Measuring Colour

Wiley-IS&T Series in Imaging Science and Technology

Series Editor:
Michael A. Kriss

Consultant Editor:
Lindsay W. MacDonald

Reproduction of Colour (6th Edition)
R. W. G. Hunt

Colour Appearance Models (2nd Edition)
Mark D. Fairchild

Colorimetry: Fundamentals and Applications
Noburu Ohta and Alan R. Robertson

Color Constancy
Marc Ebner

Color Gamut Mapping
Ján Morovič

Panoramic Imaging: Sensor-Line Cameras and Laser Range-Finders
Fay Huang, Reinhard Klette and Karsten Scheibe

Digital Color Management (2nd Edition)
Edward J. Giorgianni and Thomas E. Madden

The JPEG 2000 Suite
Peter Schelkens, Athanassios Skodras and Touradj Ebrahimi (Eds.)

Color Management: Understanding and Using ICC Profiles
Phil Green (Ed.)

Fourier Methods in Imaging
Roger L. Easton, Jr.

Measuring Colour (4th Edition)
R.W.G. Hunt and M.R. Pointer

Published in association with the Society for Imaging Science and Technology

Measuring Colour

Fourth Edition

R.W.G. Hunt

Independent Colour Consultant, UK

M.R. Pointer

Independent Colour Consultant and Visiting Professor, University of Leeds, UK

A John Wiley & Sons, Ltd., Publication

Previous Editions:
1st Edition ISBN 0 7458 01250 0, Ellis Horwood, Chichester, 1987
2nd Edition ISBN 0 13 567686 X, Ellis Horwood, Chichester, 1991
3rd Edition ISBN 0 86343 387 1, Fountain Press, Kingston-upon-Thames, 1998

Registered office
John Wiley & Sons Ltd, The Atrium, Southern Gate, Chichester, West Sussex, PO19 8SQ, United Kingdom

For details of our global editorial offices, for customer services and for information about how to apply for permission to reuse the copyright material in this book please see our website at www.wiley.com.

Library of Congress Cataloging-in-Publication Data

Hunt, R.W.G. (Robert William Gainer), 1923-
Measuring colour / R.W.G. Hunt, M.R. Pointer. – 4th ed.
p. cm. – (The Wiley-IS&T series in imaging science and technology)
Includes bibliographical references and index.
ISBN 978-1-119-97537-3 (hardback)
1. Colorimetry. I. Pointer, Michael, Ph. D. II. Title.
QC495.H84 2011
535.6028′7 – dc23

2011018730

A catalogue record for this book is available from the British Library.

Print ISBN: 978-1-119-97537-3
ePDF ISBN: 978-1-119-97573-1
oBook ISBN: 978-1-119-97559-5
ePub ISBN: 978-1-119-97840-4
Mobi ISBN: 978-1-119-97841-1

Set in 10/12pt Times by Laserwords Private Limited, Chennai, India.
Printed and bound in Malaysia by Vivar Printing Sdn Bhd

Prologue

This is the story of Mister Chrome
who started out to paint his home.
The paint ran out when half way through
so to the store he quickly flew
to buy some more of matching hue,
a delicate shade of egg-shell blue.
But when he tried this latest batch,
he found it simply didn't match.
No wonder he was in a fix,
for of the colours we can mix,
the major shades and those between,
ten million different can be seen.

You foolish man, said Missis Chrome,
you should have taken from the home
a sample of the colour done;
you can't remember every one.
Taking care that she had got
a sample from the early pot,
she went and bought her husband more
of better colour from the store.
Before she paid, she checked the shade,
and found a perfect match it made.
In triumph now she took it home,
and gave it straight to Mister Chrome.
He put it on without delay,
and found the colour now okay.

But, after dark, in tungsten light,
they found the colour still not right.
So to the store they both went now,
with samples clear, and asked them how

a paint that matched in daylight bright
could fail to match in tungsten light.
The man's reply to their complaint
was that the pigments in the paint
had been exchanged, since they had bought,
for others of a different sort.
To solve the problem on their wall,
he gave them paint to do it all
from just one batch of constant shade,
and then at last success was made.

To compensate them for their trouble,
the store sent to them curtains double.
They hung them up with great delight;
they matched in tungsten and daylight.
A neighbour then did make a call
and fixed his eye upon the wall;
the paint, he said was all one colour,
but clearly saw the curtains duller!

Though colours strange at times appear,
the moral of this tale is clear:
to understand just what we see,
object, light, and eye, all three,
must colour all our thinking through
of chromic problems, old or new!

Contents

About the Authors xv

Series Preface xvii

Preface xix

Acknowledgements xxi

1 Colour Vision 1

1.1 Introduction . . . 1
1.2 The spectrum . . . 1
1.3 Construction of the eye . . . 3
1.4 The retinal receptors . . . 4
1.5 Spectral sensitivities of the retinal receptors . . . 5
1.6 Visual signal transmission . . . 8
1.7 Basic perceptual attributes of colour . . . 9
1.8 Colour constancy . . . 10
1.9 Relative perceptual attributes of colours . . . 11
1.10 Defective colour vision . . . 13
1.11 Colour pseudo-stereopsis . . . 15
References . . . 16
General References . . . 17

2 Spectral Weighting Functions 19

2.1 Introduction . . . 19
2.2 Scotopic spectral luminous efficiency . . . 19
2.3 Photopic spectral luminous efficiency . . . 21
2.4 Colour-matching functions . . . 26
2.5 Transformation from R, G, B to X, Y, Z . . . 32
2.6 CIE colour-matching functions . . . 33
2.7 Metamerism . . . 38
2.8 Spectral luminous efficiency functions for photopic vision . . . 39
References . . . 40
General References . . . 40

3 Relations between Colour Stimuli **41**
3.1 Introduction . . . 41
3.2 The Y tristimulus value . . . 41
3.3 Chromaticity . . . 42
3.4 Dominant wavelength and excitation purity . . . 44
3.5 Colour mixtures on chromaticity diagrams . . . 46
3.6 Uniform chromaticity diagrams . . . 48
3.7 CIE 1976 hue-angle and saturation . . . 51
3.8 CIE 1976 lightness, L^* . . . 52
3.9 Uniform colour spaces . . . 53
3.10 CIE 1976 colour difference formulae . . . 57
3.11 CMC, CIE94, and CIEDE2000 color difference formulae . . . 61
3.12 An alternative form of the CIEDE2000 colour-difference equation . . . 64
3.13 Summary of measures and their perceptual correlates . . . 64
3.14 Allowing for chromatic adaptation . . . 65
3.15 The evaluation of whiteness . . . 66
3.16 Colorimetric purity . . . 67
3.17 Identifying stimuli of equal brightness . . . 67
3.18 CIEDE2000 worked example . . . 69
References . . . 71
General References . . . 72

4 Light Sources **73**
4.1 Introduction . . . 73
4.2 Methods of producing light . . . 74
4.3 Gas discharges . . . 74
4.4 Sodium lamps . . . 75
4.5 Mercury lamps . . . 76
4.6 Fluorescent lamps . . . 78
4.7 Xenon lamps . . . 81
4.8 Incandescent light sources . . . 82
4.9 Tungsten lamps . . . 86
4.10 Tungsten halogen lamps . . . 87
4.11 Light emitting diodes . . . 88
4.12 Daylight . . . 89
4.13 Standard illuminants and sources . . . 91
4.14 CIE standard illuminant A . . . 91
4.15 CIE illuminants B and C . . . 92
4.16 CIE sources . . . 93
4.17 CIE illuminants D . . . 94
4.18 CIE indoor daylight . . . 94
4.19 Comparison of commonly used sources . . . 96
References . . . 97
General References . . . 97

5 Obtaining Spectral Data and Tristimulus Values **99**
5.1 Introduction . . . 99
5.2 Radiometry and photometry . . . 99
5.3 Spectroradiometry . . . 100
5.4 Tele-spectroradiometry . . . 100
5.5 Spectroradiometry of self-luminous colours . . . 101
5.6 Spectrophotometry of non-self-luminous colours . . . 101
5.7 Reference whites and working standards . . . 102
5.8 Geometries of illumination and viewing . . . 103
5.9 CIE Geometries of illumination and measurement . . . 104
5.10 Spectroradiometers and spectrophotometers . . . 108
5.11 Choice of illuminant . . . 110
5.12 Calculation of tristimulus values from spectral data . . . 111
5.13 Colorimeters using filtered photo-detectors . . . 114
References . . . 115
General References . . . 115

6 Metamerism and Colour Constancy **117**
6.1 Introduction . . . 117
6.2 The cause of metamerism . . . 117
6.3 The definition of metamerism . . . 118
6.4 Examples of metamerism in practice . . . 119
6.5 Degree of metamerism . . . 121
6.6 Index of metamerism for change of illuminant . . . 122
6.7 Index of metamerism for change of observer . . . 122
6.8 Index of metamerism for change of field size . . . 124
6.9 Colour matches and geometry of illumination and measurement . . . 124
6.10 Correcting for inequalities of tristimulus values . . . 125
6.11 Terms used in connection with metamerism . . . 126
6.12 Colour inconstancy . . . 127
6.13 Chromatic adaptation transforms . . . 129
6.14 The Von Kries transform . . . 130
6.15 The CAT02 transform . . . 131
6.16 A colour inconstancy index . . . 134
6.17 Worked examples . . . 135
References . . . 141

7 Colour Rendering by Light Sources **143**
7.1 Introduction . . . 143
7.2 The meaning of colour rendering . . . 144
7.3 CIE colour rendering indices . . . 145
7.4 Spectral band methods . . . 147
7.5 Other methods for assessing the colour rendering of light sources . . . 150

7.6 Comparison of commonly used sources 151
References 152
General References 154

8 Colour Order Systems **155**
8.1 Introduction 155
8.2 Variables 155
8.3 Optimal colours 157
8.4 The Munsell System 159
8.5 The Munsell Book of Color 164
8.6 Unique hues and colour opponency 168
8.7 The Natural Colour System (NCS) 170
8.8 Natural Colour System Atlas 172
8.9 The DIN System 179
8.10 The Coloroid System 182
8.11 The Optical Society of America (OSA) System 183
8.12 The Hunter Lab System 187
8.13 The Tintometer 190
8.14 The Pantone System 191
8.15 The RAL System 191
8.16 Advantages of colour order systems 192
8.17 Disadvantages of colour order systems 192
References 194
General References 195

9 Precision and Accuracy in Colorimetry **197**
9.1 Introduction 197
9.2 Sample preparation 198
9.3 Thermochromism 199
9.4 Geometry of illumination and measurement 199
9.5 Reference white calibration 200
9.6 Polarisation 200
9.7 Wavelength calibration 202
9.8 Stray light 202
9.9 Zero level and linearity 202
9.10 Use of secondary standards 203
9.11 Bandwidth 203
9.12 Correcting for errors in the spectral data 204
9.13 Calculations 207
9.14 Precautions to be taken in practice 214
References 215

10 Fluorescent Colours **219**
10.1 Introduction 219
10.2 Terminology 219
10.3 Use of double monochromators 220

10.4 Illumination with white light 221
10.5 Correcting for differences between an actual and the desired source 222
10.6 Two-monochromator method 224
10.7 Two-mode method 225
10.8 Filter-reduction method 226
10.9 Luminescence-weakening method 226
10.10 Practical considerations 227
References 230

11 RGB Colorimetry **231**
11.1 Introduction 231
11.2 Choice and specification of matching stimuli 231
11.3 Choice of units 233
11.4 Chromaticity diagrams using r and g 233
11.5 Colour-matching functions in RGB systems 234
11.6 Derivation of XYZ from RGB tristimulus values 235
11.7 Using television and computer displays 239
References 240
General Reference 240

12 Colorimetry with Digital Cameras **241**
12.1 Introduction 241
12.2 Camera characterisation 242
12.3 Metamerism 244
12.4 Characterisation methods 244
12.5 Practical considerations in digital camera characterisation 249
12.6 Practical example 251
12.7 Discussion 254
References 255
General References 256

13 Colorant Mixtures **257**
13.1 Introduction 257
13.2 Non-diffusing colorants in a transmitting layer 257
13.3 Non-diffusing colorants in a layer in optical contact with a diffusing surface 259
13.4 Layers containing colorants which diffuse and absorb light 262
13.5 The use of multi-spectral analysis to reduce metamerism in art restoration 264
References 265
General References 265

14 Factors Affecting the Appearance of Coloured Objects **267**
14.1 Introduction 267
14.2 Measuring optical properties 267
14.3 Colour 268
14.4 Gloss 271

14.5 Translucency 279
14.6 Surface texture 281
14.7 Conclusions 289
References 289

15 The CIE Colour Appearance Model CIECAM02 293
15.1 Introduction 293
15.2 Visual areas in the observing field 294
15.3 Chromatic adaptation in CIECAM02 294
15.4 Spectral sensitivities of the cones in CIECAM02 295
15.5 Cone dynamic response functions in CIECAM02 297
15.6 Luminance adaptation in CIECAM02 297
15.7 Criteria for achromacy and for constant hue in CIECAM02 299
15.8 Effects of luminance adaptation in CIECAM02 300
15.9 Criteria for unique hues in CIECAM02 303
15.10 Redness-greenness, a, and yellowness-blueness, b, in CIECAM02 303
15.11 Hue angle, h, in CIECAM02 305
15.12 Eccentricity factor, e, in CIECAM02 305
15.13 Hue quadrature, H, and hue composition, H_c, in CIECAM02 306
15.14 The achromatic response, A, in CIECAM02 308
15.15 Correlate of lightness, J, in CIECAM02 308
15.16 Correlate of brightness, Q, in CIECAM02 309
15.17 Correlate of chroma, C, in CIECAM02 310
15.18 Correlate of colourfulness, M, in CIECAM02 311
15.19 Correlate of saturation, s, in CIECAM02 311
15.20 Comparison of CIECAM02 with the natural colour system 311
15.21 Testing model CIECAM02 312
15.22 Filtration of projected slides and CIECAM02 314
15.23 Comparison of CIECAM02 with CIECAM97s 315
15.24 Uniform colour space based on CIECAM02 315
15.25 Some problems with CIECAM02 316
15.26 Steps for using the CIECAM02 model 316
15.27 Steps for using the CIECAM02 model in reverse mode 319
15.28 Worked example for the model CIECAM02 321
References 322

16 Models of Colour Appearance for Stimuli of Different Sizes 325
16.1 Introduction 325
16.2 Stimuli of different sizes 325
16.3 Room colours 325
16.4 A model for predicting room colours 326
16.5 Steps in using the model for predicting room colours 327
References 328

17 Model of Colour Appearance for Unrelated Colours in Photopic and Mesopic Illuminances **329**

17.1 Introduction 329
17.2 A model for predicting unrelated colours 330
17.3 Input data required for the model 331
17.4 Steps in using the model for unrelated colours 332
17.5 Worked example in the model for predicting unrelated colours 333
References 334

Appendices **335**

Appendix 1 Radiometric and Photometric Terms and Units **337**

A1.1 Introduction 337
A1.2 Physical detectors 337
A1.3 Photometric units and terms 338
A1.4 Radiant and quantum units and terms 340
A1.5 Radiation sources 340
A1.6 Terms for measures of reflection and transmission 341
A1.7 Other spectral luminous efficiency functions 343
A1.8 Mesopic photometry 343
Reference 344

Appendix 2 Spectral Luminous Efficiency Functions **345**

Appendix 3 CIE Colour-Matching Functions **347**

Appendix 4 CIE Spectral Chromaticity Co-Ordinates **351**

Appendix 5 Relative Spectral Power Distributions of Illuminants **355**

A5.1 Introduction 355
A5.2 CIE illuminants 355
A5.3 Representative fluorescent lamps 359
A5.4 Planckian radiators 368
A5.5 Gas discharge lamps 371
A5.6 Method of calculating D illuminant distributions 374

Appendix 6 Colorimetric Formulae **379**

A6.1 Chromaticity relationships 379
A6.2 CIELUV, CIELAB, and U*V*W* relationships 379

Appendix 7 Calculation of the CIE Colour Rendering Indices **383**
A7.1 Spectral radiance factors of test colours 383
A7.2 Worked example of the CIE colour rendering indices 388

Appendix 8 Illuminant-Observer Weights for Calculating Tristimulus Values **393**

Appendix 9 Glossary of Terms **431**
Reference 453

Index **455**

About the Authors

Dr Robert W. G. Hunt received his Ph.D and DSc from the University of London. He was a research scientist at the Kodak Research Laboratories, where he worked on factors affecting the quality of colour images, and devices for making reflection prints from both negative and positive images on film; he was finally Assistant Director of Research. Since 1982 he has worked as an independent colour consultant, and has taken a leading role in the development of colour appearance models. He has written over 100 papers on colour vision, colour reproduction, and colour measurement, and his other book, *The Reproduction of Colour*, is now in its sixth edition. He has been awarded the Newton Medal of the Colour Group (Great Britain) (1974), the Progress Medal of the Royal Photographic Society (1984), the Judd-AIC Medal of the International Colour Association (1987), the Gold Medal of the Institute of Printing (1989), the Johann Gutenberg Prize of the Society for Information Display (2002), the Godlove Award of the Inter-Society Color Council, U.S.A (2007), and Honorary Fellowship of the Society of Dyers and Colourists (2009). In 2009 he was appointed an Officer of the British Empire (OBE) for 'services to the field of colour science and to young people through Crusaders'.

Dr Michael R. Pointer received his Ph.D from Imperial College, London, working with David Wright. He then worked in the Research Division of Kodak Limited on fundamental issues of colour science applied to the photographic system. After periods at the University of Westminster and the National Physical Laboratory, he is now a Visiting Professor at the University of Leeds, as well as working as a consultant scientist. In 1997, he received the Fenton Medal, The Royal Photographic Society's award for services to the Society. In 2004, he received a Silver Medal from the Society of Dyers and Colourists for 'contributions to colour science'. He has authored over 100 scientific papers, is a Fellow of The Royal Photographic Society and the Institute of Physics, Secretary of CIE Division 1 Vision & Colour, and UK Associate Editor of the journal, *Color Research & Application.*

Series Preface

Imagine Alice in Wonderland saying this: 'I wonder if I've changed ***colour*** in the night? Let me think. Was I the same ***hue*** when I got up this morning? I almost think I can remember feeling a little ***less saturated***. But if I'm not the same ***x-y value***, the next question is '***What Lab value*** in the world am I?' Ah, that's the great puzzle!'

The fourth edition of *Measuring Colour* by Dr Robert W.G. Hunt and Dr Michael R. Pointer is the eleventh book in the Wiley-IS&T Series in Imaging Science and Technology. This excellent text, while not solving the complex puzzle of colour, provides readers with the means to solve their colour puzzle.

The 17-chapter book starts with the basic concepts of colour vision then covers the methodology of converting a spectrum to CIE values (*XYZ* or *Lab*) so one can match colours and detect metamers. Visual models are then used to indicate how colour changes under different viewing conditions and to explain why surface characteristics influence the perception of a given spectrum. The details of using digital cameras to measure colour are an important addition in the fourth edition, as the authors recognise that the CCD and CMOS sensors in digital cameras, together with colour filter arrays and digital signal processing, present a new opportunity to measure spatial variation in colour.

Human beings are very sensitive to colour changes or differences and find it difficult to decide, from a set of colours, for example on a paint palette, which one is wanted. People have a strong sense of memory of preference for the colour of green grass, blue skies or pink sunsets. They notice when a photographic image (from a film or digital camera) of a red tablecloth comes out wrong or when the sweater that was bought in a shopping mall under tungsten (fluorescent) light looks different in daylight.

Neural scientists can use Functional Magnetic Resonance Imaging to locate where in the brain the perceptions of the colours of the visible spectrum are located. Colour scientists know that each colour has an exact spectral power distribution which can be measured to a high degree of accuracy. Why, then, is colour such a puzzle?

What Alice did not know when she fell down the rabbit hole was that the human visual system can play a lot of tricks on how we perceive colour. The perception of colour depends not just on its native spectrum but also on the spectra of the direct illumination, the ambient illumination and the near and far surround colours. Geometrical patterns can cause local colour changes as seen by the observer, as can adaptation to a uniform colour field. So how can the puzzle called colour be solved when there are so many variables and boundary conditions?

The Mock Turtle might have said: *'What is the use of* ***studying*** *all that* ***colour*** *stuff, if you* ***can't measure*** *it as you go on? It's by far the most confusing thing I ever heard!'*

But then, he was not privy to the fourth edition of *Measuring Colour* which provides a welcome and major contribution to the continuing understanding of the puzzle that is colour.

MICHAEL A. KRISS
Formerly of the Eastman Kodak Research Laboratories and the University of Rochester, USA

Preface

To the First Edition

This book is intended to provide the reader with the basic facts needed to measure colour. It is a book about principles, rather then a guide to instruments. With the continual advances in technology, instruments are being improved all the time, so that any description of particular colorimeters, spectroradiometers, or spectrophotometers is likely to become out of date very quickly. For such information, manufacturers' catalogues are a better source of information than books. But the principles of measuring colour are not subject to rapid change, and are therefore appropriate for treatment in the more permanent format offered by books.

Recommendations about the precise way in which the basic principles of colour measurement should be applied have for over 50 years been the province of the International Commission on Illumination (CIE). The second edition of its publication *No. 15, Colorimetry*, includes several new practices, and it is therefore timely to restate the basic principles of colorimetry together with these latest international recommendations on their application; this is the aim of *Measuring Colour*.

Colour is, of course, primarily a sensation experienced by the individual. For this reason, the material has been set in the context of the colour vision properties of the human observer: the first chapter is a review of our current knowledge of colour vision; and the last chapter provides a description of a model of colour vision that can be used to extend colour measurement, beyond the territory covered by the CIE at present, to the field of colour appearance.

To the Second Edition

The second edition contains all the material of the first edition, together with four new chapters. Two of these chapters provide entirely new material: one is on light sources and the other is on precision and accuracy in colorimetry. The other two new chapters provide expanded treatments of metamerism and of the colorimetry of fluorescent materials. Extensive revisions have been made to the chapter on the model of colour vision, so as to present it in its latest version. Finally, minor revisions have been made to the rest of the book to improve the treatment in various respects.

To the Third Edition

The following changes have been made to this third edition. The chapter on metamerism has been expanded to include a discussion of corresponding colours, colour constancy and a description of a colour inconstancy index. The material in Chapter 11 of the second edition, entitled 'Miscellaneous topics', has been included at the end of Chapter 3. Chapter 11 now provides a discussion of the way in which the colours of colorant mixtures can be evaluated. Chapter 12 has been updated to provide the colour appearance model adopted internationally, designated CIECAM97s. Two new Appendices have been added: Appendix 7 provides illuminant-observer weights for band-pass corrected data, and Appendix 8 provides illuminant-observer weights for band-pass uncorrected data. In addition, various minor changes have been made to update the text.

To the Fourth Edition

For this fourth edition Dr Michael R. Pointer has become a joint author. Much of the book is concerned with CIE procedures, and, as current secretary of CIE Division 1 Colour and Vision, Dr Pointer has enabled the important features of the latest CIE publications to be covered; he has also provided new chapters on 'Factors affecting the appearance of coloured objects' and 'Colorimetry with digital cameras'. The first of these chapters covers the important topics of gloss, translucency and texture, which were not previously included; the second of these chapters covers the technology that enables colorimetry to be carried out on objects with complicated shapes or patterns, and which has been developed since publication of the third edition. The important topic of colour rendering by light sources now has its own chapter, and this includes descriptions of alternatives to the current CIE Colour Rendering Index. Additions to the Appendices include the recent CIE procedure for mesopic photometry, and the spectral reflectance factors for the Munsell colours used in the CIE Colour Rendering Index. For the current state of CIE publications see www.CIE.co.at. As with previous editions, various minor changes have been made to update the text. An important change in this fourth edition is the availability of colour printing on every page; this has made it possible to improve the clarity of many figures, and to position colour reproductions at their appropriate positions in the text, instead of being grouped into a section of colour plates. Since the publication of the third edition in 1998, the measurement of colour has become increasingly important in many areas, including science, medicine and manufacturing, and this fourth edition provides a more up-to-date and comprehensive treatment of this fascinating subject.

Acknowledgements

For the First Edition

I am most grateful to Dr M. R. Pointer of Kodak Limited for kindly making many helpful comments on the text, for providing some of the numerical data, and for help with the proof reading. My grateful thanks for help are also due to Dr A. Hård in connection with the section on the NCS, to Dr H. Terstiege with that on the DIN system, and to Dr A. Nemcsics with that on the Coloroid system. For permission to reproduce figures, my thanks are due to the Institute of Physics for Figure 3.5; to John Wiley and Sons for Figures 7.5, 7.12, 7.19 and 8. 1; to Dr A. Hård for Figure 7.12; to Dr H. Terstiege for Figure 7.19; and to Academic Press for Figures 9.1, 9.2 and 9.3. I would also like to thank Dr J. Schanda for kindly supplying me with copies of recent CIE documents.

With regard to the colour plates, my thanks are due to the following for kindly supplying the originals: Dr A.A. Clarke and Dr M.R. Luo, of Loughborough University of Technology, for Plates 2 and 3; Dr A. Hård for Plate 5; Dr H. Terstiege for Plate 8; and Mr R. Ingalls for Plates 1, 6, and 7. I would also like to thank the Munsell Corporation for permission to reproduce Plate 4.

I am also most grateful to Mr A.J. Johnson, and some of his colleagues, of Crosfield Electronics Limited, for kindly supplying the separations for the colour illustrations.

Finally my grateful thanks are due to my wife for editorial assistance and for help with the proof reading.

For the Second Edition

I am very grateful for help that has kindly been given to me by experts on the subject matter of the new material in this second edition. Dr F.W. Billmeyer has made many suggestions for improving the new chapters on metamerism, on precision and accuracy in colorimetry, and on the colorimetry of fluorescent materials. Miss M.B. Halstead, Mr D.O. Wharmby and Dr M.G. Abeywickrama have helped with the new chapter on light sources. Dr R.F. Berns has helped with the section on correcting for errors in spectral data, and Dr W.H. Venable with the section on the computation of tristimulus values. I am indebted to Mr J.K.C. Kempster for the data on which Figure 6.2 is based. Once again, I am most grateful to Dr M.R. Pointer for general comments, for help with computations, and for proof reading, and to my wife for editorial help and for proof reading.

For the Third Edition

I am most grateful to Dr M. R. Pointer for kindly suggesting that publication of this third edition be assisted by The Tintometer Limited, a company which has been continuously involved with colour measurement for over a hundred years; in this connection Miss Nicola Pointer's word-processing help is much appreciated. As with the earlier editions, Dr Pointer has also provided much expert help by means of general comments, and proof reading, for which I am most grateful. The tables given in Appendices 7 and 8 were originally published by the American Society for Testing Materials in their Standard ASTM E 308 - 95, *Standard Practice for Computing the Colours of Objects by Using the CIE System*; their permission to reproduce these tables is acknowledged with thanks. For preparing the final text in such a helpful way, I am very grateful to Mr Dennis Shearman of Priory Publications, who also performed the same task for my other book *The Reproduction of Colour*. Finally my best thanks are due to my wife for editorial help and for proof reading.

For the Fourth Edition

We are grateful to Janos Schanda for help with Chapter 7, to Jan and Peter Morovic for help with Chapter 12, and to Ronnier Luo and Changjun Li with Chapter 15. References to Munsell® in this publication are used with permission from X-Rite Inc. References to NCS in this publication are used with permission from NCS Color AB.

For preparing the final text in such a helpful way, we are very grateful to the Wiley staff at Chichester, who also performed the same task for Dr Hunt's *The Reproduction of Colour*.

1

Colour Vision

1.1 INTRODUCTION

Ten million! That is the number of different colours that we can distinguish, according to one reliable estimate (Judd and Wyszecki, 1975). It is, therefore, no wonder that we cannot remember colours well enough to identify a particular shade. People are thus well advised to take samples of their clothing colours with them when purchasing accessories that are intended to match. They are also usually well aware that it is not enough to examine the colour match in just one type of light in a shop, but to see it in daylight as well as in artificial light. Finally, a second opinion about the match, expressed by a friend or a shop assistant, is often wisely sought.

The above activity involves the three basic components of colour: sources of light, objects illuminated by them and observers. Colour, therefore, involves not only material sciences, such as physics and chemistry, but also biological sciences, such as physiology and psychology; and, in its applications, colour involves various applied sciences, such as architecture, dyeing, paint technology, and illuminating engineering. Measuring colour is, therefore, a subject that has to be broadly based and widely applied.

Without observers possessing the faculty of sight, there would be no colour. Hence it is appropriate to start by considering the nature of the colour vision provided by the human eye and brain. Before doing this, however, a brief description must be given of the way in which it is necessary to characterise the nature of the light which stimulates the visual system.

1.2 THE SPECTRUM

It is fair to say that understanding colour finds its foundations in the famous experiments performed by Isaac Newton in 1666. Before this date, opinions on the nature of colours and the relationships between them were most vague and of very little scientific use, but, after Newton's work became known, a way was open for progress based on experimental facts.

The historic experiments were performed in Trinity College, Cambridge, when Newton made a small hole, a third of an inch in diameter, in the shutter of an otherwise entirely

Measuring Colour, Fourth Edition. R.W.G. Hunt and M.R. Pointer.

dark room; through this hole, the direct rays of the sun could shine and form an image of the sun's disc on the opposite wall of the room, like a pin-hole camera. Then, taking a prism of glass, and placing it close to the hole, he observed that the light was spread out fan-wise into what he was the first to call a *spectrum*: a strip of light, in this case about ten inches long, and coloured red, orange, yellow, green, blue, indigo, and violet, along its length. The natural conclusion, which Newton was quick to draw, was that white light was not the simple homogeneous entity which it was natural to expect it to be, but was composed of a mixture of all the colours of the spectrum.

The next question which arose was whether these spectral colours themselves, red, green, etc., were also mixtures and could be spread out into further constituent colours. A further experiment was performed to test this suggestion. A card with a slit in it was used to obscure all the light of the spectrum, except for one narrow band. This band of light, say a yellow or a green, was then made to pass through a second prism, but the light was then seen not to be spread out any further, remaining exactly the same colour as when it emerged from the slit in the card. It was, therefore, established that the spectral colours were in fact the basic components of white light.

The inclusion by Newton of indigo in the list of spectral colours is rather puzzling since, to most people, there appears to be a gradual transition between blue and violet with no distinct colour between them, as there is in the case of orange between red and yellow. Several explanations of the inclusion of indigo have been suggested, but the most likely is that Newton tried to fit the colours into a scale of tones in a way analogous to the eight-tone musical scale; to do this he needed seven different colours to correspond to the seven different notes of the scale (McLaren, 1985).

In Figure 1.1, the main bands of colour in the spectrum are shown against a scale of the *wavelength* of the light. Light is a form of electro-magnetic radiation, as is also the case for x-rays, radar, and radio waves, for instance, and the property of this radiation that gives it particular characteristics is its wavelength. Radio waves have quite long wavelengths, typically in the range from about a metre to several kilometres, whereas x-rays

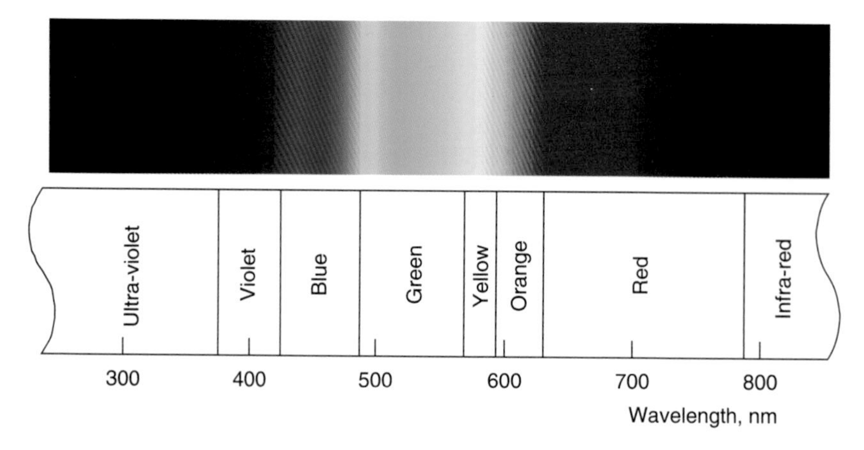

Figure 1.1 The colour names usually given to the main regions of the spectrum. Because of the limitations of printing, the colours of the spectrum cannot be shown accurately, but their general layout is displayed adequately

have extremely short wavelengths, typically about a millionth of a millimetre or shorter. Light waves have wavelengths in between, ranging from slightly above to slightly below a half a millionth of a metre. To obtain convenient numbers for the wavelengths of light, the unit used for expressing them is the *nanometre* (abbreviation, *nm*), which is a millionth of a millimetre, or 10^{-9} of a metre; this is the unit used in Figure 1.1. It must be emphasised that the colour names and wavelength boundaries given in Figure 1.1 are only intended as a rough guide; each colour gradually merges into the next so that there is really no exact boundary; moreover, the colour appearance of light of a given wavelength depends on the viewing conditions, and is also liable to be slightly different from one observer to another. Even so, the names given in Figure 1.1 are useful to bear in mind when considering data that are presented as functions of wavelength. Radiation having wavelengths longer than those of the visible spectrum and less than about 1 mm is called *infrared*; and that having wavelengths shorter than those of the visible spectrum and longer than about 100 nm is called *ultraviolet*. These radiations can provide radiant energy that tans the skin or warms the body, for instance, but they cannot normally be seen as light. In colour science, although it is the long-established practice to identify different parts of the spectrum by using wavelength, it would be more fundamental to use *frequency*. This is because, for light from any part of the spectrum, as it passes through a medium, its wavelength decreases by being divided by the refractive index of that medium; however, the velocity also decreases in the same proportion, so that the frequency (the velocity divided by the wavelength) remains constant. The values of wavelength quoted are usually as measured in air, and, although those measured in vacuum would be more fundamental, they differ by only about 3 parts in 10 000. (The velocity of light in vacuum is about 2.998×10^8 metres per second.)

1.3 CONSTRUCTION OF THE EYE

A diagrammatic representation of a cross-section of the human eye is given in Figure 1.2. Most of the optical power is provided by the curved surface of the *cornea*, and the main function of the *lens* is to alter that power by changing its shape, being thinner for viewing

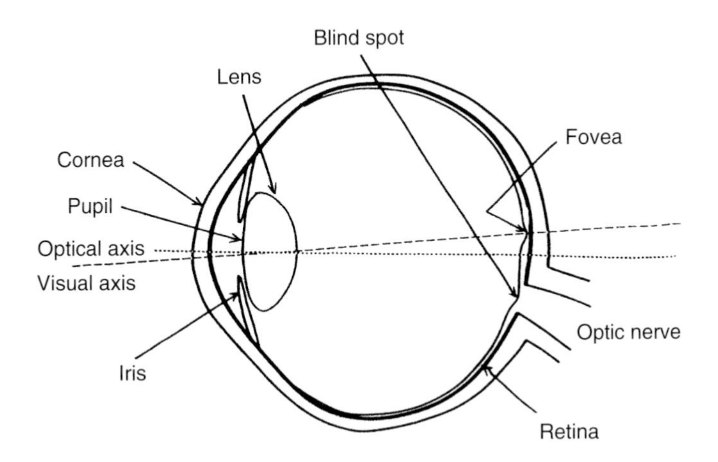

Figure 1.2 Cross-sectional diagram of the human eye

distant objects and thicker for near objects. The cornea and lens acting together form a small inverted image of the outside world on the *retina*, the light-sensitive layer of the eye. The *iris*, the annular-shaped coloured part of the eye that we see from the outside, changes its shape, having a central aperture that is only about 2 mm in diameter in bright light, but which is larger in dim light, having a maximum diameter of about 8 mm. The aperture referred to is called the *pupil*, and is the area through which the light passes. The iris, by changing its diameter, provides some compensation for changes in the level of illumination under which objects are seen; however, this compensation only amounts to a factor of about 8 to 1, rather than the 16 to 1 to be expected from the ratio of the squares of the diameters, because rays that pass through the edge of the pupil are less effective in stimulating the retina than those that pass through the centre, a property known as the *Stiles-Crawford effect*.

The retina lines most of the interior of the approximately spherically-shaped eye-ball, and this provides the eye with a very wide field of view. However, the retina is far from being uniform in sensitivity over its area. Colour vision is very limited for stimuli seen beyond about 40° off the visual axis (Hurvich, 1981), and in this area vision is used mainly for the detection of movement. Within the 40° on either side of the eye's axis, the ability to see both colour and fine detail gradually increases as the eye's axis is approached, the area of sharpest vision being termed the *fovea*, which comprises approximately the central 1.5° diameter of the visual field. An area within this, termed the *foveola*, corresponds to a field of about 1°. A curious feature of the fovea and foveola is that they are not centred on the *optical axis* of the eye, but lie about 4° to one side as shown in Figure 1.2, thus resulting in the visual axis being offset by this amount. About 10° to the other side of the optical axis (equivalent to about 14° from the fovea) is the *blind spot*, where the nerve fibres connecting the retina to the brain pass through the surface of the eye-ball, and this area has no sensitivity to light at all. There is also an area covering part of the fovea, called the *yellow spot* or *macula lutea*, containing a yellowish pigment. In addition to these spatial variations in the retina, there are changes in the types of light receptors present in different areas. In the foveola, the receptors are all of one type, called *cones*; outside this area, there is, in addition, another type, called *rods*. The ratio of cones to rods varies continuously from all cones and no rods in the foveola to nearly all rods and very few cones beyond about 40° from the visual axis. Finally, the individual cones and rods are connected to the brain by nerve fibres in very different ways, depending on their position: in the foveola, there are about the same number of nerve fibres as cones; but, as the angle from the visual axis increases, the number of nerve fibres decreases continuously until as many as several hundred rods and cones may be served by each nerve fibre.

1.4 THE RETINAL RECEPTORS

The function of the rods in the retina is to give monochromatic vision under low levels of illumination, such as moonlight and starlight. This *scotopic* form of vision operates when the stimuli have luminances of less than some hundredths of a candela per square metre (cd m^{-2}; for a summary of photometric terms and units, see Appendix 1).

The function of the cones in the retina is to give colour vision at normal levels of illumination, such as daylight and typical indoor artificial light. This *photopic* form of vision operates when stimuli have luminances of several cd m^{-2} or more.

There is a gradual change from photopic to scotopic vision as the illumination level is lowered, and, for stimuli having luminances between several cd m^{-2} to some hundredths of a cd m^{-2}, both cones and rods make significant contributions to the visual response, and this is called *mesopic* vision. The wavelengths of the light to which the rods are most sensitive are shorter than is the case for most of the cones, and, as a result, as the illumination level falls through the mesopic range, the relative brightnesses of red and blue colours change. This can often be seen in a garden at the end of the day; red flowers that look lighter than blue flowers in full daylight look darker than the blue ones as the light fades. This is known as the *Purkinje phenomenon*.

The rods and cones are so named because of their shapes, but they are all very small, being typically about a five-hundredth of a millimetre in diameter, with a length of around a twenty-fifth of a millimetre. They are packed parallel to one another and face end-on towards the pupil of the eye so that the light is absorbed by them as it travels along their length. They are connected to nerve fibres via an extremely complicated network of cells situated immediately on the pupil-side of their ends. The nerve fibres then travel across the pupil-side of the retina to the blind spot where they are collected together to form the *optic nerve* which connects the eye to the brain. Hence, before the light reaches the receptors, it has to pass through the cells and nerve fibres, which are largely transparent. In each eye, there are about 6 million cones, 100 million rods, and 1 million nerve fibres.

1.5 SPECTRAL SENSITIVITIES OF THE RETINAL RECEPTORS

The rods and the cones are not equally sensitive to light of all wavelengths. In the case of the rods, the initial step in the visual process is the absorption of light in a photosensitive pigment called *rhodopsin*. This pigment absorbs light most strongly in the blue-green part of the spectrum, and decreasingly as the wavelength of the light becomes either longer or shorter. As a result, the spectral sensitivity of the scotopic vision of the eye is as shown by the broken curve of Figure 1.3. This curve is obtained by having observers adjust the strength of a beam of light of one wavelength until the perception it produces has the same intensity as that produced by a beam of fixed strength of a reference wavelength. If the strength of the variable beam had to be, for example, twice that of the fixed beam, then the scotopic sensitivity at the wavelength of the variable beam would be regarded as a half of that at the wavelength of the fixed beam. These relative sensitivities are then plotted against wavelength to obtain the broken curve of Figure 1.3, the maximum value being made equal to 1.0 by convention. To obtain a sensitivity curve representing scotopic vision, it is necessary to use beams of sufficiently low intensity to be entirely in the scotopic range, and the curve of Figure 1.3 was obtained in this way. It is based on results obtained from about 70 observers (22 in a study by Wald, 1945; and 50 in a study by Crawford, 1949) and represents scotopic vision of observers under 30 years of age; above this age, progressive yellowing of the lens of the eye makes the results rather variable. The curve represents the scotopic sensitivity for light incident on the cornea, and thus the effects of any absorption in the ocular media are included. The strengths of the beams can be evaluated in various ways, but the convention has been adopted to use the amount of power (energy per unit time) per small constant-width wavelength interval. If the beams used have the same small width of wavelength throughout the spectrum, then all that is required is to know the relative power

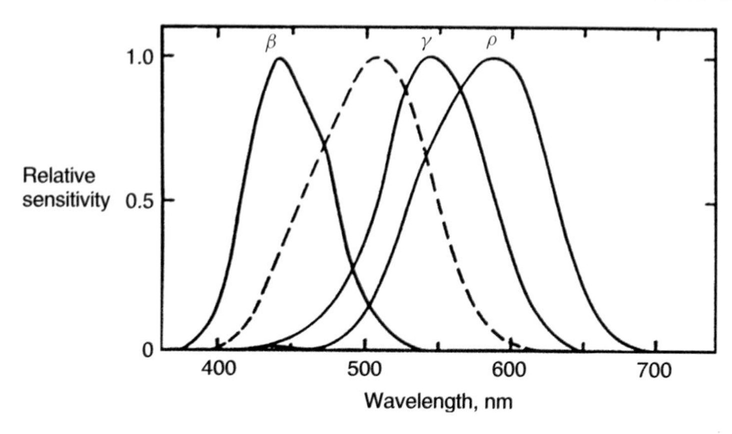

Figure 1.3 Broken line: the spectral sensitivity of the eye for scotopic (rod) vision. Full lines: spectral sensitivity curves representative of those believed to be typical of the three different types of cones, ρ, γ, and β, of the retina that provide the basis of photopic vision. The sensitivities are for equal power per small constant-width wavelength interval

in each beam. However, if the beams have different widths of wavelength, then the values of the relative power per unit wavelength interval have to be determined for each beam.

A system having a single spectral sensitivity function, such as that shown by the broken line in Figure 1.3, cannot, on its own, provide a basis for colour vision. Thus, although, for example, light of wavelength 500 nm would result in a response about 30 times as great as the same strength of light of wavelength 600 nm, the two responses could be made equal simply by increasing the strength of the 600 nm beam by a factor of about 30 times. The system is thus not able to distinguish between changes in wavelength and changes in intensity, and this is what is needed to provide a basis for colour vision. Scotopic vision therefore provides only shades of whites, greys, and blacks, as occur in moonlight.

In the case of the cones of the human retina, it has proved difficult to extract the photosensitive pigments, and our knowledge of them has had to be obtained largely by indirect means. These include very careful measurements of the light absorbed at each wavelength of the spectrum by individual cones removed from eyes that have become available for study (Dartnall, Bowmaker, and Mollon, 1983), and deductions from experiments on colour matching together with data (Estévez, 1979) on colour defective vision (to be discussed in Section 1.10). Also the genes for the pigments have been expressed in tissue cultures, enabling the pigments to be produced for study (Nathans, Merbs, Sung, Weitz, and Wang, 1992). As a result of these studies, sets of curves typified by those shown by the full lines of Figure 1.3 have been obtained. (See also Stockman, Sharpe and Fach, 1999; Stockman and Sharpe, 2000; Stockman, 2008.)

The exact shapes of the curves that best typify the spectral sensitivities of the cones are still a matter of some debate, and in some sets the right-hand curve peaks at about 565 nm (Smith and Pokorny, 1972) instead of at about 585 nm as in Figure 1.3; but the set shown in Figure 1.3 shows all the important features of any reasonably plausible set, and is adequate for our present descriptive purposes. These curves represent the spectral sensitivities for light incident on the cornea, and allowance has thus again been made for any absorptions in the ocular media; these types of curve are often referred to as *action spectra*. For convenience, they have been plotted so that their maxima are all equal to 1.0.

The full curves of Figure 1.3 have been labelled ρ, γ, and β, to distinguish them. If Figure 1.3 is compared to Figure 1.1, it is clear that the ρ curve has a maximum sensitivity in the yellow-orange part of the spectrum, the γ curve in the green part, and the β curve in the blue-violet part. Various designations have been used by different authors for the three types of cone to which the three curves refer, including L, M, S (Long, Medium, and Short wavelength), π_5, π_2, π_1 (after the work of Stiles), and R, G, B (sensitive mainly to the Reddish, Greenish, and Bluish thirds of the spectrum). The L, M, S and R, G, B designation are perhaps the most widely used, but *L* is used for luminance, *M* for correlates of colourfulness, and *s* for correlates of saturation, and it is convenient to keep R, G, B to represent red, green, and blue lights and colours; hence the similar, but distinctive, Greek symbols, ρ, γ, β, have been adopted instead.

It is clear that there are *three* different curves for the cones, and they correspond to three different types of cone, each containing a different photosensitive pigment. We now have a basis for colour vision. For, if we consider again lights of wavelengths 500 and 600 nm, it is clear that the 500 nm light will produce about twice as much γ response as ρ response, whereas the 600 nm light will produce about twice as much ρ response as γ response. If the strengths of the beams are altered, the ρ and γ responses will also alter, but their ratios will always remain typical of those for their respective wavelengths. Hence, in this case, the strengths of the signals can indicate the intensities of the lights, and the ratios the wavelengths of the lights. We can thus distinguish between changes in the intensity, and changes in the wavelength, of the lights, and hence a basis for colour vision exists. Most colours consist of a mixture of many wavelengths of the spectrum, not just a single wavelength as considered so far; but the above argument is quite general, and changes in spectral composition of such spectrally complex colours will cause changes in the ratios of the cone responses, and changes in the amount of light will cause changes in the strengths of the responses. Of course, in general, there will also be changes in the ratios of the β to γ and β to ρ responses as the spectral composition is changed, and these further assist in the discrimination of colours.

The different types of cone, ρ, γ, and β are distributed more or less randomly in the retinal mosaic of receptors on which the light falls (Mollon and Bowmaker, 1992; Hofer, Carroll, Neitz, Neitz, and Williams, 2005; Williams, 2006). The relative abundance of the three types of cone varies considerably from one observer to another (Williams, 2006), but it is always found that there are many fewer β cones than ρ and γ; one estimate of their relative abundances is that they are, on average, in the ratios of 40 to 20 to 1 for the ρ, γ, and β cones, respectively (Walraven and Bouman, 1966). This rather asymmetrical arrangement is, in fact, very understandable. Because the eye is not corrected for chromatic aberration, it cannot simultaneously focus sharply the three regions of the spectrum in which the ρ, γ, and β cones are most sensitive, that is, wavelengths of around 580 nm, 540 nm, and 440 nm, respectively. The eye focuses light of wavelength about 560 nm, so both the ρ and γ responses correspond to images that are reasonably sharp; the β cones then have to receive an image that is much less sharp, and hence it is unnecessary to provide such a fine network of β cones to detect it.

1.6 VISUAL SIGNAL TRANSMISSION

When light is absorbed in a receptor in the retina, the molecules of its photosensitive pigment are excited, and, as a result, a change in electrical potential is produced. This change then travels through a series of relay cells, and eventually results in a series of voltage pulses being transmitted along a nerve fibre to the brain. The rates at which these pulses are produced provide the signal modulation, a higher rate indicating a stronger signal, and a lower rate a weaker signal. However, zero signal may be indicated by a *resting rate*, and rates lower than this can then indicate an opposite signal. The pulses themselves are all of the same amplitude, and it is only their frequency that carries information to the brain. The frequencies involved are typically from a few per second to around 400 per second.

It might be thought that, as there are four different types of receptor, the rods and the three different types of cone, there would be four different types of signal, transmitted along four different types of nerve fibre, each indicating the strength of the response from one of the four receptor types. However, there is overwhelming evidence that this is not what happens (Mollon, 1982). While much still remains unknown about the way in which the signals are encoded for transmission, the simple scheme shown in Figure 1.4 can be regarded as a plausible framework for incorporating some of the salient features of what is believed to take place. The strengths of the signals from the cones are represented by the symbols, ρ, γ, and β. These strengths will depend on the amount of radiation usefully absorbed by the three different types of cone, and on various other factors (as will be discussed in more detail in Chapter 15).

The rod and cone receptors are shown in Figure 1.4 to be connected to *neurons* (nerve cells) that eventually result in just three, not four, different types of signal in the nerve fibres. One of these signals is usually referred to as an *achromatic signal*; its neurons collect inputs from both rods and all three types of cone. Because of the different abundances of the ρ, γ, and β cones, the cone part of its signal is represented as:

$$2\rho + \gamma + (1/20)\beta$$

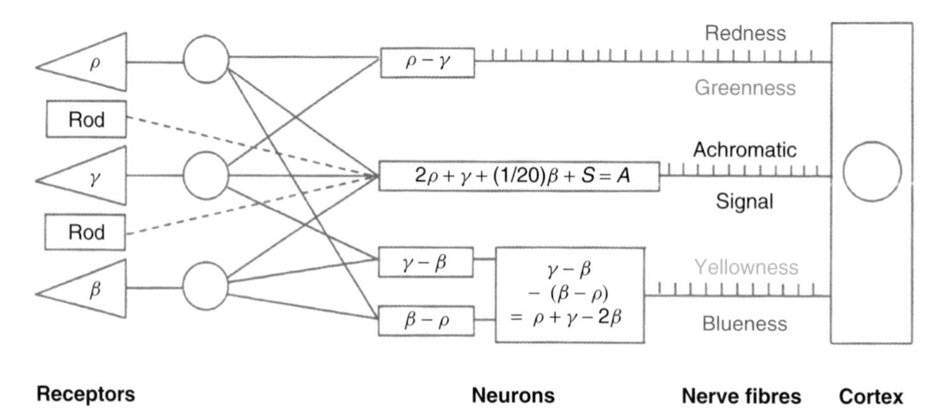

Figure 1.4 Greatly over-simplified and hypothetical diagrammatic representation of possible types of connections between some retinal receptors and some nerve fibres

(The 1/20 factor in the above expression represents a very small contribution from the β cones to the achromatic signal; some studies suggest that there is no such contribution.)

If the scotopic contribution from the rods is represented by S, then the total achromatic signal is:

$$2\rho + \gamma + (1/20)\beta + S = A$$

The other two signals in the nerve fibres are usually referred to as *colour-difference* signals. Three basic difference signals are possible:

$$\rho - \gamma = C_1$$

$$\gamma - \beta = C_2$$

$$\beta - \rho = C_3$$

To transmit all three of these signals would be to include redundancy, since, if two of them are known, the third can be deduced from the fact that $C_1 + C_2 + C_3 = 0$. In fact, there is various evidence to suggest that the signals actually transmitted resemble:

$$C_1 = \rho - \gamma$$

and

$$C_2 - C_3 = \gamma - \beta - (\beta - \rho) = \rho + \gamma - 2\beta$$

Behavioural studies have shown the presence of well-developed colour vision in various non-human species; in many cases, physiological experiments on these species have revealed signals of three general types that are broadly similar to the signals, A, C_1, and $C_2 - C_3$, proposed above.

1.7 BASIC PERCEPTUAL ATTRIBUTES OF COLOUR

We shall now consider some of the perceptual attributes of colour, in the context of these visual signals. There are three basic attributes, brightness, hue and colourfulness; they are defined as follows:

Brightness

Attribute of a visual perception according to which an area appears to exhibit more or less light. (Adjectives: *bright* and *dim*.)

Hue

Attribute of a visual perception according to which an area appears to be similar to one, or to proportions of two, of the perceived colours red, yellow, green, and blue.

Colourfulness

Attribute of a visual perception according to which an area appears to exhibit more or less of its hue.

The achromatic channel is largely responsible for providing a basis for the attribute of brightness: every colour has a brightness, and, as this channel collects responses from all types of receptor, it could indicate an overall magnitude of response for all colours. Hence we could have:

A	large	Bright colours
A	small	Dim colours

There is some evidence that the magnitude of the colour-difference signals may also make a contribution to brightness; if this is so, then the stimulation of the β cones could contribute to brightness even if they did not contribute to the achromatic signal.

If we assume that, for white, grey, and black colours, $\rho = \gamma = \beta$, then the colour difference signals, C_1, C_2, and C_3 would be zero for these colours. The hues of colours could then be indicated thus:

C_1	positive	Reddish colours
C_1	negative	Greenish colours
$C_2 - C_3$	positive	Yellowish colours
$C_2 - C_3$	negative	Bluish colours

The particular hue of any colour could then be indicated by the ratio of C_1 to $C_2 - C_3$, which corresponds to C_1, C_2, and C_3 being in constant ratios to one another; and the colourfulness of colours could then be indicated by the strengths of the signals C_1 and $C_2 - C_3$, zero indicating zero colourfulness (that is, white, grey, or black, the *achromatic colours*), and signals increasingly different from zero indicating the degree to which the hue is exhibited in colours possessing a hue (the *chromatic colours*). The four hues, red, green, yellow, and blue are known as *unique hues*, or *unitary hues*.

1.8 COLOUR CONSTANCY

One of the most important practical uses of colour is as an aid to the recognition of objects. Objects can, however, be illuminated under a very wide range of conditions; in particular, the level and colour of the illumination can vary very considerably. Thus bright sunlight represents a level of illumination that is typically about a thousand times that inside a living room; and electric tungsten-filament lighting is much yellower than daylight. However, the human visual system is extremely good at compensating for changes in both the level and the colour of the lighting; as a result of this *adaptation*, objects tend to be recognized as having nearly the same colour in very many conditions, a phenomenon known as *colour constancy*. Colour constancy is only approximate, and considerable changes in colour appearance can sometimes occur, as in the tendency for colours that appear purple in daylight to appear distinctly redder in tungsten light; but colour constancy is, nevertheless, an extremely powerful and important effect in colour perception.

We can, for the moment, regard colour constancy as corresponding to the ρ, γ, and β responses for whites, greys, and blacks, being approximately equal to one another and constant, no matter what the level and colour of the illuminant. (This will be discussed more fully in Sections 3.13 and 6.12, and in Chapter 15.)

1.9 RELATIVE PERCEPTUAL ATTRIBUTES OF COLOURS

Let us now consider, as an example, a white and a grey patch seen side by side on a piece of paper. If we observe the patches in bright sunlight they will look very bright, and if we take them into the shade, or indoors, they will look less bright: but the white will still look white, and the grey will still look grey. By means which are not fully understood, the eye and brain subconsciously allow for the fact that the lower brightnesses are not caused by changes in the objects, but by changes in the illumination. The same is also true for changes in the colour of the illuminant over the range of typical 'white light' illuminants. Thus, in tungsten light, the patches still look approximately white and grey, and certainly not yellow and brown. (It is to explain these phenomena that the *Retinex Theory* was produced by Edwin Land; see, for example, Land and McCann, 1971.) This is such an important phenomenon that certain relative perceptual attributes of colours are given separate names.

The term *lightness* is used to describe the brightness of objects relative to that of a similarly illuminated white. Thus, whereas brightness could depend on the magnitude of an achromatic signal such as A, lightness could depend on a signal such as A/A_n, where the subscript n indicates that the signal is for an appropriately chosen reference white. Changes in the level of illumination would tend to change A and A_n in the same proportions, thus tending to keep A/A_n constant; hence lightness would tend to remain constant for a given colour. Whites and greys could then be recognised as such by their *lightnesses*, independent of their *brightnesses*.

Just as it is possible to judge brightness relative to that of a white, so it is also possible to judge colourfulness in proportion to the brightness of a white, and the relative colourfulness then perceived is called *chroma*. It is well known that, as the illumination level falls, the colourfulness of objects decreases. Thus, in bright daylight, a scene may look very colourful, but it will look less so under dark clouds; and as the illumination level falls in the evening the colourfulness gradually reduces to zero when scotopic levels are reached. Over most of the photopic range of illumination levels however, the colours of objects are recognised as approximately constant. Let us take, as an example, a red tomato on a white plate. In bright daylight, the red tomato looks very colourful: its red hue is exhibited very strongly. If we then view it at a much lower level of illumination it will look less colourful (its hue will not be exhibited so strongly); but the white plate will also look less bright, and the visual system then subconsciously judges that the lower colourfulness in the dimmer light is caused by the lower level of illumination characterised by the lower brightness of the white. The colourfulness judged in proportion to the brightness of the white is then seen to be unchanged, and this relative colourfulness is the chroma. Thus, whereas colourfulness could depend on the magnitudes of signals such as C_1 and $C_2 - C_3$, chroma could depend on the magnitudes of signals such as C_1/A_n and $(C_2 - C_3)/A_n$ where the subscript n again indicates that the signal is for the white.

Lightness and chroma are therefore defined as follows:

Lightness

The brightness of an area judged relative to the brightness of a similarly illuminated area that appears to be white or highly transmitting. (Adjectives: *light* and *dark*.)

Chroma

The colourfulness of an area judged in proportion to the brightness of a similarly illuminated area that appears to be white or highly transmitting. (Adjectives: *strong* and *weak*.)

Because these two attributes are defined with reference to a 'similarly illuminated area' they apply only to *related colours*, that is, colours perceived to belong to areas seen in relation to other colours. They do not apply to *unrelated colours*, that is, colours perceived to belong to areas seen in isolation from other colours. Self-luminous colours, such as light sources, are usually perceived as unrelated colours; colours produced by objects reflecting light in ordinary viewing conditions are usually perceived as related colours. A television display is of itself self-luminous, but, within the picture area, related colours can be seen if the portrayal is of illuminated objects. Equally the colours of objects within the picture area may be seen as related to the colours of objects within the room; this is especially true when the level of illumination in the room approaches, or even exceeds, that of the television display. A transmitting colour can be perceived as a related colour if seen in suitable relationship to other areas, as in a stained glass window in a church or in a photographic transparency; it is for this reason that the words 'highly transmitting' are included in the above definitions.

If we consider the case of the tomato on the plate again, because the tomato is a solid object, the level of illumination will vary considerably over its surface, being high where the light falls on it perpendicularly, low where it falls on it at a glancing angle, and even lower for those parts in shadow. The brightness of a similarly illuminated white can then be readily judged only for a few areas, and hence lightness and chroma can only be evaluated in these areas. It is also possible however, to judge colourfulness relative to the brightness of the same area, instead of relative to that of a similarly illuminated white; when this is done the attribute is called *saturation*. This attribute can be readily judged at all parts of the tomato; hence, saturation, together with hue, can then be used to judge the uniformity of colour over the surface of the tomato. Saturation could depend on signals such as C_1/A and $(C_2 - C_3)/A$, where A is the achromatic signal for the same area as that of the colour, rather than for the white. Saturation is then defined as follows:

Saturation

Colourfulness of an area judged in proportion to its brightness.

Because the judgement of saturation does not require the concept of a similarly illuminated white, it is applicable to both related and unrelated colours. Consider, as an example of an unrelated colour, a red traffic-light signal seen first directly, and then reflected in a piece of plane glass, such as a shop window. When seen directly, the red signal will usually look quite bright and colourful: but its reflection will look both less bright and less colourful. However, it will still look red, not pink, because its lower colourfulness will be judged in proportion to its lower brightness and it will be perceived to have the same saturation. Thus recognition of the colour of the traffic-light will depend on its hue and saturation, not on its hue and colourfulness.

In photopic conditions of fairly high levels of illumination, the rod response, S, is usually regarded as negligibly small; in this case, when ρ, γ, and β are in constant ratios to one another, $C_1/(C_2 - C_3)$ will be constant, and this implies constant hue; and C_1/A, and $(C_2 - C_3)/A$ will be constant, and this implies constant saturation; but the brightness will not be constant (which is in accord with the discussion of the 500 nm and 600 nm lights in Section 1.5).

1.10 DEFECTIVE COLOUR VISION

It has long been known that some observers have colour vision that is markedly different from the average of most observers. Such people are popularly known as 'colour blind', but a more appropriate term is *colour defective*, because, in most cases, what is involved is a reduction in colour discrimination, not its complete loss.

There are various types of colour deficiency, and their exact causes are still a matter of some debate (Ruddock and Naghshineh, 1974; Nathans, Merbs, Sung, Weitz, and Wang, 1992; Birch, 2001; Carroll, Neitz, Hofer, Neitz, and Williams, 2004). However, the most likely causes for the various categories are given below in brackets:

Protanopia

No discrimination of the reddish and greenish contents of colours, with reddish colours appearing dimmer than normal. (ρ pigment missing.)

Deuteranopia

No discrimination of the reddish and greenish contents of colours, without any colours appearing appreciably dimmer than normal. (γ pigment missing; the similarity of the shapes of the ρ and γ spectral sensitivity curves on the short wavelength side of their peaks preserves approximately normal brightness of colours.)

Tritanopia

No discrimination of the bluish and yellowish contents of colours, without any colours appearing appreciably dimmer than normal. (β pigment missing; the very small contribution of the β cones to the achromatic signal preserves approximately normal brightness of colours.)

Cone monochromatism

No colour discrimination, but approximately normal brightnesses of colours. (γ and β pigments missing; that is, a combination of deuteranopia and tritanopia.) Or no colour discrimination, and brightness confined mainly to the blue part of the spectrum (ρ and γ pigments missing, that is, a combination of protanopia and deuteranopia).

Rod monochromatism

No colour discrimination, and brightnesses typical of scotopic vision. (No cones present.)

Protanomaly

Some reduction in the discrimination of the reddish and greenish contents of colours, with reddish colours appearing dimmer than normal. (ρ cones having a spectral sensitivity curve shifted along the wavelength axis towards that of the normal γ cones, because of molecular changes in the ρ pigment.)

Deuteranomaly

Some reduction in the discrimination of the reddish and greenish contents of colours, without any colours appearing abnormally dim. (γ cones having a spectral sensitivity curve shifted along the wavelength axis towards that of the normal ρ cones, because of molecular changes in the γ pigment.)

Tritanomaly

Some reduction in the discrimination of the bluish and yellowish contents of colours, without any colours appearing appreciably dimmer than normal. (β cones having a spectral sensitivity curve shifted along the wavelength axis towards that of the normal γ cones, because of molecular changes in the β pigment.)

The degree of reduction in colour discrimination in the cases of protanomaly, deuteranomaly, and tritanomaly (referred to as *anomalous trichromatism*) varies from only slight differences from normal observers, to nearly as great a loss as occurs in protanopia, deuteranopia, and tritanopia (referred to as *dichromatism*). Colour matches made by normal observers are accepted by dichromats, but not always by anomalous trichromats.

The occurrence of these different types of defective colour vision varies enormously, and is different for men and for women. The following figures are estimates for Western races, based on various surveys.

Type	Men %	Women %
Protanopia	1.0	0.02
Deuteranopia	1.1	0.01
Tritanopia	0.002	0.001
Cone monochromatism	Very rare	Very rare
Rod monochromatism	0.003	0.002
Protanomaly	1.0	0.02
Deuteranomaly	4.9	0.38
Tritanomaly	Rare	Rare
Total	8	0.4

The nature of the colour confusions likely to be made by colour defective observers will be described in terms of colorimetry in Section 3.6. Detection of colour deficiency is usually carried out by means of confusion charts, such as the widely used *Ishihara* charts (Dain, 2004), the *City University Colour Vision Test* (Barbur, Harlow, and Plant, 1994), and the *Cambridge (HRR) Colour Vision Test* (Bailey, Neitz, Tait, and Neitz, 2004), or the more diagnostic *Farnsworth-Munsell 100 Hue Test* (Farnsworth, 1943); but accurate classification of the type of deficiency usually requires the use of other test methods (see, for instance, Fletcher and Voke, 1985; Birch, 2001). Most cases of colour deficiency are hereditary, but progressive tritanomaly is also acquired with certain diseases.

1.11 COLOUR PSEUDO-STEREOPSIS

When saturated red and blue lettering is viewed on a dark or black background, some observers perceive the red letters as standing out in front of the plane of the paper, and the blue letters lying behind it, even though the letters are all actually in the plane of the paper. However, although a majority of observers see this phenomenon, there is a minority for whom the reverse effect occurs, with the red letters receding and the blue letters advancing; and there is a third, and still smaller group, who see all the letters in the same plane as the paper. This effect is caused by a combination of the chromatic aberration of the eye and the fact that the pupils of the eyes are not always central with respect to their optical axes. This is illustrated in Figure 1.5. In the left hand diagram, the pupils are displaced outwards relative to the optical axes, A, so that the rays from an object at O are dispersed by refraction with images of blue (B) light in the two eyes being closer to one another than in the case of red (R) light. The red light therefore appears to emanate from an object that is closer than an object from which blue light appears to emanate. In the right hand diagram, the pupils are displaced inwards relative to the optical axes, A, with the result that a blue object now appears nearer than a red object. In the centre diagram, the pupils are concentric with the optical axis, A, and no effect occurs.

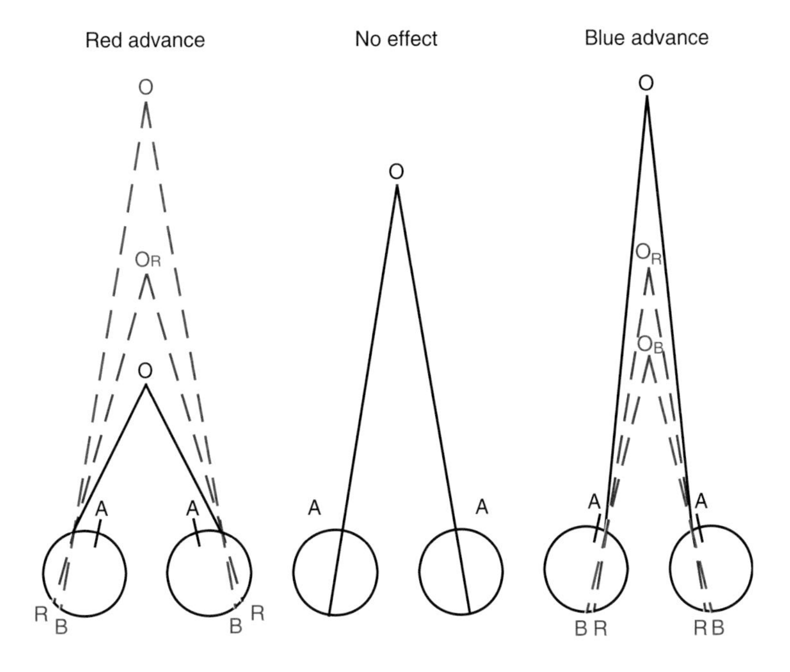

Figure 1.5 Illustration (not to scale) of the basis for colour pseudo-stereopsis

REFERENCES

Bailey, J.R., Neitz, M., Tait, D.M., and Neitz, J., Evaluation of an updated HRR color vision test, *Visual Neuroscience*, **21**, 431–436 (2004).

Barbur, J.L., Harlow, A.J., and Plant, G.T., Insights into the different exploits of colour in the visual cortex, *Proc Roy. Soc. London B*, **258**, 327–334 (1994).

Birch, J., *Diagnosis of Defective Colour Vision*, 2nd *Ed*., Butterworth-Heinemann, Edinburgh, Scotland (2001).

Crawford, B.H., The scotopic visibility function, *Proc. Phys. Soc. B*, **62**, 321–334 (1949).

Carroll, J., Neitz, M., Hofer, H., Neitz, J., and Williams, D.R., Functional photoreceptor loss revealed with adaptive optics: an alternate cause of color blindness, *Proceedings of the National Academy of Sciences, U.S.A.*, **101**, 22, 8461–8466 (2004).

Dain, S.J., Clinical colour vision tests, *Clinical Experimental Optometry*, **87**, 276–293 (2004).

Dartnall, H.J.A., Bowmaker, J.K., and Mollon, J.D., Human visual pigments: microspectrophotometric results from the eyes of seven persons, *Proc. Roy. Soc. Lond. B*, **220**, 115–130 (1983).

Estévez, O., *On the Fundamental Data-Base of Normal and Dichromatic Colour Vision*, Ph.D. Thesis, University of Amsterdam, Holland (1979).

Farnsworth, D., The Farnsworth-Munsell 100-hue and dichotomous tests for color vision, *J. Opt. Soc. Am*., **33**, 568–574 (1943).

Fletcher, R., and Voke, J., *Defective Colour Vision*, Hilger, Bristol, England (1985).

Hofer, H., Carroll, J., Neitz, J., Neitz, M., and Williams, D. R., Organization of the human trichromatic cone mosaic, *J. Neurosci*., **25**(42), 9669–9679 (2005).

Hurvich, L.M., *Color Vision*, p.21, Sinauer Associates, Sunderland, Mass., U.S.A. (1981).

Judd, D.B. and Wyszecki, G., *Color in Business Science and Industry*, *3rd Ed*., p.388, Wiley, New York, NY, U.S.A. (1975).

Land, E.H., and McCann, J.J., Lightness and Retinex theory, *J. Opt. Soc. Amer*., **61**, 1–11 (1971).

McLaren, K., Newton's indigo, *Color Res. Appl*., **10**, 225–229 (1985).

Mollon, J.D., Color vision, *Ann. Rev. Psychol*., **33**, 41–85 (1982).

Mollon, J.D., and Bowmaker, J.K., The spatial arrangement of cones in the primate fovea, *Nature*, **360**, 677–679 (1992).

Nathans, J., Merbs, S.L., Sung, C.-H., Weitz, C.J., and Wang, Y., Molecular genetics of human visual pigments, In Campbell, A., ed., *Annual Review Genetics*, **26**, 403–424, Annual Reviews Inc., Palo Alto, CA., U.S.A. (1992).

Ruddock, K.H., and Naghshineh, S., Mechanisms of red-green anomalous trichromacy: hypothesis and analysis, *Mod. Prob. Ophthal*., **13**, 210–214 (1974).

Smith, V.C., and Pokorny, J., Spectral sensitivity of colorblind observers and the cone pigments, *Vision Res*., **12**, 2059–2071 (1972).

Stockman, A., Physiologically-based color matching functions, *IS&T and SID's* 16th *Color Imaging Conference: Color Science and Engineering Systems, Technologies, and Applications*, pp. 1–5, IS&T Springfield, VA, U.S.A. (2008).

Stockman, A., Sharpe, L. T., and Fach, C. C., The spectral sensitivity of the human short-wavelength cones, *Vision Res*., **39**, 2901–2927 (1999).

Stockman, A., and Sharpe, L. T., Spectral sensitivities of the middle- and long-wavelength sensitive cones derived from measurements in observers of known genotype, *Vision Res*., **40**, 1711–1737 (2000).

Wald, G., Human vision and the spectrum, *Science*, **101**, 653–658 (1945).

Walraven, P.L. and Bouman, M.A., Fluctuation theory of colour discrimination of normal trichromats, *Vision Res.*, **6**, 567–586 (1966).

Williams, D., Color and the cone mosaic, *IS&T and SID's 14th Color Imaging Conference: Color Science and Engineering Systems, Technologies, and Applications*, pp. 1–2, IS&T Springfield, VA., U.S.A. (2006).

GENERAL REFERENCES

Bass, M., ed., *Handbook of Optics, Vol. III, Vision and Vision Optics*, McGraw Hill, New York, NY, U.S.A. (2010).

Hunt, R.W.G., Colour terminology, *Color Res. Appl.*, **3**, 79–87 (1978).

Gregory, Richard L., *Eye and Brain: The Psychology of Seeing*, 5th *revised Ed.*, Oxford University Press, Oxford, England (1998).

Kaiser, P., and Boynton, R.M., *Human Color Vision*, 2nd *Ed.*, Optical Society of America, Washington DC, U.S.A. (1996).

Mollon, J.D., Pokorny, J., and Knoblauch, K., (editors) *Normal and Defective Colour Vision*, *revised Ed.*, Oxford University Press, Oxford, England (2003).

Valberg, A., *Light Vision and Color*, Wiley, Chichester, England (2005).

Wandell, Brian A., *Foundations of Vision: Behaviour, Neuroscience and Computation*, Sinauer Associates, Sunderland, MA., U.S.A. (1995).

2

Spectral Weighting Functions

2.1 INTRODUCTION

When white light is passed through a prism, or some other suitable device, to form a spectrum, it normally appears to be brightest somewhere near the middle, around the green part. As the part considered is increasingly displaced from the brightest part, the brightness decreases continually until, at the two ends, it merges into complete darkness. For most sources of white light these changes in brightness along the length of the spectrum are not usually caused mainly by changes in the power present, but by changes in the sensitivity of the eye to different wavelengths.

2.2 SCOTOPIC SPECTRAL LUMINOUS EFFICIENCY

For the scotopic type of vision that occurs at low light levels, it was explained, in Section 1.5, that the photosensitive pigment, rhodopsin, on which the rods depend, absorbs light most strongly in the blue-green part of the spectrum and less strongly in the other parts. As a result, the sensitivity, throughout the spectrum, of the eye in scotopic vision is as shown by the broken line in Figure 2.1 (which was also shown as the broken line in Figure 1.3). This function was obtained by having observers adjust the power in a beam of each wavelength of the spectrum until its apparent brightness matched that of a beam of fixed power of a reference wavelength (see Section 1.5).

But what is the situation if light of other wavelengths is present in the same stimulus? At each wavelength, the power of the light can be multiplied by the heights, at those wavelengths, of the broken curve in Figure 2.1, and the resulting products added together; then, if this sum is equal for any two stimuli, experiment shows that they will look equally bright if viewed under the same scotopic conditions. The broken curve of Figure 2.1 can, therefore, be thought of as a function representing the different efficiencies with which different parts of the spectrum excite the scotopic visual system. By convention, the maximum efficiency is given a value of 1.0, and, for scotopic vision, this occurs at a wavelength of approximately 510 nm. It is clear that, for wavelengths above and below this, the efficiency falls until it reaches approximately zero for wavelengths greater than about 620 nm, and

Measuring Colour, Fourth Edition. R.W.G. Hunt and M.R. Pointer.

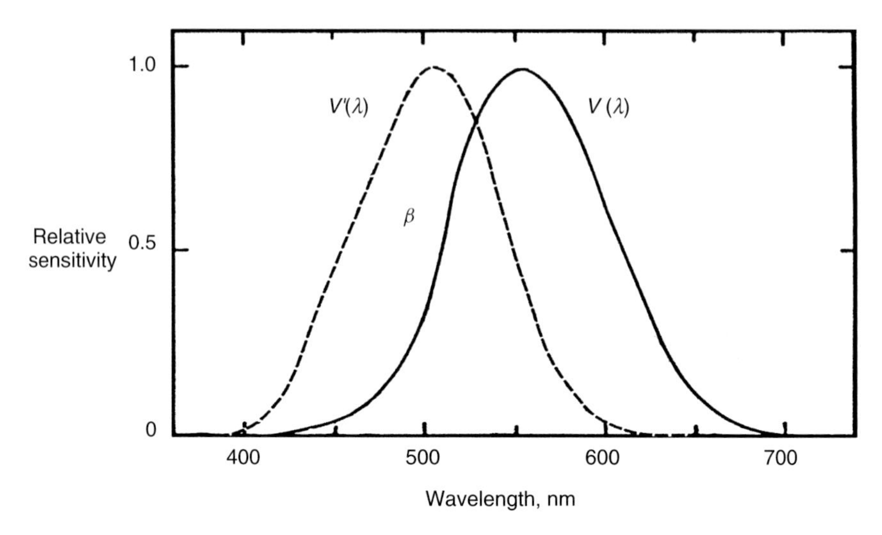

Figure 2.1 The spectral luminous efficiency curves, $V'(\lambda)$, for low level (scotopic) vision, and $V(\lambda)$, for high level (photopic) vision

less than about 400 nm. The broken curve of Figure 2.1, which has been standardised internationally by the CIE (Commission Internationale de l'Éclairage), is known as the *spectral luminous efficiency for scotopic vision*, and is given the symbol $V'(\lambda)$. Its values are given in full in Appendix 2. An observer having a relative spectral sensitivity function that is the same as the $V'(\lambda)$ function is known as the *CIE standard scotopic photometric observer*.

The $V'(\lambda)$ function can therefore be used as a weighting function to determine which of any two lights, whatever their spectral composition, will appear, under the same scotopic viewing conditions, to have the greater brightness, or whether they have the same brightness. This is done by using the following formula:

$$L' = K'_{\mathrm{m}}(P_1V'_1 + P_2V'_2 + P_3V'_3 + \ldots\ldots..P_{\mathrm{n}}V'_{\mathrm{n}})$$

In this formula K'_{m} is a constant, P_1, P_2, P_3, etc., are the amounts of power per small constant-width wavelength interval throughout the visible spectrum, and V'_1, V'_2, V'_3, etc., are the values of the $V'(\lambda)$ function at the central wavelength of each interval. The effect of this formula is to evaluate the radiation by giving most weight to its components having wavelengths near 510 nm, and progressively less and less weight as the wavelength differs more and more from 510 nm, until at the two ends of the spectrum no weight is given at all. The result is an evaluation of the radiation, not in terms of its total power, but in terms of its ability to stimulate the eye (at low levels of illumination). The use of weighting functions in this way is a very important procedure in photometry and colorimetry, and different weighting functions are used to represent different visual properties. The $V'(\lambda)$ function is used to represent the relative spectral sensitivity of the eye in scotopic conditions.

In the above formula, if K'_{m} is put equal to 1700, and P is the radiance in watts per steradian and per square metre, L' is the scotopic luminance expressed in scotopic candelas per square metre. (A steradian is a unit of solid angle defined as the solid angle that, having its vertex in the middle of a sphere, cuts off an area on the surface of the sphere equal to that of a square with side of length equal to that of the radius of the sphere.) If P is in some other radiant measure, then L' will be the corresponding scotopic photometric measure; for example, if P is the irradiance in watts per square metre, then L' will be the scotopic illuminance in scotopic lux. (For details of these corresponding photometric and radiometric measures, see Appendix 1, Table A1.1).

It is important to realise the limitations of photometric measures. They can tell us whether two lights are equally bright; or, if one is brighter, they can tell us by what factor the other must be increased in radiant power (keeping the relative spectral composition the same) for the two brightnesses to appear equal. Photometric measures, however, do not tell us how bright a given light will look, because this depends on the conditions of viewing. For example, a motor car headlight that appears dazzlingly bright at night, appears to have a much more modest brightness when seen in bright sunlight: however, in both situations the values of radiance are the same. Moreover, even for a fixed set of viewing conditions, the relationship between photometric measures and brightness is not one of simple proportionality. (This topic will be discussed more fully in Chapter 15.)

Let us now consider, as an example, two lights, B and R, whose spectral radiances, P_B and P_R, per small constant-width wavelength interval throughout the spectrum, are as shown in Table 2.1. (To obtain convenient numbers, the powers are expressed in microwatts, μW.) The table shows how the $V'(\lambda)$ function is used to obtain the weighted spectral power, PV', at each wavelength. These products are then summed, and multiplied by 1700 (and by 10^{-6} to allow for the use of microwatts) to obtain the scotopic luminance, L', in scotopic candelas per square metre. It is seen that the scotopic luminances are 0.00806 for light B and 0.00680 for light R; we therefore conclude that the scotopic luminance of light B is greater than that of light R by a factor of 806/680, that is by 1.18 times. Hence the powers in light R would have to be increased by a factor of 1.18 to ensure that light R appeared as bright as light B when seen under the same scotopic viewing conditions. In Table 2.1, data at every 10 nm have been used; this interval is adequate for this example, but data at 5 nm or even 1 nm may be required for some applications.

2.3 PHOTOPIC SPECTRAL LUMINOUS EFFICIENCY

If we wish to compare the brightnesses of stimuli seen under photopic conditions of vision, we have to use, not the $V'(\lambda)$ function shown by the broken line of Figure 2.1, but a different function, the *photopic spectral luminous efficiency function*, $V(\lambda)$, shown by the full line in Figure 2.1. Its values are given in full in Appendix 2. It is clear that the $V(\lambda)$ function, although of similar general shape as the $V'(\lambda)$ function, peaks at a wavelength of about 555 nm instead of at about 510 nm. This represents a change in relative spectral sensitivity towards an increase for reddish colours and a decrease for bluish colours, which, as mentioned in Section 1.4, is referred to as the *Purkinje phenomenon*.

The experimental basis for obtaining the $V(\lambda)$ function is not as simple as that used for the scotopic $V'(\lambda)$ function. Because the different colours of the spectrum are not seen at scotopic levels, the observer's task of adjusting the power at each wavelength to obtain equality of apparent brightness with a reference beam is a comparatively simple one. But, if the same technique is used at photopic levels, the large difference in colour between the light of most of the wavelengths and that of the reference beam makes the task very much more difficult. However, if the two beams are seen, not side by side, but alternately, it has

Table 2.1 Calculation of scotopic values using the CIE $V'(\lambda)$ function

		Power in microwatts μW, in interval $\lambda - 5$ nm to $\lambda + 5$ nm		Scotopic light	
Wavelength, λ	$V'(\lambda)$	P_B	P_R	$P_B V'$	$P_R V'$
380 nm	0.0006	1.10	0.00	0.0007	0.0000
390	0.0022	1.05	0.00	0.0023	0.0000
400	0.0093	1.00	0.00	0.0093	0.0000
410	0.0348	0.95	0.00	0.0331	0.0000
420	0.0966	0.90	0.00	0.0869	0.0000
430	0.1998	0.85	0.05	0.1698	0.0100
440	0.3281	0.80	0.10	0.2625	0.0328
450	0.4550	0.75	0.15	0.3413	0.0683
460	0.5670	0.70	0.20	0.3969	0.1134
470	0.6760	0.65	0.25	0.4394	0.1690
480	0.7930	0.60	0.30	0.4758	0.2379
490	0.9040	0.55	0.35	0.4972	0.3164
500	0.9820	0.50	0.40	0.4910	0.3928
510	0.9970	0.45	0.45	0.4487	0.4487
520	0.9350	0.40	0.50	0.3740	0.4675
530	0.8110	0.35	0.55	0.2839	0.4461
540	0.6500	0.30	0.60	0.1950	0.3900
550	0.4810	0.25	0.65	0.1203	0.3127
560	0.3288	0.20	0.70	0.0658	0.2302
570	0.2076	0.15	0.75	0.0311	0.1557
580	0.1212	0.10	0.80	0.0121	0.0970
590	0.0655	0.05	0.85	0.0033	0.0557
600	0.0332	0.00	0.90	0.0000	0.0299
610	0.0159	0.00	0.95	0.0000	0.0151
620	0.0074	0.00	1.00	0.0000	0.0074
630	0.0033	0.00	1.05	0.0000	0.0035
640	0.0015	0.00	1.10	0.0000	0.0017
650	0.0007	0.00	1.15	0.0000	0.0008
660	0.0003	0.00	1.20	0.0000	0.0004
670	0.0001	0.00	1.25	0.0000	0.0001
680	0.0001	0.00	1.30	0.0000	0.0001
690	0.0000	0.00	1.35	0.0000	0.0000
700	0.0000	0.00	1.40	0.0000	0.0000
710	0.0000	0.00	1.45	0.0000	0.0000
720	0.0000	0.00	1.50	0.0000	0.0000
730	0.0000	0.00	1.55	0.0000	0.0000
740	0.0000	0.00	1.60	0.0000	0.0000
750	0.0000	0.00	1.65	0.0000	0.0000
760	0.0000	0.00	1.70	0.0000	0.0000
770	0.0000	0.00	1.75	0.0000	0.0000
780	0.0000	0.00	1.80	0.0000	0.0000
Totals				4.7402	4.0022
$\times 1700$				8058.30	6803.74
$\times 10^{-6}$				0.00806	0.00680

been found that a rate of alternation can be set that is too fast for colour differences to be detected, but slow enough for brightness differences still to be visible. Using this rate of alternation, it is then a somewhat easier task to set the brightnesses of the two beams so that no flicker can be seen, and this then corresponds to equality of brightness. This technique is known as *flicker photometry*, and it was used to obtain the $V(\lambda)$ function shown by the full line of Figure 2.1. An observer having a relative spectral sensitivity function that is the same as the $V(\lambda)$ function is known as the *CIE standard photometric observer*. (Flicker photometry is not the only experimental method of obtaining the $V(\lambda)$ function, as will be discussed in Section 2.8, where other methods, and other photopic functions, will be briefly reviewed.)

The $V(\lambda)$ function is used in the same way as the $V'(\lambda)$ function for evaluating the relative brightnesses of stimuli, but using the formula:

$$L = K_{\mathrm{m}}(P_1V_1 + P_2V_2 + P_3V_3 + \ldots\ldots P_{\mathrm{n}}V_{\mathrm{n}})$$

where K_{m} is another constant, P_1, P_2, P_3, etc., are the amounts of power per small constant-width wavelength interval throughout the visible spectrum, and V_1, V_2, V_3, etc., are the values of the $V(\lambda)$ function at the central wavelength of each interval over the wavelength range considered. (The official CIE equation for computing L uses integration, but, because the $V(\lambda)$ function is not defined as a continuous function, summation has to be used in practice.) If K_{m} is put equal to 683 and P is the radiance in watts per steradian and per square metre, L is the *luminance* expressed in candelas per square metre (cd m^{-2}). As before, if other radiant measures are used for P, the corresponding photometric measures are obtained (see Appendix 1, Table A1.1, for details).

In Table 2.2, two stimuli CB and CR are evaluated using $V(\lambda)$ as a weighting function; they have the same relative spectral power distributions as the stimuli B and R, respectively, of Table 2.1, but the powers are high enough to be in the photopic range (the units are watts instead of microwatts). It is seen that now the photopic luminances are 1595 for light CB and 5116 for light CR; we therefore conclude that the photopic luminance for light CR is greater than that for light CB by a factor of 5116/1595, that is by 3.21 times. Hence the powers in light CB would have to be increased by a factor of 3.21 to ensure that it appeared as bright as light CR when seen in the same viewing conditions and judged by flicker photometry; but in Table 2.1 light B had a higher scotopic luminance than light R. This illustrates the Purkinje phenomenon described in Section 1.4, lights B and CB being bluish, and lights R and CR being reddish. In Table 2.2, data at every 10 nm have been used again.

Once again, these photometric measures do not tell us how bright a stimulus will look. As with scotopic measures, the brightness is greatly affected by the viewing conditions, and, in any one set of viewing conditions, brightness and photometric measures are not proportional. But there is an additional factor operating in the case of photopic vision. For colours of the same luminance, there is a tendency for the brightness to increase gradually as the colourfulness of the colour increases (this phenomenon is known as the *Helmholtz-Kohlrausch effect*). Thus, if a white light and a colourful red light of the same luminance are compared side by side, the red usually looks brighter than the white. Similarly, if a colourful blue-green light and a white light have the same luminance, then the blue-green light usually looks brighter. But, if we add the red and blue-green lights together, and compare this mixture with that of the two white lights added together, we find that the

Table 2.2 Calculation of photopic values using the CIE $V(\lambda)$ function

		Power in watts W, in interval $\lambda - 5$ nm to $\lambda + 5$ nm		Photopic light	
Wavelength, λ	$V(\lambda)$	P_{CB}	P_{CR}	$P_{CB}V$	$P_{CR}V$
380 nm	0.0000	0.000	0.00	0.0000	0.0000
390	0.0001	0.000	0.00	0.0000	0.0000
400	0.0004	1.000	0.00	0.0004	0.0000
410	0.0012	0.950	0.00	0.0011	0.0000
420	0.0040	0.900	0.00	0.0036	0.0000
430	0.0116	0.850	0.05	0.0099	0.0006
440	0.0230	0.800	0.10	0.0184	0.0023
450	0.0380	0.750	0.15	0.0285	0.0057
460	0.0600	0.700	0.20	0.0420	0.0120
470	0.0910	0.650	0.25	0.0592	0.0228
480	0.1390	0.600	0.30	0.0834	0.0417
490	0.2080	0.550	0.35	0.1144	0.0728
500	0.3230	0.500	0.40	0.1615	0.1292
510	0.5030	0.450	0.45	0.2264	0.2264
520	0.7100	0.400	0.50	0.2840	0.3550
530	0.8620	0.350	0.55	0.3017	0.4741
540	0.9540	0.300	0.60	0.2862	0.5724
550	0.9950	0.250	0.65	0.2488	0.6468
560	0.9950	0.200	0.70	0.1990	0.6965
570	0.9520	0.150	0.75	0.1428	0.7140
580	0.8700	0.100	0.80	0.0870	0.6960
590	0.7570	0.050	0.85	0.0379	0.6435
600	0.6310	0.000	0.90	0.0000	0.5679
610	0.5030	0.000	0.95	0.0000	0.4779
620	0.3810	0.000	1.00	0.0000	0.3810
630	0.2650	0.000	1.05	0.0000	0.2783
640	0.1750	0.000	1.10	0.0000	0.1925
650	0.1070	0.000	1.15	0.0000	0.1231
660	0.0610	0.000	1.20	0.0000	0.0732
670	0.0320	0.000	1.25	0.0000	0.0400
680	0.0170	0.000	1.30	0.0000	0.0221
690	0.0082	0.000	1.35	0.0000	0.0111
700	0.0041	0.000	1.40	0.0000	0.0057
710	0.0021	0.000	1.45	0.0000	0.0030
720	0.0010	0.000	1.50	0.0000	0.0015
730	0.0005	0.000	1.55	0.0000	0.0008
740	0.0003	0.000	1.60	0.0000	0.0005
750	0.0001	0.000	1.65	0.0000	0.0002
760	0.0001	0.000	1.70	0.0000	0.0002
770	0.0000	0.000	1.75	0.0000	0.0000
780	0.0000	0.000	1.80	0.0000	0.0000
Totals				2.3360	7.4904
×683				1595.49	5115.93

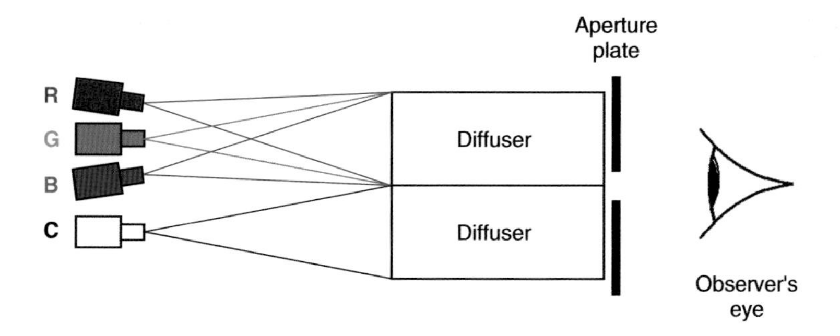

Figure 2.2 Principle of trichromatic colour matching by additive mixing of lights. R, G, and B are sources of red, green, and blue light, whose amounts can be adjusted; C is the light whose colour is to be matched. The diffusers result in uniform fields for viewing by the observer

brightnesses of the two mixtures are now similar. This arises because the mixture of the red and blue-green results in a whitish colour, and the additional brightnesses associated with the high colourfulnesses of the red and the blue-green colours have disappeared. This state of affairs is sometimes described by saying that luminances are additive, but brightnesses are not. (These phenomena could be explained if brightness depended not only on the magnitude of the signal in the achromatic channel but also on contributions from the colour-difference signals.) The type of mixing of colours referred to above is called *additive* to distinguish it from *subtractive* mixing, in which the light passes through colorants which subtract part of the light in different parts of the spectrum by absorption. Additive mixture can be either by superimposing beams of light on a diffuser as shown in Figure 2.2 (or on other types of beam combiners), or by viewing beams in succession at a frequency high enough to remove all sense of flicker, or by viewing them in adjacent areas that are too small to resolve (as in typical television displays). Subtractive mixture can be either by mixing pigments together in a suitable vehicle (as artists often do) or by having dyes or inks in superimposed thin layers (as in colour photography and printing).

The equations given for evaluating L and L' imply that the contributions of light of all the various wavelengths are exactly additive after weighting by the $V(\lambda)$, or $V'(\lambda)$, function, as appropriate. In the case of the $V'(\lambda)$ function, this is to be expected because only a single photosensitive pigment is involved. But, in the case of the $V(\lambda)$ function, the photopic achromatic signal, on which it presumably depends, is composed of contributions from the three different photosensitive pigments in the three types of cone. Moreover, there is considerable evidence that the output signals from the cones are not proportional to the light absorbed, but more nearly to its square-root (or some similar function, as will be discussed in Chapter 15). However, the contribution of the β cones is very small, and the spectral sensitivities of the γ and ρ cones overlap considerably (see Figure 1.3); the result of this is that a single function, such as the $V(\lambda)$ function, can represent the effective spectral sensitivity of the photopic achromatic signal quite closely. This is borne out by experiments that show that luminances are additive. By this is meant that, if two colour stimuli, A and B, are perceived to be equally bright, as judged by flicker photometry, and two other stimuli, C and D, are similarly perceived to be equally bright, then the additive mixtures of A with C and B with D will also be similarly perceived to be equally bright. If the criterion is equality of brightness in side by side comparisons, as already explained, additivity does not hold. However, the discrepancies from brightness additivity can be small for many stimuli, especially those of low colourfulness. When brightness additivity occurs it is sometimes referred to as *Abney's Law*. (A method of allowing for the contribution of colourfulness towards brightness will be discussed in Section 3.17.)

For most practical applications, photopic levels of illumination apply, and the convention has been adopted that, unless otherwise indicated, all photometric measures are assumed to be based on the $V(\lambda)$ function and to represent photopic vision. In those cases where the $V'(\lambda)$ function has been used, the adjective scotopic precedes the photometric term (for example, scotopic luminance, scotopic cd m^{-2}), and a prime is added to the symbol (for example, L').

2.4 COLOUR-MATCHING FUNCTIONS

Colour vision is basically a function of three variables. There are three different types of cone, and, although the rods provide a fourth spectrally different receptor, there is overwhelming evidence that, at some later stage in the visual system, the number of variables is reduced once again to three, as indicated in Figure 1.4. Hence, it is to be expected that the evaluation of colour from spectral power data should result in three different measures. At levels of illumination that are high enough for colour vision to be operating properly, there is evidence that the output from the rods is in some way rendered ineffective. At levels where both cones and rods are operating together, colour vision must be based on the four different spectral sensitivities of the cones and the rods, but, at these levels, colour discrimination is not very good, and any disturbing influence of the rods is usually not of appreciable practicable importance (although rod activity can be detected by special techniques (Trezona, 1973; Palmer, 1981), and its effects will be considered in Chapter 17).

In view of the small effects of rod intrusion, it might be thought that a good basis for evaluating colour from spectral power data would be the use of the three cone spectral sensitivity curves shown by the full lines in Figure 1.3. However, this is not what was done in establishing the internationally accepted method of evaluating colour, and the reason is that the curves of the type shown in Figure 1.3 are not known with sufficient precision. Instead, the basis chosen was three-colour matching or *trichromatic matching*, as it is usually called.

In Figure 2.2, the basic experimental arrangement for trichromatic matching is shown diagrammatically. The test colour to be matched is seen in one half of the field of view, and, in the other half, the observer sees an additive mixture of beams of red, green, and blue light. The amounts of red, green, and blue light are then adjusted until the mixture matches the test colour in brightness, hue, and colourfulness. The fact that the colours can be matched in this way, using just three *matching stimuli* in the mixture is a consequence of the fact that there are only three spectrally different types of cone in the retina.

For trichromatic matches to form a proper basis for a system of colour measurement, various parts of the experimental system must be precisely specified. Because, as we saw in Section 1.3, the retina varies considerably in its properties from one part to another, it is necessary to specify the angular size of the matching field. Standards now exist for two field sizes, 2° and 10°, but the original system was based on the 2° field size, so this will be considered first. Then the precise colours of the red, green, and blue matching stimuli need to be specified. In the original work that led to the 2° data, two separate experimental arrangements were used. John Guild (Guild 1931), at the National Physical Laboratory at Teddington, England, used a tungsten lamp and coloured filters; W. David Wright (Wright, 1969), at Imperial College, London, England, used monochromatic bands of light isolated from a spectrum formed by a system of prisms. In order to combine the two sets of results, each set of data was transformed mathematically to what would have been obtained if the following monochromatic matching stimuli had been used:

Red	700 nm
Green	546.1 nm
Blue	435.8 nm

The green and blue stimuli were chosen to coincide with two prominent lines in the mercury discharge spectrum, because this facilitated wavelength calibration; the red stimulus was chosen to be in a part of the spectrum where hue changes vary slowly with wavelength, so as to reduce the effects of any errors in wavelength calibration. The results were also transformed to what would have been obtained if the amounts of red, green, and blue had been measured using, not units of luminance, but units that result in a white being matched by equal amounts of the three matching stimuli. Compared with the unit used for the amount of red, the photometric value of the unit used for the green was rather larger, and that for the blue was very much smaller. The reason for the very considerable change in the case of the blue is that, because, as we saw in Section 1.5, there are many fewer β cones in the retina than γ or ρ cones, the blue component of white light, which stimulates mainly the β cones, results in smaller contributions to the achromatic signal than the red or green components; hence a typical beam of white light is matched by a red, green, and blue mixture in which the luminance of the blue is much less than that of the green or red. For example, to match 5.6508 cd m^{-2} of a white light consisting of equal amounts of power per small constant-width wavelength interval throughout the spectrum (a stimulus known as the *equi-energy stimulus*, S_E), the amounts of these red, green, and blue matching stimuli are:

Red	1.0000 cd m^{-2}
Green	4.5907 cd m^{-2}
Blue	0.0601 cd m^{-2}
Mixture	5.6508 cd m^{-2}

But, because white is a colour that is not perceptually biassed to either red, green, or blue, it was decided to measure the amount of green in a unit that was 4.5907 cd m^{-2}, and the amount of blue in a unit that was 0.0601 cd m^{-2}. The match of 5.6508 cd m^{-2} of S_E is then represented by:

Red	1.0000 cd m^{-2}
Green	1.0000 new green units
Blue	1.0000 new blue units

Of course, changing the units does not change the amount of matching stimulus in the match; it merely expresses it by using a more convenient number, just as, if an article cost 10 000 cents, it is more convenient to change the unit, and to speak of it as costing

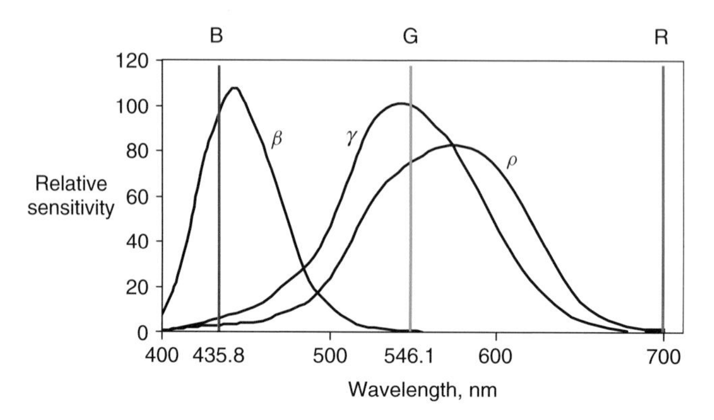

Figure 2.3 Representative spectral sensitivity curves of the three different types, ρ, γ, and β, of cone in the retina, together with the wavelengths, R, G, and B used for defining the CIE 1931 Standard Colorimetric Observer, 700 nm, 546.1 nm, and 435.8 nm, respectively

100 dollars. The amounts of the three matching stimuli, expressed in the units adopted for them, are known as *tristimulus values*.

In Figure 2.3 the probable sensitivity curves of the three types of cone are reproduced from Figure 1.3. Superimposed on this diagram are shown the wavelengths of the three matching stimuli, R, G, B. Although the ρ curve is very low at 700 nm it is not zero, and we can take it that the R stimulus will excite the ρ cones, but it will not excite the γ or β; the B stimulus will excite mainly the β cones, but with some very small excitations of the γ and ρ cones; but the G stimulus, although exciting the γ cones most, and the β cones very little, clearly excites the ρ cones quite strongly. From the point of view of trichromatic colour matching, the R stimulus can give us all the ρ excitation that we need, so the excitation of the ρ cones by the G stimulus is unwanted, and in fact leads to a complication in colour matching.

For simplicity, let us assume that the ordinate scale in Figure 2.3 shows the cone responses for unit power (in the chosen wavelength interval) at each wavelength. It is then clear from Figure 2.3 that one power unit of light of wavelength 490 nm will produce responses of approximately the following magnitudes:

$$\rho = 20 \qquad \gamma = 40 \qquad \beta = 20$$

For simplicity, let us also assume that 100 units of B produce cone responses as shown by the ordinates on Figure 2.3, that 100 units of G also produce cone responses as shown by the ordinates, and that 100 units of R produce cone responses equal to 100 times the ordinates (which are taken as 1 for ρ, and zero for γ and β; the reason why this large factor appears in the case of R is that, as can be seen from Figure 2.3, whereas G and B are near the peaks of the curves for the γ and β cones, respectively, R is displaced well away from the peak of that for the ρ cones. Figure 2.3 then shows that:

100 units of B produce	$\beta = 100$	$\gamma = 5$	$\rho = 4$
100 units of G produce	$\beta = 1$	$\gamma = 100$	$\rho = 75$
100 units of R produce	$\beta = 0$	$\gamma = 0$	$\rho = 100$

To match one power unit of 490 nm light we must produce:

$$\beta = 20 \qquad \gamma = 40 \qquad \rho = 20$$

We proceed as follows (keeping to one place of decimals for simplicity):

19.6 of B produces	$\beta = 19.6$	$\gamma = 1$	$\rho = 0.8$
39 of G produces	$\beta = 0.4$	$\gamma = 39$	$\rho = 29.2$

So together:

19.6 of B plus 39 of G produce	$\beta = 20$	$\gamma = 40$	$\rho = 30$

We have thus produced a ρ response of 30, without having added any R stimulus. The only way in which a match can be made is to add some of the R stimulus to the 490 nm light, thus:

1 unit of 490 nm light produces	$\beta = 20$	$\gamma = 40$	$\rho = 20$
10 of R produces	$\beta = 0$	$\gamma = 0$	$\rho = 10$

So together:

1 of 490 nm plus 10 of R produce	$\beta = 20$	$\gamma = 40$	$\rho = 30$

We thus have that:

1 of 490 nm plus 10 of R is matched by 39 of G plus 19.6 of B

and by convention this is written as:

1 of 490 nm is matched by -10 of R + 39 of G + 19.6 of B

The use of the negative sign does not, of course, mean that there is any such thing as negative light! It is only a convenient way of expressing the experimental fact that one of the matching stimuli, in this case the red, had to be added to the colour being matched instead of to the mixture.

If the light of 490 nm was then additively mixed with light of say 600 nm, we might have:

1 of 490 nm is matched by	-10 of R + 39 of G + 19.6 of B
1 of 600 nm is matched by	95 of R + 30 of G + 0 of B

So together:

1 of 490 nm + 1 of 600 nm is matched by	85 of R + 69 of G + 19.6 of B

The negative amount of R in the match on 490 nm has been subtracted from the positive amount in the match on 600 nm; and experiment shows that this correctly predicts the amounts of R, G, and B needed to match this mixture. The result is quite general, and the amounts of matching stimuli required in matches can be added and subtracted using the ordinary rules of algebra. This is sometimes referred to as a consequence of *Grassmann's Laws* (see the entry for *Grassmann's Laws* in Appendix 9) (CIE, 2009).

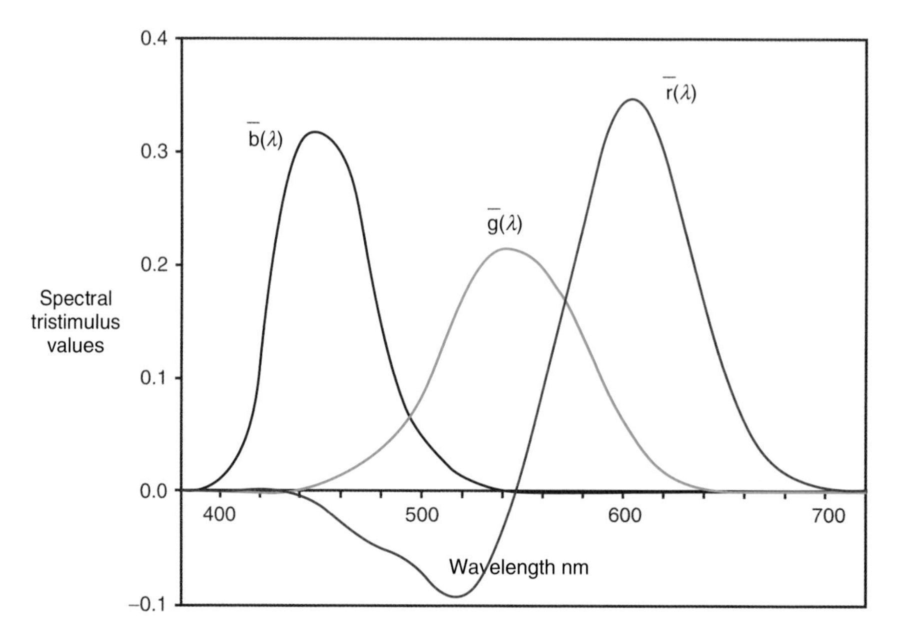

Figure 2.4 The colour-matching functions for the CIE 1931 Standard Colorimetric Observer, expressed in terms of matching stimuli, R, G, and B, consisting of monochromatic stimuli of wavelengths 700 nm, 546.1 nm, and 435.8 nm, respectively, using units such that equal amounts of them are needed to match the equi-energy stimulus, S_E

In Figure 2.4 the amounts of R, G, and B needed to match a constant amount of power per small constant-width wavelength interval at each wavelength of the spectrum are shown. Curves of this type are called *colour-matching functions*, and are designated by the symbols $\bar{r}(\lambda)$, $\bar{g}(\lambda)$, and $\bar{b}(\lambda)$. The $\bar{r}(\lambda)$ curve shows the amount of R needed for a match to be made for each wavelength of the spectrum; and the $\bar{g}(\lambda)$ and $\bar{b}(\lambda)$ curves similarly show the amounts of G and B, respectively. The $\bar{r}(\lambda)$ curve is strongly positive in the orange part of the spectrum, and negative in the blue-green part, with a small positive part in the violet. The $\bar{g}(\lambda)$ curve is strongly positive in the greenish part of the spectrum, with a small negative part in the violet; this negative part arises because these spectral colours have to be matched by R and B only with a small amount of G added to the test colour. The $\bar{b}(\lambda)$ curve is strongly positive in the bluish part of the spectrum, with a small negative part in the yellow; this negative part arises because these spectral colours have to be matched by R and G only, with a small amount of B added to the test colour. As is to be expected, light of 700 nm is matched by R only, light of 546.1 nm by G only, and light of 435.8 nm by B only, the other two curves being zero at each of these wavelengths.

If the matching stimuli had consisted of monochromatic lights of different wavelengths, then the positions at which two of the three curves were zero would move to these different wavelengths, and the sizes and positions of the positive and negative parts of the curves would also have been different. But all sets of matching stimuli, no matter what their colours, give rise to colour-matching functions having some negative parts to their curves; this is because the ρ and β curves overlap, so that it is never possible to stimulate the γ

cones on their own. The areas under the three curves of Figure 2.4 are equal (subtracting the area of the negative part from the sum of the areas of the positive parts in the case of each curve), and this arises from the choice of the units used being such that equal amounts are required to match the equi-energy stimulus, S_E.

The amounts of R, G, and B needed to match the colours of the spectrum, represented by the curves of Figure 2.4, were based on the experimental results obtained by 10 observers in Wright's investigation, and by 7 in Guild's. The two investigations produced similar results, and the average obtained, which is what is shown in Figure 2.4, therefore represents the average colour matching properties of the eyes of 17 British observers (see Fairman, Brill, and Hemmendinger, 1997 and 1998). Subsequent investigations have shown that this average result represents the normal human colour vision in 2° fields adequately (see for instance, Stiles, 1955), apart from the values being lower than they should be below 450 nm because of the use of the $V(\lambda)$ function which has this same feature; but the effect of this discrepancy on the colour-matching functions is negligible in most practical situations.

As has already been mentioned, experiment shows that colour matches are additive. Hence, if 1 power unit of light at one wavelength, λ_1, is matched by

$$\bar{r}_1 \text{ of R } + \bar{g}_1 \text{ of G } + \bar{b}_1 \text{ of B,}$$

and one power unit of light of wavelength λ_2 by

$$\bar{r}_2 \text{ of R } + \bar{g}_2 \text{ of G } + \bar{b}_2 \text{ of B,}$$

then the additive mixture of the two lights, λ_1 and λ_2, is matched by

$$(\bar{r}_1 + \bar{r}_2) \text{ of R } + (\bar{g}_1 + \bar{g}_2) \text{ of G } + (\bar{b}_1 + \bar{b}_2) \text{ of B}$$

This means that the colour-matching functions of Figure 2.4 can be used as weighting functions to determine the amounts of R, G, and B needed to match any colour, if the amount of power per small constant-width wavelength interval is known for that colour throughout the spectrum. These tristimulus values, R, G, and B, are evaluated using the formulae:

$$R = k(P_1\bar{r}_1 + P_2\bar{r}_2 + P_3\bar{r}_3 + \ldots\ldots P_n\bar{r}_n)$$
$$G = k(P_1\bar{g}_1 + P_2\bar{g}_2 + P_3\bar{g}_3 + \ldots\ldots P_n\bar{g}_n)$$
$$B = k(P_1\bar{b}_1 + P_2\bar{b}_2 + P_3\bar{b}_3 + \ldots\ldots P_n\bar{b}_n)$$

where k is a constant; the values of P are the amounts of power per small constant-width wavelength interval throughout the spectrum, and $\bar{r}$, $\bar{g}$, and $\bar{b}$, are the heights of the colour-matching functions, at the central wavelength of each interval.

The luminance of a colour matched by those amounts, R, G, and B, of the three matching stimuli is given by:

$$L = 1.0000R + 4.5907G + 0.0601B$$

the expressions on the right being the amounts of the matching stimuli expressed in units of luminance again. The constant k can be chosen so that, if P is in watts per steradian and per square metre, L is in candelas per square metre (cd m^{-2}); but if P is in some other radiant measure, then L will be in the corresponding photometric measure (see Appendix 1, Table A1.1 for details).

The fact that these tristimulus values can be *calculated* is of very great practical importance; if colorimetry required the visual matching of samples, it would be a very slow, observer-dependant, operation; but the calculation of tristimulus values merely requires the availability of the amounts of power throughout the spectrum, and this only requires purely physical measurements which modern equipment can provide accurately and quickly.

2.5 TRANSFORMATION FROM R, G, B TO X, Y, Z

The values of R, G, and B, calculated in the above way, could provide a system of colour specification which is precise, is based on representative human colour vision, and could be used to calculate colour specifications from spectral power data. But it is not used in the above form. In 1931, the CIE drew up the system of colour specification that has been adopted internationally; it was felt at that time that the presence of negative values in the colour-matching functions, and the fact that, for some colours, the value of one of the tristimulus values could be negative, would militate against the system being used without error and adopted without misgivings.

Therefore, to avoid negative numbers in colour specifications, the CIE recommended that the tristimulus values R, G, B, should be replaced by a new set of tristimulus values, X, Y, Z, that are obtained by means of the equations:

$$X = 0.49R + 0.31G + 0.20B$$

$$Y = 0.17697R + 0.81240G + 0.01063B$$

$$Z = 0.00R + 0.01G + 0.99B$$

The numbers in these equations were carefully chosen so that X, Y, and Z would always be all positive for all colours. That this is possible can be illustrated by considering a blue-green spectral colour. Such a colour will have a negative value of R, but positive values of G, and B; in fact for a certain amount of light of wavelength 500 nm, $R = -0.07173$, $G = 0.08536$, and $B = 0.04776$. The corresponding value of X is given by:

$$0.49 \times (-0.07173) + 0.31 \times (0.08536) + 0.20 \times (0.04776)$$

$$= -0.035148 + 0.026462 + 0.009552$$

$$= +0.000866$$

The values of Y and Z are also both positive for this colour. Hence the use of X, Y, and Z instead of R, G, and B has eliminated the presence of negative numbers in this colour specification, and in fact eliminates them for all colours.

There is, moreover, another advantage in using X, Y, and Z. It is clear from the equations used to evaluate them that fairly simple numbers have been used for the coefficients in the case of X and Z; but, in the case of Y, the coefficients are given to five places of decimals. The reason for this is that these coefficients have been carefully chosen so that they are in the same ratios as those of the luminances of the units used for measuring the amounts of the matching stimuli R, G, and B; that is to say:

$$0.17697, 0.81240, \text{ and } 0.01063$$

are in the same ratios as:

$$1.0000, 4.5907, \text{ and } 0.0601$$

This means that the value of Y is proportional to L, the luminance of the colour being specified. Hence, the ratio of the values of Y for any two colours, Y_1 and Y_2, is the same as the ratio of their luminances, L_1 and L_2:

$$Y_1/Y_2 = L_1/L_2$$

In the equations given above relating X, Y, Z and R, G, B, the coefficients in the equation for Y are equal to 1.0000, 4.5907, and 0.0601, each divided by the sum of these numbers, 5.6508; the coefficients in the equation for Y therefore sum to unity. The coefficients in the equation for X also sum to unity, as is also the case for the equation for Z. This means that, when $R = G = B$, it is the case that $X = Y = Z$. Hence, for the equi-energy stimulus, S_E, since $R = G = B$, it is also the case that $X = Y = Z$, and the equations were designed to achieve this result.

2.6 CIE COLOUR-MATCHING FUNCTIONS

The colour-matching functions, $\bar{r}(\lambda)$, $\bar{g}(\lambda)$, and $\bar{b}(\lambda)$, represent the amounts of R, G, and B needed to match a constant amount of power per small constant-width wavelength interval throughout the spectrum. The equivalent values of X, Y, and Z, which are denoted by the symbols $\bar{x}(\lambda)$, $\bar{y}(\lambda)$, and $\bar{z}(\lambda)$ can be calculated thus:

$$\bar{x}(\lambda) = 0.49\bar{r}(\lambda) + 0.31\bar{g}(\lambda) + 0.20\bar{b}(\lambda)$$

$$\bar{y}(\lambda) = 0.17697\bar{r}(\lambda) + 0.81240\bar{g}(\lambda) + 0.01063\bar{b}(\lambda)$$

$$\bar{z}(\lambda) = 0.00\bar{r}(\lambda) + 0.01\bar{g}(\lambda) + 0.99\bar{b}(\lambda)$$

These values are called the *CIE colour-matching functions* and they define the colour matching properties of the *CIE 1931 Standard Colorimetric Observer*, often referred to as the 2° *Observer*. These $\bar{x}(\lambda)$, $\bar{y}(\lambda)$ and $\bar{z}(\lambda)$ functions are shown in Figure 2.5 by the full lines. It is clear that, as is to be expected, these functions have no negative parts; and, because for S_E, $X = Y = Z$, the areas under the three curves are equal. The figure also illustrates another advantage of these functions: the $\bar{y}(\lambda)$ function is exactly the same shape as the $V(\lambda)$ function of Figure 2.1, and this means that, for photopic vision, instead of having four spectral functions, $\bar{r}(\lambda), \bar{g}(\lambda), \bar{b}(\lambda)$, and $V(\lambda)$, only three are required, $\bar{x}(\lambda)$, $\bar{y}(\lambda) = V(\lambda)$, and $\bar{z}(\lambda)$. The $\bar{y}(\lambda)$ function has to be the same shape as the $V(\lambda)$ function because the values of Y are proportional to luminance; by making it have a

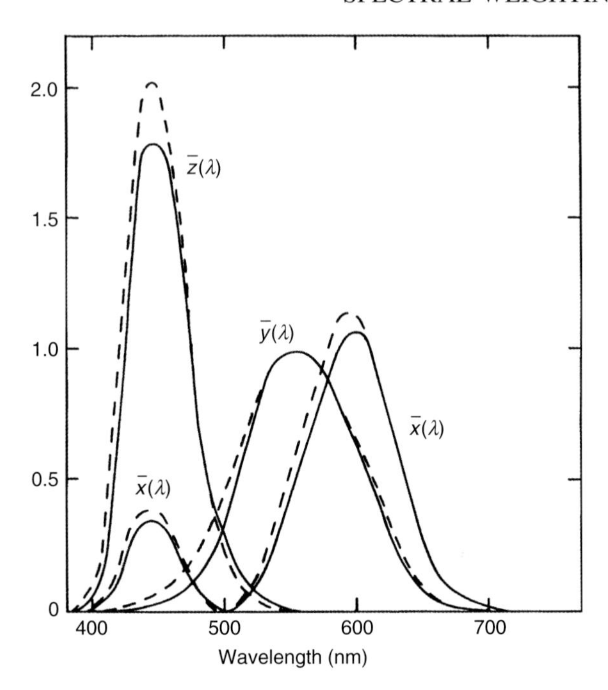

Figure 2.5 The CIE colour-matching functions for the 1931 Standard Colorimetric Observer (full lines), and for the 1964 Standard Colorimetric Observer (broken lines)

maximum value of 1.0, the $\bar{y}(\lambda)$ function is made, not only the same shape as, but also identical to, the $V(\lambda)$ function. The $\bar{r}(\lambda)$, $\bar{g}(\lambda)$, and $\bar{b}(\lambda)$ curves shown in Figure 2.4 are scaled such that $1.0000\bar{r}(\lambda) + 4.5907\bar{g}(\lambda) + 0.0601\bar{b}(\lambda)$ equals $V(\lambda)$ at each wavelength. To make the $\bar{y}(\lambda)$ function identical to $V(\lambda)$ function, the $\bar{r}(\lambda)$, $\bar{g}(\lambda)$, and $\bar{b}(\lambda)$ curves are all multiplied by 5.6508 before using them in the equations to generate the $\bar{x}(\lambda)$, $\bar{y}(\lambda)$, and $\bar{z}(\lambda)$ functions. Thus, in the example given earlier for 500 nm, it is necessary to multiply the result, 0.000866 by 5.6508 to obtain the correct result of 0.0049 for $\bar{x}(\lambda)$ at this wavelength.

The $\bar{x}(\lambda)$, $\bar{y}(\lambda)$, and $\bar{z}(\lambda)$ functions can be used as weighting functions to enable X, Y, and Z to be evaluated from spectral power data by using the formulae:

$$X = K(P_1\bar{x}_1 + P_2\bar{x}_2 + P_3\bar{x}_3 + \ldots\ldots P_n\bar{x}_n)$$

$$Y = K(P_1\bar{y}_1 + P_2\bar{y}_2 + P_3\bar{y}_3 + \ldots\ldots P_n\bar{y}_n)$$

$$Z = K(P_1\bar{z}_1 + P_2\bar{z}_2 + P_3\bar{z}_3 + \ldots\ldots P_n\bar{z}_n)$$

where K is a constant; the values of P are the amounts of power per small constant-width wavelength interval throughout the spectrum, and the values of $\bar{x}$, $\bar{y}$ and $\bar{z}$ are the heights of the CIE colour-matching functions at the central wavelength of each interval. (The official CIE equations for computing X, Y, and Z, use integration, but, because the colour-matching functions are not defined as continuous functions, summations have to be used in practice.) If P is in watts per steradian and per square metre and K is put equal to 683, then Y is the luminance in candelas per square metre (cd m^{-2}); when this is the case it is convenient to use the symbols X_L, Y_L, and Z_L for these luminance-related tristimulus values. If other radiometric measure are used for P, the corresponding photometric measures are obtained (see Appendix 1, Table A1.1 for details).

It is much more customary, however, to express the luminance separately from the colour specification, and to choose the constant K so that $Y = 100$ for a *perfect reflecting (or transmitting) diffuser* similarly illuminated and viewed.

A perfect diffuser reflects (or transmits) all the light that is incident on it at every wavelength throughout the visible spectrum, and does so in such a manner that it looks equally bright for all directions of viewing. The value of Y then gives the percentage luminous reflection or transmission of the colour. For non-diffusing samples a perfectly reflecting mirror, or a perfectly transmitting specimen may be used instead.

Specific terms are defined to denote different ways in which the light from the sample and the perfect diffuser may be received and evaluated. *Reflectance factor* and *transmittance factor* are used when the beams reflected or transmitted by the sample and by the perfect diffuser lie within a defined cone. If this cone is a hemisphere, the terms *reflectance* and *transmittance* are used. If the light is monochromatic, the adjective *spectral* is used; for example, *spectral reflectance factor:* if the light is evaluated using the $V(\lambda)$ or $\bar{y}(\lambda)$ function as a weighting function, the adjective *luminous* is used; for example, *luminous reflectance factor*. If the cone in which the light is reflected or transmitted is very small, the radiant term used is *radiance factor*, and the luminous term used is *luminance factor*. Y is then equal to whichever is the appropriate one of these luminous measures. (See Appendix 1, Section A1.6.)

The CIE colour-matching functions illustrated in Figure 2.5 are the most important spectral functions in colorimetry. They are used to obtain X, Y, and Z tristimulus values from spectral power data. If two colour stimuli have the same tristimulus values, they will look alike, when viewed under the same photopic conditions, by an observer whose colour vision is not significantly different from that of the CIE 1931 Standard Colorimetric Observer; conversely, if the tristimulus values are different, the colours may be expected to look different in these circumstances.

In Table 2.3, an example is given of how the tristimulus values X, Y, and Z, are obtained for two samples, A and B, having different spectral power distributions. The samples, in this case, are both reflecting surfaces, so that their spectral data are recorded as their spectral reflectance factors $R_A(\lambda)$ and $R_B(\lambda)$.They are assumed to be viewed under the same light source, whose spectral power distribution per small constant-width wavelength interval, $S(\lambda)$, is given in the table. In colorimetry, it is usual to work with *relative spectral power distributions*, and in this case $S(\lambda)$ is normalised by having an arbitrary value of 100 at 560 nm, the values at all the other wavelengths having the correct ratios relative to this value. Also listed are the values for the CIE colour-matching functions $\bar{x}(\lambda)$, $\bar{y}(\lambda)$, and $\bar{z}(\lambda)$. The relative spectral power distributions of the two samples under this light source are then obtained by multiplying, at each wavelength, the value of R_A by S and R_B by S; such spectral products are referred to as *relative colour stimulus functions*, $\phi(\lambda)$; if the absolute spectral power of the illuminant is used, the products are referred to as *colour stimulus functions*, $\phi_\lambda(\lambda)$. Each of these relative spectral power distributions is then weighted by each of the CIE colour-matching functions, and these weighted products summed to obtain:

Table 2.3 Calculation of tristimulus values from CIE colour-matching functions

Wave-length nm	$\bar{x}(\lambda)$	$\bar{y}(\lambda)$	$\bar{z}(\lambda)$	$S(\lambda)$	$S(\lambda)$ x $\bar{y}(\lambda)$	$R_A(\lambda)$	$R_B(\lambda)$	X_A/K	Y_A/K	Z_A/K	X_B/K	Y_B/K	Z_B/K
400	0.0143	0.0004	0.0679	15	0.0	0.20	0.20	0.0	0.0	0.2	0.0	0.0	0.2
420	0.1344	0.0040	0.6456	21	0.1	0.40	0.10	1.1	0.0	5.4	0.3	0.0	1.4
440	0.3483	0.0230	1.7471	29	0.7	0.40	0.11	4.0	0.3	20.3	1.1	0.1	5.6
460	0.2908	0.0600	1.6692	38	2.3	0.10	0.32	1.1	0.2	6.3	3.5	0.7	20.3
480	0.0956	0.1390	0.8130	48	6.7	0.10	0.27	0.5	0.7	3.9	1.2	1.8	10.5
500	0.0049	0.3230	0.2720	60	19.4	0.20	0.20	0.1	3.9	3.3	0.1	3.9	3.3
520	0.0633	0.7100	0.0782	72	51.1	0.40	0.10	1.8	20.4	2.3	0.5	5.1	0.6
540	0.2904	0.9540	0.0203	86	82.0	0.40	0.10	10.0	32.8	0.7	2.5	8.2	0.2
560	0.5945	0.9950	0.0039	100	99.5	0.10	0.39	5.9	10.0	0.0	23.2	38.8	0.2
580	0.9163	0.8700	0.0017	114	99.2	0.10	0.36	10.4	9.9	0.0	37.6	35.7	0.1
600	1.0622	0.6310	0.0008	129	81.4	0.20	0.20	27.4	16.3	0.0	27.4	16.3	0.0
620	0.8544	0.3810	0.0002	144	54.9	0.40	0.11	49.2	21.9	0.0	13.5	6.0	0.0
640	0.4479	0.1750	0.0000	158	27.6	0.40	0.10	28.3	11.1	0.0	7.1	2.8	0.0
660	0.1649	0.0610	0.0000	172	10.5	0.10	0.75	2.8	1.0	0.0	21.3	7.9	0.0
680	0.0468	0.0171	0.0000	185	3.1	0.10	0.49	0.9	0.3	0.0	4.2	1.5	0.0
700	0.0114	0.0041	0.0000	198	0.8	0.20	0.20	0.5	0.2	0.0	0.5	0.2	0.0
Totals					539.3			144.0	129.0	42.4	144.0	129.0	42.4

For sample A,

$$X_A = 144.0(100/539.3) = 26.7$$
$$Y_A = 129.0(100/539.3) = 23.9$$
$$Z_A = 42.4(100/539.3) = 7.9$$

For sample B

$$X_B = 144.0(100/539.33) = 26.7$$
$$Y_B = 129.0(100/539.33) = 23.9$$
$$Z_B = 42.4(100/539.33) = 7.9$$

In this example, only values at 20 nm intervals, over the range of wavelengths from 400 to 700 nm, are included in the calculation; this is the minimum number of values that can be used, and was done only to keep the example as simple as possible. For most applications, it is necessary to take values either at every 10 nm, or preferably at every 5 nm, and to use a range of wavelengths from 380 nm to 780 nm. (For methods of improving the accuracy of 20 nm calculations, see Section 9.13.)

$$X = K(R_1S_1\bar{x}_1 + R_2S_2\bar{x}_2 + R_3S_3\bar{x}_3 + \ldots\ldots R_nS_n\bar{x}_n)$$

$$Y = K(R_1S_1\bar{y}_1 + R_2S_2\bar{y}_2 + R_3S_3\bar{y}_3 + \ldots\ldots R_nS_n\bar{y}_n)$$

$$Z = K(R_1S_1\bar{z}_1 + R_2S_2\bar{z}_2 + R_3S_3\bar{z}_3 + \ldots\ldots R_nS_n\bar{z}_n)$$

In Table 2.3, these products and summations are listed as $X/K, Y/K, Z/K$.

The value of the constant K that results in Y being equal to 100 for the perfect diffuser is given by

$$K = 100(S_1\bar{y}_1 + S_2\bar{y}_2 + S_3\bar{y}_3 + \ldots\ldots S_n\bar{y}_n)$$

This arises because, for the perfect diffuser, the reflectance factor $R(\lambda)$, is equal to 1 at all wavelengths, so that its Y tristimulus value is given by:

$$Y = 100 = K(S_1\bar{y}_1 + S_2\bar{y}_2 + S_3\bar{y}_3 + \ldots\ldots S_n\bar{y}_n)$$

The products $S\bar{y}$, and their summation, are therefore also evaluated in Table 2.3.

The tristimulus values for the two samples, X_A, Y_A, Z_A, and X_B, Y_B, Z_B are then obtained, and it is seen from the table that $X_A = X_B$, $Y_A = Y_B$, and $Z_A = Z_B$. The samples will therefore look alike, when seen under the same photopic viewing conditions, by an observer whose colour vision is not significantly different from the CIE 1931 Standard Colorimetric Observer, for the particular light source, $S(\lambda)$, used.

In Table 2.3, the summation is carried out using data at every 20 nm, from 400 nm to 700 nm. This is the minimum number of values that can be used. For most applications it is necessary to take values either at every 10 nm, or preferably at every 5 nm, and to use a range of wavelengths from 380 nm to 780 nm. In Appendix 3, the values of the $\bar{x}(\lambda)$, $\bar{y}(\lambda)$, and $z(\lambda)$ functions are given in full at every 5 nm (CIE, 2004); values at every 1 nm are also available in various publications (CIE, 2006a; Grum and Bartleson, 1980; Wyszecki and Stiles, 1982) or by interpolation (see Section 9.13).

As already mentioned, if the spectral data available are in terms of an absolute measure of a radiant quantity per small constant-width wavelength interval throughout the spectrum, it is referred to, not as the *relative colour stimulus function*, but as the *colour stimulus function*; in this case, tristimulus values, X_L, Y_L, Z_L, can be obtained for which Y_L is equal to the absolute value of a photometric quantity; the constant K must then be equal to 683 lumens per watt, and the radiometric quantity must correspond to the photometric quantity required (see Appendix 1, Table A1.1). However, this is seldom done, even for light sources; these are usually evaluated using convenient arbitrary values for K.

As mentioned in Section 1.3, the retina varies considerably in its properties from one point to another, and matches made with a 2° field may not remain matches if the field size is altered. If the field size is reduced, the ability to discriminate one colour from another becomes less marked, and hence 2° matches usually appear to remain matches; but if the field size is increased, colour discrimination becomes more pronounced, and 2° matches tend to break down. For this reason, in 1964, the CIE recommended a different set of colour-matching functions for samples having field sizes greater than 4°. These supplementary colour-matching functions, $\bar{x}_{10}(\lambda)$, $\bar{y}_{10}(\lambda)$, and $\bar{z}_{10}(\lambda)$, are shown in Figure 2.5 by the broken lines, the full lines showing the $\bar{x}(\lambda)$, $\bar{y}(\lambda)$, and $\bar{z}(\lambda)$, functions for comparison. It can be seen from the figure that the two sets of functions are similar, but the differences are large enough to be significant. Values for these functions are given in Appendix 3 at every 5 nm (CIE 1986a); 1 nm values are also available in various publications (CIE, 2006a; Grum and Bartleson, 1980; Wyszecki and Stiles, 1982) or by interpolation (see Section 9.13). The $\bar{x}_{10}(\lambda)$, $\bar{y}_{10}(\lambda)$, and $\bar{z}_{10}(\lambda)$ functions can be used as weighting functions to obtain tristimulus values X_{10}, Y_{10}, and Z_{10}, by means of procedures analogous to those adopted to obtain X, Y, and Z. All measures obtained using $\bar{x}_{10}(\lambda)$, $\bar{y}_{10}(\lambda)$, and $\bar{z}_{10}(\lambda)$, colour-matching functions as a basis are distinguished by the presence of a subscript 10. It has been arranged that, for the equi-energy stimulus, S_E, just as $X = Y = Z$, so also $X_{10} = Y_{10} = Z_{10}$.

CIE Publication 165 defines the $\bar{y}_{10}(\lambda)$ function as a spectral luminous efficiency function. However, photometric measures are not additive in 10° fields to the extent that they are in 2° fields (CIE, 2005).

The calculations of tristimulus values are covered by one of a series of CIE Standards that have also been adopted by the International Organisation for Standardisation, ISO. The primary reference of the CIE Standard Observer colour-matching functions is:

> ISO 11664-1: 2008(E)/CIE S 014-1/E:2006:Colorimetry - Part 1: CIE Standard Colorimetric Observers. This contains the 1 nm spectral data. In addition, 5 nm data are obtained in CIE Publication 15:2004: Colorimetry, 3rd ed.

The calculation of XYZ tristimulus values will be described in CIE S 014-3E:2011: Colorimetry – Part 3: Calculation of CIE tristimulus values. (CIE, 2011).

As already mentioned in connection with the 1931 Standard Colorimetric Observer, in these standards, integrals are used to compute tristimulus values; but, because the colour-matching functions are not defined continuously, in practice summations have to be used.

Estimates are available for cone spectral sensitivity functions for normal observers, ranging in viewing angle from 1° to 10°. By correcting these functions for the absorption of the ocular media and the macular pigment, taking into account the optical densities of the cone visual pigments, and incorporating the relationship of the absorption of the lens as a function of age, colour-matching functions for field sizes between 1° and 10°, for observers of different ages, have been obtained (CIE, 2006b); these sets of colour-matching functions provide useful data for projects in vision research; but, for the application of colorimetry in most commercial and practical situations, a limited number of representative functions, such as the two CIE Standard Observers, is more convenient.

2.7 METAMERISM

When two colours having different spectral compositions match one another, they are said to be *metameric*, or *metamers*, and the phenomenon is called *metamerism*. A characteristic of metamerism is that, in general, metameric matches are upset if a different observer is used, or, in the case of reflecting or transmitting samples, if a different illuminant is used. Thus, for example, a change of field size from 2° to 10° necessitating a change from the CIE 1931 Standard Colorimetric Observer to the CIE 1964 Standard Colorimetric Observer, will usually indicate that metameric matches have been upset. Real observers will also often see that metameric matches are upset when field sizes are changed. In the case of illuminants, if a change is made from daylight to tungsten filament lighting, for example, then again, in general, both colorimetric computations and real observers will indicate that metameric matches have been upset. These effects are of great practical importance, and will be discussed further in Chapter 6.

2.8 SPECTRAL LUMINOUS EFFICIENCY FUNCTIONS FOR PHOTOPIC VISION

As mentioned in Section 2.4, the values of the $V(\lambda)$ function at wavelengths below about 450 nm are lower than they should be, but the resulting errors in deriving photopic photometric measures are negligible in most practical cases. The $V(\lambda)$ function has therefore been retained both for photometric purposes, and in the form of the $\bar{y}(\lambda)$ function in the CIE X, Y, Z system of colorimetry. A set of values that is more correct was proposed by Deane B. Judd in 1951 (CIE, 1951), and has been referred to as the *Judd correction*; in 1988 the CIE (CIE, 1988b) provided such a set of values known as the *CIE 1988 modified two degree spectral luminous efficiency function for photopic vision*, $V_M(\lambda)$; these values are given in Appendix 2. Another similar function, $V^*(\lambda)$ has also been proposed (Sharpe, Stockman, Jagla, and Jägle, 2005). It has been found that the $V_M(\lambda)$ function can be used to represent the brightnesses of *point sources*, by which is meant those of less than 10 minutes of arc in angular size.

The $V(\lambda)$ function was originally obtained by the method of flicker photometry. It has been found since that similar results for photopic efficiency can be obtained by three other methods:

a. Step by step comparisons involving pairs of intermediate stimuli having small colour differences.

b. Minimum-distinct-border evaluation, in which the amounts of two stimuli are judged equal when the border between them is minimally distinct when they are viewed with a negligible gap between them.

c. Visual acuity evaluation, in which the amounts of two stimuli are judged equal to one another when the visibility of fine detail in each is judged equally clear.

When the brightnesses are judged in other ways, such as by equality when seen side by side in the presence of large colour differences, or by subjective scaling of apparent brightness, or by threshold amounts for detection in a given set of viewing conditions, then the functions obtained are usually broader than the $V(\lambda)$ function. Sets of values for such functions have been provided by the CIE (CIE, 1988a) for both 2° and 10° field sizes, known as the *spectral luminous efficiency functions based upon brightness matching for monochromatic 2° and 10° fields*, $V_{b,2}(\lambda)$ and $V_{b,10}(\lambda)$, which are given in Appendix 2. The reasons for the differences between these broader functions and the $V(\lambda)$ function are not known, but perhaps the $V(\lambda)$ function represents activity in the achromatic channel only, whereas brightness also has some contributions from the colour-difference signals; this could also explain the Helmholtz-Kohlrausch effect (see Section 2.3). It is important to note that the $V_{b,2}(\lambda)$ and $V_{b,10}(\lambda)$ functions cannot be used as weighting functions for stimuli consisting of a variety of wavelengths, because additivity for lights of different wavelengths does not generally hold in the case of brightness.

It must be emphasised that, in colorimetry, none of the above functions is used, except the $V(\lambda)$ and $V_{10}(\lambda)$ functions.

REFERENCES

CIE Proceedings, 12th Session, Stockholm, Vol. 1, Committee No.7, Colorimetry, *Compte Rendu*, pp. 11–52, Bureau Central de la CIE, Paris, France (1951).

CIE Publication 75:1988, *Spectral luminous efficiency functions based upon brightness matching for monochromatic point sources, 2°, and 10° fields*, Commission Internationale de l'Éclairage, Vienna, Austria (1988a).

CIE Publication 86:1988, *CIE 1988 2° spectral luminous efficiency functions for photopic vision*, Commission Internationale de l'Éclairage, Vienna, Austria (1988b).

CIE Publication 165:2005, *CIE 10 degree photopic photometric observer*, Commission Internationale de l'Éclairage, Vienna, Austria (2005).

CIE S 014-1E:2006, *Colorimetry – Part 1: CIE standard colorimetric observer*, Commission Internationale de l'Éclairage, Vienna, Austria (2006a).

CIE Publication 170-1:2006, *Fundamental chromaticity diagram with physiological axes - Part 1*, Commission Internationale de l'Éclairage, Vienna, Austria (2006b).

CIE Publication 185:2009, *Reappraisal of colour matching and Grassmann's Laws*, Commission Internationale de l'Éclairage, Vienna, Austria (2009).

CIE S 014-3E:2011, *Colorimetry – Part 3: Calculation of CIE tristimulus values*, Commission Internationale de l'Éclairage, Vienna, Austria (2011).

Fairman, H.S., Brill, M.H., and Hemmendinger, H., How the CIE 1931 color-matching functions were derived from Wright-Guild data, *Color Res Appl*., **22**, 11–23 (1997).

Fairman, H.S., Brill, M.H., and Hemmendinger, H., How the CIE 1931 color-matching functions were derived from Wright-Guild data, *Color Res Appl*., **23**, 259 (1998).

Grum, F., and Bartleson, C.J., *Color Measurement*, pp. 48–68, Academic Press, New York, NY, USA (1980).

Guild, J., The colorimetric properties of the spectrum, *Phil. Trans. Roy. Soc., (London), A* **230**, 149 (1931).

Palmer, D.A., Nonadditivity in color matches with four instrumental stimuli, *J. Opt. Soc. Amer*., **71**, 966–969 (1981).

Sharpe, L.T., Stockman, A., Jagla, W., and Jägle, H., A luminous efficiency function, $V^*(\lambda)$, for daylight adaptation, *J. of Vision*, **5**, 948–968 (2005).

Stiles, W.S., 18th Thomas Young Oration; the basic data of colour matching, *Phys. Soc. Year Book*, pp. 44–65 (1955).

Trezona, P.W., The tetrachromatic colour match as a colorimetric technique, *Vision Res*., **13**, 9–25 (1973).

Wyszecki, G. and Stiles, W.S., *Color Science*, *2nd Ed*., pp. 725–747, Wiley, New York, NY, USA (1982).

GENERAL REFERENCES

Berns, R.S., *Billmeyer and Saltzman's Principles of Color Technology*, 3rd *Ed*., Wiley, New York, NY, USA (2000).

CIE Publications 15, 15.2, and 15:2004, *Colorimetry*, Commission Internationale de l'Éclairage, Vienna, Austria (1971, 1986a and 2004).

Judd, D.B., and Wyszecki, G., *Color in Business, Science, and Industry*, 2nd *Ed*., Wiley, New York, NY, USA (1975).

Wright, W.D., *The Measurement of Colour*, 4th *Ed*., Hilger, Bristol, England (1969).

Wyszecki, G., and Stiles, W.S., *Color Science*, 2nd *Ed*., Wiley, New York, NY, USA (1982 and 2000).

3

Relations Between Colour Stimuli

3.1 INTRODUCTION

We have seen in the previous chapter that, if two colours have the same tristimulus values, then they will look alike when seen under the same photopic viewing conditions by an observer whose colour vision is not significantly different from the CIE 1931 or 1964 Standard Colorimetric Observer, according to whether the field size is about 2° or about 10°, respectively. But, for colours in general, the tristimulus values differ over a wide range. In this chapter we see how such differences can be related to various colour attributes.

3.2 THE Y TRISTIMULUS VALUE

If the Y tristimulus value is evaluated on an absolute basis as Y_L, in candelas per square metre, for example, it represents the luminance of the colour. This provides a basis for a correlation with the perceptual attribute of brightness. As has already been explained, the correlation is complicated by the effect of the viewing conditions, by adaptation, by the non-linear relationship between brightness and luminance, and by the partial dependence of brightness on colourfulness. These factors will be discussed in some detail in Chapter 15. For the moment, however, we shall simply regard luminance as an approximate correlate of brightness.

When, as is customary (see in Section 2.6), Y is evaluated such that, for the similarly illuminated and viewed perfect diffuser, $Y = 100$, then Y is equal to the *reflectance factor*, or *transmittance factor*, expressed as a percentage. When viewed by the human eye, the cone of light collected from samples is very small, and for small cones of collection these factors are approximately the same as the *luminance factor*. Hence it is customary to regard Y as representing the percentage luminance factor, and this is an approximate correlate of the perceptual attribute of lightness. The Y values then range from 100 for white, or transparent objects, that absorb no light, to zero for objects that absorb all the light. (Values of Y greater than 100 can occur for objects that fluoresce; see Chapter 10). It is sometimes convenient to use the ratio Y/Y_n, where Y_n is the value of Y for a suitable

Measuring Colour, Fourth Edition. R.W.G. Hunt and M.R. Pointer.

chosen reference white or reference transparent specimen; in this case, if the reference object absorbs some light, as is usually the case, Y/Y_n will be slightly more than 1 for the perfect diffuser (this will be discussed further in Section 5.7).

3.3 CHROMATICITY

If the Y tristimulus value correlates approximately with brightness or, more usually, with lightness, with what do the X and Z tristimulus values correlate? The answer is that they do not correlate, even approximately, with any perceptual attributes, and it is necessary to derive from the tristimulus values other measures that can provide such correlates.

Important colour attributes are related to the *relative* magnitudes of the tristimulus values. It is therefore helpful to calculate a type of relative tristimulus values called *chromaticity co-ordinates*, as follows:

$$x = X/(X + Y + Z)$$

$$y = Y/(X + Y + Z)$$

$$z = Z/(X + Y + Z)$$

The chromaticity co-ordinates, for which lower-case letters are always used, thus represent the relative amounts of the tristimulus values. For example, if $X = 8$, $Y = 48$, $Z = 24$, then $X + Y + Z = 80$, and $x = 8/80 = 0.1$; $y = 48/80 = 0.6$; $z = 24/80 = 0.3$. This indicates that, for this particular colour, there is in its specification 10% of X, 60% of Y, and 30% of Z. If, for the moment, we regard the equi-energy stimulus, S_E, as a white, we can deduce from these values that the colour considered has a higher y value, and a lower x value than the white, for which x, y, and z are all 1/3, because X, Y, and Z are equal to one another for S_E. Since the $\bar{x}(\lambda)$, $\bar{y}(\lambda)$, and $\bar{z}(\lambda)$ curves correspond very approximately to the $\bar{r}(\lambda)$, $\bar{g}(\lambda)$, and $\bar{b}(\lambda)$ curves, respectively, we can deduce that the colour considered is probably greener and less red (that is, more blue-green), than the white. This, together with the fact that the Y value is 48, suggests that we have a bluish green of lightness about midway between white and black. These deductions are only partly correct, and need to be refined, but they show the way in which chromaticity co-ordinates are useful in understanding the implications of tristimulus values.

It is clear from the way in which x, y, and z are calculated that

$$x + y + z = 1$$

and hence, if x and y are known, z can always be deduced from $1 - x - y$. With only two variables, such as x and y, it becomes possible to construct two-dimensional diagrams, or *chromaticity diagrams* as they are usually called, an example of which is shown in Figure 3.1. In this figure, y is plotted as ordinate against x as abscissa, and this important diagram is usually referred to as the *x, y chromaticity diagram*. This diagram provides a sort of colour map on which the chromaticities of all colours can be plotted. As already mentioned, the equi-energy stimulus, S_E, has tristimulus values that are equal to one another, so that its chromaticity co-ordinates are $x = 1/3$, $y = 1/3$, and $z = 1/3$, and it is marked at this position in Figure 3.1.

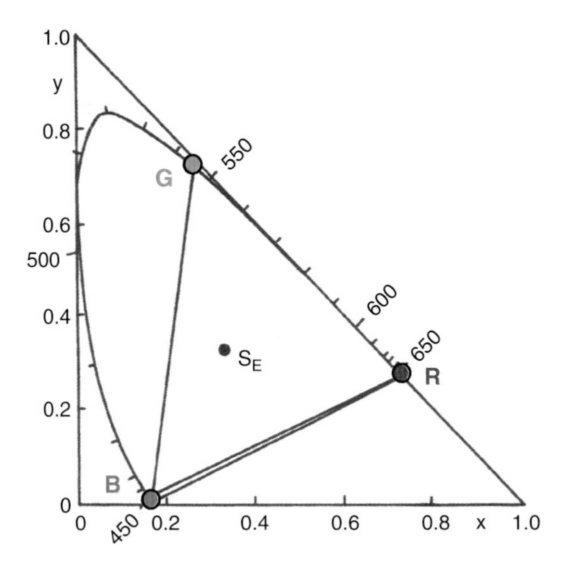

Figure 3.1 The CIE x, y chromaticity diagram, showing the spectrum locus and equi-energy stimulus, S_E, and the CIE red, green, and blue matching stimuli, R, G, B

The curved line in the diagram shows where the colours of the spectrum lie and is called the *spectrum locus*; the wavelengths are indicated in nanometres along the curve, and the corresponding values of x and y are listed at 5 nm intervals in Appendix 4.

If two colour stimuli are additively mixed together, then the point representing the mixture is located in the diagram by a point that always lies on the line joining the two points representing the two original colours (this is a consequence of the additivity of colour matches, see Section 2.4). This means that, if the two ends of the spectrum are joined by a straight line, that line represents mixtures of light from the two ends of the spectrum; as these colours are mixtures of red and violet, this line is known as the *purple boundary*. The area within the spectrum locus and the purple boundary encloses the domain of all colours; this is because the spectrum locus consists of a continuously convex boundary so that all mixtures of wavelengths must lie inside it. The positions of the three matching stimuli, R, G, and B, having wavelengths of 700 nm, 546.1 nm, and 435.8 nm, respectively, are also shown in the diagram. The line joining the points R and G represents all the colours formed by the additive mixtures of various amounts of the R and G stimuli; and the lines joining G and B, and B and R, have similar relationships to the appropriate mixtures of the other pairs of stimuli. The area within the triangle formed by the three lines represents all the colours that can be matched by additive mixtures of these three stimuli. The location of the spectrum locus outside this triangle, especially in the blue-green part of the spectrum, is a consequence of negative amounts of one of the three matching stimuli being necessary in order to match the colours of the spectrum (as explained in Section 2.4).

If, now, points are considered between the point representing S_E and a given point on the spectrum locus, these must represent colours formed by additive mixtures of S_E and light of the particular wavelength considered. If the mixture consists mainly of the spectral colour, then the corresponding point will lie near the spectrum locus and will tend to represent a saturated colour of that particular hue. If, on the other hand, the mixture consists mainly of S_E, then the corresponding point will lie near S_E and will tend to represent a pale colour of that hue. The chromaticity diagram thus represents a continuous gradation of stimuli from those like S_E to those like spectral colours, in the space between S_E and the spectrum locus. It is very important to remember, however, that chromaticity diagrams only show the *proportions* of the tristimulus values; hence, bright and dim lights, or brightly lit and dimly lit surface colours, or light and dark surface colours, having tristimulus values in the same ratios to one another, all plot at the same point. For this reason, the point S_E, for instance, represents all levels of the equi-energy stimulus; and, if this stimulus is regarded as an illuminant, the same point also represents whites, greys, and blacks illuminated by it at any level. Similarly, a single point can represent orange and brown surface colours illuminated at any level.

It is also very important to remember that chromaticity diagrams are maps of relationships between colour stimuli, not between colour perceptions. Although it is often helpful to consider the approximate nature of the colour perceptions that are likely to be represented by various chromaticities, it must be remembered that the exact colour perceptions will depend on the viewing conditions, and on the adaptation and other characteristics of the observer. An example of the importance of the viewing conditions is the fact that the appearance of a colour can be greatly affected by surrounding colours, a phenomenon termed *simultaneous contrast* (or *chromatic induction*); the effect is usually to change the appearance to be less like the surrounding colour, but if the stimuli are very small the appearance may become more like the surrounding colour (*assimilation*). The appearances of colours are also affected by previous viewing of different colours, an effect known as *successive contrast*. These effects are discussed in Section 14.3.

A similar chromaticity diagram is derived from the X_{10}, Y_{10}, and Z_{10} tristimulus values, by evaluating:

$$x_{10} = X_{10}/(X_{10} + Y_{10} + Z_{10})$$

$$y_{10} = Y_{10}/(X_{10} + Y_{10} + Z_{10})$$

$$z_{10} = Z_{10}/(X_{10} + Y_{10} + Z_{10})$$

The values of x_{10}, and y_{10} for the spectrum locus at 5 nm intervals are listed in Appendix 4. In Figure 3.2, the spectral loci for the x, y and the x_{10}, y_{10} chromaticity diagrams are superimposed; it is clear that they are broadly similar in shape, but the locations of the points representing some of the individual wavelengths are appreciably different (see, for instance, those for 480 nm). The equi-energy stimulus, S_E, is located at the same point in both diagrams, because for this stimulus $X = Y = Z$, and $X_{10} = Y_{10} = Z_{10}$.

The following sections describe various colorimetric measures in terms of tristimulus values for the 1931 Observer; they can all also be evaluated using tristimulus values for the 1964 Observer, in which case they would all be distinguished by having a subscript 10.

3.4 DOMINANT WAVELENGTH AND EXCITATION PURITY

Chromaticity diagrams can be used to provide measures that correlate approximately with the perceptual colour attributes, hue and saturation. In Figure 3.3, the x, y chromaticity

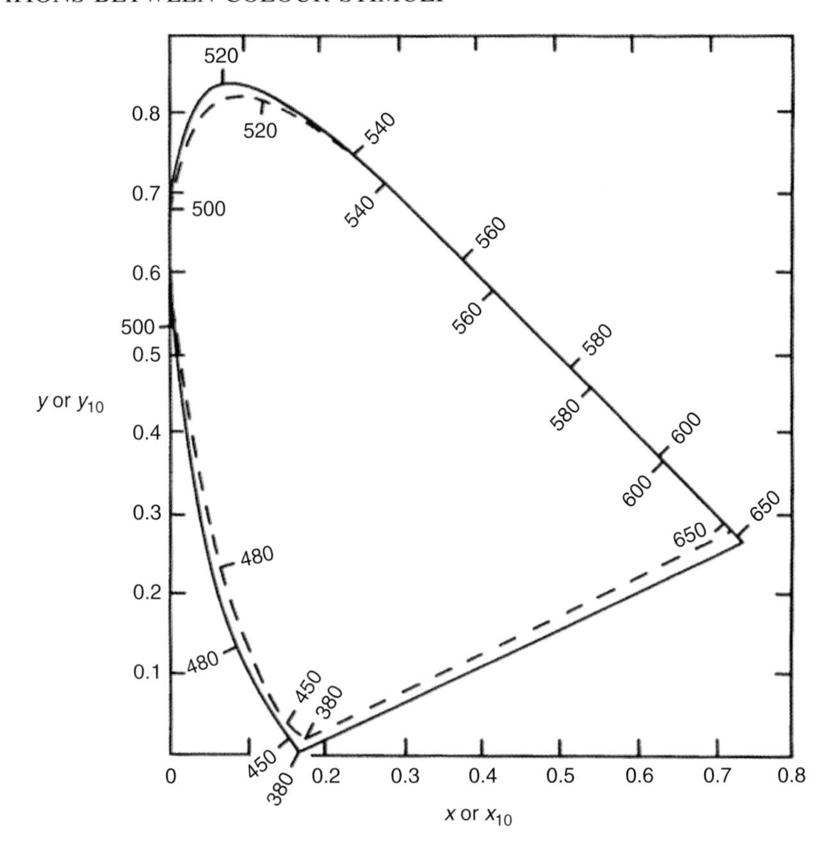

Figure 3.2 The CIE x_{10}, y_{10} chromaticity diagram (broken lines), compared with the CIE x, y chromaticity diagram (full lines)

diagram is shown again. The point, C, represents a colour being considered. The point, N, represents a suitably chosen reference white (or grey), the *achromatic stimulus*, which is almost always different from S_E, the equi-energy stimulus. For reflecting surface colours, it is common practice to consider the perfect diffuser, illuminated by the light source in use, to be the reference white. A line is then drawn from N through C, to meet the spectrum locus at a point, D. The wavelength on the spectrum locus that corresponds to D, is then called the *dominant wavelength*, λ_d, of the colour, C, relative to the white point, N. It is clear from the diagram that the colour, C, can be thought of as an additive mixture of the colour, D, and the white, N; in this sense, dominant wavelength is a very descriptive name for this measure. If, for a colour, C′, the point, D′ is on the purple boundary, then NC′ is extended in the opposite direction, as shown in Figure 3.3, to meet the spectrum locus at a point, D_c, which defines the *complementary wavelength*, λ_c. Dominant (and complementary) wavelength may be considered to be approximately correlated with the hue of colours, but, in addition to the usual factors that make this inexact, loci of constant hue (for a given set of viewing conditions) are not straight lines, like ND, (the Abney phenomenon) but are slightly curved (this will be discussed in more detail in Chapter 15).

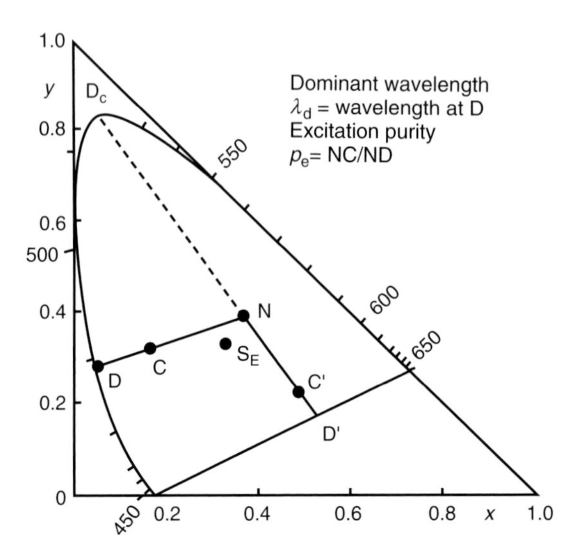

Figure 3.3 The derivation on the x,y chromaticity diagram of dominant wavelength, λ_d, complementary wavelength, λ_c, at D_c, and excitation purity, p_e, equal to NC/ND or NC′/ND′

The ratio of the length of the lines NC and ND (or NC′ and ND′) provides a measure called *excitation purity*, p_e:

$$p_e = \mathrm{NC/ND} \text{ or } \mathrm{NC'/ND'}$$

If p_e is nearly unity, the point is near the spectrum locus or purple boundary, and will tend to represent a colour that is highly saturated; but, if p_e is nearly zero, the point is near the reference white, and will tend to represent a colour that is very pale. Excitation purity may be considered to be approximately correlated with the saturation of colours, but, in addition to the usual factors that make this inexact, loci of constant saturation are not the same shape as loci of constant excitation purity (this will also be discussed in more detail in Chapter 15). The term *purity* is quite helpfully descriptive, but the adjective *excitation* serves only to distinguish this measure from another type of purity in colorimetry (as will be described in Section 3.16).

3.5 COLOUR MIXTURES ON CHROMATICITY DIAGRAMS

If two colours C_1 and C_2 are represented by points, C_1 and C_2, as shown in Figure 3.4, then, as has already been mentioned, the additive mixture of the two colours is represented by a point, such as C_3, lying on the line joining C_1 and C_2. The exact position of C_3 on the line depends on the relative luminances of C_1 and C_2, and is calculated as follows.

If the chromaticity co-ordinates of C_1 and C_2 are x_1, y_1, z_1, and x_2, y_2, z_2, respectively, and if, in the mixture, there are m_1 luminance units of C_1 and m_2 luminance units of C_2, we proceed as follows. For some amount of C_1, it will be the case that $X = x_1, Y = y_1,$

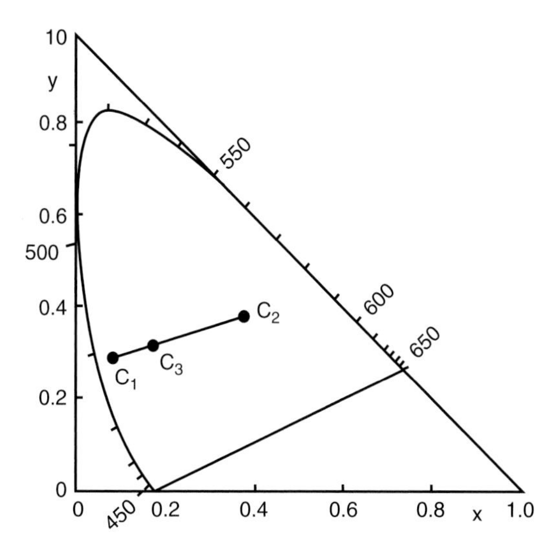

Figure 3.4 The colour, C_3, that can be matched by an additive mixture of the colours, C_1 and C_2

and $Z = z_1$. The luminance of this amount will be proportional to Y, and may therefore be expressed as $L_Y Y$, which is the same as $L_Y y_1$, where L_Y is a constant. Hence, since for $L_Y y_1$ luminance units of C_1, $X = x_1$, $Y = y_1$, and $Z = z_1$ it follows that for one luminance unit of C_1 the values of X, Y, and Z will become

$$X = x_1/L_Y y_1 \qquad Y = y_1/L_Y y_1 \qquad Z = z_1/L_Y y_1$$

and for m_1 luminance units of C_1

$$X = m_1 x_1/L_Y y_1 \qquad Y = m_1 y_1/L_Y y_1 \qquad Z = m_1 z_1/L_Y y_1$$

Similarly for m_2 luminance units of C_2

$$X = m_2 x_2/L_Y y_2 \qquad Y = m_2 y_2/L_Y y_2 \qquad Z = m_2 z_2/L_Y y_2$$

Hence for these two amounts of C_1 and C_2 additively mixed together:

$$X = (m_1 x_1)/(L_Y y_1) + (m_2 x_2)/(L_Y y_2)$$

$$Y = (m_1 y_1)/(L_Y y_1) + (m_2 y_2)/(L_Y y_2)$$

$$Z = (m_1 z_1)/(L_Y y_1) + (m_2 z_2)/(L_Y y_2)$$

Remembering that $x_1 + y_1 + z_1 = 1$ and $x_2 + y_2 + z_2 = 1$

$$X + Y + Z = m_1/(L_Y y_1) + m_2/(L_Y y_2)$$

Hence for the mixture:

$$x = [(m_1 x_1)/y_1) + (m_2 x_2)/y_2)]/[(m_1/y_1) + (m_2/y_2)]$$

$$y = [(m_1 y_1)/y_1) + (m_2 y_2)/y_2)]/[(m_1/y_1) + (m_2/y_2)]$$

The geometrical interpretation of this result on the x, y chromaticity diagram is that the point, C_3, representing the mixture, is on the line joining the points C_1, at x_1, y_1, and C_2, at x_2, y_2, in the ratio

$$C_1C_3/C_2C_3 = (m_2/y_2)/(m_1/y_1)$$

This means that C_3 is at the centre of gravity of weights m_1/y_1 at C_1 and m_2/y_2 at C_2, and hence the result is referred to as *the Centre of Gravity Law of Colour Mixture*.

3.6 UNIFORM CHROMATICITY DIAGRAMS

Although the x, y chromaticity diagram has been widely used, it suffers from a serious disadvantage: the distribution of the colours on it is very non-uniform. This is illustrated in Figure 3.5. Each of the short lines in this figure joins a pair of points representing two colours having a perceptual colour difference of the same magnitude, the luminances of all the colours being the same. Ideally these identical colour differences should be represented by lines of equal length. But it is clear that this is far from being the case, the lines being much longer towards the green part of the spectrum locus, and much shorter towards the violet part, than the average length. This is rather similar to some maps of the world. Because it is not possible to represent a curved surface accurately on a flat piece of paper, distortions occur. For example, in some maps of the world, countries near one of the poles, such as Greenland, are represented as far too large compared with those near the equator, such as India. On such maps, pairs of locations equally distant from one another on the earth's surface are represented by points that are much closer together in India than in Greenland. No map on a flat piece of paper can avoid this problem entirely, but some types of map are better than others in minimising the effect.

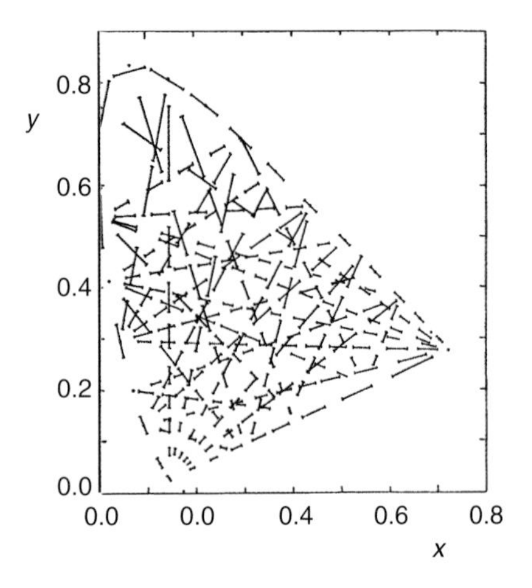

Figure 3.5 The x, y chromaticity diagram with lines representing small colour differences (after the work of Wright, 1941). Each line is three times the length of a distance representing a difference that is just noticeable in a 2° field for colours of constant luminance

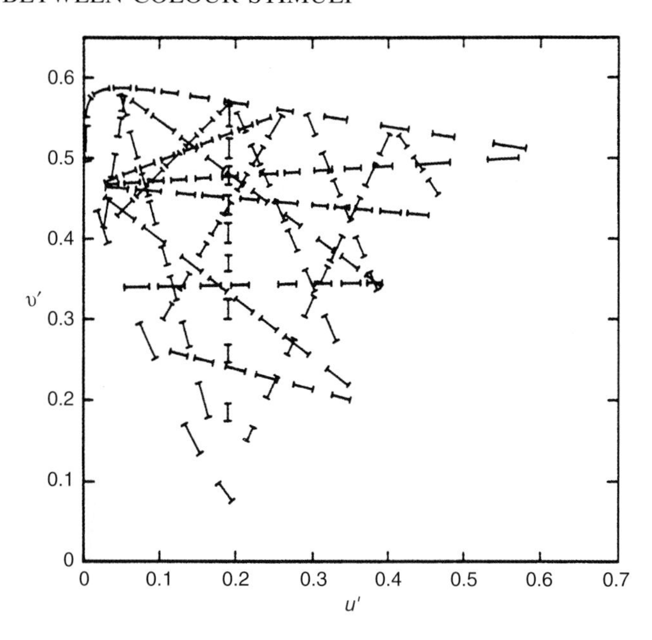

Figure 3.6 The approximately uniform CIE u', v' chromaticity diagram with a selection of lines of Figure 3.5

There is, in fact, no chromaticity diagram that can entirely avoid the problem illustrated in Figure 3.5; but, just as with maps, some chromaticity diagrams are better than others in this respect. In Figure 3.6 a different chromaticity diagram is used, in which a selection of the lines of Figure 3.5 are shown again. It is immediately clear that the variation in the length of the lines, while not eliminated, has been much reduced; in fact, the ratio of the longest to the shortest line in Figure 3.6 is only about four to one, instead of about twenty to one in Figure 3.5 (see Figure 3.7). The diagram shown in Figure 3.6 is known as the *CIE 1976 uniform chromaticity scale diagram* or the *CIE 1976 UCS diagram*, commonly referred to as the *u', v' diagram*. It is obtained by plotting v' against u', where:

$$u' = 4X/(X + 15Y + 3Z) = 4x/(-2x + 12y + 3)$$

$$v' = 9Y/(X + 15Y + 3Z) = 9y/(-2x + 12y + 3)$$

To obtain x, y from u', v' the following equations can be used:

$$x = 9u'/(6u' - 16v' + 12)$$

$$y = 4v'/(6u' - 16v' + 12)$$

The values of u' and v' for the spectrum locus are given in Appendix 4. The u′,v′ diagram was recommended by the CIE in 1976; prior to that a similar diagram, the *u,v diagram* was used in which $u = u'$, and $v = (2/3)v'$. All chromaticity diagrams, whether x,y, or u′,v′, or u,v, have the property that additive mixtures of colours are represented by points lying on the straight line joining the points representing the constituent colours. The

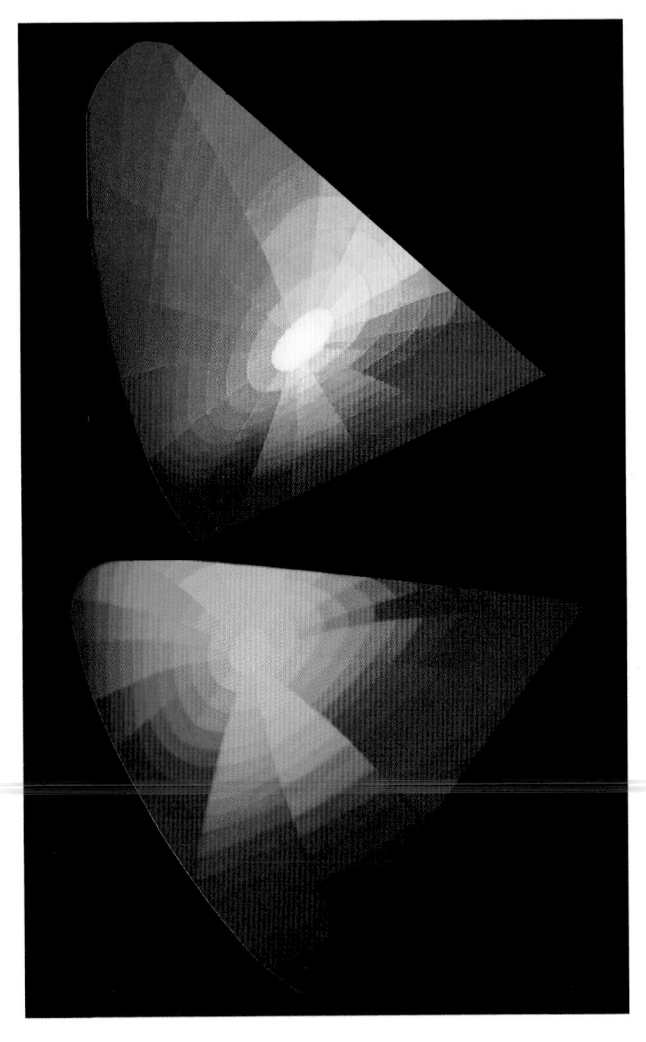

Figure 3.7 Artist's representations of the distribution of colours in chromaticity diagrams. Upper: the x,y diagram; green colours are much more spread out than violet colours. Lower: the u', v' diagram; colours are more evenly distributed than in the x,y diagram. Most of the colours near the spectrum locus are outside the gamut of the printing process, so that the colours shown are not accurate; however, the figures are useful in giving a general impression of the differences between the two chromaticity diagrams. Based on paintings by Louis M. Condax

position of the mixture point has to be calculated by the method given in Section 3.5, and in the u′, v′ diagram the weights used are m_1/v'_1 and m_2/v'_2; and in the u, v diagram the weights are m_1/v_1 and m_2/v_2.

The u′,v′ diagram is useful for showing the relationships between colours whenever the interest lies in their discriminability. For example, in Figure 3.8 are shown the loci of colours (of equal luminance) likely to be confused by colour defective observers. The loci

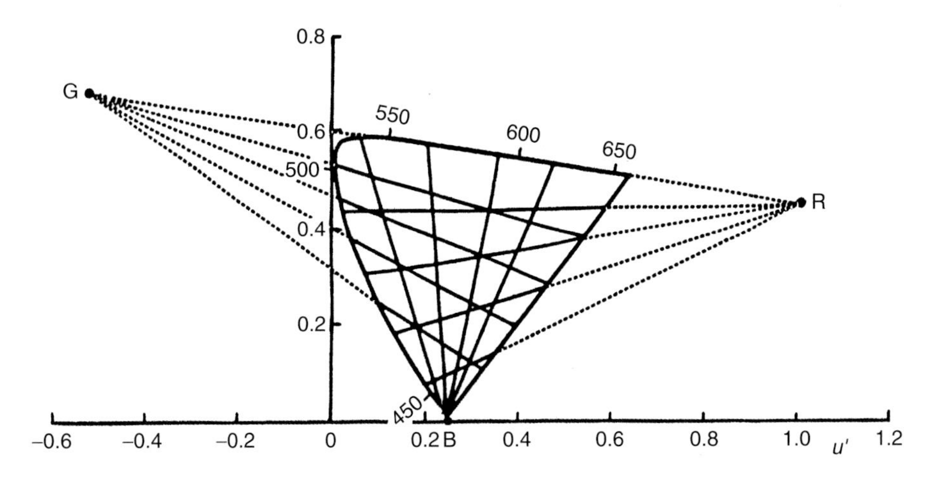

Figure 3.8 Loci of colours confused by protan, deutan, and tritan observers

for protan observers (those with protanopia or protanomaly) emanate from the point R, at $u' = 1.020$, $v' = 0.447$; those for deutan observers (those with deuteranopia or deuteranomaly) emanate from the point G, at $u' = -0.534$, $v' = 0.680$; and those for tritan observers (those with tritanopia or tritanomaly) from the point B, at $u' = 0.251$, $v' = 0$ (Estevez, 1979). Of course, the u', v' diagram represents the discriminability for normal observers; for colour deficient observers, diagrams compressed along the directions of the loci would be required. Thus, for a protanope, for example, all colours lying on any one line emanating from the point R would be confused; and, for a protanomalous observer, pairs of colours lying on any one of these lines would be seen to be less distinguishable than in the case of a normal observer, the difference depending on the degree of severity of the deficiency.

3.7 CIE 1976 HUE-ANGLE AND SATURATION

Because the x,y diagram is so non-uniform in its distribution of colours, excitation purity, p_e, does not correlate uniformly with the perception of saturation; and dominant wavelength correlates very non-uniformly with the perception of hue, because equal differences of hue correspond to very unequal differences of wavelength at different parts of the spectrum. Two new measures have therefore been provided, based on the u′,v′ diagram, that correlate with hue and saturation more uniformly. They are:

CIE 1976 u, v hue-angle, h_{uv}

$$h_{uv} = \arctan[(v' - v'_n)/(u' - u'_n)]$$

CIE 1976 u, v saturation, s_{uv}

$$s_{uv} = 13[(u' - u'_n)^2 + (v' - v'_n)^2]^{1/2}$$

where u'_n, v'_n are the values of u', v' for a suitable chosen reference white; arctan means ‘the angle whose tangent is’, and 13 is a scale factor introduced to harmonise s_{uv} with

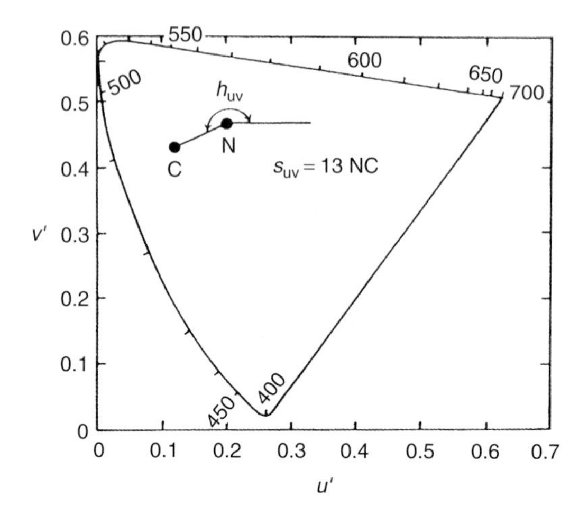

Figure 3.9 The derivation in the u', v' diagram of h_{uv} and s_{uv}

some other measures that will be considered later. h_{uv} lies between 0° and 90° if $v' - v'_n$ and $u' - u'_n$ are both positive; between 90° and 180° if $v' - v'_n$ is positive and $u' - u'_n$ is negative; between 180° and 270° if $v' - v'_n$ and $u' - u'_n$ are both negative; and between 270° and 360° if $v' - v'_n$ is negative and $u' - u'_n$ is positive.

In Figure 3.9, the geometrical meanings of h_{uv} and s_{uv} can be seen. C is the point representing the colour considered, and N that for the reference white. Then h_{uv} is the angle between the line NC and a horizontal line drawn from N to the right, the angle being measured in an anticlockwise direction; and s_{uv} is equal to 13 times the distance NC. The uniformity of correlation between h_{uv} and hue, and between s_{uv} and saturation, will still not be perfect, because of the remaining non-uniformities of the u', v' diagram. In the case of hue, further discrepancies are caused by the fact, already mentioned, that loci of constant hue are not straight lines, but are slightly curved in chromaticity diagrams; an approach that attempts to allow for this will be described in Chapter 15. But h_{uv} and s_{uv} are approximations to uniform correlates of hue and saturation that are of practical use.

3.8 CIE 1976 LIGHTNESS, L^*

A scale of greys can be constructed in which the values of the ratios $100L/L_n$ are in uniform increments, such as 10, 20, 30, 40, 50, 60, 70, 80, 90, and 100, where L denotes the luminances of the greys and L_n that of the appropriate reference white. If this is done, the apparent differences in lightness between adjacent pairs are very much less at the light end of the scale than at the dark end. In 1976 the CIE therefore recommended a non-linear empirical function to provide a measure that correlated with lightness more uniformly. It is defined in terms of the ratio of the Y tristimulus value of the colour considered to that of the reference white, that is Y/Y_n (which is equal to L/L_n), as follows:

CIE 1976 lightness, L^*

$$L^* = 116\text{f}(Y/Y_n) - 16$$

where $f(Y/Y_n) = (Y/Y_n)^{1/3}$ for $Y/Y_n > (6/29)^3$

$f(Y/Y_n) = (841/108)(Y/Y_n) + 4/29$ for $Y/Y_n \leq (6/29)^3$

The reference white, for which $Y = Y_n$, then has a value of $L^* = 100$; a perfect black, for which $Y/Y_n = 0$, has a value of $L^* = 0$ (the lower expression for L^* is included to avoid negative values occurring when Y/Y_n is very small); and a medium grey has a value of L^* of about 50. (The scale is similar in its distribution of lightnesses to the Munsell Value scale, to be considered in Section 8.4, but the numbers obtained are about 10 times as great as those for Munsell Value.)

3.9 UNIFORM COLOUR SPACES

Chromaticity diagrams have many uses, but, as they show only *proportions* of tristimulus values, and not their actual magnitudes, they are only strictly applicable to colours that all have the same luminance and luminance factor. In general, colours differ in luminance and luminance factor, as well as in chromaticity, and some method of combining these variables is therefore required. To meet this need for luminance factor (but not for luminance), the CIE has recommended the use of one of two alternative *colour spaces*. The first of these that we shall consider is the *CIE 1976 (L*u*v*) colour space* or the *CIELUV colour space*. It is produced by plotting, along three axes at rights to one another, the quantities:

$$L^* = 116f(Y/Y_n) - 16$$

$$u^* = 13L^*(u' - u'_n)$$

$$v^* = 13L^*(v' - v'_n)$$

where $f(Y/Y_n) = (Y/Y_n)^{1/3}$ for $Y/Y_n > (6/29)^3$

$f(Y/Y_n) = (841/108)(Y/Y_n) + 4/29$ for $Y/Y_n \leq (6/29)^3$

where u'_n, v'_n are values of u', v' for the appropriately chosen reference white. It is clear from the expressions for u^* and v^* that a given difference in chromaticity will be reduced in magnitude by the factor L^* as the colour becomes darker. This allows for the perceptual fact that a given difference in chromaticity represents a smaller and smaller perceived colour difference as its value of Y/Y_n is reduced. The constant 13 is included to equalise the perceptual significance of the L^*, u^*, and v^* scales.

Since $(v' - v'_n)/(u' - u'_n) = v^*/u^*$, hue-angle, h_{uv}, can also be defined as follows:

CIE 1976 u, v hue-angle, h_{uv}

$$h_{uv} = \arctan(v^*/u^*).$$

These variables u^* and v^* also enable a correlate of chroma to be evaluated:

CIE 1976 u,v chroma, C^*_{uv}

$$C^*_{uv} = (u^{*2} + v^{*2})^{1/2} = L^* s_{uv}$$

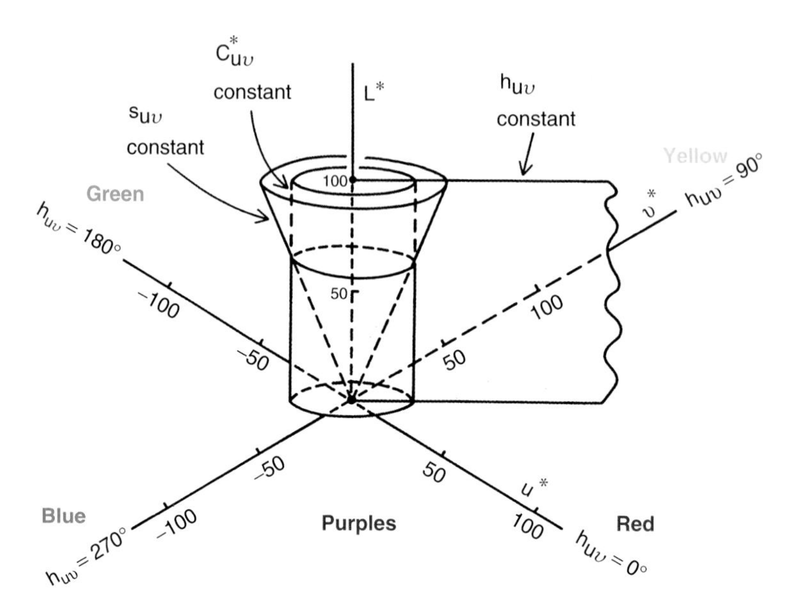

Figure 3.10 A three dimensional representation of the CIELUV space. The CIELAB space is similar, except that there is no representation of saturation

In Figure 3.10 the $L^*u^*v^*$ space is illustrated with the L^* axis being considered vertical, and the u^* and v^* axes lying in a horizontal plane. Examples of surfaces of constant CIE 1976 hue-angle, saturation, and chroma, are shown. Surfaces of constant hue-angle, h_{uv}, are planes that have the L^* axis as one edge. Surfaces of constant chroma, C^*_{uv}, are cylinders that have the L^* axis as their axes. Surfaces of constant saturation, s_{uv}, are cones that have the L^* axis as their axes and their apices at the origin of the space. If colours of a single hue-angle are considered, they will lie on a plane, with the lightest at the top and the darkest at the bottom, and with those of low chroma near the L^* axis and those of high chroma displaced farthest from it. This arrangement is very similar to the way in which the samples in the Munsell System are arranged (to be considered in Section 8.4.) A series of colours of constant chromaticity, but gradually decreasing value of Y/Y_{n} (often referred to as a *shadow series*) will fall on a line that slopes down towards the origin, on the hue-angle plane, and these will be a series of colours of constant saturation. In Figure 3.11 a reproduction is given of a representation of a horizontal section through the CIELUV system for a value of $L^* = 50$. In Figure 3.12 similar representations are given for vertical sections for red and cyan colours, for yellow and blue colours, and for green and magenta colours.

In an earlier colour space, based on the u, v chromaticity diagram, the variables used were:

$$W^* = 25Y^{1/3} - 17$$

$$U^* = 13W^*(u - u_{\mathrm{n}})$$

$$V^* = 13W^*(v - v_{\mathrm{n}})$$

where Y was L/L_{n} expressed as a percentage.

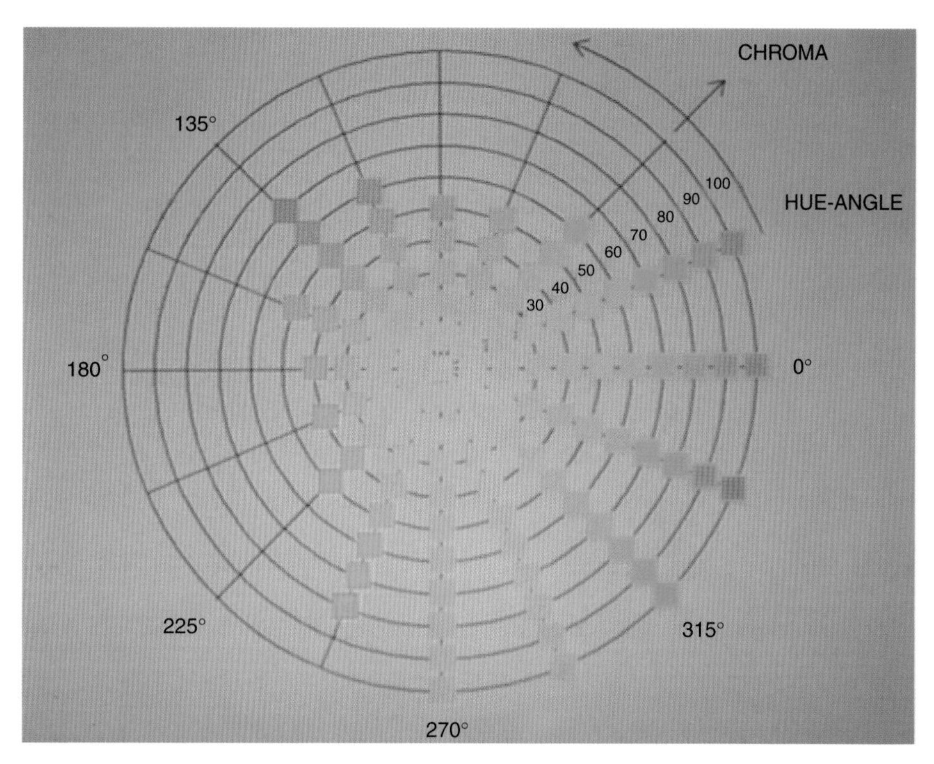

Figure 3.11 Representation of a horizontal section through the CIELUV system for a value of $L^* = 50$. Because of the limitations of printing, all the colours cannot be shown accurately, but the general layout is displayed adequately

The second space recommended by the CIE is the *CIE 1976* $(L^*a^*b^*)$ *colour space* or the *CIELAB colour space* (see Figure 3.13). It is produced by plotting, along three axes at right angles to one another, the quantities:

$$L^* = 116\mathrm{f}(Y/Y_\mathrm{n}) - 16$$

$$a^* = 500[\mathrm{f}(X/X_\mathrm{n}) - f(Y/Y_\mathrm{n})]$$

$$b^* = 200[\mathrm{f}(Y/Y_\mathrm{n}) - f(Z/Z_\mathrm{n})]$$

where $\mathrm{f}(X/X_\mathrm{n}) = (X/X_\mathrm{n})^{1/3}$ for $X/X_\mathrm{n} > (6/29)^3$

$\mathrm{f}(X/X_\mathrm{n}) = (841/108)(X/X_\mathrm{n}) + 4/29$ for $X/X_\mathrm{n} \leq (6/29)^3$

and $\mathrm{f}(Y/Y_\mathrm{n}) = (Y/Y_\mathrm{n})^{1/3}$ for $Y/Y_\mathrm{n} > (6/29)^3$

$\mathrm{f}(Y/Y_\mathrm{n}) = (841/108)(Y/Y_\mathrm{n}) + 4/29$ for $Y/Y_\mathrm{n} \leq (6/29)^3$

and $\mathrm{f}(Z/Z_\mathrm{n}) = (Z/Z_\mathrm{n})^{1/3}$ for $Z/Z_\mathrm{n} > (6/29)^3$

$\mathrm{f}(Z/Z_\mathrm{n}) = (841/108)(Z/Z_\mathrm{n}) + 4/29$ for $Z/Z_\mathrm{n} \leq (6/29)^3$

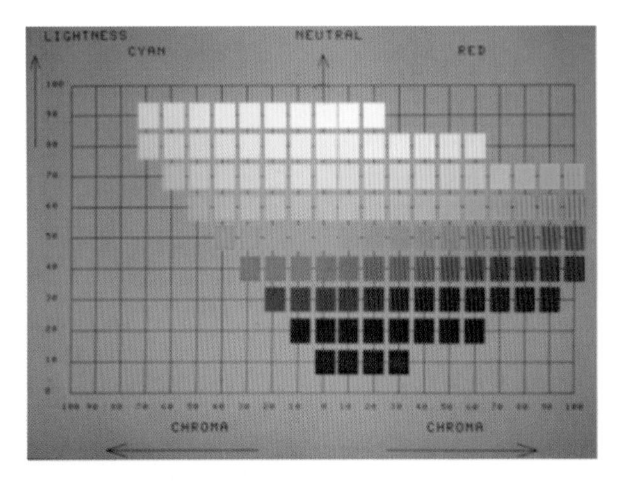

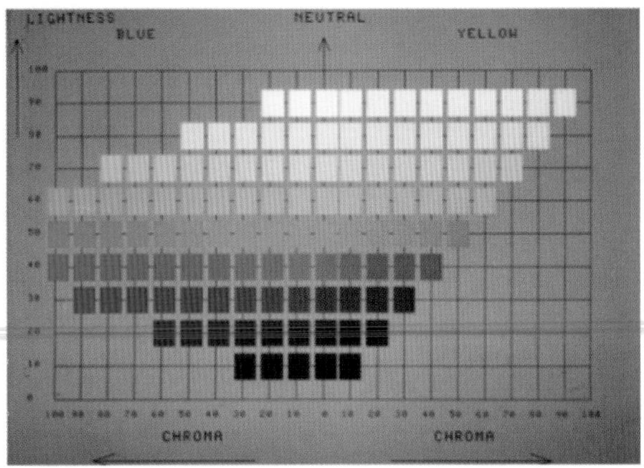

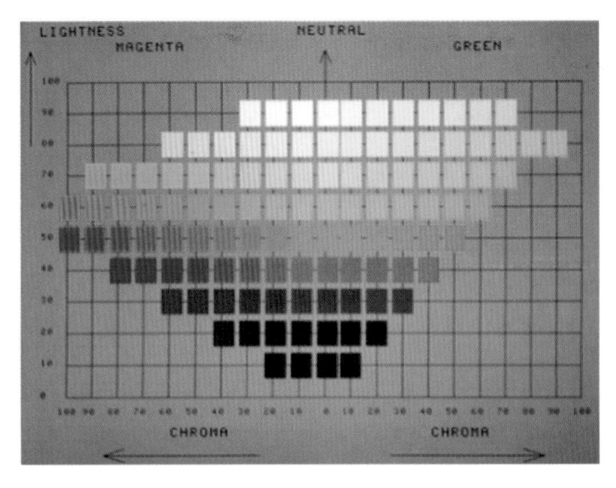

Figure 3.12 Representations similar to that shown in Figure 3.11, but for vertical sections for red and cyan colours, for yellow and blue colours, and for green and magenta colours. Because of the limitations of printing, all the colours cannot be shown accurately, but the general layout is displayed adequately

where X_n, Y_n, Z_n are the values of X, Y, Z for the appropriately chosen reference white. In these formulae, the reduced perceptual significance of a given difference in chromaticity caused by a reduction in luminance factor is incorporated by using the tristimulus ratios X/X_n, Y/Y_n, Z/Z_n instead of chromaticity co-ordinates. Because these ratios of the tristimulus values are incorporated as cube-roots, there can be no chromaticity diagram associated with the CIELAB space, and therefore no correlate of saturation. This is because the definition of a chromaticity diagram requires that any mixture of two colours falls on the straight line joining those two colours: in CIELAB space this line would not be straight because of the cube-root transformation. Correlates of hue and chroma are, however, available, and are formulated in exactly analogous ways to those used in connection with the CIELUV space:

CIE 1976 a,b hue-angle, h_{ab}

$$h_{ab} = \arctan(b^*/a^*)$$

CIE 1976 a,b chroma, C^*_{ab}

$$C^*_{ab} = (a^{*2} + b^{*2})^{1/2}$$

The CIELUV and CIELAB spaces are intended to apply to object colours of the same size and shape, viewed in identical white to mid-grey surroundings, by an observer photopically adapted to a field of chromaticity not too different from that of average daylight. If the samples considered have an angular subtense greater than 4°, the X_{10}, Y_{10}, and Z_{10} tristimulus values should be used instead of X, Y, and Z, and the resulting measures are then all distinguished by an additional subscript 10, such as L^*_{10}, $h_{uv,10}$, etc.

CIELUV and CIELAB uniform colour spaces are described in the following CIE Standards:

ISO 11664–4:2008(E)/CIE S 014–4/E 2007: Colorimetry – Part 4: CIE 1976 L*a*b* colour space.

ISO 11664–5:2009(E)/CIE S 014–5/E:2009: Colorimetry – Part 5: CIE 1976 L*u*v* colour space and u′,v′ uniform chromaticity scale diagram.

Uniform colour spaces based on the CIECAM02 Colour Appearance Model (to be described in Chapter 15) have also been devised; see Section 15.24 (Luo, Cui, and Li, 2006).

3.10 CIE 1976 COLOUR DIFFERENCE FORMULAE

If the differences between two colours in L^*, u^*, and v^* are denoted by ΔL^*, Δu^*, and Δv^*, respectively, then the total colour difference may be evaluated as:

CIE 1976 ($L^*u^*v^*$) colour difference or *CIELUV colour difference*

$$\Delta E^*_{uv} = [(\Delta L^*)^2 + (\Delta u^*)^2 + (\Delta v^*)^2]^{1/2}$$

Thus ΔE^*_{uv} is equal to the distance between the two points representing the colours in the CIELUV space. A just noticeable difference is equal to about one ΔE^*_{uv} unit for samples

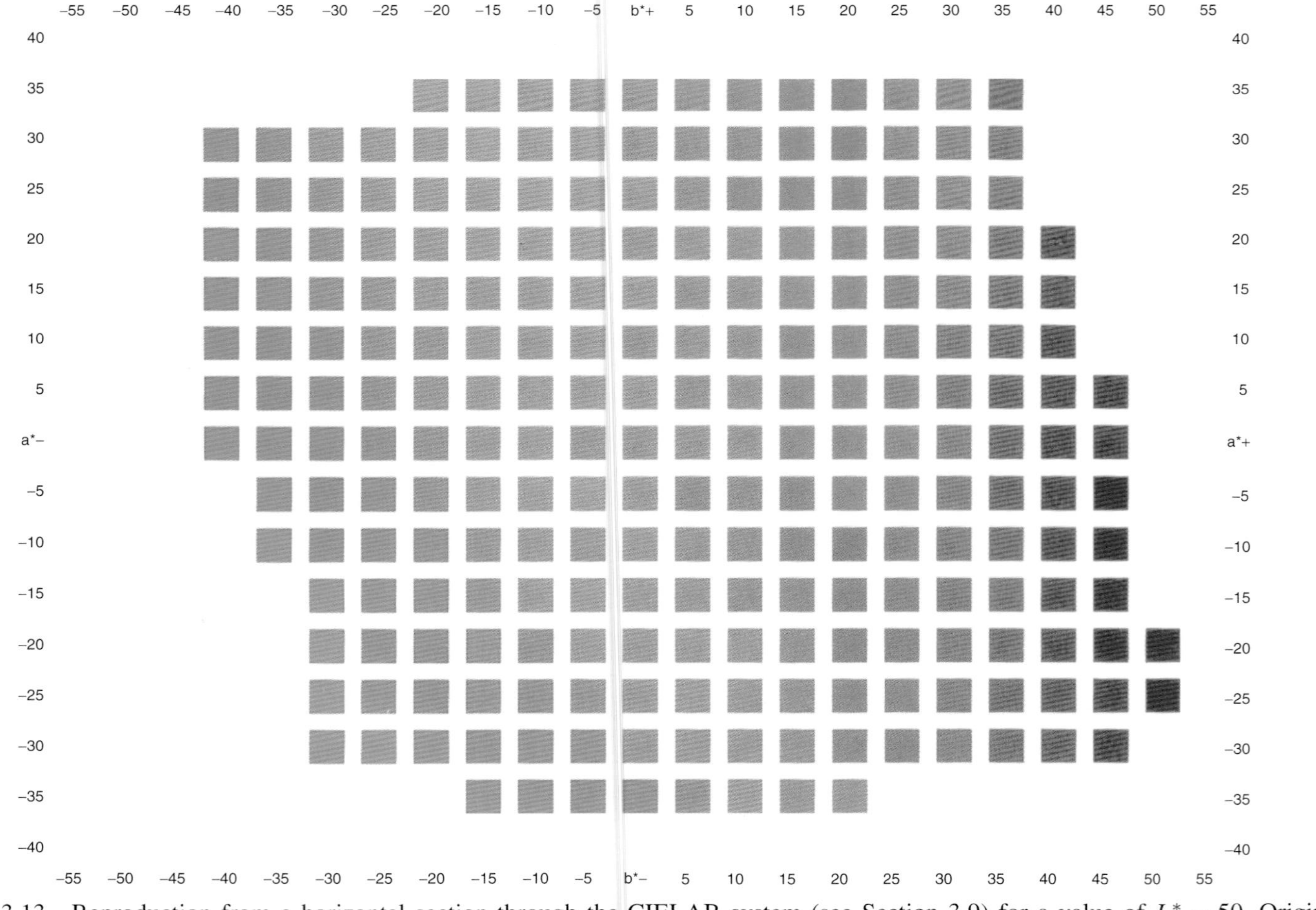

Figure 3.13 Reproduction from a horizontal section through the CIELAB system (see Section 3.9) for a value of $L^* = 50$. Original produced on Kodak Ektacolor paper by Richard Ingalls of Target Color Technology. Because of the limitations of printing, all the colours cannot be shown accurately, but the general layout is displayed adequately

seen side by side, but when samples are separated in space or time the just noticeable difference can be much larger.

It is convenient to be able to express the same colour difference in terms of differences in L^*, C^*, and a measure that correlates with hue difference. To do this, h_{uv} is not used because, being an angular measure, it cannot be combined with L^* and C^* easily. What is used instead is:

CIE1976 u,v hue-difference, ΔH^*_{uv}

$$\Delta H^*_{uv} = [(\Delta E^*_{uv})^2 - (\Delta L^*)^2 - (\Delta C^*_{uv})^2]^{1/2}$$

where ΔC^*_{uv} is the difference in C^*_{uv} between the two colours. It follows that:

$$\Delta E^*_{uv} = [(\Delta L^*)^2 + (\Delta H^*_{uv})^2 + (\Delta C^*_{uv})^2]^{1/2}$$

For small colour differences away from the L^* axis, the geometry is such that $\Delta H^*_{uv} = C^*_{uv}\Delta h_{uv}(\pi/180)$. The rule is that ΔH^*_{uv} is to be regarded as positive if indicating an increase in h_{uv} and negative if indicating a decrease in h_{uv}. In some applications different weightings for ΔL^*, ΔC^*_{uv}, and ΔH^*_{uv} may be necessary.

The following similar measures can also be evaluated in connection with the CIELAB space:

*CIE 1976(L*a*b*) colour difference* or *CIELAB colour difference*

$$\Delta E^*_{ab} = [(\Delta L^*)^2 + (\Delta a^*)^2 + (\Delta b^*)^2]^{1/2}$$

$$\Delta E^*_{ab} = [(\Delta L^*)^2 + (\Delta H^*_{ab})^2 + (\Delta C^*_{ab})^2]^{1/2}$$

CIE 1976 a,b hue-difference, ΔH^*_{ab}

$$\Delta H^*_{ab} = [(\Delta E^*_{ab})^2 - (\Delta L^*)^2 - (\Delta C^*_{ab})^2]^{1/2}$$

ΔH^*_{ab} can also be defined as follows:

$$\Delta H^*_{ab} = 2(C^*_{ab,1} - C^*_{ab,2})^{1/2}\sin(\Delta h_{ab}/2)$$

where $C^*_{ab,1}$ and $C^*_{ab,2}$ refer to the values of C^* of the two samples being compared, and Δh_{ab} is the difference in the values of the hue angles (in radians) of those two samples. For small colour differences away from the achromatic axis the above equation can be written as:

$$\Delta H^*_{ab} = 2(C^*_{ab,1} - C^*_{ab,2})^{1/2}\Delta h_{ab}$$

Corresponding equations can be applied to CIE 1976 CIELUV colour difference calculations.

It was unfortunate that in 1976 the CIE could not recommend a single colour space and associated colour difference formula to meet all needs. At that time, some representatives from the colorant industries argued strongly in favour of a difference formula that was similar to one known as the Adams-Nickerson (AN40) formula. The CIELAB formula fulfils this requirement, its colour difference evaluations being on average (but not uniformly throughout its space) about 1.1 times those produced by the AN40 formula. On the other hand, representatives from the television industry preferred a space having an associated chromaticity diagram because of the very simple way in which additive colour mixtures (such as occur in television display devices) are represented on it. Since 1976, evaluations of the CIELUV and CIELAB colour spaces and difference formulae have shown that they are about equally good (or bad) in representing the perceptual sizes of colour differences, the ratios of the maximum to the minimum values of ΔE^* corresponding to a given perceptual difference being about six to one (Robertson, 1977).

The CIELUV space has been used in television and the CIELAB space in the colorant industries. However, in the colorant industries, there has been a continuous search for better colour difference formulae, and, in some applications, these have now superseded the CIELAB formula. These more advanced formulae are usually non-Euclidean, in the sense that colour differences cannot be represented as simple distances in an ordinary three-dimensional space. Three of these advanced formulae will be described in the next section although the use of the CIE94 formula is not now recommended by CIE.

Some differences between the u′,v′ diagram, and diagrams in which v^* is plotted against u^*, or b^* is plotted against a^*, are illustrated in Figure 3.14. What is shown in each diagram is the gamut of reproducible colours on a typical television display, when one of the three primaries is at the same strength as that used for white, another primary is at zero, and the third primary ranges between zero and the strength used for white. On the u′,v′ diagram, the gamut is the triangle formed by the points representing the chromaticities of the three primaries; on the u*, v* and a*, b* diagrams the colours intermediate between the primary colours are further from the white point than a straight line joining the points representing

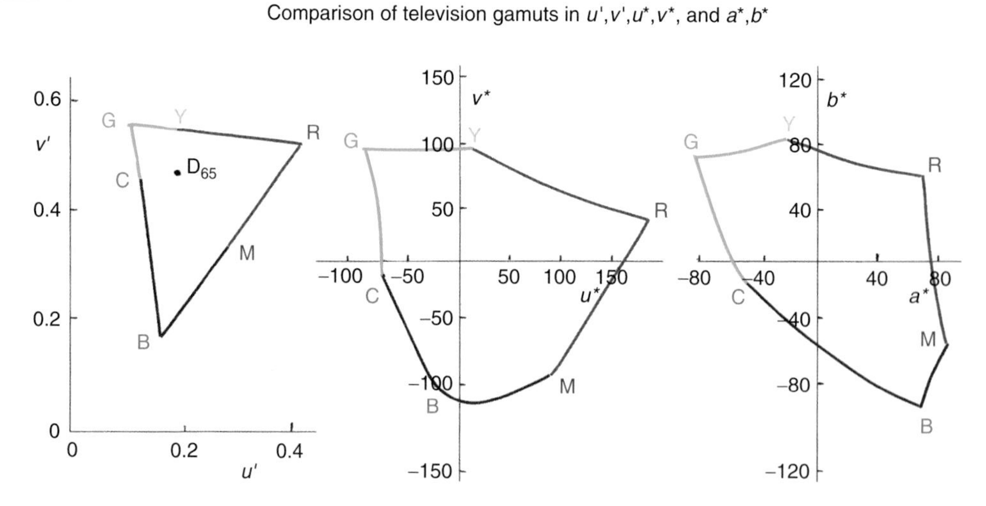

Figure 3.14 Gamut of reproducible colours on a typical colour television display plotted on u′,v′, u*, v*, and a*, b* diagrams

the primaries, because the higher luminance factor of these colours, caused by the addition of two primaries, increases the value of L^*, and hence the values of u^*, v^* and a^*, b^*. Also, the edges of the gamuts between the primaries are curved in the u*, v* and a*, b* diagrams. There is no simple relationship between the pattern of points on these three diagrams, and data can only be transferred from one diagram to another by transforming back to X, Y, Z tristimulus values and then using a second transformation into the required diagram.

3.11 CMC, CIE94, AND CIEDE2000 COLOR DIFFERENCE FORMULAE

A more complicated colour difference formula which has been shown to give better correlation than the CIELAB formula for small colour differences in the colorant industries is known as the CMC(l:c) formula (Clarke, McDonald, and Rigg, 1980; McLaren, 1986; BS 6923, 1988). In this formula the colour difference, $\Delta E^*_{\mathrm{CMC}}$, is evaluated as:

$$\Delta E^*_{\mathrm{CMC}}(\mathrm{l{:}c}) = \left[\left(\frac{\Delta L^*}{l S_{\mathrm{L}}}\right)^2 + \left(\frac{\Delta C^*_{\mathrm{ab}}}{c S_{\mathrm{C}}}\right)^2 + \left(\frac{\Delta H^*_{\mathrm{ab}}}{S_{\mathrm{H}}}\right)^2\right]^{1/2}$$

where

$$S_{\mathrm{L}} = 0.0409075L^*/(1 + 0.01765L^*) \qquad \text{unless } L^* < 16 \text{ when } S_{\mathrm{L}} = 0.511$$

$$S_{\mathrm{C}} = 0.0638C^*_{\mathrm{ab}}/(1 + 0.0131C^*_{\mathrm{ab}}) + 0.638$$

$$S_{\mathrm{H}} = (fT + 1 - f)S_{\mathrm{C}})$$

where

$$f = \left[\frac{(C^*_{\mathrm{ab}})^4}{(C^*_{\mathrm{ab}})^4 + 1900}\right]^{1/2}$$

and

$$T = 0.36 + |0.4\cos(h_{\mathrm{ab}} + 35)|$$

unless h_{ab} is between 164° and 345° when

$$T = 0.56 + |0.2\cos(h_{\mathrm{ab}} + 168)|$$

The vertical lines (| and |) enclosing some of the expressions indicate that, for these, the absolute value is to be taken, i.e. the value is always to be taken as positive whatever the numerical result obtained by the initial calculation.

It is clear that what this formula does is to vary the relative weightings of the contributions of the differences in L^*, C^*_{ab}, and H^*_{ab} according to the position of the colour in the CIELAB space. The values of l and c are chosen to give the most appropriate weighting of differences in lightness and chroma, respectively, relative to differences in hue. For predicting the *perceptibility* of colour differences, l and c are both set equal to unity, and this is referred to as the CMC(1:1) formula. For predicting the *acceptability* of colour differences, it is sometimes preferable to set l and c at values greater than unity.

In 1994 the CIE (CIE, 1995) introduced a colour difference formula, CIE94, that is similar to the CMC formula in that it is an elaboration of the CIELAB formula, but is less complicated providing only one of the features of the CMC formula; the decreasing weight given to differences in ΔC^*_{ab} and ΔH^* with increasing C^*_{ab}. In this formula the colour difference is evaluated as

$$\Delta E^*_{\mathrm{CIE94}} = \left[\left(\frac{\Delta L^*}{k_L S_L}\right)^2 + \left(\frac{\Delta C^*_{ab}}{k_C S_C}\right)^2 + \left(\frac{\Delta H^*_{ab}}{k_H S_H}\right)^2\right]^{0.5}$$

$$S_L = 1$$

$$S_C = 1 + 0.045(C^*_{abS}) \quad \text{or} \quad S_C = 1 + 0.045(C^*_{abS1}C^*_{abS2})^{0.5}$$

$$S_H = 1 + 0.015(C^*_{abS}) \quad \text{or} \quad S_H = 1 + 0.015(C^*_{abS1}C^*_{abS2})^{0.5}$$

These expressions involve asymmetry if standard, S, and batch, B are interchanged. | These alternative expressions avoid asymmetry between colours 1 and 2.

For most applications $k_L, k_C, k_H = 1$ but for the textile industry k_L is often set equal to a value of 2.

The use of the CIE94 colour difference formula is no longer recommended by the CIE.

Subsequent work by the CIE led to the recommendation of the CIEDE2000 formula for colour difference evaluation (CIE, 2001). This formula includes improvements on the dividing terms used in the CIE94 formula, all based on new experimental data, together with a hue-rotation term to allow for an interaction between the hue and chroma terms in the blue region of colour space. It has been found that most ellipses that represent colour differences tilt towards the origin of the CIELAB colour space; the ellipse tilt in the blue region, however, is in a counter-clockwise direction and away from the direction of constant hue angle. To account for this effect, a rotation function is applied to the weighted hue and chroma difference that only has an effect on high chroma, blue, colour differences.

In this formula, the colour difference is evaluated as:

$$\Delta E_{00} = [(\Delta L'/k_L S_L)^2 + (\Delta C'/k_C S_C)^2 + (\Delta H'/k_H S_H)^2 + R_T(\Delta C'/k_C S_C)(\Delta H'/k_H S_H)]^{0.5}$$

where

$$L' = L^*$$

$$\Delta L' = L'_b - L'_s$$

$$\Delta C' = C'_b - C'_s$$

$$\Delta H' = 2(C'_b C'_s)^{1/2}\sin(\Delta h'/2)$$

and

$$\Delta h' = h'_b - h'_s$$

the subscripts b and s indicate that the value is for the batch (or test) colour and for the standard colour, respectively.

The CIELAB axes are transformed such that:

$$L' = L^*$$
$$a' = a^*(1 + G)$$
$$b' = b^*$$

where

$$G = 0.5\{1 - [\bar{C}_{ab}^{*7}/(\bar{C}_{ab}^{*7} + 25^7)]^{0.5}\}$$

The bar sign over the symbol C^* indicates that the value is the arithmetic mean of the values for a pair of colours, and similarly when used over L', h', and C', below.

Modified chroma, C' and hue angle h' values are calculated using:

$$C' = (a'^2 + b'^2)^{0.5}$$
$$h' = \tan^{-1}(b'/a')$$

The weighting functions S_L, S_C, S_H are given by

$$S_\mathrm{L} = 1 + [0.015(\bar{L}' - 50)^2]/[20 + (\bar{L}' - 50)^2]^{0.5}$$
$$S_\mathrm{C} = 1 + 0.045\bar{C}'$$
$$S_\mathrm{H} = 1 + 0.015\bar{C}'T$$

where

$$T = 1 - 0.17\cos(\bar{h}' - 30^\circ) + 0.24\cos(2\bar{h}') + 0.32\cos(3\bar{h}' + 6^\circ) - 0.20\cos(4\bar{h}' - 63^\circ)$$

The last step is to deal with the anomalies of CIELAB colour space in the blue region where the main axes of tolerance ellipses do not point towards the origin of the space. A rotation parameter, R_T, is defined as follows.

$$R_\mathrm{T} = -\sin(2\Delta\theta)R_\mathrm{C}$$

with

$$\Delta\theta = 30\exp\{-[(\bar{h}' - 275^\circ)/25)]^2\}$$
$$R_\mathrm{C} = 2[\bar{C}'^7/(\bar{C}'^7 + 25^7)]^{0.5}$$

A set of worked examples for the DE2000 colour difference formula is given at the end of the chapter.

3.12 AN ALTERNATIVE FORM OF THE CIEDE2000 COLOUR-DIFFERENCE EQUATION

In many applications of a colour-difference formula, for example in apportioning the colour difference into hue, chroma, and lightness, components for shade sorting, or for attributing size and direction to a specific difference in dye recipe formulation, it is more convenient to have the formula defined with only three terms pertaining to lightness, chroma, and hue respectively. It has been shown that this is possible and the procedure for calculating the terms is given below (Nobbs, 2002).

Step 1 Calculate the CIELAB L^*, a^*, b^*, and C^* as usual

Step 2 Calculate a', b', C' and h' as for CIEDE2000 above

Step 3 Calculate $\Delta L'$, $\Delta C'$ and $\Delta H'$ as for CIEDE2000 above

Step 4 Calculate the weighting functions and rotation term as for CIEDE2000 above

Step 5 Convert to three terms

$$\tan(2\phi) = R_{\mathrm{T}} \frac{(k_{\mathrm{C}} S_{\mathrm{C}})(k_{\mathrm{H}} S_{\mathrm{H}})}{(k_{\mathrm{H}} S_{\mathrm{H}})^2 - (k_{\mathrm{C}} S_{\mathrm{C}})^2}$$

$$\Delta C'' = \Delta C' \cos(\phi) + \Delta H' \sin(\phi)$$

$$\Delta H'' = \Delta H' \cos(\phi) - \Delta C' \sin(\phi)$$

$$S''_{\mathrm{C}} = (k_{\mathrm{C}} S_{\mathrm{C}}) \sqrt{\frac{2(k_{\mathrm{H}} S_{\mathrm{H}})}{2(k_{\mathrm{H}} S_{\mathrm{H}}) + R_{\mathrm{T}}(k_{\mathrm{C}} S_{\mathrm{C}}) \tan(\phi)}}$$

$$S''_{\mathrm{H}} = (k_{\mathrm{H}} S_{\mathrm{H}}) \sqrt{\frac{2(k_{\mathrm{C}} S_{\mathrm{C}})}{2(k_{\mathrm{C}} S_{\mathrm{C}}) - R_{\mathrm{T}}(k_{\mathrm{H}} S_{\mathrm{H}}) \tan(\phi)}}$$

$$\Delta L_{00} = \frac{\Delta L'}{k_{\mathrm{L}} S_{\mathrm{L}}}$$

$$\Delta C_{00} = \frac{\Delta C''}{S''_{\mathrm{C}}}$$

$$\Delta H_{00} = \frac{\Delta H''}{S''_{\mathrm{H}}}$$

Step 6 Calculate CIEDE2000 ΔE_{00}

3.13 SUMMARY OF MEASURES AND THEIR PERCEPTUAL CORRELATES

A summary of the colorimetric measures described in this chapter, together with their perceptual correlates, is given in the first three columns of Table 3.1. The fourth column

Table 3.1 Colorimetric measures and their perceptual correlates (for the Colour Appearance Model CIECAM02 see Chapter 15)

Non-uniform measures	Approximately uniform measures	Approximate perceptual correlates	Correlates in CIECAM02
Luminance L	Only in Colour Appearance Model	Brightness	Q
Chromaticity x, y	CIE 1976 chromaticity u', v'	Hue and saturation	None
Dominant wavelength λ_d	CIE 1976 hue-angle h_{uv} or h_{ab}	Hue	h or H
Excitation purity p_e	CIE 1976 saturation s_{uv}	Saturation	s
Luminance factor L/L_n	CIE 1976 lightness L^*	Lightness	J
None	CIE 1976 chroma C^*_{uv} or C^*_{ab}	Chroma	C
None	Only in Colour Appearance Model	Colourfulness	M

lists the relevant measures provided by the CIE colour appearance model, CIECAM02, to be described in Chapter 15.

The attributes lightness and chroma apply only to related colours, such as surface colours seen in normal surroundings. Hence, for unrelated colours, such as light sources, luminance factor, and CIE 1976 lightness and chroma, do not apply.

Sets of measures that are particularly useful in practice include:

for unrelated colours u', v', L

for related colours X, Y, Z, L

u', v', Y/Y_n, L

h_{uv}, C^*_{uv}, L^*, L

h_{ab}, C^*_{ab}, L^*, L where L is the luminance.

3.14 ALLOWING FOR CHROMATIC ADAPTATION

At the end of Section 3.9, it was pointed out that the CIELUV and CIELAB systems were intended for application to conditions of adaptation to fields of chromaticity not too different from daylight; this is also the case for the CMC and CIEDE2000 colour difference formulae. For illuminants of other chromaticities, one of two alternative procedures is usually adopted to allow for chromatic adaptation to them.

First, the CIELUV and CIELAB systems (and the CMC, CIE94, and CIEDE2000 elaborations of CIELAB) are applied as they stand. This is the simplest approach, and, because the formulae in the systems are normalised for the reference white, they include an allowance for chromatic adaptation that results in the reference white (and greys of the same chromaticity) always having C^* equal to zero, related adjustments being made to the values of C^* and h for other colours. However, this procedure is only approximately valid, the approximation being much worse for CIELUV than for CIELAB.

The second procedure is to use a chromatic adaptation transform to change the set of tristimulus values for each colour in the non-daylight illuminant to a set for *corresponding colours* that have the same appearance when seen in daylight. The colour difference formulae are then used with the corresponding colours. Several different chromatic adaptation transforms have been proposed; they are discussed in Section 6.13.

3.15 THE EVALUATION OF WHITENESS

Whiteness is an important attribute of colours in certain industries, such as paper making, textiles, laundering, and paint manufacture. The higher the luminance factor, the whiter a sample will look, although if luminance factors of over 100% can be achieved the sample may have an appearance of fluorescence rather than of whiteness. But whiteness does not depend only on luminance factor; chromaticity also has an effect. If two whites have the same luminance factor and one is slightly bluer than the other it will look whiter. It is for this reason that *optical brightening agents* are often added to paper; such agents absorb ultra-violet radiation and re-emit it as blue light.

To promote uniformity of practice in the evaluation of whiteness the CIE has recommended that the formulae for whiteness, W or W_{10}, and for tint, T_W or T_{W10}, given below, be used for comparisons of the whiteness of samples evaluated for CIE Standard Illuminant D65 (a standard daylight type illuminant to be described in Section 4.17). The application of the formulae should be restricted to samples that are called 'white' commercially, that do not differ much in colour and fluorescence, and that are measured on the same instrument at nearly the same time; within these restrictions, the formulae provide relative, but not absolute, evaluations of whiteness, that are adequate for commercial use, when employing measuring instruments having suitable modern and commercially available facilities.

$$W = Y + 800(x_{\mathrm{p}} - x) + 1700(y_{\mathrm{p}} - y)$$

$$W_{10} = Y_{10} + 800(x_{\mathrm{p},10} - x_{10}) + 1700(y_{\mathrm{p},10} - y_{10})$$

$$T = 1000(x_{\mathrm{p}} - x) - 650(y_{\mathrm{p}} - y)$$

$$T_{\mathrm{W},10} = 900(x_{\mathrm{p},10} - x_{10}) - 650(y_{\mathrm{p},10} - y_{10})$$

where Y is the Y tristimulus value of the sample, x and y are the x,y chromaticity co-ordinates of the sample, and x_{p}, y_{p} are the chromaticity co-ordinates of the perfect reflecting diffuser, all for the CIE 1931 Standard Colorimetric Observer; the subscript 10 indicates similar values for the CIE 1964 Standard Colorimetric Observer.

The higher the value of W or W_{10}, the greater is the indicated whiteness. The more positive the value of T_W or $T_{W,10}$, the greater is the indicated greenishness; the more negative the value of T_W or $T_{W,10}$, the greater is the indicated reddishness. However, these variables do not provide uniform scales of whiteness, greenishness, or reddishness. For the perfect reflecting diffuser, W and W_{10} are equal to 100, and T_W or $T_{W,10}$ are equal to zero.

The formulae are only applicable to samples whose values of W and W_{10} and T_W or $T_{W,10}$ lie within the following limits:

W or W_{10} greater than 40 and less than $5Y - 280$ or $5Y_{10}-280$

T_W or $T_{W,10}$ greater than -3 and less than $+3$.

The formulae for T_W and $T_{W,10}$ are based on the empirical fact that lines of equal tint in whites are approximately parallel to lines of dominant wavelength 466 nm.

The colorimetric evaluation of whiteness is an active area of research.

3.16 COLORIMETRIC PURITY

As explained in Section 3.4, a measure, *excitation purity*, p_e, that correlates approximately with saturation can be obtained from the x,y chromaticity diagram by dividing the distance on this diagram from the white point to the colour considered, by the distance from the white point to the point on the spectrum locus having the same dominant wavelength; this measure is adversely affected by the non-uniformity of the x, y chromaticity diagram (see Section 3.6). An alternative measure of a similar type is called *colorimetric purity*, p_c. It is defined as:

$$\text{Colorimetric purity, } p_c = L_d/(L_n + L_d)$$

where L_d and L_n are, respectively, the luminances of the spectral (monochromatic) stimulus and of the reference white that match the colour stimulus in an additive mixture.

Unlike excitation purity, colorimetric purity is independent of the use of any particular chromaticity diagram. In the case of stimuli characterised by complementary wavelength, suitable mixtures of the light from the two ends of the spectrum are used instead of the monochromatic stimuli. Colorimetric, p_c, and excitation, p_e, purities are related by the expression:

$$p_c = p_e y_d / y$$

where y_d and y are the y chromaticity co-ordinates, respectively, of the monochromatic stimulus and the colour stimulus considered. Similar measures, $p_{c,10}$ and $p_{e,10}$, for the CIE 1964 Standard Colorimetric Observer can also be calculated and are related by a similar expression.

3.17 IDENTIFYING STIMULI OF EQUAL BRIGHTNESS

As mentioned in Sections 2.3 and 3.2, colours of equal luminance, even if seen under the same photopic viewing conditions, will not necessarily look equally bright, there being a tendency for brightness to increase with colour saturation. Experimental work on this effect has shown it to vary very considerably from one study to another, and amongst observers in a given study. But some guidance is desirable for applications where the effect is large enough to be important, and this is particularly the case when displays are designed that utilise saturated self-luminous colours. An empirical formula has therefore been developed

(Cowan and Ware, 1986; Kaiser, 1986) that makes it possible to identify stimuli that, on average, may be expected to look equally bright. A factor F is evaluated from the x, y chromaticity co-ordinates of the colour as follows:

$$F = 0.256 - 0.184y - 2.527xy + 4.656x^3y + 4.657xy^4$$

Then, if two stimuli have luminances, L_1, L_2, and factors, F_1, F_2, the two stimuli are equally bright if:

$$\log(L_1) + F_1 = \log(L_2) + F_2$$

If these two expressions are not equal, then whichever is greater indicates the stimulus having the greater brightness. In Figure 3.15, loci of equal values of F are shown. It is clear that, although F tends to increase with saturation, it does so from a minimum that does not occur at the point representing the equi-energy stimulus (for which $u' = 0.2105$ and $v' = 0.4737$), but from a point that is displaced from it towards more yellowish colours.

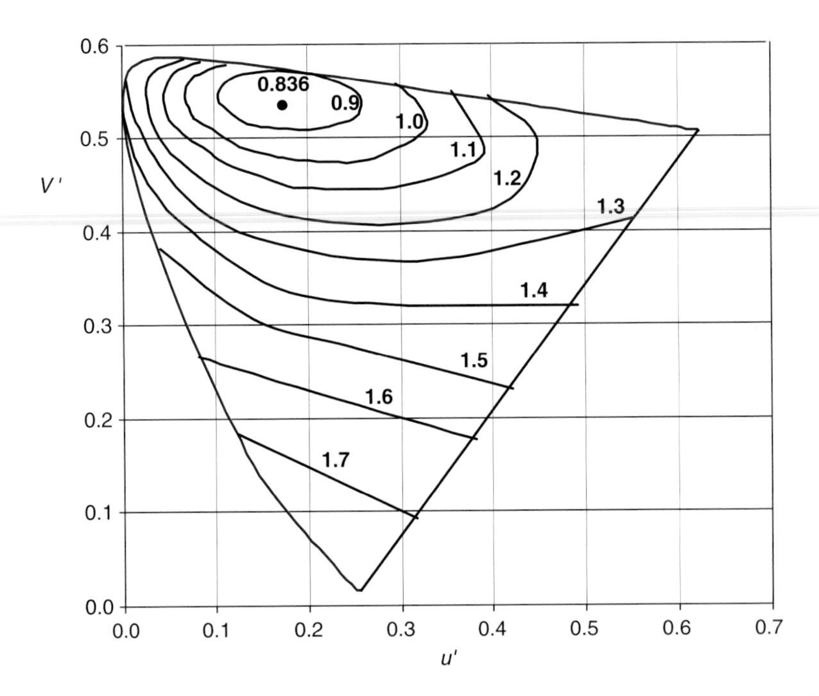

Figure 3.15 Contours of constant factor F representing the difference in log luminances of colours that, on average, appear equally bright under the same viewing conditions. The figures on the curves are for 10^F. For example, if $F = 0.2$, then $10^{0.2} = 1.6$. The figure on each contour therefore shows the factor by which a stimulus whose chromaticity lies on the 1.0 contour would have to be increased (or, for yellowish colours, decreased) to appear to have the same brightness as a colour situated on each contour when seen under the same viewing conditions. (After Cowan and Ware, 1986)

3.18 CIEDE2000 WORKED EXAMPLE

How to calculate the colour difference, D_i, between a reference sample R_i and a test sample T_i.

The ten sets of input test data have been taken from Color Research and Application, 26(5), 349 (2001): the tristimulus values are defined for the CIE 1964 Standard Observer and CIE Standard Illuminant D65.

	Illuminant D65		
	$X_{n,10}$	$Y_{n,10}$	$Z_{n,10}$
	94.811	100.000	107.304

	X_{10}	Y_{10}	Z_{10}	$f(X_{10}/X_{n,10})$	$f(Y_{10}/Y_{n,10})$	$f(Z_{10}/Z_{n,10})$
***R*1**	19.4100	28.4100	11.5766	0.5894	0.6574	0.4761
***T*1**	19.5525	28.6400	10.5791	0.5908	0.6592	0.4620
***R*2**	22.4800	31.6000	38.4800	0.6189	0.6811	0.7105
***T*2**	22.5833	31.3700	36.7901	0.6199	0.6795	0.6999
***R*3**	28.9950	29.5800	35.7500	0.6737	0.6663	0.6932
***T*3**	28.7704	29.7400	35.6045	0.6720	0.6675	0.6923
***R*4**	4.1400	8.5400	8.0300	0.3521	0.4404	0.4214
***T*4**	4.4129	8.5100	8.6453	0.3597	0.4399	0.4319
***R*5**	4.9600	3.7200	19.5900	0.3740	0.3338	0.5673
***T*5**	4.6651	3.8100	17.7848	0.3664	0.3365	0.5493
***R*6**	15.6000	9.2500	5.0200	0.5480	0.4523	0.3603
***T*6**	15.9148	9.1500	4.3872	0.5516	0.4506	0.3445
***R*7**	73.0000	78.0500	81.8000	0.9165	0.9207	0.9135
***T*7**	73.9351	78.8200	84.5156	0.9204	0.9237	0.9235
***R*8**	73.9950	78.3200	85.3060	0.9207	0.9218	0.9264
***T*8**	69.1762	73.4000	79.7130	0.9003	0.9021	0.9057
***R*9**	0.7040	0.7500	0.9720	0.1958	0.1963	0.2085
***T*9**	0.6139	0.6500	0.8510	0.1883	0.1885	0.1997
***R*10**	0.2200	0.2300	0.3250	0.1560	0.1558	0.1615
***T*10**	0.0933	0.1000	0.1453	0.1456	0.1457	0.1485

	L^*_{10}	a^*_{10}	b^*_{10}	C^*_{10}	$\bar{L}'_{10}$	$\bar{C}^*_{10}$
***R*1**	60.2574	−34.0099	36.2677	49.7194	60.3600	50.9525
***T*1**	60.4626	−34.1751	39.4387	52.1857		
***R*2**	63.0109	−31.0961	−5.8663	31.6447	62.9148	30.8591
***T*2**	62.8187	−29.7946	−4.0864	30.0735		
***R*3**	61.2901	3.7196	−5.3901	6.5490	61.3597	5.9982
***T*3**	61.4292	2.2480	−4.9620	5.4474		
***R*4**	35.0831	−44.1164	3.7933	44.2792	35.0532	42.1911
***T*4**	35.0232	−40.0716	1.5901	40.1031		
***R*5**	22.7233	20.0904	−46.6940	50.8326	22.8782	47.9757
***T*5**	23.0331	14.9730	−42.5619	45.1188		

(Continued)

(Continued)

R6	36.4612	47.8580	18.3852	51.2680	36.3664	53.0262
T6	36.2715	50.5065	21.2231	54.7844		
R7	90.8027	−2.0831	1.4410	2.5329	90.9778	2.0885
T7	91.1528	−1.6435	0.0447	1.6441		
R8	90.9257	−0.5406	−0.9208	1.0677	89.7819	1.1108
T8	88.6381	−0.8985	−0.7239	1.1538		
R9	6.7748	−0.2908	−2.4247	2.4421	6.3231	2.3364
T9	5.8715	−0.0985	−2.2286	2.2308		
R10	2.0776	0.0795	−1.1350	1.1378	1.4904	0.8464
T10	0.9033	−0.0636	−0.5514	0.5550		

	G_{10}	a'_{10}	C'_{10}	$\bar{C}'_{10}$	h'_{10}	h'_{10}	$\bar{h}'_{10}$
R1	0.0017	−34.0678	49.7590	50.9914	133.21	133.21	132.08
T1		−34.2333	52.2238		130.96	130.96	
R2	0.0490	−32.6195	33.1428	32.3315	−169.80	190.20	188.82
T2		−31.2542	31.5202		−172.55	187.45	
R3	0.4966	5.5668	7.7488	6.8719	−44.08	315.92	310.03
T3		3.3643	5.9950		−55.86	304.14	
R4	0.0063	−44.3939	44.5557	42.4554	175.12	175.12	176.43
T4		−40.3237	40.3550		177.74	177.74	
R5	0.0026	20.1424	50.8532	47.9924	−66.67	293.33	291.38
T5		15.0118	45.1317		−70.57	289.43	
R6	0.0013	47.9197	51.3256	53.0850	20.99	20.99	21.88
T6		50.5717	54.8444		22.77	22.77	
R7	0.4999	−3.1244	3.4407	2.9531	155.24	155.24	167.10
T7		−2.4651	2.4655		178.96	178.96	
R8	0.5000	−0.8108	1.2269	1.3784	−131.37	228.63	218.44
T8		−1.3477	1.5298		−151.76	208.24	
R9	0.4999	−0.4362	2.4637	2.3486	−100.20	259.80	263.00
T9		−0.1477	2.2335		−93.79	266.21	
R10	0.5000	0.1192	1.1412	0.8504	−84.01	275.99	268.09
T10		−0.0954	0.5595		−99.82	260.18	

	T_{10}	$\Delta h'_{10}$	$\Delta L'_{10}$	$\Delta C'_{10}$	$\Delta H'_{10}$	$S_{L,10}$	$S_{C,10}$	$S_{H,10}$
D1	1.3010	−2.25	0.2052	2.4648	−2.0018	1.1427	3.2946	1.9951
D2	0.9402	−2.75	−0.1922	−1.6226	−1.5490	1.1831	2.4549	1.4560
D3	0.6952	−11.79	0.1391	−1.7538	−1.3995	1.1586	1.3092	1.0717
D4	1.0168	2.63	−0.0599	−4.2007	1.9430	1.2148	2.9105	1.6476
D5	0.3636	−3.91	0.3098	−5.7215	−3.2653	1.4014	3.1597	1.2617
D6	0.9239	1.78	−0.1897	3.5189	1.6444	1.1943	3.3888	1.7357
D7	1.1546	23.72	0.3501	−0.9751	1.1972	1.6110	1.1329	1.0511
D8	1.3916	−20.39	−2.2876	0.3029	−0.4850	1.5930	1.0620	1.0288
D9	0.9556	6.41	−0.9033	−0.2302	0.2622	1.6517	1.1057	1.0337
D10	0.7827	−15.81	−1.1743	−0.5817	−0.2198	1.7246	1.0383	1.0100

(Continued)

(*Continued*)

	$\Delta\theta_{10}$	$R_{T,10}$	$\Delta E_{00,10}$
***D*1**	0.0000	0.0000	1.2644
***D*2**	0.0002	0.0000	1.2630
***D*3**	4.2110	−0.0032	1.8731
***D*4**	0.0000	0.0000	1.8645
***D*5**	19.5282	−1.2537	2.0373
***D*6**	0.0000	0.0000	1.4146
***D*7**	0.0000	0.0000	1.4440
***D*8**	0.1794	0.0000	1.5381
***D*9**	23.8303	−0.0004	0.6378
***D*10**	27.7922	0.0000	0.9082

To convert to three terms:

	ϕ_{10}	$\Delta L''_{10}$	$\Delta C''_{10}$	$\Delta H''_{10}$	$S'_{C,10}$	$S'_{H,10}$
***D*1**	6.4012E-15	0.1796	2.4648	−2.0018	3.2946	1.9951
***D*2**	6.1324E-06	−0.1624	−1.6226	−1.5490	2.4549	1.4560
***D*3**	3.9555E-03	0.1201	−1.7593	−1.3925	1.3092	1.0717
***D*4**	1.5271E-07	−0.0493	−4.2007	1.9430	2.9105	1.6476
***D*5**	2.6858E-01	0.2211	−6.3828	−1.6300	4.1926	1.2204
***D*6**	2.1856E-45	−0.1589	3.5189	1.6444	3.3888	1.7357
***D*7**	3.2173E-11	0.2173	−0.9751	1.1972	1.1329	1.0511
***D*8**	3.8731E-06	−1.4360	0.3029	−0.4850	1.0620	1.0288
***D*9**	1.3933E-03	−0.5469	−0.2298	0.2625	1.1057	1.0337
***D*10**	1.0839E-04	−0.6809	−0.5817	−0.2198	1.0383	1.0100

	$\Delta L_{00,10}$	$\Delta C_{00,10}$	$\Delta H_{00,10}$	$\Delta E_{00,10}$
***D*1**	0.1796	0.7481	−1.0034	1.2644
***D*2**	−0.1624	−0.6610	−1.0639	1.2630
***D*3**	0.1201	−1.3438	−1.2994	1.8731
***D*4**	−0.0493	−1.4433	1.1793	1.8645
***D*5**	0.2211	−1.5224	−1.3356	2.0373
***D*6**	−0.1589	1.0384	0.9474	1.4146
***D*7**	0.2173	−0.8608	1.1389	1.4440
***D*8**	−1.4360	0.2853	−0.4714	1.5381
***D*9**	−0.5469	−0.2078	0.2539	0.6378
***D*10**	−0.6809	−0.5603	−0.2176	0.9082

REFERENCES

CIE Publication 119:1995, *Industrial colour difference evaluation*, Commission Internationale de l'Éclairage, Vienna, Austria (1995).

CIE Publication 142:2001, *Improvement to industrial colour difference evaluation*, Commission Internationale de l'Éclairage, Vienna, Austria (2001).

Clarke, F.J.J., McDonald, R., and Rigg, B., Modification to the JPC79 color-difference formula, *J. Soc. Dyers and Colourists*, **100**, 128–132 (1980).

BS 6923:1988, *Method for calculation of small colour differences*, British Standards Institution, London, England (1988).

Cowan, W.B., and Ware, C., unpublished communication (1986).

Estévez, O., *On the Fundamental Data-Base of Normal and Dichromatic Colour Vision*, PhD Thesis, University of Amsterdam, Holland (1979).

Kaiser, P.K., Models of heterochromatic brightness matching, *CIE Journal* **5**, 57–59 (1986).

Luo, M.R., Cui, G., and Li, C., Uniform colour spaces based on CIECAM02 Colour Appearance Model, *Color Res. Appl.*, **31**, 320–330 (2006).

McLaren, K., *The Colour Science of Dyes and Pigments*, *2nd Ed.*, p.143, Hilger, Bristol, England (1986).

Nobbs, J.H., A lightness, chroma and hue splitting approach to CIEDE2000 colour differences, *Advances in Colour Science and Technology*, **5**(2), 46–53 (2002).

Robertson, A.R., The CIE 1976 color-difference formulae, *Color Res. Appl.* **2**, 7–11 (1977).

Wright, W.D., The sensitivity of the eye to small colour differences, *Proc. Phys. Soc.* **53**, 93–112 (1941).

GENERAL REFERENCES

CIE Publication 15:2004, *Colorimetry*, 3rd *Ed.*, Commission Internationale de l'Éclairage, Vienna, Austria (2004).

4

Light Sources

4.1 INTRODUCTION

Without light there is normally no colour. Light sources therefore play a very important part in colorimetry. If the colour is self-luminous, such as in the case of fireworks, for example, then the light source itself is the colour. But, more often, colours are associated with objects that, instead of being self-emitting, reflect or transmit the light emitted by light sources. That the nature of these light sources can have a profound effect on the appearance of coloured objects is a well known experience. For instance, to take an extreme example, objects that normally appear red, become dark brown or black when seen under the yellow sodium lamps that are widely used in street lighting. It is also well known that, compared to their appearance in daylight, red and yellow objects seen under tungsten filament light, or candle light, look lighter and more colourful, while blue objects look darker and less colourful. If the visual system did not adapt to the level and colour of the prevailing illumination, these changes would be even greater (see Section 1.8; a quantitative treatment of adaptation will be given in Chapter 15).

We have already seen that light sources are involved in colorimetry in various ways. First, for reflecting and transmitting objects, the Y tristimulus value is usually evaluated so that $Y = 100$ for the perfect reflecting diffuser, or the perfect transmitter, similarly illuminated and viewed (Section 2.6); the similar illumination implies the use of a particular light source. Second, the evaluation of dominant wavelength and excitation purity requires the adoption of an achromatic stimulus, and this is usually taken as the perfect diffuser or transmitter illuminated by a particular light source (Section 3.4). Third, the variables in the CIELUV and CIELAB systems require the use of a reference white, and this is normally defined in terms of a specified white or a highly transmitting object illuminated by a particular light source (Section 3.9). Fourth, pairs of colours that are a match under one light source may not match under another having different spectral composition (Section 2.7). Fifth, samples that fluoresce will usually vary in colour as light sources of different spectral compositions are used, particularly when variations in the ultraviolet content of the light sources occur (to be discussed in Chapter 10). Thus, in all these cases, the colour specifications relate to a particular light source. In fact, for reflecting and transmitting objects, X,

Measuring Colour, Fourth Edition. R.W.G. Hunt and M.R. Pointer.

Y, and Z tristimulus values, and any colorimetric measures derived from them, are largely meaningless unless the light source used is also specified.

It is therefore clear that the proper specification of light sources is an essential part of colorimetry. There are many different light sources, however, and their choice and specification therefore warrant this chapter devoted to them.

4.2 METHODS OF PRODUCING LIGHT

Light can be produced by a variety of methods. These include:

> *Incandescence:* solids and liquids emit light when their temperatures are above about 1000 K (where K indicates the temperature in kelvin, which is equal to the Celsius temperature plus 273).
>
> *Gas discharges:* gases can emit light when an electric current is passed through them; the spectral distribution of the light is characteristic of the chemicals present in the gas.
>
> *Electroluminescence:* certain solids, such as semi-conductors or phosphors, can emit light when an electric current is passed through them.
>
> *Photoluminescence:* when radiation is absorbed by some substances, light is emitted with a change of wavelength; if the emission is immediate it is termed *fluorescence*; if the emission continues appreciably after the absorbing radiation is removed, it is termed *phosphorescence*.
>
> *Cathodoluminescence:* phosphors emit light when they are bombarded with electrons.
>
> *Chemiluminescence:* certain chemical reactions result in light being emitted without necessarily generating heat.

Of the above methods, the most widely occurring in light sources are incandescence (in the sun, and in tungsten filament lamps), gas discharges (in sodium and mercury street lamps, in fluorescent lamps, in mercury-based lamps for stadia and studios, and to some extent on the surface of the sun), photoluminescence (in fluorescent lamps), electroluminescence (in light-emitting diodes), and cathodoluminescence (in cathode ray tubes, as used in older oscilloscopes, televisions, and computer displays).

An important feature of light sources is the success with which they use the power they consume to produce their light; the term *efficacy* is used to characterise this, and can conveniently be expressed as lumens per watt, lm W^{-1} (as discussed in Appendix 1). This term is distinct from the term *efficiency*, which is used when two different amounts of the same entity are compared, such as in luminous efficiency (see Sections 2.2 and 2.3).

4.3 GAS DISCHARGES

Gases are normally insulators, so that electric currents cannot pass through them. But, in some circumstances, their insulating properties break down, a current passes, and, when it does, light may be emitted. One example of this is when a switch in an electrical circuit is opened: frequently a spark can be seen, and this is caused by an electric current being able to flow through the air across the very small gap between the contacts of the switch immediately after it is opened. Another example is lightning; in this case, a current flows because of very high voltages produced by friction in clouds, with moisture in the air facilitating the conduction. To produce light in gases continually in a controlled manner, it is necessary to increase their conductivity, and one way of doing this is by having them at low pressures. When the pressure of a gas is low, its atoms are less concentrated, and the passage of electric current may then be facilitated.

When electrodes are used to produce an electric field across a gas, a current will still not flow, even with the gas at low pressure, unless some of its molecules are broken into electrons and ions; this ionization can be achieved in various ways, one of which is to have glowing filaments at each end of the electric field. With some electrons and ions present, a high voltage can then start a current to flow, and the consequent movement of the electrons and ions causes collisions which produce more ionization, so that the current continues to increase. This means that suitable circuits have to provide a high starting voltage and a current limiting device, the latter often being in the form of a choke (a coil having a high impedance to alternating current).

Light is produced in a gas discharge as a result of electrons colliding with atoms of the gas and exciting them to energy states that are higher than their ground states; the excited atoms then return to their ground states and, in doing so, emit light. The light emitted by this process can have the very distinctive property of being confined to narrow lines in the spectrum.

Lamps producing light by the process of gas-discharge contain the gas in a transparent envelope of glass, fused silica, or other ceramic, usually in the form of a tube, with the electrodes at the two ends.

4.4 SODIUM LAMPS

When low pressure sodium is used for the gas in a gas-discharge lamp, the visible spectrum consists almost entirely of two spectral lines at wavelengths 589.0 nm and 589.6 nm. This spectral power distribution is shown in Figure 4.1; the lines at about 770 nm are too

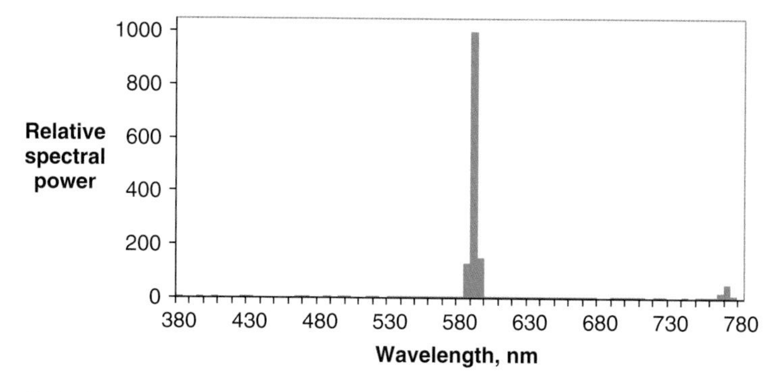

Figure 4.1 Relative spectral power distribution of a low pressure sodium lamp. The power is represented as a histogram in blocks of wavelength 5 nm wide

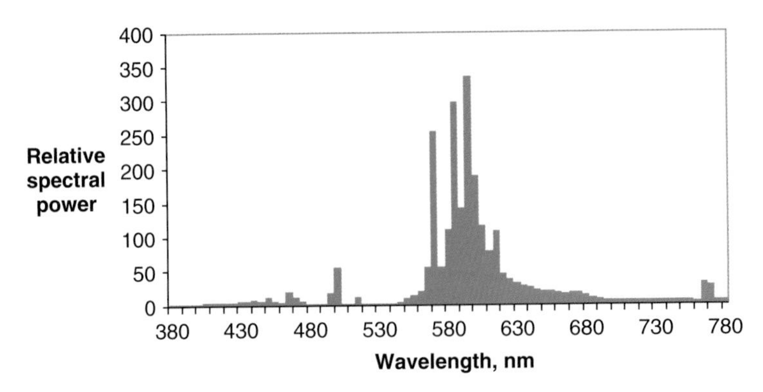

Figure 4.2 Same as Figure 4.1, but for a high pressure sodium lamp

far towards the infrared to have any appreciable visual effect. The theoretical maximum efficacy for any radiation is 683 lumens per watt (lm W^{-1}), and this occurs only for monochromatic radiation of wavelength 555 nm, this wavelength being at the maximum of the $V(\lambda)$ function (Section 2.3). The maximum possible efficacy for light of wavelength 589 nm is $525\,\text{lm W}^{-1}$, but in practice low-pressure sodium lamps have efficacies of only about $160\,\text{lm W}^{-1}$, because of wastage of power in operating the lamps. However, this efficacy is greater than for most other types of lamp, and low-pressure sodium lamps have therefore historically been used in street lighting, where high efficacy is very important; but the colour rendering of these lamps is extremely poor. The relative spectral power distribution of a typical low-pressure sodium lamp is given in Appendix 5.5.

If the pressure of the sodium gas in the envelope is increased, the proximity of neighbouring atoms affects their energy levels and the lines of emission become broader; in addition to this effect, the sodium gas in the tube is sufficiently concentrated to absorb the primary radiation at 589.0 nm and 589.6 nm, with the result that the spectral power distribution is as shown in Figure 4.2. Although this type of spectral power distribution is still far from giving good colour rendering, it is much improved over that of the low pressure lamp, is very useful for floodlighting, and is widely used for street lighting; the colour of the light appears a pleasant pale pink-orange, instead of the deep yellow of the low pressure lamps. These high pressure lamps have an efficacy of $100–150\,\text{lm W}^{-1}$. The relative spectral power distribution of a typical high pressure sodium lamp is given in Appendix 5.5. Further improvement in colour rendering is achieved in colour-enhanced high pressure sodium lamps, for which a typical spectral power distribution is shown in Figure 4.3; the corresponding numerical values are given in Appendix 5.5.

4.5 MERCURY LAMPS

When low pressure mercury is used for the gas in a gas-discharge lamp, the spectrum consists mainly of a series of lines, the more prominent of which are at wavelengths of 253.7 nm, 365.4 nm, 404.7 nm, 435.8 nm, 546.1 nm, and 578.0 nm, together with a very low level continuous radiation throughout the spectrum. If the pressure is increased, the lines tend to broaden and the continuum tends to represent a greater proportion of the light output.

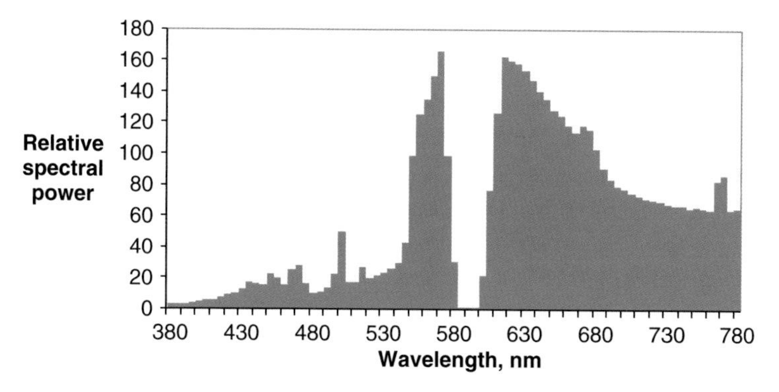

Figure 4.3 Same as Figure 4.1 but for a colour-enhanced high pressure sodium lamp

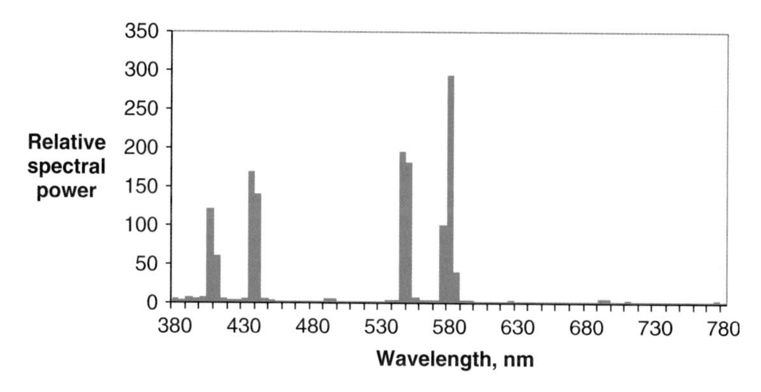

Figure 4.4 Same as Figure 4.1, but for a high pressure mercury lamp type MB

The absence of lines at longer wavelengths than 578.0 nm results in the typical blue-green colour appearance of mercury lamps; and the absence of light at the long wavelength end of the spectrum results in the colour rendering being very poor, especially of Caucasian skin tones. The radiation at 253.7 nm is in the ultraviolet and cannot be seen, but is used in fluorescent lamps to excite phosphors that emit in the visible part of the spectrum (this will be discussed in Section 4.6; the line at 365.4 nm can also be used to excite some phosphors). In Figure 4.4, the relative spectral power distribution is shown for a typical high pressure mercury lamp (MB). Figure 4.5 shows the distribution for a high pressure mercury lamp with a red-emitting phosphor coated on the inside of the envelope (MBF); and Figure 4.6 shows the distribution for a similar lamp using a tungsten filament ballast (MBTF); the red-emitting phosphor in these lamps improves their colour rendering appreciably. The relative spectral power distributions of typical lamps of the types shown in Figures 4.4, 4.5, and 4.6 are given in Appendix 5.5. Typical efficacies range from about 50 lm W^{-1} (for MB and MBF lamps) to about 20 lm W^{-1} (for MBTF lamps).

By adding metal halides to the mercury vapour, extra emission lines can be provided in the spectrum, and the colour rendering can then be improved until it is comparable with that from large area, lower luminance, types, such as fluorescent lamps (to be discussed

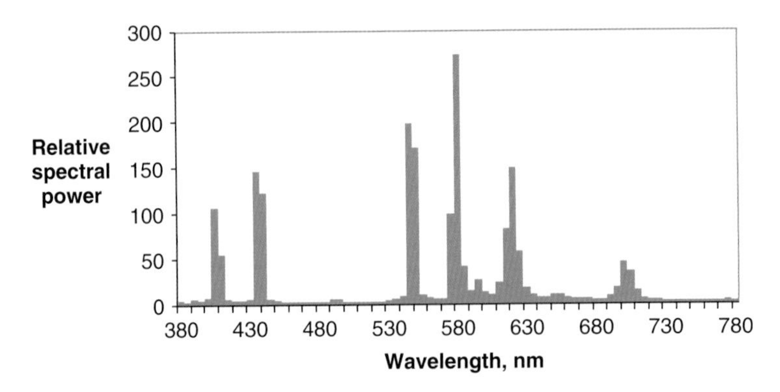

Figure 4.5 Same as Figure 4.4, but for a high pressure mercury lamp type MBF, which has a red-emitting phosphor coated on the inside of the envelope

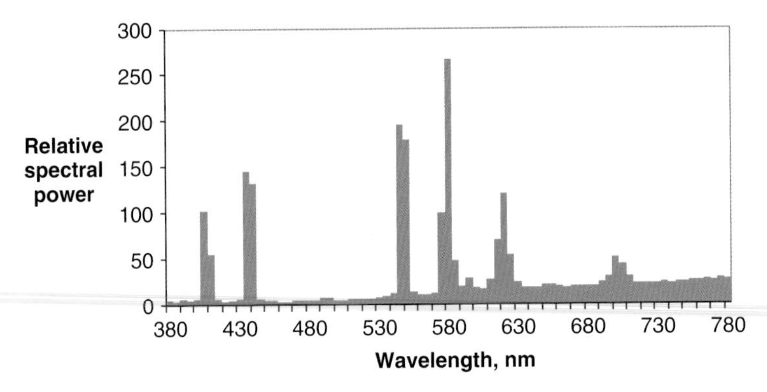

Figure 4.6 Same as Figure 4.5, but for a high pressure mercury lamp type MBTF, which is an MBF lamp with tungsten filament ballast

in Section 4.6). These metal-halide discharge lamps are useful for lighting large areas, for example car parks and the insides of industrial warehouses, and for floodlighting sports stadia where, being compact, they can be used in reflectors to beam the light towards the scene of the action. Lamps can be made that give light of various colours, including the range from daylight to tungsten filament light. One important member of this series of lamps is the HMI lamp, which was developed for use in television to supplement daylight. In Figure 4.7 the relative spectral power distribution of the HMI lamp is shown, and corresponding numerical data are given in Appendix 5.5. Typical efficacies for metal halide mercury lamps are around $80\,\mathrm{lm\,W^{-1}}$.

4.6 FLUORESCENT LAMPS

Fluorescent lamps are very widely used for general lighting, especially in industrial and commercial environments where high levels of illumination, high efficacies, and good colour rendering are required. These lamps consist of a glass tube containing low-pressure

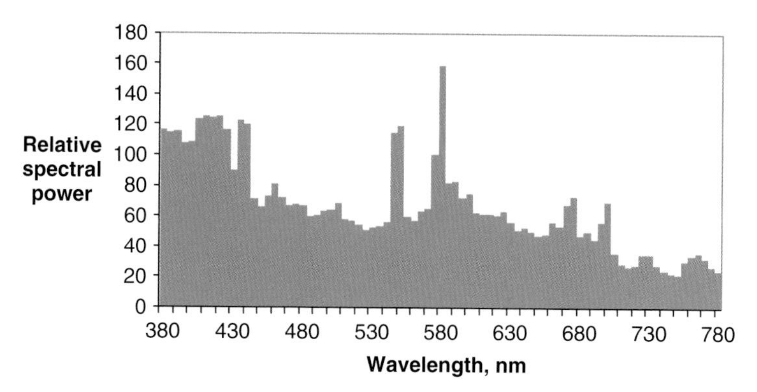

Figure 4.7 Same as Figure 4.4, but for an HMI (Mercury, Medium arc, Iodides) lamp, which has iodides added to the high pressure mercury gas, and is used to supplement daylight for television production

mercury gas, in which a gas-discharge is produced; and the inside of the tube is coated with phosphors which are excited by the emission of the ultraviolet lines of the mercury spectrum, particularly that at 253.7 nm, to produce additional light. The phosphors are carefully chosen to supplement the light from the gas-discharge, special attention being paid to the need to provide additional light in the long wavelength parts of the spectrum where there are no lines in the mercury spectrum. Thus the light from these lamps comes partly from the gas-discharge, but mainly from the phosphors. There is a wide range of different types of fluorescent lamp, according to the phosphors used; these vary from those having high efficacy but poor colour rendering, to those having lower efficacy but good colour rendering. Typical efficacies range from about 45 lm W^{-1} to 95 lm W^{-1}.

In Appendix 5.3, spectral power distributions are given for twelve different types of fluorescent lamp, and these have been chosen to illustrate some of the main groups. The first six are designated *normal*; they have been designed to have a reasonably high efficacy, and their colour rendering is adequate for many purposes; but it is not very good for reddish colours, because there is some deficiency in emission at the long wavelength end of the spectrum.

The second group is designated *broadband*; they have been designed to have very good colour rendering, and their emission at the long wavelength end of the spectrum is appreciably greater than in the case of the normal group; however, their efficacy is lower.

The third group is designated *three-band*. As their name implies, their emission tends to be concentrated in three bands of the spectrum, and these bands are quite narrow, and are designed to occur around wavelengths of approximately 610 nm, 545 nm, and 435 nm. Lamps in this group tend to have relatively high efficacies and fairly good colour rendering. They tend to increase the saturation of most colours, and this makes them attractive for some purposes, such as lighting goods in stores; but the appearance of some colours can be somewhat distorted, so that they are less suitable for critical evaluation of colours in general.

There are no CIE standard illuminants representing fluorescent lamps, but three of the distributions given in Appendix 5.3, F2, F7, and F11, one from each group, are chosen as deserving priority over the others when a few typical illuminants are to be selected. These

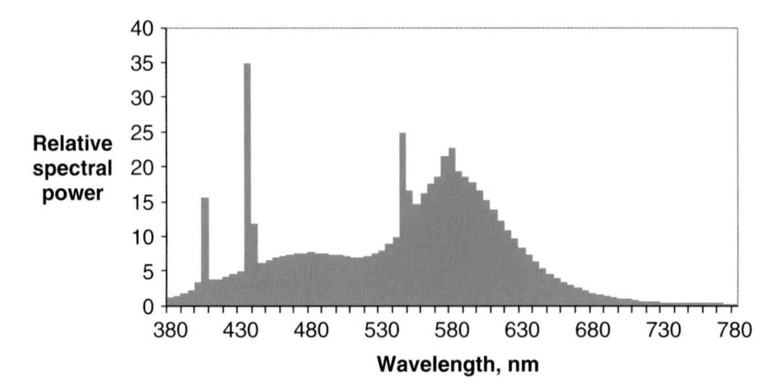

Figure 4.8 Relative spectral power distribution of a fluorescent lamp representative of the *normal* type (F2). The power is represented as a histogram in blocks of wavelengths 5 nm wide

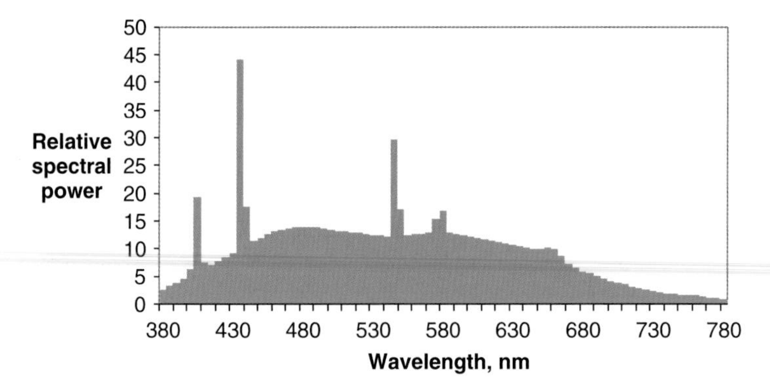

Figure 4.9 Same as Figure 4.8, but for a fluorescent lamp representative of the *broadband* type (F7)

three distributions are illustrated in Figures 4.8, 4.9, and 4.10 respectively. In Figure 4.11 the distribution is shown for a broadband type of lamp typical of those often used for critical evaluation in the graphic arts industry, together with that for a daylight (D50 to be described in Section 4.17) of the same chromaticity; it is clear that these two spectral power distributions, although of the same chromaticity, are very different.

In addition, CIE Publication 15:2004 (CIE, 2004) includes tables of the spectral power distributions of three standard halophosphate fluorescent lamps, three DeLuxe type lamps, six three-band lamps and three multiband lamps; these are also reproduced in Appendix 5.3.

With an increasing requirement to reduce energy consumption, the use of fluorescent lamps is being encouraged over the more traditional tungsten lamp, especially for home use. Indeed some countries have now banned the sale of some tungsten lamps for both domestic and commercial use. This has led to the design of compact fluorescent lamps of a variety of shapes, many of which can be used as direct replacements in traditional

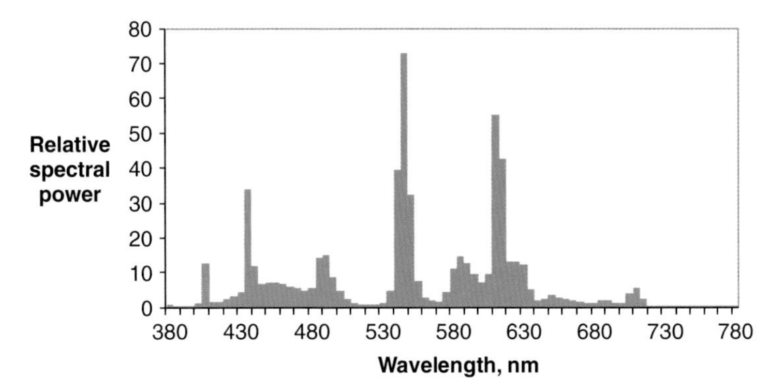

Figure 4.10 Same as Figure 4.8, but for a fluorescent lamp representative of the *three-band* type (F11)

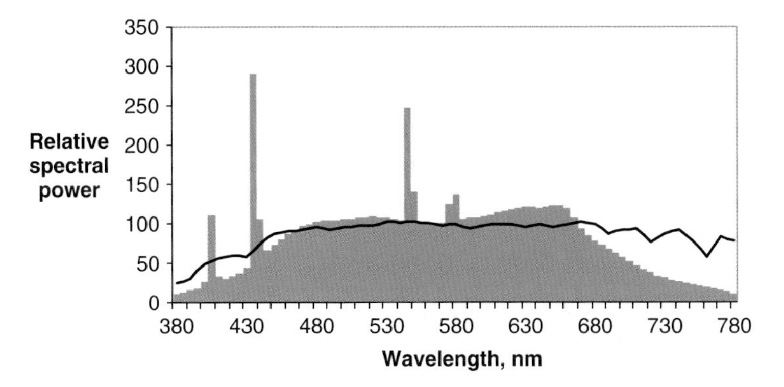

Figure 4.11 Same as Figure 4.8, but for a fluorescent lamp representative of the *broadband* type (F8) having the same chromaticity as that of the daylight spectral power distribution (D50 to be described in Section 4.17) shown for comparison

tungsten lamp fittings; their higher cost is offset by their greater efficacy and longer life. A significant advance in these fluorescent lamps is the integration of the electronic ballast into the lamp, which removes most of the flicker and slow starting usually associated with fluorescent lighting, and can provide a dimming facility.

4.7 XENON LAMPS

An electric current can be made to pass through xenon gas by using a high voltage pulse to cause ionisation. The electrodes in the gas can either be separated by only a few millimetres (compact source xenon), or by about 10 centimetres or more (linear xenon). The operation can be either continuous, or in the form of a series of short pulses at each half-cycle of mains frequency (pulsed xenon), or as isolated flashes (xenon flash tubes). The cold-filling pressure of the gas varies from about 10 atmospheres for some compact source lamps, to about a tenth of an atmosphere for linear lamps.

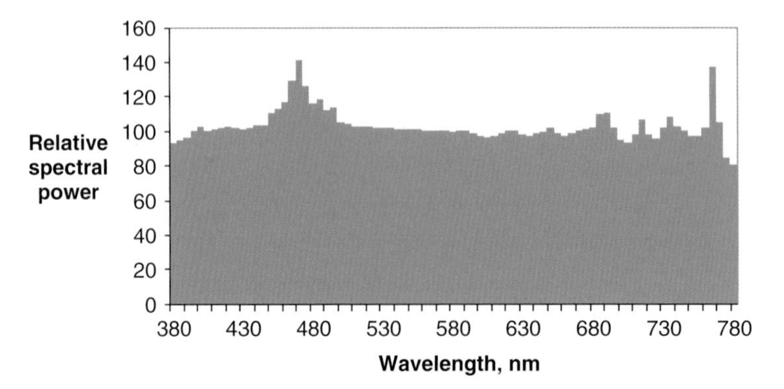

Figure 4.12 Relative spectral power distribution typical of a xenon lamp. The power is represented as a histogram in blocks of wavelengths 5 nm wide

At low pressures, the xenon emission spectrum is comprised of many discrete lines, but, at the pressures used in xenon lamps, the lines broaden to provide the type of continuum shown in Figure 4.12. The exact spectral power distribution depends on the pressure of the gas and on other operating characteristics of the lamp. The colour of the light emitted is quite similar to that of average daylight, but slightly more purple because of relatively greater emissions at the two ends of the spectrum. The relative spectral power distribution of a typical xenon lamp is given in Appendix 5.5. Efficacies for xenon lamps range from about 20 lm W^{-1} to about 50 lm W^{-1}.

Compact source xenon lamps are used in projectors in cinemas, and in lighthouses; continuously run and pulsed xenon lamps are used for floodlighting, for accelerated fading tests, and for copy-board lighting in the graphic arts industry; xenon flash tubes are used extremely widely for flash photography. The near daylight quality of the spectral power distribution makes xenon lamps a useful source for artificial daylight, and this is especially so for fluorescent materials because of their adequate emission in the ultraviolet part of the spectrum (to be discussed in Chapter 10).

4.8 INCANDESCENT LIGHT SOURCES

We have already seen that, in gas discharges, when the gas pressure is raised, the spectral lines in the emission are broadened because of the effects of the neighbouring atoms in the gas. In solids and liquids, the atoms are much more closely packed than in gases, and the line-broadening effects are usually so great that the emission is continuous throughout the spectrum. Solids and liquids can be excited to produce light in various ways, including the passage of an electric current (electroluminescence), exposure to radiation (photo-luminescence), and bombardment with electrons (cathodoluminescence). The most important way is by heating them to a temperature above about 1000 K (incandescence). The most important natural incandescent light source is the sun. The most important man-made incandescent source is the tungsten filament lamp; but sources depending, at least partly, on incandescence that have been important in the past include *limelight* (calcium oxide heated by coal, gas, or hydrogen), gas mantles (meshes of thorium heated by combustible gases), and carbon arcs (carbon electrodes heated by arcs in air across small gaps).

For incandescent sources, their temperature is the most important factor determining the magnitude and spectral composition of the light they radiate, but various materials do show some differences in their facility to radiate, both generally and at different wavelengths. These differences in *emissivity* modify the magnitude and the spectral composition of the radiation somewhat, but there is one class of light sources for which the nature of the radiation depends only on their temperature. They consist of heated enclosures, whose radiation escapes through an opening whose area is small compared to that of the interior surface of the enclosures; they are important theoretical sources, and in the past have been variously referred to as *black-body radiators*, or *full radiators*; but they are now generally referred to as *Planckian radiators*, after Max Planck, who originated a formula for deriving the power radiated at each wavelength from a knowledge of the temperature of the source. Planck's formula is as follows:

$$M_{\mathrm{e}} = \frac{c_1}{\lambda^5}\left(\frac{1}{\mathrm{e}^{c_2/\lambda T} - 1}\right)$$

where M_{e} is the spectral concentration of radiant excitance, in watts per square metre per wavelength interval ($\mathrm{W\,m^{-3}}$), as a function of wavelength, λ, in metres, and temperature, T, in kelvins, and where $c_1 = 3.74183 \times 10^{-16}\,\mathrm{W\,m^{-2}}$ and $c_2 = 1.4388 \times 10^{-2}\,\mathrm{mK}$; and $e = 2.718282$.

The unit used for the temperature of Planckian radiators is always the *kelvin, K* (which, as already mentioned, is equal to the Celsius temperature plus 273). In Figure 4.16 the locus of chromaticities of Planckian radiators, the *Planckian locus*, is shown in the u′,v′ chromaticity diagram. Relative spectral power distributions and chromaticity co-ordinates for a selection of temperatures are given in Appendix 5.5.

The amount of power radiated by incandescent sources increases as their temperatures increase; this is illustrated in Figure 4.13. It can also be seen from this figure that, at

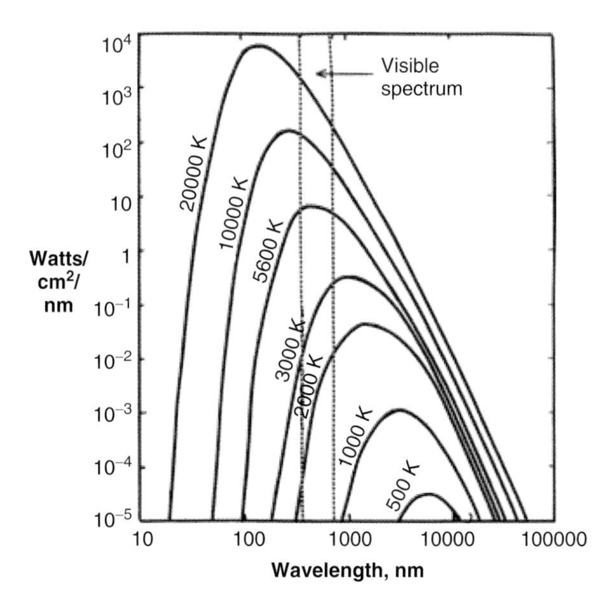

Figure 4.13 Spectral power distribution of Planckian radiators at different temperatures

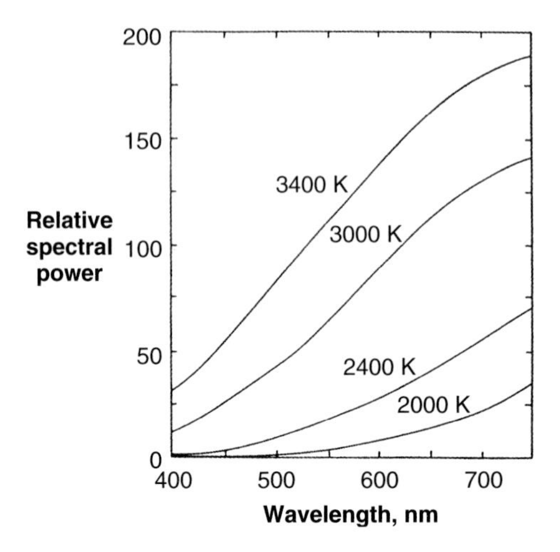

Figure 4.14 Spectral power distribution of Planckian radiators at four temperatures

temperatures below about 5600 K, more power is radiated at the long, rather than at the short, wavelength end of the visible spectrum; and at temperatures above about 5600 K more power is radiated at the short, rather than at the long, wavelength end. In Figure 4.14, data are shown, for the visible spectrum only, for temperatures of 2000 K, 2400 K, 3000 K, and 3400 K (a range of values of particular importance for tungsten filament lamps); this figure shows that, as the temperature is raised from 2000 K to 3400 K, more light is emitted, and the ratio of short to long wavelength light gradually increases. In Figure 4.15, data

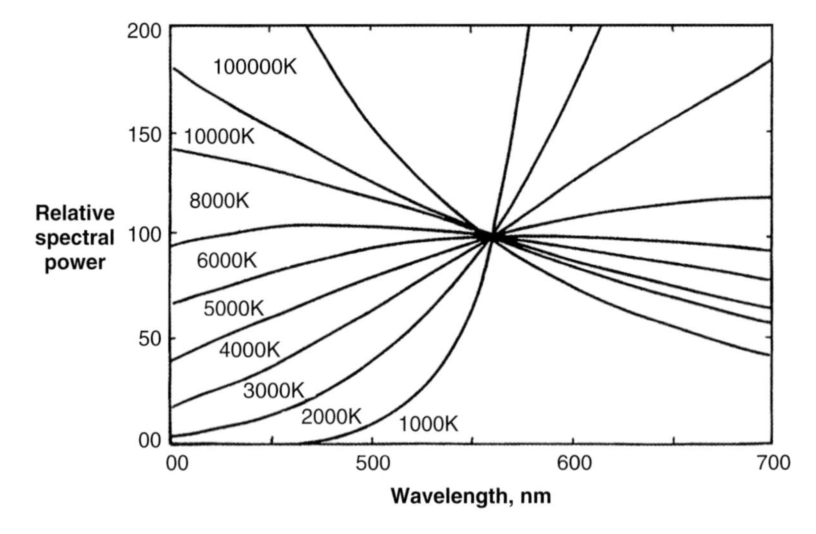

Figure 4.15 Relative spectral power distributions of Planckian radiators at various temperatures, normalised at 560 nm

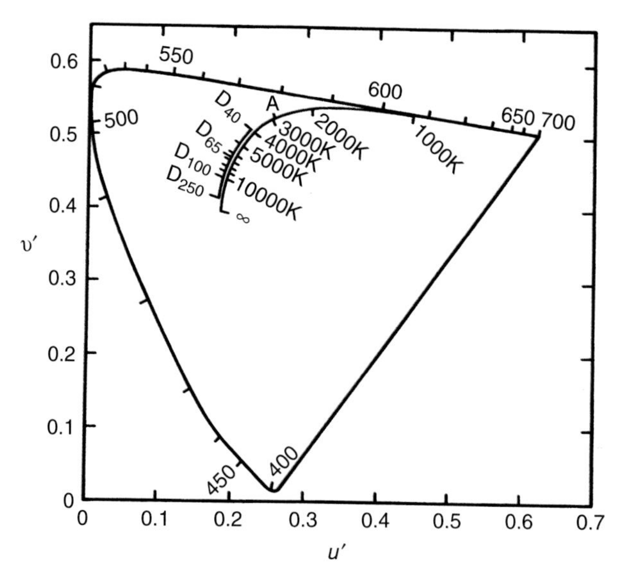

Figure 4.16 The loci of the chromaticities of Planckian radiators (the *Planckian locus*) and CIE D Illuminants (the *daylight locus*) in the u′, v′ diagram

for the visible spectrum are shown, normalised for a wavelength of 560 nm; it can be seen that the amount of light from the two ends of the spectrum becomes similar when the temperature is between about 5000 K and 6000 K.

Planckian radiators are not usually available as practical sources, but many incandescent sources emit light whose spectral composition or colour bears a particular relationship to a Planckian radiator. A source whose emissivity is constant with wavelength, but less than that of a Planckian radiator, has the same spectral power distribution, but at a lower level. It is convenient, in this case, to characterise the spectral power distribution of the radiation by quoting the temperature of the Planckian radiator having the same relative spectral power distribution, and this is called the *distribution temperature* of the radiation.

Distribution temperature, T_{D}

The temperature of a Planckian radiator whose relative spectral power distribution is the same as that of the radiation considered in the spectral range of interest.

Many sources have relative spectral power distributions that are not quite the same as those of Planckian radiators, but are very similar. Tungsten filament lamps fall into this category. Such stimuli usually have chromaticities that are on the Planckian chromaticity locus, and it is convenient, in this case, to characterise the colour of the stimulus by quoting the temperature of the Planckian radiator having the same chromaticity, and this is called the *colour temperature*.

Colour temperature, T_{C}

The temperature of a Planckian radiator whose radiation has the same chromaticity as that of a given stimulus.

For stimuli whose chromaticities are near, but not exactly on, the Planckian locus, it is useful to characterise the colour of the stimulus by quoting the temperature of the Planckian radiator that has the most similar perceived colour, and this is called the *correlated colour temperature*.

Correlated colour temperature, T_{cp}

The temperature of the Planckian radiator whose perceived colour most closely resembles that of a given stimulus seen at the same brightness and under specified viewing conditions. The recommended method of calculating the correlated colour temperature of a stimulus is to determine on the u,v (not the u′,v′) chromaticity diagram the temperature corresponding to the point on the locus of Planckian radiators that is nearest to the point representing the stimulus.

The u,v diagram (as described in Section 3.6) is obtained by plotting:

$$u = 4X/(X + 15Y + 3Z) = u'$$

$$v = 6Y/(X + 15Y + 3Z) = (2/3)v'$$

The u,v diagram is used instead of the u′,v′ diagram only for historical reasons. Correlated colour temperatures are used in calculating CIE Colour Rendering Indices (to be described in Chapter 7), and these were introduced before the u′,v′ diagram was available; changing to the u′,v′ diagram would involve changes to Colour Rendering Indices, and the consequent inconvenience has so far prevented this change being made. The concept of correlated colour temperature is only valid for stimuli whose chromaticities are reasonably close to the Planckian locus and accepted practice is to quote correlated colour temperatures for commonly used 'white' light sources, including tungsten lamps, tungsten halogen lamps, fluorescent lamps, and the various aspects of daylight. However, in the 2011 revision of the International Lighting Vocabulary the CIE recommends that the concept of correlated colour temperature should not be used if the chromaticity of a test source differs on the u,v chromaticity diagram by more than 5×10^{-2} from the Planckian locus (this distance is given by $[(u'_t - u'_P)^2 + (4/9)(v'_t - v'_P)^2]^{1/2}$, where u'_t, v'_t refer to the test source, u'_P, v'_P to the Planckian radiator (CIE, 2011).

The most widely used method of calculating colour temperature is that due to Robertson (Robertson, 1968; see also Wyszecki and Stiles, 2000; and Krystek, 1985).

4.9 TUNGSTEN LAMPS

The simplest way of providing an incandescent source is to heat a conductor by passing an electric current through it. If this is done in air, when the temperature is high enough to provide a reasonable amount of light, rapid oxidation usually occurs and the material soon burns away. But, if the conductor is enclosed in a transparent envelope, such as glass, that contains a vacuum or only inert gases, then a stable and convenient light source can be obtained. The earliest lamps of this type used carbon filaments in vacuum; but, although carbon has a high emissivity, it tends to evaporate and produce a dark deposit on the inside of the envelope.

The most suitable material found so far for the filament is tungsten. The spectral emissivity of tungsten is nearly constant in the visible spectrum, but decreases slightly with increasing wavelength; consequently, the spectral power distributions of tungsten filaments are almost identical to those of Planckian radiators whose temperatures are about 50 K greater than those of the corresponding filaments. The melting point of tungsten is 3410°C, that is 3683 K, and the filament must obviously be run below this temperature. It is desirable to operate at as high a temperature as possible, because then, as can be seen from Figure 4.14, the amount of light emitted increases, and the ratio of short, relative to long, wavelength light becomes less imbalanced; but higher temperatures tend to result in earlier failure of lamps because of greater evaporation of tungsten from the filaments leading to their breakage. This evaporation can be reduced by filling the envelope with inert gases, and such *gas filled* lamps can be run at higher temperatures for a given life, as compared to vacuum lamps. One disadvantage of the gas filling is that it provides a means for heat being removed from the filament thus reducing the efficacy. However, this can be minimised by having the filament as compact as possible by making it in the form of a tight coil, which is then often made into a larger *coiled coil*. A reasonable life can be obtained for lamps with thick filaments (about a fifth of a millimetre in diameter) operating at colour temperatures of about 3200 K; but with thin filaments (about a fiftieth of a millimetre in diameter) colour temperatures of only about 2500 K can be achieved because of the fragility of the filament and the serious weakening resulting from any evaporation from it. Thick filaments result in low resistance, and hence can only be used with low voltage or high wattage lamps. Typical efficacies for tungsten filament lamps are about 25 lm W^{-1} for lamps operating at colour temperatures of 3200 K, and about 10 lm W^{-1} at 2650 K.

4.10 TUNGSTEN HALOGEN LAMPS

In an ordinary tungsten lamp, tungsten gradually evaporates from the filament (thus weakening it) and forms a dark deposit on the envelope (which reduces the light output). In tungsten halogen lamps, the envelope is made of pure fused silica (often referred to as 'quartz'), and is much smaller than in an ordinary lamp; its envelope contains a halogen gas, and, when tungsten evaporates from the filament, it combines with the halogen gas at the hot envelope wall to form tungsten halide (a gas). This gas then migrates back to the filament, where it decomposes to deposit the tungsten back on to the filament and the halide back into the mixture of gases in the envelope. The halide is thus available for further use, and can provide a continuous cycle of combination with, and dissociation from, tungsten from the filament. Thus dark deposits on the envelope are avoided. The compact shape of the lamp results in the gas pressure being higher than in a conventional lamp, and this reduces the evaporation of tungsten, and hence the filament is weakened less. This means that the lamp can be run at a higher colour temperature with better efficacy. Tungsten halogen lamps can be run at colour temperatures in the range from about 2850 K to about 3300 K and have efficacies from about 15 to about 35 lumens per watt.

The compact nature of the tungsten-halogen lamp makes it very suitable for use with mirrors and lenses in optical systems, and it has been very widely used in slide projectors and overhead projectors; these projector lamps may operate on a supply of less than 30 volts in order to keep them as compact as possible. Low voltage tungsten-halogen lamps are also widely used for automobile headlamps. Mains voltage tungsten-halogen lamps are used for floodlighting, and for studio lighting for television and filming.

4.11 LIGHT EMITTING DIODES

A light-emitting diode (LED) is an inorganic semiconductor that can be made to emit narrow-band incoherent radiation, an effect which is a form of electroluminescence. The colour of the emitted radiation is a function of the composition of the semi-conducting material used and can be in the infrared, visible, or near ultraviolet region of the spectrum. Most LEDs emit light of a narrow band of wavelengths, typically about 40 nm in width; and some sources are available with more than one LED in the package. Thus an LED emitting white light can be made using a combination of a red-, a green- and a blue-emitting LED, all in the same package, see Figure 4.17. Many white LEDs in production, however, emit blue light of wavelengths centred between 450 nm and 470 nm and the inside of the package is coated with a phosphor that emits yellow light, usually cerium-doped yttrium aluminium garnet. This is activated by the blue light to emit radiation centred on approximately 580 nm (see Figure 4.18), giving an overall appearance of white light. Varying the thickness of the phosphor, or substituting the cerium with other rare earth elements, enables some tuning of the correlated colour temperature of the emitted radiation. An alternative method of producing white light is to use an ultraviolet emitting LED with europium based red- and blue-emitting phosphors plus a green-emitting copper-doped zinc sulphide. This can offer better colour rendering than the blue-LED based device. LEDs are finding application, for example, in architectural and decorative lighting, as signal lights on automobiles, and for traffic lights. They are also being used to provide the backlight required in liquid-crystal computer displays and television receivers.

The key advantage of LED based lighting is the high efficacy in terms of light output per unit input of power, typically four times that of incandescent lamps. They usually operate at low voltage and, because of their efficacy emit relatively little heat. They are often operated in clusters to provide greater intensity.

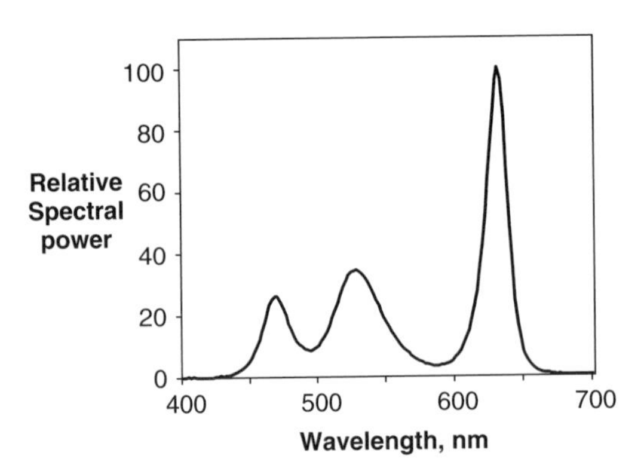

Figure 4.17 Spectral power distribution of a 'white' LED composed of separate red-, green-, and blue-emitting LEDs. The correlated colour temperature in this case is approximately 3700 K

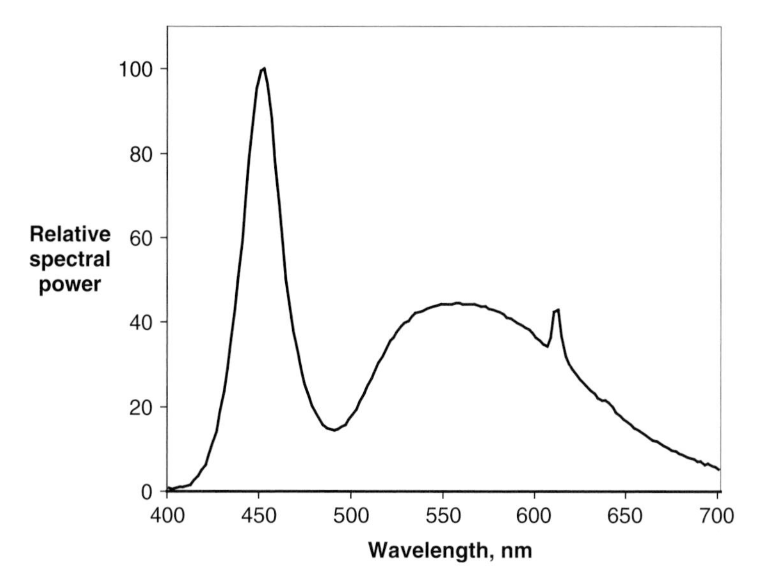

Figure 4.18 Spectral power distribution of a 'white' LED comprising a blue-emitting LED and a phosphor. The correlated colour temperature in this case is approximately 6500 K

4.12 DAYLIGHT

All daylight originates from the sun, the temperature of which is millions of degrees at its centre; but, at its surface, it is only about 5800 K. Because the light from the sun has to pass through both its own atmosphere and that of the earth, its correlated colour temperature as seen from the surface of the earth is not 5800 K, but is about 5500 K when high in the sky, and much lower when near the earth's horizon. A spectral power distribution that is typical for sun and sky light received at the earth's surface is shown in Figure 4.19. Also shown in this figure is the spectral power distribution of the Planckian radiator having the same correlated colour temperature. It is clear that the sun and sky light combination does not have a smooth distribution like that of the Planckian radiator, and this is the result of the selective absorptions of the light by the sun's, and by the earth's, atmospheres. The absorptions by the sun's atmosphere are mostly in the form of very narrow absorption lines (Fraunhofer lines). The most significant absorptions by the earth's atmosphere are of two types. First, light is lost at the short wavelength end of the spectrum as a result of the scattering of blue and ultraviolet radiation back into space; this is indicated by the rapid downward slope of the full curve at the left-hand end in Figure 4.19. Second, light is absorbed by molecules of various gases, particularly water vapour and ozone; this is the cause of the considerable undulation of the full curve at the right-hand end in Figure 4.19.

In sunlit areas in clear sunny weather, about 90% of the illumination comes directly from the sun and only about 10% from the sky. Because the sky is normally blue in these conditions, the addition of the sky light makes the light slightly bluer, and raises the correlated colour temperature somewhat. The correlated colour temperature of clear sunlight plus skylight thus depends on the altitude of the sun in the sky, and on the colour

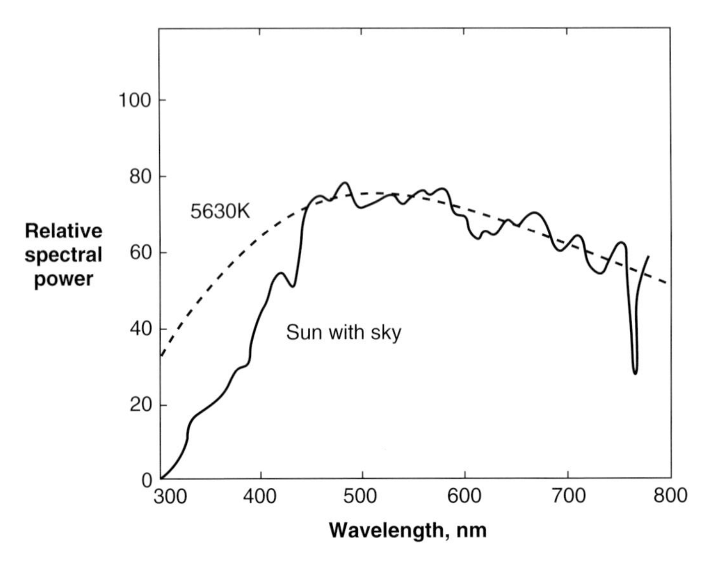

Figure 4.19 Relative spectral power distribution of a typical daylight (full curve), and of the Planckian radiator of the same correlated colour temperature, in this case 5630 K (broken line)

of the sky; a correlated colour temperature of 5500 K is usually taken as a representative figure for sunny daylight for solar altitudes not lower than about 30°.

In fully overcast weather, the light from the sun and sky above the clouds is diffused by them to produce a much more uniform light. If the clouds are low, the light usually has a correlated colour temperature similar to that of sunlight plus skylight on a clear day, that is, about 5500 K. But, if the clouds are very high, the blue light scattered by the atmosphere can be reflected back down to the earth, and the lighting can become of higher correlated colour temperature.

There are many types of weather between clear sun and fully overcast, and the resulting daylight can vary in spectral power distribution very considerably according to the solar altitude and the nature of the cloud formation and atmospheric conditions.

In Figure 4.20 are shown relative spectral power distributions for three examples of daylight, having correlated colour temperatures of 5610, 7140, and 10350 K. It is clear that the shape of these distributions remains much the same, except for a general tilting of the curve.

Indoor daylight varies according to the nature of the sky (and sunlight, if any), the geometry of the windows relative to the sky, the spectral transmission of the windows, and the nature of the interior of the room. Indoor daylight can therefore be even more variable than outdoor daylight, and, in particular, the decor of the room can have a large effect. For instance, if a room has a red carpet, and pink walls, the effective colour of the daylight in the room can be considerably redder than outside. A representative correlated colour temperature for indoor daylight is generally taken as 6500 K; this is higher than the value of 5500 K taken for sun and sky light, because, when indoor situations are averaged, the sky plays a more important part than the sun for providing illumination.

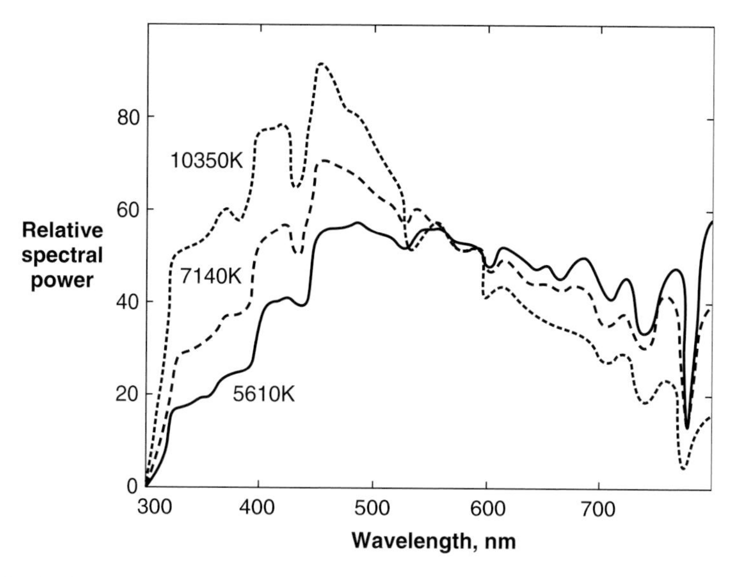

Figure 4.20 Relative spectral power distributions of three examples of daylight, having correlated colour temperatures of 5610, 7140, and 10350 K

From the point of view of colorimetry, daylight presents three problems. First, the spectral power distribution is complicated. Second, it is variable. Third, it has correlated colour temperatures much higher than can be achieved by artificial incandescent sources.

4.13 STANDARD ILLUMINANTS AND SOURCES

Enough has been said in this chapter to show that commonly used light sources provide a great variety of spectral power distributions. The result is that, in spite of the adaptation processes of the visual system, coloured objects undergo appreciable changes in colour appearance as the light source is changed. Furthermore, pairs of colours that match under one light source will not necessarily match under another. To reduce the complexity of dealing with this situation, and yet retain some simplified representation of it, the CIE has introduced some standardisation (CIE, 2006). The CIE distinguishes between *illuminants*, which are defined in terms of spectral power distributions, and *sources*, which are defined as physically realisable producers of radiant power (CIE, 2010).

4.14 CIE STANDARD ILLUMINANT A

Historically, the most common domestic artificial light source is the tungsten filament lamp. Tungsten lamps are widely used because they are inexpensive, compact, and usually considered flattering to the human complexion. It is thus appropriate that one of the standard illuminants adopted by the CIE should represent the light from tungsten lamps. The tungsten lamp is also a very convenient standard illuminant, because the spectral power distribution

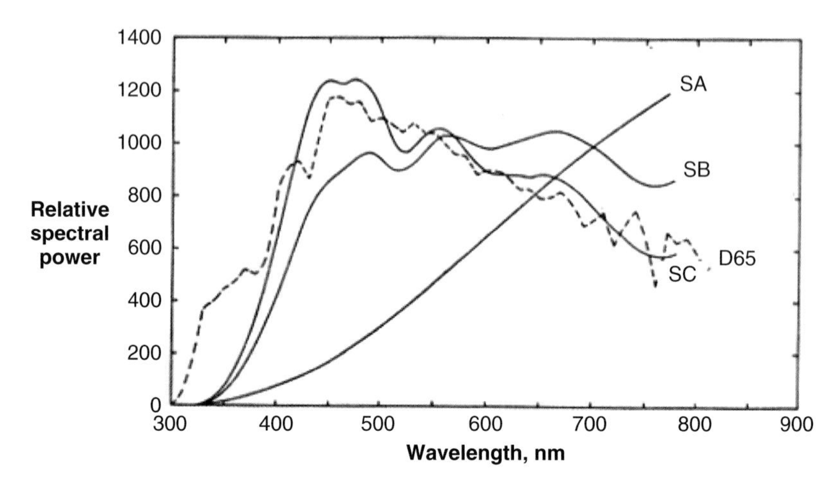

Figure 4.21 Relative spectral power distributions of CIE Standard Illuminants A (SA) and D65, and of CIE Illuminants B (SB) and C (SC)

of its light is almost entirely dependent on just one variable, the temperature of its filament. The CIE standard tungsten illuminant is defined as:

Standard Illuminant A

An illuminant having the same relative spectral power distribution as a Planckian radiator at a temperature of about 2856 K.

The temperature is quoted as 'about 2856 K' because the temperature associated with a particular relative spectral power distribution depends on the value adopted for one of the constants, c_2, in the formula for Planck's law. The value assigned to this constant has changed over the years since this Standard Illuminant was first defined in 1931, and, to keep the power distribution unchanged, the temperature is now defined as (1.4388/1.4350)2848, which is approximately equal to 2856, using the 'International Practical Temperature Scale, 1968'. The relative spectral power distribution of Standard Illuminant A (SA) is illustrated in Figure 4.21, and its values are given at every 5 nm throughout the visible spectrum in Appendix 5.2 (CIE, 1986a, 2004). The values are also available at every 1 nm in published tables (CIE, 1971; CIE, 1986b; CIE, 2006) or by interpolation (see Section 9.13). In the case of CIE Standard Illuminant A, 2856 K is both its distribution temperature and its colour temperature.

4.15 CIE ILLUMINANTS B AND C

Of even more practical importance than tungsten light is daylight, and standard illuminants representing daylight are therefore essential; they are, however, more complicated to define. In 1931 the CIE established two different illuminants to represent daylight. They were:

CIE Illuminant B

An illuminant having the relative spectral power distribution given in Appendix 5.2, Column B.

CIE Illuminant C

An illuminant having the relative spectral power distribution given in Appendix 5.2, Column C.

These distributions are shown in Figure 4.21. CIE Illuminant B (SB) has a correlated colour temperature of about 4874 K and was intended to represent sunlight. CIE Illuminant C (SC) has a correlated colour temperature of about 6774 K and was intended to represent average daylight. CIE Illuminant B is now obsolete (but a source approximating it is still used in the European brewing scale); and the CIE no longer recommends the use of CIE Illuminant C, but its wide use in the past has resulted in its involvement in much data in the literature.

4.16 CIE SOURCES

In the case of Standard Illuminant A, its radiation can be produced by CIE Source A, which is defined as:

CIE Source A

A gas-filled tungsten filament lamp operating at a correlated colour temperature of 2856 K ($c_2 = 1.4388 \times 10^{-2}$ mK). A lamp with a fused-quartz envelope or window is recommended if the spectral power distribution of the ultraviolet radiation of Illuminant A is to be realised more accurately.

The radiation of CIE Illuminant C can be produced by CIE Source C, which is defined as:

CIE Source C

CIE Source A combined with a filter consisting of a layer, 1 cm thick, of each of two solutions C_1 and C_2, contained in a double cell made of colourless optical glass, the solutions to be as follows:

Solution C_1:

Copper Sulphate ($CuSO_4$. $5H_2O$)	3.412 g
Mannite [$C_6H_8(OH)_6$]	3.412 g
Pyridine (C_5H_5N)	30.0 ml
Distilled water to make	1000.0 ml

Solution C_2:

Cobalt Ammonium Sulphate [$CoSO_4.(NH_4)_2SO_4.6H_2O$]	30.58 g
Copper Sulphate ($CuSO_4.5H_2O$)	22.52 g
Sulphuric Acid (density 1.835 $g.ml^{-1}$)	10.0 ml
Distilled water to make	1000.0 ml

The use of these liquid filters is inconvenient in practice, and considerable care is necessary in making them up to obtain the correct result (Billmeyer and Gerrity, 1986). As a result, glass filters are sometimes used instead, but they do not exactly reproduce the required spectral transmission characteristics.

4.17 CIE ILLUMINANTS D

It has been mentioned that daylight has a lower ultraviolet content than Planckian radiators of the same correlated colour temperatures. But CIE Illuminants B and C have too little power in the ultraviolet region. This is not significant in the case of colours that do not fluoresce, but, for those that do, these illuminants result in less fluorescence than occurs in real daylight. With the extensive use of fluorescing agents in white colorants, often referred to as *optical brightening agents* or *fluorescent whitening agents* there was a need for a Standard Illuminant that was more representative of daylight in the ultraviolet region. Therefore, in 1963, the CIE recommended a new Standard Illuminant, D65, to represent average daylight throughout the visible spectrum and into the ultraviolet region as far as 300 nm (CIE, 2006). The spectral power distribution of this illuminant is also shown in Figure 4.21, and it can be seen that below about 380 nm it has considerably more power than Illuminant C. Standard Illuminant D65 has a correlated colour temperature of about 6504 K and is one of a series of CIE D Illuminants representing daylights of different correlated colour temperatures; these are designated D50, D55, etc., for example, for daylights having correlated colour temperatures of about 5000 K, 5500 K, respectively. In Appendix 5.2 the spectral power distributions are given at 5 nm intervals for illuminants D50, D55, and D65 (CIE, 2004). The method of calculating the spectral power distributions of these D illuminants is given in Appendix 5.6. If the values of the spectral power distributions are required at closer intervals than every 5 nm, they are interpolated linearly from the 5 nm values. For D65 these have been made available at every 1 nm (CIE, 1971, 1986b, and 2004). The locus of the chromaticities of the D illuminants (the *daylight locus*) is given in Figure 4.16; it can be seen that the D illuminants lie slightly on the green side of the Planckian locus.

There are no CIE sources that realise the D illuminants, and their main use is in computing tristimulus values and other colorimetric data. However, the simulation of D illuminants remains a need in colorimetry, and various attempts have been made to provide such sources (Terstiege, 1989). The degree to which a practical source approximates a D illuminant in relative spectral power distribution for colorimetric purposes can be assessed by using carefully chosen samples and a colour difference formula (CIE, 1981; ISO/CIE, 2004/2005). The method presented is especially useful in evaluating a light source in a viewing booth used in, for example, a quality control environment in the graphic arts (D50) and dyeing industries (D65) (CIE, 2010).

The primary reference for CIE Standard Illuminants A and D65 is:

CIE S 014-2/E:2006:Colorimetry – Part 2: CIE standard illuminants.

This standard is also available as ISO Standard: ISO 11664 – 2:2008. These standards both contain the 1 nm spectral data. In addition, 5 nm data are available in:

CIE Publication 15:2004: Colorimetry, 3rd ed.

4.18 CIE INDOOR DAYLIGHT

In 2008, the CIE recommended two new daylight illuminants ID50 and ID65 that correspond to phases of indoor daylight of approximately 5000 K and 6500 K correlated colour

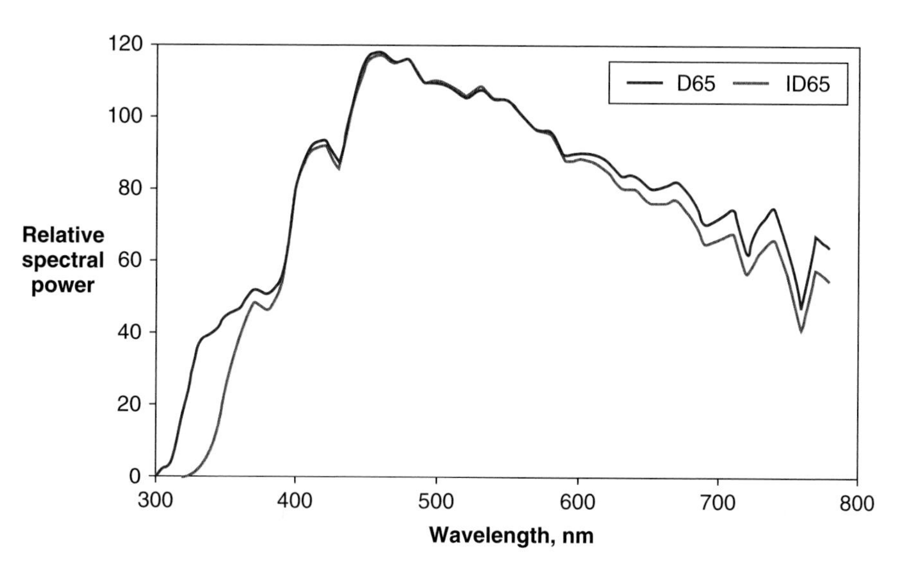

Figure 4.22 The relative spectral power distributions of CIE standard illuminant D65 and CIE ID65 indoor-daylight illuminants

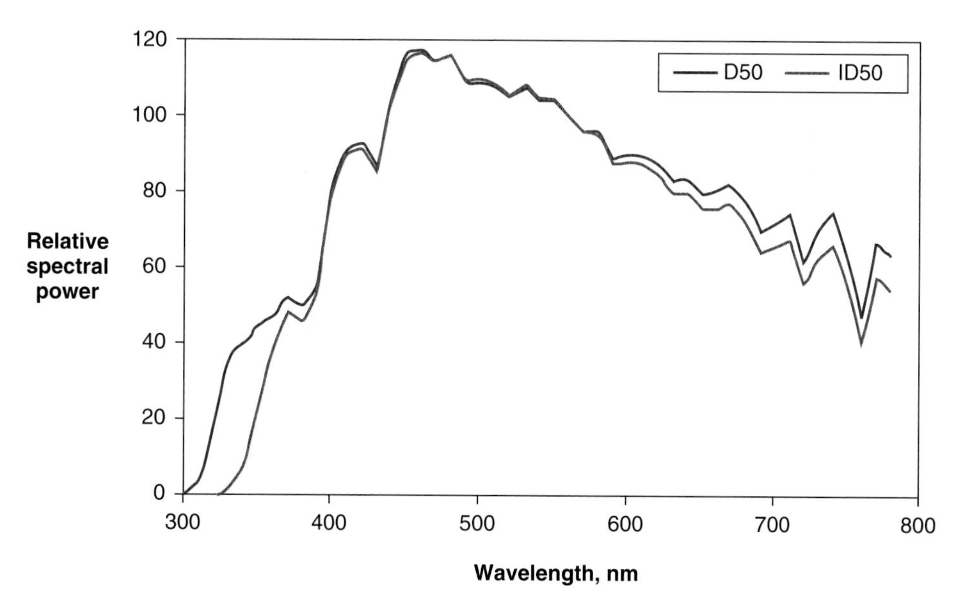

Figure 4.23 The relative spectral power distributions of CIE standard illuminant D50 and CIE ID50 indoor-daylight illuminants

temperature respectively. As described above, daylight illuminants contain ultraviolet radiation in proportions similar to those found in natural daylight. Indoors this daylight is filtered by the transmission of window glass and illuminants ID50 and ID65 take this into consideration (CIE 2008). Figures 4.22 and 4.23 show the spectral power distributions of these two new illuminants compared with those of D50 and D65 respectively.

4.19 COMPARISON OF COMMONLY USED SOURCES

In Table 4.1 are given the correlated colour temperatures of a selection of commonly used light sources, together with their reciprocal temperatures multiplied by 10^6. These values of reciprocal megakelvins (the use of the previous name, *mireds*, is now deprecated) are useful because, quite fortuitously, they represent an approximately uniform scale of colour differences.

Table 4.1 Correlated Colour Temperatures of some typical light sources

Source	K	10^6 K^{-1}
North-sky light	7500	133
Average daylight	6500	154
Fluorescent lamps (Northlight, Colour-matching)	6500	154
xenon (electronic flash or continuous)	6000	167
Sunlight plus skylight	5500	182
Blue flash bulbs	5500	182
Carbon arc for projectors	5000	200
Sunlight at solar altitude 20°	4700	213
Fluorescent lamps (Cool White)	4200	238
Sunlight at solar altitude 10°	4000	250
Clear flash-bulbs	3800	263
Fluorescent lamps (White or Natural)	3500	286
Photoflood tungsten lamps	3400	296
Tungsten-halogen lamps (short life)	3300	303
Projection tungsten lamps	3200	312
Studio tungsten lamps	3200	312
Tungsten-halogen lamps (short life)	3000	333
Fluorescent lamps (Warm White)	3000	333
Tungsten lamps for flood-lighting	3000	333
Tungsten lamps (domestic, 100 watts)	2800	357
Tungsten lamps (domestic, 40 watts)	2700	370
Sunlight at sunset	2000	500
Candle flame	1900	526

REFERENCES

Billmeyer, F.W. and Gerrity, A., Redetermination of CIE standard source C, *Color Res. Appl*. **8**, 90–100 (1983).

CIE Publication 51, *A method for assessing the quality of daylight simulators for colorimetry*, Commission Internationale de l'Éclairage, Vienna, Austria (1981).

CIE Publication 15:2004, *Colorimetry, 3rd Ed*., Commission Internationale de l'Éclairage, Vienna, Austria (2004).

CIE Publication 184:2009, *Indoor daylight illuminants*, Commission Internationale de l'Éclairage, Vienna, Austria (2009).

CIE Publication 192:2010, *Practical daylight sources for colorimetry*, Commission Internationale de l'Éclairage, Vienna, Austria (2010).

ISO/CIE 23603:2005(E)/CIE S 012/E:2004, *Standard method of assessing the spectral quality of daylight simulators for visual appraisal and measurement of colour*, Commission Internationale de l'Éclairage, Vienna, Austria (2004/2005).

ISO/CIE 11664-2:2008(E)/CIE S 014-2/E:2006, *Colorimetry – Part 2: CIE standard illuminants*, Commission Internationale de l'Éclairage, Vienna, Austria (2006).

Krystek, M., An algorithm to calculate correlated color temperature, *Color Res. Appl*., **10**, 38–40 (1985).

Robertson, A.R., Computation of correlated color temperature and distribution temperature, *J. Opt. Soc. Amer*., **58**, 1528–1535 (1968).

Terstiege, H., Artificial daylight for measurement of optical properties of materials, *Color Res. Appl*., **14**, 131–138 (1989).

Wyszecki. G. and Stiles, W.S., *Color Science, 2nd Ed*., pp. 224–229, Wiley, New York, NY, USA (2000).

GENERAL REFERENCES

Cayless, M.A., and Marsden, A.M., *Lamps and Lighting. 4th Ed*., Arnold, London, England (1997).

Henderson, S.T. *Daylight and its Spectrum, 2nd Ed*., Hilger, Bristol, England (1977).

Society for Light and Lighting: Lighting Handbook, Society for Light and Lighting, Balham, England (2009).

5

Obtaining Spectral Data and Tristimulus Values

5.1 INTRODUCTION

CIE X, Y, Z tristimulus values can be obtained by one of three main methods: visual matching, calculation from spectral power data, or direct measurement with filtered photo-detectors.

The visual matching may be done using either additive mixtures of red, green, and blue light (as described in Section 2.4, and as will be further described in Chapter 11), or by comparing colours with those in colour order systems (as will be described in Chapter 8). However, visual matching tends to be time-consuming, and a rather difficult task, and therefore the vast majority of tristimulus value determinations use one of the other two methods. In this chapter we shall consider these two methods, together with some related topics.

The second method, calculation from spectral power data, is the most widely used, and is therefore of the greatest importance. The X, Y, Z tristimulus values are obtained from the spectral power data by using CIE colour-matching functions as weighting functions, as described in Section 2.6. Measuring the spectral power data involves spectroradiometry or spectrophotometry, and there are various factors that can affect the results. Different methods of calculation can also affect the results. Different considerations also apply to the measurement of self-luminous samples, and non-self-luminous (reflecting or transmitting) samples.

The third method, measurement with filtered photodetectors, is quick and uses inexpensive equipment, but is usually less accurate than calculation from spectral power data.

5.2 RADIOMETRY AND PHOTOMETRY

In *radiometry*, the amount of *radiant power* in a stimulus is measured; hence the detectors that are used have to be sensitive to all required wavelengths. Measurements are usually made by comparison with a standard source of known power output.

Measuring Colour, Fourth Edition. R.W.G. Hunt and M.R. Pointer.

In *photometry*, the amount of *light* in a stimulus is compared with that of a standard stimulus, and the ratio of the two amounts then provides a photometric measure. The amounts of light may be compared by visual assessment (and this was historically the case). But it is much more common for them to be measured using filtered photo-detectors whose spectral sensitivities approximate the $V(\lambda)$ or the $V'(\lambda)$ function, according to whether photopic or scotopic levels of illumination are involved (see Sections 2.2 and 2.3). Alternatively the comparison can be made by calculations from spectral power data using either the $V(\lambda)$ or the $V'(\lambda)$ function as a weighting function (see Tables 2.1 and 2.2).

5.3 SPECTRORADIOMETRY

In *spectroradiometry*, the amounts of radiation are measured using narrow bands of wavelengths situated at regular intervals throughout the spectrum, and the measures obtained are then referred to as *spectral*; the corresponding measuring instruments are known as *spectroradiometers*.

5.4 TELE-SPECTRORADIOMETRY

If we want to obtain tristimulus values that accurately represent a colour that was seen in a given set of viewing conditions, then ideally we should set up an instrument at the same position as was occupied by the observer's eye, and direct it at the colour while it is illuminated by the same source in the same surroundings. The instrument for doing this is called a *tele-spectroradiometer, tele* because it incorporates a telescope, or other suitable imaging optics to collect the light from the colour at the observing position, *spectro* because it analyses the light throughout the spectrum, and *radiometer* because the measurement it makes is of the radiant power. Such instruments are commercially available, and they should be used when it is important that the measurements relate to a particular set of viewing conditions that occurs in practice. If measurements of *absolute* radiant power are required, the instrument must be calibrated to give this result; such a calibration is usually most conveniently done by using a standard light source whose absolute spectral power distribution is known to the required accuracy. Such a calibration is usually offered by the instrument manufacturer as part of the instrument package. Then the instrument is able to measure absolute radiant power in watts per steradian per square metre per nanometer ($\mathrm{W\,sr^{-1}\,m^{-2}\,nm^{-1}}$). If, as is often the case, only *relative* spectral power data are required, then it is only necessary to know the *relative* spectral power distribution of the standard source.

If it is required to evaluate variables that depend on values for a reference white (as described in Chapter 3), it is also necessary to measure with the tele-spectroradiometer the spectral power distribution of a suitably chosen reference white under exactly the same conditions as were used for the colours.

The above procedures can be somewhat tedious, and, for general work, it is usually more convenient to adopt a standard geometry for the illuminating and viewing conditions. We shall now therefore consider these simpler procedures.

5.5 SPECTRORADIOMETRY OF SELF-LUMINOUS COLOURS

For self-luminous colours, such as light sources or television and computer displays, the light from the sample has to be made to enter the spectroradiometer by some suitable means.

In the case of light sources, one possible way is to let the source illuminate a stable white reflecting surface, whose spectral power distribution is then measured. Similar measurements are then repeated using a standard source, whose spectral power distribution is known, such as CIE Standard Source A. The ratio of the two sets of measurements at each wavelength can then be used to modify the spectral power distribution of the known source to give that of the unknown source.

In the case of self-luminous colours that are not light sources, such as television and computer displays, the amount of light available is not usually sufficient to adopt the above procedure, and the light emitted has to be passed directly into the spectroradiometer. The standard reflecting surface (whose spectral reflectance factor data must now be known) is then only used with the standard source for calibration purposes.

It must be borne in mind that the geometry of the collection of the light from self-luminous colours may affect the results; two instruments having different collection angles will only give the same results if the spectral composition of the light emitted by the colour is independent of the angle at which it leaves its surface.

In general, if F_{M1} is the measurement of a known spectral radiance F_{R1}, and F_{M2} is a measurement of an unknown spectral radiance F_{R2}, then $F_{M1}/F_{R1} = F_{M2}/F_{R2}$ so that $F_{R2} = F_{R1}(F_{M2}/F_{M1})$.

5.6 SPECTROPHOTOMETRY OF NON-SELF-LUMINOUS COLOURS

Non-self-luminous colours are required to be illuminated by a source that emits light throughout the spectrum. A comparison is then made, throughout the spectrum, between the radiant power of a beam of light after it has been reflected from, or transmitted by, the sample, and the radiant power of a similar beam after it has been reflected from, or transmitted by, a calibrated working standard. Although the comparison is still of radiant power, the instruments involved are then usually called *spectrophotometers*, and the procedure referred to as *spectrophotometry*, implying a comparison, not of radiant power, but of light (radiant power weighted according to the $V(\lambda)$ or $V'(\lambda)$function); however, in the narrow bands of wavelengths usually involved in spectral measurements, there is generally little or no difference between ratios of radiant power and ratios of light.

In the case of diffusing samples, the working standard is usually a white reflecting surface or a white transmitting diffuser. In the case of non-diffusing objects, for reflecting samples, a calibrated mirror can be used; for transmitting samples, either no sample (that is, air) can be used, or, in the case of liquids, a cell of the same type as used for containing the sample, filled with a suitable liquid such as deionised water.

5.7 REFERENCE WHITES AND WORKING STANDARDS

In measuring spectral reflectance (or transmittance) factor, the CIE recommends, as the reference white, the adoption of the *perfect reflecting (or transmitting) diffuser*. As already mentioned in Section 2.6, this is defined as:

The perfect reflecting (or transmitting) diffuser
An ideal isotropic diffuser with a reflectance (or transmittance) equal to unity.

By isotropic is meant that the spatial distribution of the reflected (or transmitted) radiation is such that the radiance (or luminance) is the same in all directions in which the radiation is reflected (or transmitted); hence, the perfect reflecting diffuser is completely matt and is entirely free from any gloss or sheen. The reflectance (or transmittance) of the perfect diffuser is equal to unity at all wavelengths, so that the diffusion is independent of wavelength, that is, it is *non-selective*.

The perfect diffuser is easy to use in making colorimetric calculations, but, for practical measurements, it is not available. It is therefore necessary to use a working standard that has been calibrated against the perfect diffuser. These calibrations are usually obtained from national standardising laboratories, and are often supplied with the instruments. The results obtained with these working standards are then corrected to obtain those that would have been obtained if the perfect diffuser had been used instead. (This topic will be discussed further in Section 9.5.)

As in tele-spectroradiometry, to evaluate variables that depend on values for a reference white (see Chapter 3), it is necessary to include the reference white in the samples considered. One choice for the reference white is the perfect diffuser, and this may be appropriate for general use in evaluating products such as textiles and paints. However, in other applications, the perfect diffuser may not be appropriate.

For instance, in evaluating printing inks, the paper being used is generally a better choice. This is because, if the paper were slightly yellowish, for example, a non-selective neutral grey ink would appear to be yellowish relative to the perfect diffuser; but the ink itself is not yellowish. The ink is therefore better evaluated using the unprinted paper as the reference white; but the perfect diffuser is appropriate for evaluating the paper.

In the case of photographic reflection prints, the perfect diffuser is appropriate for evaluating the paper in areas where there is no image; but, for image areas, an area representing the reproduction of a typical white in the scene is appropriate. This may not only be of a different colour, but also appreciably darker than the perfect diffuser. By comparison, the perfect diffuser would then have a value of Y/Y_n (its Y tristimulus value divided by that for the reference white) that was appreciably higher than unity, indicating that its lightness was greater than that of a white, which indeed would be its appearance if it were incorporated into the picture. Exactly similar arguments apply in evaluating transparent areas in photographic film: Y_n should be the value of Y for a typical white in the picture, in which case Y/Y_n for an area without any image would again be in excess of unity, and this is appropriate.

When self-luminous areas are being evaluated, as in television or computer displays, the perfect diffuser is not relevant because there is no illuminant (in the usual sense of the word). The use of the peak white of the television or computer system is not usually appropriate, because whites used in the displays are usually reproduced at levels lower than that of the peak white. Again, what has to be done is to select an area representing a typical white in the scene. In the case of pictorial images this can usually be done, although not always without some difficulty; but, in the case of non-pictorial images, such as data presentations, it may be very difficult to define a stimulus that corresponds to an area that appears perceptually white. Indeed, in some cases the display may appear to be entirely self-luminous, in which case only brightness, hue, colourfulness, and saturation will be perceived, and correlates of lightness and chroma are inappropriate. In this case, Y_n has no meaning, but values of u'_n, v'_n are required, and these can either be those of the nominal 'white' for the display (often D65 or SC), or those of the equi-energy stimulus (S_E).

When images are being evaluated, important practical problems are often caused by lack of uniformity of the illumination at various stages in the process of forming the image. For example, the lenses used in cameras, enlargers, and projectors, invariably give images that have lower luminances at the edges, than at the centre, of the picture (Hunt, 2003). The human visual system, however, is extraordinarily adept at discounting the effects of this. For example, a white reproduced at the corner of a picture may have a lower luminance than a light grey at the centre of the picture; but the white and grey are usually correctly recognised without difficulty. Realistic evaluations of colours in images therefore often require that meticulous corrections be made for the effects of non-uniformities of illumination.

It must be stressed that Y tristimulus values are always to be evaluated such that $Y = 100$ for the perfect diffuser. It is Y_n, the value of Y for the appropriately chosen reference white, that must often be different from 100, when the ratio Y/Y_n is being used to calculate correlates of perceptual attributes.

5.8 GEOMETRIES OF ILLUMINATION AND VIEWING

If an object with a very glossy reflecting surface is viewed, its appearance is greatly affected by the angle of view relative to the angles at which the illuminating light falls on the surface. If the light comes from just one direction, then it is possible, by tilting the object appropriately, to ensure that the mirror-like image of the light source is avoided, and the colour of the object can then be seen clearly. But, if the object is illuminated from many different directions, as may occur, for instance, in a room lit by large windows or many artificial lights, then it may be impossible to find a direction in which mirror-like reflections of light sources are completely avoided. Furthermore, if the object is viewed under a very large source of light, such as an overcast sky, or a uniform ceiling light, then the colour of the object will always be seen in the presence of mirror-like reflections of part of the light source. These mirror-like (or *specular* or *regular*, as they are often referred to) reflections are produced by the top surface of objects, and, unless the objects are metals, the light reflected is not coloured by the surface and is therefore the same colour as the light source. In the usual case when the colour of the illuminant is white, the specular reflections add white light to the colour of the object, and, unless it is itself white or grey, the effect is to desaturate the colour. This is why glossy objects look more saturated in directional, than in diffuse, illumination.

In the case of completely matt objects, every beam of incident light, no matter what its angle of incidence on the surface, will result in some light entering the eye that has not penetrated the surface and which is therefore not affected by the colorant (except in the case of metals). Hence, when matt objects are viewed in white light, they are always desaturated by these top surface reflections. It is for this reason that matt objects do not usually appear as saturated as glossy objects, unless the glossy objects are lit very diffusely.

5.9 CIE GEOMETRIES OF ILLUMINATION AND MEASUREMENT

The CIE recommends that colorimetric specifications of opaque specimens be given so as to correspond to measurements made with either directional illumination and directional collection, or with directional illumination and diffuse collection, or with diffuse illumination and direct collection; these alternatives are illustrated in Figure 5.1. With the 0°:45° and 45°:0° arrangements, the zero direction is usually slightly offset to prevent inter-reflections between the sample and the light source or the detector. With the diffuse arrangements similar offsets are used, having a usual magnitude of 8°, hence the designations 8°:d and d:8°; in these cases the offsets enable gloss traps (to be described later) to be provided. Details of these conditions are given in Table 5.1; in the table the symbols before the colon refer to the illumination, and those after the colon to the measurement. The diffuse illumination or measurement is normally provided in instruments by integrating spheres (hollow spheres that are coated white inside). Baffles are provided in the integrating spheres. In the

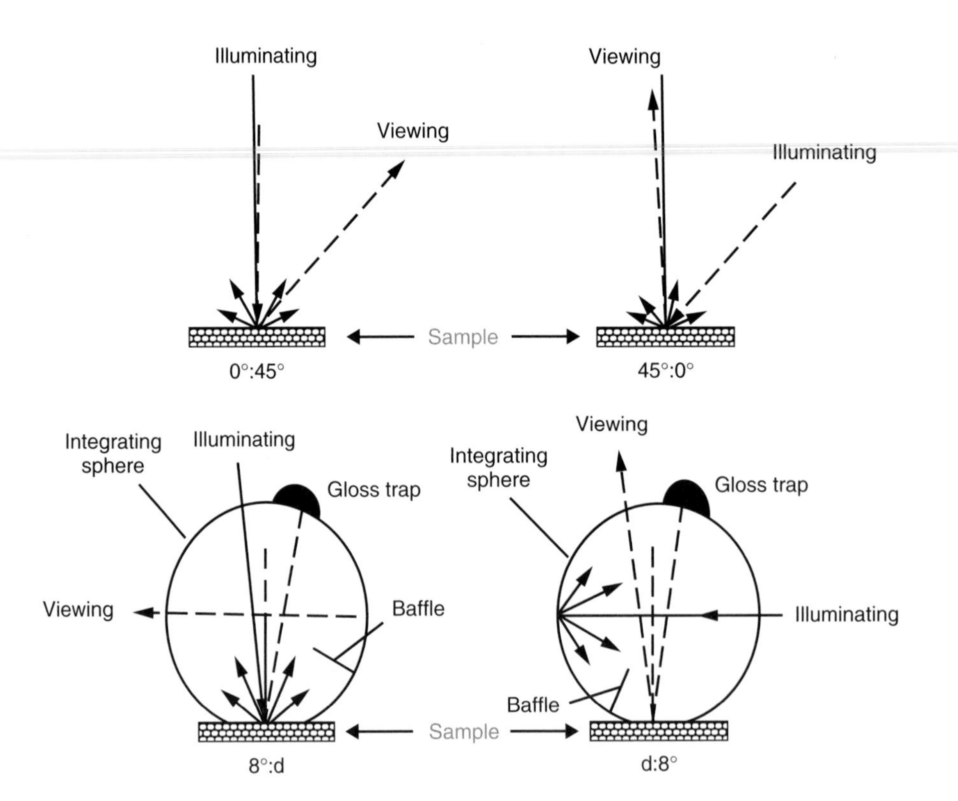

Figure 5.1 Diagrammatic representation of four of the CIE recommended geometries of illumination and measurement

Table 5.1 CIE geometries of illumination and measurement for reflection

Geometry	Illumination		Measurement		Specular reflection	Approximate CIE measure
	Axis	Spread	Axis	Spread		
45°a : 0°	45° ± 2°	±8°	±10°	±8°	Excluded	Radiance factor, 45°a : 0°
0° : 45°a	±10°	±8°	45° ± 2°	±8°	Excluded	Radiance factor, 0° : 45°a
45°x : 0°	45° ± 2°	±8°	±10°	±8°	Excluded	Radiance factor, 45°x : 0°
0° : 45°x	±10°	±8°	45° ± 2°	±8°	Excluded	Radiance factor, 0° : 45°x
di : 8°	Diffuse	Ports ≤ 10%	8°	±5°	Included	Radiance factor, di : 8°
8° : di	8°	±5°	Diffuse	Ports ≤ 10%	Included	Reflectance
de : 8°	Diffuse	Gloss trap	8°	±5°	Excluded	Radiance factor, de : 8°
8° : de	8°	±5°	Diffuse	Gloss trap	Excluded	Diffuse Reflectance

Note 1. The letter 'a' in the designation 45°a : 0 indicates the use of annular illumination which may be achieved by the use of a small source and an elliptical ring reflector or aspheric optics. This geometry is also sometimes approximated by the use of a number of light sources in a ring or a number of fibre bundles illuminated by a single source and terminated in a ring. Such an approximation to annular geometry is called *circumferential* geometry, 45°c : 0°.

Note 2. The letter 'x' in the designation 45°x : 0 indicates that the direction of the incident beam is in the same plane as that of the reflected beam.

Note 3. The axis figures denote the angle between the central ray of the beam and the normal to the surface. The spreads denote the maximum permitted deviation of the rays of the beam from the central ray.

Note 3. Integrating spheres are usually used for the diffuse conditions. When they are used, the total area of their ports should not exceed 10% of the internal reflecting sphere area.

Note 4. For specular excluded conditions, details of the gloss traps should be given.

Note 5. It is important that the particular illuminating and viewing conditions used should be specified even if they are within the range of one of these recommended conditions. Measurements of some types of specimens (for example retro-reflective materials) may require different geometries or smaller tolerances.

Note 6. When integrating spheres are used, they should be fitted with white-coated baffles to prevent light passing directly between the specimen and the spot on the sphere wall irradiated or measured. When the regular component of reflection is to be included, the sphere efficiency for that part of the sphere wall that receives the regularly reflected radiation component should be of the highest value possible.

Note 7. When fluorescent samples are measured, it is preferable not to use integrating spheres, because the samples then modify the spectral distribution of the illuminant (unless small ports are used, see Section 10.10).

Note 8. Where single-beam integrating spheres are used, a correction for the reduction of sphere efficiency caused by sample absorption is necessary. Without such a correction, the instrument will give a non-linear output. The corrected reflectance is given by:

$$\rho_S = R(\lambda)\frac{1-\rho_w(\lambda)\cdot(1-\Sigma_i f_i)}{1-\rho_w(\lambda)\cdot(1-\Sigma_i f_i)-f_S\cdot(\rho_r(\lambda)-R(\lambda))},$$

where $R(\lambda)$ is the uncorrected reflectance of the sample referred to the perfect reflecting diffuser, $\rho_w(\lambda)$ is the diffuse/diffuse spectral reflectance of the sphere wall, f_i is the fractional area of the ith port, f_S is the fractional area of the sample port, and $\rho_r(\lambda)$ is the reflectance of the reference standard. The above equation assumes ideal integrating characteristics for the sphere and that the other ports have zero effective reflectance.

Note 9. CIE standards on geometric tolerances are regularly updated. Users should check with the CIE list of publications for the latest recommendations.

layout shown in Figure 5.1, the illumination is by a parallel beam of light and the viewing is by imaging the sphere wall or the sample on to the detector; in this case the baffles prevent light travelling directly from the sample to the part of the sphere from which light is collected by the detector (8°:d), or directly from the part of the sphere illuminated by the input beam to the sample (d:8°). However, in some instruments the input window of the sphere is covered with a diffuser, and the detector is placed at the exit window; in this case the baffles are on the opposite sides to those shown in Figure 5.1, to prevent any diffuse illumination falling on to the sample without hitting the sphere wall first (d:8°), and to ensure that the detector only sees light that has been reflected from the sphere wall (8°:d). The integrating spheres can often be used with *gloss traps*, which may consist of cavities designed to absorb all the light incident on the sphere wall at angles near those corresponding to specular reflections for the incident or measurement beams. Hence, if gloss traps are used, the specularly reflected light is excluded (sometimes referred to as SPEX); if gloss traps are not used, the specularly reflected light is included (sometimes referred to as SPIN or SPINC). CIE recommendations concerning geometric conditions for colorimetry are given in Table 5.1.

Most objects are neither completely matt nor highly glossy. The effects of the geometry of the illuminating and viewing conditions on these objects are therefore intermediate to those described above, and their colour saturations tend to be less than those of glossy, but more than those of matt, samples.

These differences in surface properties greatly affect the appearance of objects, and the particular degree of sheen is the characteristic that gives rise to surface finishes that may be described by terms such as satin, egg-shell, lustre, etc.

It is clear, therefore, that the illumination and viewing conditions play a very significant part in colour appearance. Hence, when it comes to colour measurement, if telespectroradiometry is not used, and simplified procedures are adopted, it is very important that these procedures represent the viewing conditions that are of major interest in the particular application. The CIE has recommended several alternative geometries of illumination and measurement for colorimetry and these will now be described. (The use of the term 'viewing' with reference to measuring instruments is deprecated because it blurs the important distinction between instrumental measurements and visual observations.)

Although there are eight different conditions listed in Table 5.1, for practical purposes each pair, such as 45°a : 0° and 0° : 45°a, can be regarded as equivalent, so that only four need to be considered as potentially giving different results (but this is not true for polarising samples, see Section 9.6). The 45°a : 0° and 0° : 45°a, and the 45°x : 0° and 0° : 45°x both represent typical viewing of surfaces in directional light; the letter 'a' in these designations indicates the use of annular illumination, while the letter 'x' indicates that the direction of the incident beam is in the same plane as that of the reflected beam. The di : 8° and 8° : di represent viewing in a completely diffuse illumination; the letter 'i' in these designations indicates that the specularly reflected light is included. The 8° denotes the angle between the measuring or illuminating direction and the normal to the surface of the sample being measured; as already mentioned, this angle is used (instead of zero) to minimise inter-reflections between the sample and the illuminating or measuring apertures in the integrating spheres, and to enable gloss traps to be fitted if required. The de : 8° and 8° : de represent viewing in a diffuse illumination that has a black area in the direction corresponding to a mirror image of the direction of measurement or illumination; the letter 'e' in these designations indicates that the specularly reflected light is excluded. This may sound a rather artificial circumstance, but practical viewing situations often involve light sources having quite limited areas and viewers usually adjust their angle of viewing so as to exclude the specular reflections as much as possible. It is, therefore, a mode that is widely used for practical measurements. However, if it is required to compare measurements made on different integrating-sphere instruments, it is better to use the specular included mode, because the efficiency of gloss traps varies considerably from one instrument to another. It has also been found that, for computer match prediction, the specular included mode is to be preferred, because it gives more accurate results.

Use of the different CIE geometries can give very different results for the same specimen, unless it is completely matt. This is illustrated in Figure 5.2, where four glossy samples have been measured with and without the use of a gloss trap. The use of the gloss trap increases the purity; it also decreases the value of L^*, very considerably for dark colours, so that the CIELUV or CIELAB colour differences between the two types of measurement range from about 5 for the yellow sample (which is light), to about 18 for the red and blue samples (which are dark). For highly glossy samples, it is preferable to use either 45°a : 0°, 0° : 45°a, 45°x : 0°, 0° : 45°x, di : 8°, or 8° : de (but not 8° : di or de : 8°); however, both of the 45° : 0° geometries may give rise to problems of polarisation (see Section 9.6). For partly glossy, or textured, samples, the preferred geometry is the one that minimises the surface effects (such as specular included for textiles), and this may correspond to that which gives the lowest luminance factor and highest excitation purity (ASTM, 2009.)

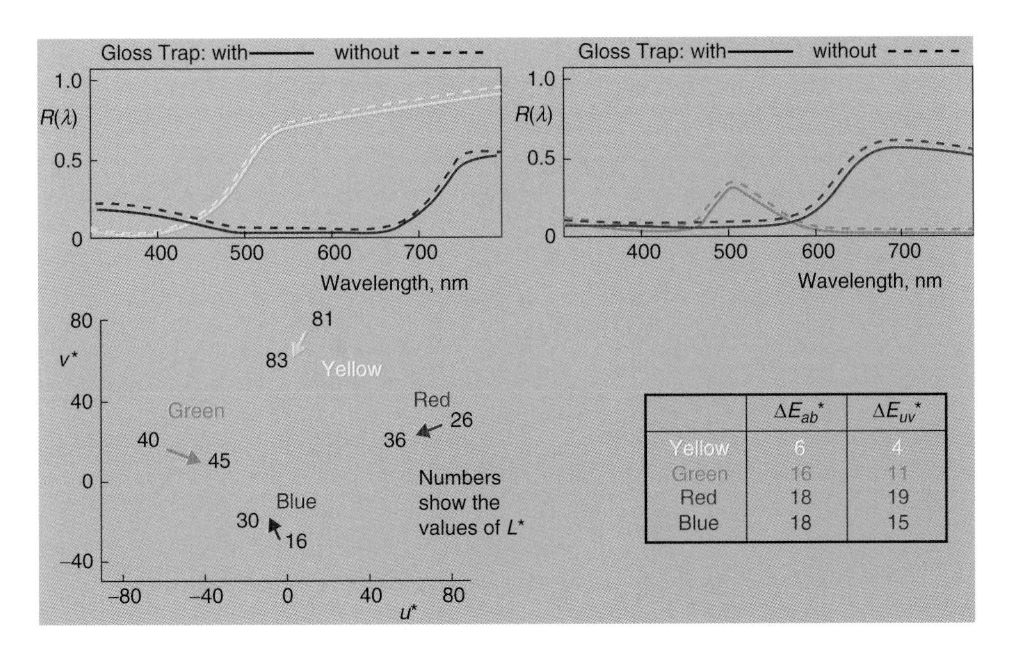

Figure 5.2 Top: spectral reflectance factors, $R(\lambda)$, for blue, yellow, green, and red glossy samples measured with (full line) and without (broken line) gloss traps. Bottom: u^*, v^* plot, with gloss trap (points) and without gloss trap (arrow-heads), and the colour differences, ΔE^*, between the two types of measurement. The numbers on the figures are the L^* values

For opaque samples, all the measures are of *reflectance factor*, the ratio of the radiant flux reflected by the sample into a defined cone to the radiant flux similarly reflected by the perfect diffuser. But they also correspond approximately to more particular CIE measures, and these are listed in Table 5.1. All except the diffuse measurement cases give approximations to the measurement of *radiance factor* because the size of the cone of rays collected is fairly small (it would have to be negligibly small to give radiance factor exactly). In the two cases of diffuse measurement, 8° : di approximates *reflectance*, and 8° : de approximates *diffuse reflectance* (the reflectance attributable to diffusion alone). In the colorant industries *reflectance factor* is often abbreviated to *reflectance* (See Table 5.1).

A similar set of conditions for transmitting samples has also been recommended by the CIE, and these are listed in Table 5.2. As with opaque samples, the similar pairs, 8° : di, di : 8° and 8° : de, de : 8°, give similar results in practice. (Definitions for the CIE measures referred to in Tables 5.1 and 5.2 are given in Appendix A1.6.)

5.10 SPECTRORADIOMETERS AND SPECTROPHOTOMETERS

Spectroradiometers and spectrophotometers of various designs have been produced. They all require a means of dispersing the light into a spectrum, and a means of measuring the radiation from samples and from the calibration beams. The light can be dispersed by prisms, but diffraction gratings are widely used because, unlike prisms, their spectra have an approximately uniform scale of wavelength; some instruments use a series of interference filters, each transmitting only a narrow band of wavelengths. For detection and comparison of the light, photoelectric cells are used; either a single cell can be used sequentially for the different wavelength bands throughout the spectrum; or an array of photosensitive elements (such as a photo-diode or CCD array) can be used to make all the measurements throughout the spectrum simultaneously.

In the case of spectrophotometers, a light source must also be provided. Many instruments use tungsten-halogen light sources, but other sources, such as xenon flash lamps, are also useful. The use of xenon flash lamps, together with simultaneous detection throughout the spectrum by an array of detectors, can produce the results extremely rapidly, and this is advantageous when many samples have to be measured. The results, in the form of reflectance factor or transmittance factor, are often produced automatically as plots against wavelength, or as lists of numerical data, often with computations of CIE tristimulus values and other useful measures calculated from them.

Some older instruments scan through the spectrum continuously and produce a continuously varying graph of radiant power emitted, of reflectance factor, or of transmittance factor, against wavelength; but when tristimulus values are calculated these data have to be sampled at intervals throughout the spectrum. Instruments now generally use multiple photo-detectors to sample the whole spectrum simultaneously, and can only do so at a given number of wavelengths. For most purposes, it is usually considered sufficient to sample the spectrum at 5 nm intervals; but in some cases 10 nm, or even 20 nm, intervals are used, particularly for simultaneous type instruments. The CIE colour-matching functions tabulated at 1 nm intervals cover a range of wavelengths from 360 nm to 830 nm; but, for most colorimetric purposes, it is sufficient to use a range from 380 nm to 780 nm, which is what is used for these functions tabulated at 5 nm intervals. Some instruments,

Table 5.2 CIE geometries of illumination and measurement for transmission

Geometry	Illumination		Measurement		Specular	Approximate CIE measure
	Axis	Spread	Axis	Spread		
$0^\circ : 0^\circ$	$\pm 5^\circ$	$\pm 5^\circ$	$\pm 5^\circ$	$\pm 5^\circ$	Included	Regular transmittance
di : 0°	Diffuse	Ports $\leq 10\%$	$\pm 5^\circ$	$\pm 5^\circ$	Included	Radiance factor, di : 0°
0° : di	$\pm 5^\circ$	$\pm 5^\circ$	Diffuse	Ports $\leq 10\%$	Included	Diffuse transmittance
de : 0°	Diffuse	Ports $\leq 10\%$	$\pm 5^\circ$	$\pm 5^\circ$	Excluded	Diffuse transmittance
0° : de	$\pm 5^\circ$	$\pm 5^\circ$	Diffuse	Ports $\leq 10\%$	Excluded	Diffuse transmittance
di : di	Diffuse	Ports $\leq 10\%$	Diffuse	Ports $\leq 10\%$	Included	Doubly diffuse transmittance

Note 1. The axis figures denote the angle between the central ray of the beam and the normal to the specimen. The spreads denote the maximum permitted deviation of the rays of the beam from the central ray.

Note 2. Integrating spheres are usually used for the diffuse conditions. When they are used, the total area of their ports should not exceed 10% of the internal sphere area.

Note 3. For specular excluded conditions, details of the light traps should be given.

Note 4. Errors may be caused by multiple reflections between the specimen and the incident beam optics if the incident beam is normal to the specimen surface; such errors can be eliminated by slightly tilting the specimen.

Note 5. The construction of an instrument for 0/0 measurements shall be such that the flux incident on the specimen and the flux reaching the detector when there is no specimen in place shall be equal.

Note 6. It is important that the particular illuminating and viewing conditions used should be specified even if they are within the range of one of these recommended conditions. Measurements of some types of specimens may require different geometries or smaller tolerances.

Note 7. When integrating spheres are used, they should be fitted with white-coated baffles to prevent radiation passing directly from the source to sample or reference in the case of diffuse irradiation or directly from sample or reference to detector in the case of diffuse collection.

Note 8. When fluorescent samples are measured, it is preferable not to use integrating spheres, because the samples then modify the spectral distribution of the illuminant (unless small ports are used, see Section 9.10).

Note 9. Where single-beam integrating spheres are used, a correction for the reduction of sphere efficiency caused by sample absorption is necessary. Without such a correction, the instrument will give a non-linear output. The corrected reflectance is given by:

$$\rho_S = R(\lambda)\frac{1 - \rho_w(\lambda)\cdot(1 - \Sigma_i f_i)}{1 - \rho_w(\lambda)\cdot(1 - \Sigma_i f_i) - f_S\cdot(\rho_r(\lambda) - R(\lambda))},$$

where $R(\lambda)$ is the uncorrected reflectance of the sample referred to the perfect reflecting diffuser, $\rho_w(\lambda)$ is the diffuse/diffuse spectral reflectance of the sphere wall, f_i is the fractional area of the ith port, f_S is the fractional area of the sample port, and $\rho_r(\lambda)$ is the reflectance of the reference standard. The above equation assumes ideal integrating characteristics for the sphere and that the other ports have zero effective reflectance.

Note 10. CIE standards on geometric tolerances are regularly updated. Users should check with the CIE list of publications for the latest recommendations.

particularly the simultaneous types, may use a smaller range of wavelengths, such as from 400 nm to 700 nm. (The effects of these differences in wavelength interval and range will be discussed in Sections 5.12 and 9.13.)

The illuminating and measuring geometry of spectrophotometers is usually intended to comply with one of the configurations described in Tables 5.1 and 5.2. However, the scope for variation within the limits given in each configuration is such that different instruments employing the same nominal configuration may give significantly different results. It is therefore always good practice to state, not only the configuration used, but also the particular realisation adopted, and this can be done in terms of the particular make and model of the instrument, or of the precise geometry actually employed. (This will be discussed further in Section 9.4.)

Other characteristics of spectroradiometers and spectrophotometers that can significantly affect the results include: the amount of stray light; the photometric linearity; the bandwidth of wavelengths used for the measurements; polarisation of the light; and the translucency, thermochromism, and fluorescence of the samples. (These topics will be discussed in Chapters 9 and 10.)

5.11 CHOICE OF ILLUMINANT

Spectral reflection or transmission data are independent of the illuminant, but tristimulus values can only be obtained from such data for a specified illuminant. In the case of fluorescent samples, a specified illuminant is necessary even to obtain the proper spectral data (as will be discussed in Chapter 10).

For the highest accuracy in colorimetry, when spectral data are used to obtain tristimulus values, the illuminant adopted should be the same as that used in the practical situation that the colorimetry is intended to represent. Thus, if samples are to be inspected in a viewing booth, then the illuminant should be the one actually used in the booth; hence its relative spectral power distribution is required. Either this can be measured with a spectroradiometer, or published data can be used.

But, for general work, the use of a large number of different illuminants is inconvenient, and makes comparisons of results impossible. The CIE has therefore recommended that, wherever possible, either Standard Illuminant A (SA) or Standard Illuminant D65 should be used. SA is a very convenient representation of tungsten lamps because it is also available as a source. D65 is a very good representation of average daylight, but has the disadvantage of not being available as a source (see Section 4.17). Visual inspection of samples whose colorimetry has been evaluated using D65 is therefore difficult; real daylight is variable, and artificial daylight sources usually differ significantly in their spectral power distributions from D65. However, experience has shown that D65 colorimetry is useful in predicting colour matches for samples viewed under sensibly chosen real and artificial daylight sources, but the possibility of the source having a disturbing influence should always be borne in mind.

For some applications, SA and D65 are not appropriate, and other sources have to be used. Sometimes a different D Illuminant (see Section 4.17) can be used: for instance, in photography, the sunlight plus skylight widely used for amateur picture taking is well represented by D55; and in the graphic arts industry D50 is used as a representative viewing illuminant. The CIE indoor illuminants ID50 and ID 65 can also find practical application in the representation of daylight as viewed through typical window glass (see Section 4.18).

5.12 CALCULATION OF TRISTIMULUS VALUES FROM SPECTRAL DATA

Tristimulus values are calculated from spectral data by the method given in Section 2.6. The essence of this procedure is to weight the spectral power distribution of a sample by the CIE colour-matching functions $\bar{x}(\lambda)$, $\bar{y}(\lambda)$, and $\bar{z}(\lambda)$, or by the functions $\bar{x}_{10}(\lambda)$, $\bar{y}_{10}(\lambda)$, and $\bar{z}_{10}(\lambda)$ if the field size has an angular subtense greater than 4°.

Spectral data consist of quantities at a selection of wavelengths throughout the spectrum; for colorimetry, the number of wavelengths chosen is at least 16 (from 400 nm to 700 nm at 20 nm intervals), and usually many more than 16; for example 81 if 380 nm to 780 nm at 5 nm data are used. But tristimulus values consist of only three quantities. The reduction of spectral data to tristimulus values is therefore an irreversible process; it is not possible, from tristimulus values, to deduce the correct corresponding spectral data.

For light sources and self-luminous colours, such as television or computer displays, the spectral power distribution provides the natural starting data. But the data for reflecting and transmitting samples usually consist of measurements of spectral reflectance factor or spectral transmittance factor; to convert these measurements to spectral power data requires their values to be multiplied, wavelength by wavelength, by the spectral power distribution of a chosen illuminant as shown in Figure 5.3.

If the values of the spectral power for a self-luminous sample, or of spectral power for the chosen illuminant and of spectral reflectance (or transmittance) factor for a non-self-luminous sample, are available at every 5 nm from 380 nm to 780 nm, then the calculation is quite straightforward. But it often happens that an interval of greater than 5 nm has been used (abridged data), or that some of the values at the ends of the spectrum are not available (truncated data).

The most common form of abridged data is when a measuring interval of 10 nm or 20 nm has been used. If such data are weighted simply by those values in the tabulations of the CIE functions that correspond to the nominal wavelengths used at the 10 nm or 20 nm intervals, ignoring the rest, then significant errors can occur. If the size of these errors is too great to be tolerated, then either the measurements should be repeated using a 5 nm interval, or the measured data should be interpolated to be at 5 nm intervals, or corrections should be provided either by deconvoluting the data, or by using optimised weights (these latter two options will be discussed in Section 9.13).

The most common form of truncated data is when a range of wavelengths of only 400 nm to 700 nm has been used. There are two different ways of dealing with this situation. The first way is to use only those values in the 5 nm tabulations of the CIE functions that occur within this range of wavelengths. The second is to estimate the missing values by extrapolation so as to compile data down to 380 nm and up to 780 nm (in the absence of any data that indicates otherwise, the extrapolation can be done by setting the missing values equal to the nearest measured value of the appropriate quantity in the truncated data). The second method is preferable because it is more accurate (this will be discussed further in Section 9.13).

Table 5.3 Tristimulus values and chromaticity co-ordinates for Standard Illuminants D65 and A

CIE Standard Illuminant D65 – 2 degree Standard Observer						
	1 nm 360 nm 830 nm	1 nm 380 nm 780 nm	5 nm 380 nm 780 nm	10 nm 380 nm 780 nm	10 nm 400 nm 700 nm	20 nm 400 nm 700 nm
X	95.04705587	95.04226741	95.04296694	95.01739696	94.94009232	95.56383488
Y	100.00000000	100.00000000	100.00000000	100.00000000	100.00000000	100.00000000
Z	108.88287364	108.86103689	108.88005470	108.81276377	108.70912221	109.66852364
x	0.31272687	0.31273851	0.31272053	0.31273194	0.31266372	0.31308553
y	0.32902321	0.32905203	0.32903069	0.32913125	0.32932738	0.32761926
u'	0.19783997	0.19783724	0.19783275	0.19780294	0.19768195	0.19861863
v'	0.46833638	0.46835351	0.46833944	0.46839488	0.46848953	0.46763707
L^*	100.0000	100.0000	100.0000	100.0000	100.0000	100.0000
u^*	0.0000	−0.0035	−0.0094	−0.0481	−0.2054	1.0123
v^*	0.0000	0.0223	0.0040	0.0760	0.1991	−0.9091
ΔE^*_{uv}		0.023	0.010	0.090	0.286	1.361
L^*	100.0000	100.0000	100.0000	100.0000	100.0000	100.0000
a^*	0.0000	−0.0084	−0.0072	−0.0520	−0.1876	0.9045
b^*	0.0000	0.0134	0.0017	0.0429	0.1064	−0.4799
ΔE^*_{ab}		0.0158	0.0074	0.0674	0.2157	1.0240

CIE Standard Illuminant D65 – 10 degree Standard Observer						
	1 nm 360 nm 830 nm	1 nm 380 nm 780 nm	5 nm 380 nm 780 nm	10 nm 380 nm 780 nm	10 nm 400 nm 700 nm	20 nm 400 nm 700 nm
X_{10}	94.81106006	94.81073156	94.81178687	94.82495748	94.78112394	95.15495748
Y_{10}	100.00000000	100.00000000	100.00000000	100.00000000	100.00000000	100.00000000
Z_{10}	107.30466954	107.30398114	107.32410766	107.38071589	107.35174388	108.37059248
x_{10}	0.31382365	0.31382362	0.31380511	0.31377623	0.31370676	0.31349900
y_{10}	0.33099899	0.33100010	0.33097689	0.33090047	0.33098021	0.32946156
u'_{10}	0.19786053	0.19786009	0.19785595	0.19786455	0.19778657	0.19821196
v'_{10}	0.46955091	0.46955150	0.46953645	0.46949162	0.46952365	0.46868489
L^*_{10}	100.0000	100.0000	100.0000	100.0000	100.0000	100.0000
u^*_{10}	0.0000	−0.0006	−0.0060	0.0052	−0.0961	0.4569
v^*_{10}	0.0000	0.0008	−0.0188	−0.0771	−0.0354	−1.1258
$\Delta E^*_{uv,10}$		0.001	0.020	0.077	0.102	1.215
L^*_{10}	100.0000	100.0000	100.0000	100.0000	100.0000	100.0000
a^*_{10}	0.0000	−0.0006	0.0013	0.0244	−0.0526	0.6038
b^*_{10}	0.0000	0.0004	−0.0121	−0.0472	−0.0292	−0.6601
$\Delta E^*_{ab,10}$		0.0007	0.0121	0.0532	0.0602	0.8946

(*Continued*)

Table 5.3 (*Continued*)

CIE Standard Illuminant A – 2 degree Standard Observer						
	1 nm 360 nm 830 nm	1 nm 380 nm 780 nm	5 nm 380 nm 780 nm	10 nm 380 nm 780 nm	10 nm 400 nm 700 nm	20 nm 400 nm 700 nm
X	109.85031527	109.84880155	109.84899312	109.83112576	109.69091286	109.77661789
Y	100.00000000	100.00000000	100.00000000	100.00000000	100.00000000	100.00000000
Z	35.58493013	35.58148799	35.58247363	35.54565668	35.54597316	35.46373695
x	0.44757351	0.44757638	0.44757502	0.44760195	0.44728554	0.44762869
y	0.40743944	0.40744767	0.40744572	0.40753652	0.40776900	0.40776323
u'	0.25597108	0.25596932	0.25596929	0.25594679	0.25564079	0.25586453
v'	0.52429065	0.52429426	0.52429330	0.52433250	0.52437504	0.52442423
L^*	100.0000	100.0000	100.0000	100.0000	100.0000	100.0000
u^*	0.0000	−0.0023	−0.0023	−0.0316	−0.4294	−0.1385
v^*	0.0000	0.0047	0.0035	0.0544	0.1097	0.1737
ΔE^*_{uv}		0.005	0.004	0.063	0.443	0.222
L^*	100.0000	100.0000	100.0000	100.0000	100.0000	100.0000
a^*	0.0000	−0.0023	−0.0020	−0.0291	−0.2420	−0.1118
b^*	0.0000	0.0064	0.0046	0.0736	0.0730	0.2273
ΔE^*_{ab}		0.0068	0.0050	0.0792	0.2527	0.2533

CIE Standard Illuminant A – 10 degree Standard Observer						
	1 nm 360 nm 830 nm	1 nm 380 nm 780 nm	5 nm 380 nm 780 nm	10 nm 380 nm 780 nm	10 nm 400 nm 700 nm	20 nm 400 nm 700 nm
X_{10}	111.14394095	111.14331583	111.14390763	111.15514934	111.06119792	110.89639627
Y_{10}	100.00000000	100.00000000	100.00000000	100.00000000	100.00000000	100.00000000
Z_{10}	35.19994369	35.19994479	35.19951777	35.19420552	35.20467936	35.18414576
x_{10}	0.45117394	0.45117254	0.45117465	0.45120942	0.45098086	0.45065081
y_{10}	0.40593660	0.40593763	0.40593736	0.40592759	0.40606519	0.40637102
u'_{10}	0.25896454	0.25896318	0.25896466	0.25899156	0.25878208	0.25843216
v'_{10}	0.52424830	0.52424849	0.52424870	0.52425013	0.52426923	0.52433838
L^*_{10}	100.0000	100.0000	100.0000	100.0000	100.0000	100.0000
u^*_{10}	0.0000	−0.0018	0.0002	0.0351	−0.2372	−0.6921
v^*_{10}	0.0000	0.0002	0.0005	0.0024	0.0272	0.1171
$\Delta E^*_{uv,10}$		0.0018	0.0005	0.0352	0.2388	0.7019
L^*_{10}	100.0000	100.0000	100.0000	100.0000	100.0000	100.0000
a^*_{10}	0.0000	−0.0009	0.0000	0.0168	−0.1241	−0.3715
b^*_{10}	0.0000	0.0000	0.0008	0.0109	−0.0090	0.0299
$\Delta E^*_{ab,10}$		0.0009	0.0008	0.0200	0.1244	0.3727

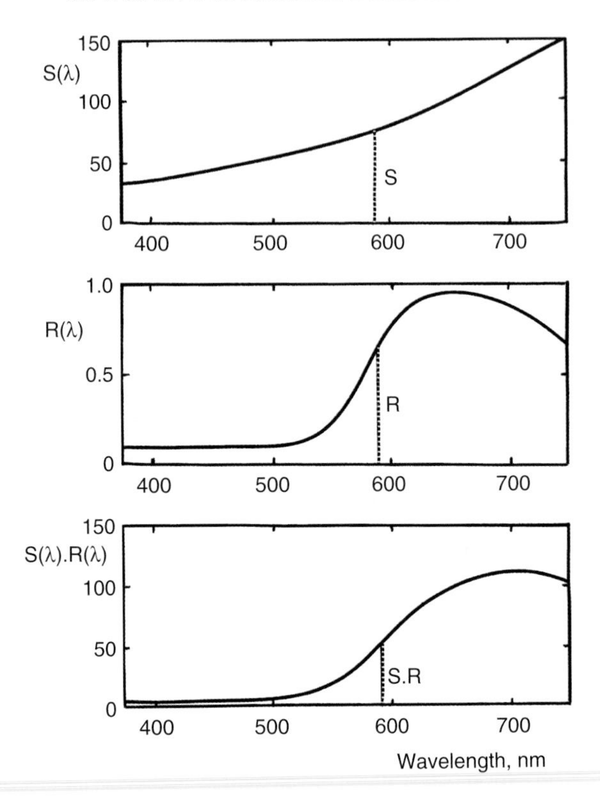

Figure 5.3 Top: spectral power distribution, $S(\lambda)$, of an illuminant. Centre: spectral reflectance factor, $R(\lambda)$, of a sample. Bottom: spectral power distribution, $S(\lambda)R(\lambda)$, of the sample in the illuminant

It is very important that the same wavelength interval, and range of wavelengths, is used for any set of calculations in which data for different colours are to be compared precisely. This is illustrated in Table 5.3 where the tristimulus values and chromaticity co-ordinates are given for CIE Standard Illuminants A and D65 evaluated using different wavelength intervals and ranges. The variations in the values for a given illuminant in Table 5.3 emphasise the importance of using self-consistent data when evaluating colour differences, or combining results for sample colours and for a reference white to obtain correlates of perceptual attributes of colour such as those included in the CIELUV and CIELAB systems.

5.13 COLORIMETERS USING FILTERED PHOTO-DETECTORS

As mentioned at the beginning of this chapter, a simple, but usually rather less accurate, method of obtaining CIE tristimulus values is to use filtered photo-detectors. The photo-detectors are covered with carefully designed colour filters so that the resulting spectral sensitivities (when combined with the spectral power distribution of the instrument source) are a close match to the CIE colour-matching functions combined with the spectral power distribution of a specified illuminant. The samples, and a suitably calibrated standard specimen, are then illuminated by the light source chosen, and the reflected or transmitted light made to fall on the filtered photo-detectors. The ratios of the sample and standard readings of the three photo-detectors are then used to derive the CIE tristimulus values. Instead of using three photo-detectors, it is also possible to use a single photo-detector and to take three readings sequentially through the three filters. Because it is difficult to find filtration that exactly matches the CIE colour-matching functions, the method is limited in accuracy, but offers a way of providing relatively inexpensive instruments. Instead of using the colour-matching functions themselves, a suitable combination of them can be used, and the readings obtained are then converted to CIE X, Y, Z tristimulus values by the use of an algebraic transformation (Erb, Krystek and Budde, 1984); this procedure can be further elaborated by using more than three filters to increase the accuracy (Wharmby, 1975). The geometry of illumination and measurement of such instruments should be in accordance with the CIE specifications given in Tables 5.1 and 5.2. These instruments can be used with different illuminants, but they cannot provide diagnostic data on metamerism (to be discussed in Chapter 6).

REFERENCES

ASTM Document E1164, *Standard practice for obtaining spectrophotometric data for object-color evaluation*, ASTM International, West Conshohocken, PA, USA (2009).

Erb, W., Krystek, M., and Budde, W., A method for improving the accuracy of tristimulus colorimeters. *Color Res. Appl.*, **9**, 84–88 (1984).

Hunt, R.W.G., The importance of being not too earnest, *IS&T and SID's 11th Color Imaging Conference: Color Science and Engineering Systems, Technologies, and Applications*, pp. 6-10, IS&T, Springfield, VA, USA (2003).

Wharmby, D.O., Improvements in the use of filter colorimeters, *J. Phys. E: Scientific Instrum.*, **8**, 41–43 (1975).

GENERAL REFERENCES

Bartleson, C.J., and Grum, F., *Color Measurement*, Academic Press, New York, NY, USA (1980).

Berns, R.S., *Billmeyer and Saltzman's Principles of Color Technology*, *3rd Ed.*, Wiley, New York, NY, USA (2000).

CIE Publication 15:2004, *Colorimetry*, Commission Internationale de l'Éclairage, Vienna, Austria (2004).

Judd, D.B., and Wyszecki, G., *Color in Business, Science, and Industry*, *2nd Ed.*, Wiley, New York, NY, USA (1975).

MacAdam, D.L., *Color Measurement*, Springer-Verlag, Berlin, Germany (1981).

McDonald, R., *Colour Physics for Industry*, *2nd Ed.*, ed Roderick McDonald, Society of Dyers and Colourists, Bradford, England (1997).

McLaren, K., *The Colour Science of Dyes and Pigments*, *2nd Ed.*, Hilger, Bristol, England (1986).

Wright, W.D., *The Measurement of Colour*, *4th Ed.*, Hilger, Bristol, England (1969).

Wyszecki, G., and Stiles, W.S., *Colour Science*, *2nd Ed.*, Wiley, New York, NY, USA (1982, 2000).

6

Metamerism and Colour Constancy

6.1 INTRODUCTION

As mentioned in Section 2.7, when two colours match one another, but are different in spectral composition, they are said to be metameric, and the phenomenon is referred to as metamerism. The phenomenon of metamerism is of great practical importance, because the greater the degree of metamerism, that is, the greater the degree of difference in spectral composition, the greater will be the likelihood that the colours will no longer match one another if one of the conditions of the match is altered, such as the spectral composition of the illumination, or the spectral sensitivity of the observer, or the size of the matching field. In the colorant industries, it is frequently necessary to work with samples that are metameric to one another, and it is important to know to what extent such colours will cease to match under the range of illuminants, observers, and field sizes, that are commonly met with in practice. It is therefore very desirable to have meaningful methods of evaluating any difference in spectral composition between two samples that are a metameric match. Such methods will be described later on in this chapter. But we shall first consider some of the more general aspects of metamerism.

6.2 THE CAUSE OF METAMERISM

As described in Section 1.5, the eye responds to light, not on a wavelength by wavelength basis, but as a result of the integrated stimulation of each of the three different cone types, ρ, γ, and β. If two stimuli result in identical ρ, γ, and β cone stimulations, then, when seen under the same conditions, they will look identical whatever their spectral compositions might be.

This is illustrated in Figure 6.1. The smooth curve is the spectral power distribution of a non-selective neutral grey stimulus, N, illuminated by a tungsten filament light, and the undulating curve is that of a selective grey stimulus, S, that is a metameric match to N

Measuring Colour, Fourth Edition. R.W.G. Hunt and M.R. Pointer.

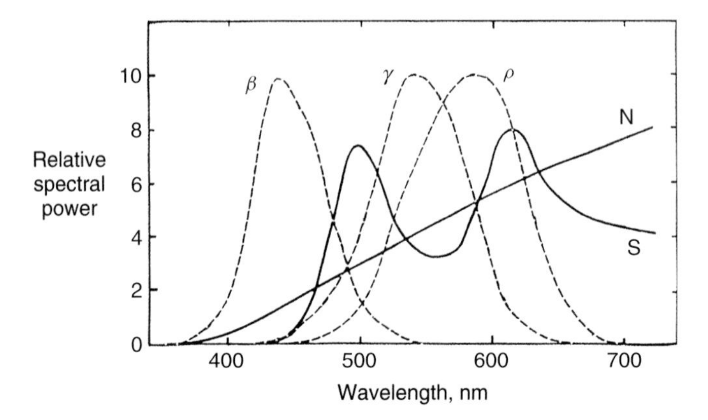

Figure 6.1 Full lines: the relative spectral power distributions of two stimuli, N and S, that are metameric for the CIE 1931 Standard Colorimetric Observer. Broken lines: spectral sensitivities representative of those believed to be typical of the three different types of cones, ρ, γ, and β, of the retina of the eye

under the same illuminant for the CIE 1931 Standard Colorimetric Observer. Also shown in the figure, by the broken lines, are the spectral sensitivity curves of the ρ, γ, and β type cones. It is clear by inspection of the figure that, although the spectral power distribution of S is greater in the long wavelength part of the ρ band, it is smaller in the short wavelength part of that band, and the total effect of the S stimulus on the ρ cones could therefore be the same as that of the N stimulus. Similarly, in the γ band the smaller power of the S stimulus in the long wavelength part of the band could be exactly compensated by its greater power in the short wavelength part. Finally, in the case of the β band, the greater power at the long wavelength part could be balanced by the smaller power at the shorter wavelengths. Hence it is clear that the stimuli, N and S, could match one another in spite of their large differences in spectral composition.

For equal ρ, γ, and β responses to result from two different spectral power distributions, it is necessary for their curves to exhibit a crossover point within each of the bands of the spectrum to which the ρ, γ, and β cones are sensitive. There can be more than one crossover point in each band, so that there may be more than three crossovers in total, but there can never be less than three. It is therefore a characteristic of metameric pairs of stimuli that their spectral power distributions exhibit three or more crossover points in the visible spectrum; the positions of the crossovers depend on the spectral compositions of the stimuli that are involved (Berns and Kuehni, 1990).

6.3 THE DEFINITION OF METAMERISM

Because the exact shapes of the ρ, γ, and β cone sensitivity curves are difficult to establish, and because they vary significantly among observers, metameric pairs of colours are defined, not in terms of equality of ρ, γ, and β responses, but as stimuli that have the same tristimulus values. This means that their spectral power distributions, when weighted by the CIE $\bar{x}(\lambda)$, $\bar{y}(\lambda)$, and $\bar{z}(\lambda)$ colour-matching functions, must produce equal results.

These colour-matching functions can be regarded as linear combinations of ρ, γ, and β spectral sensitivities, and hence the arguments about crossovers given above also apply to the colour-matching functions. It is necessary to choose colour-matching functions either for the CIE 1931 Standard Colorimetric Observer, or for the CIE 1964 Standard Colorimetric Observer, according to whether the field size is less than or equal to 4°, or greater then 4°, respectively; when the latter is the case all the measures involved are distinguished by a subscript 10.

Equality of tristimulus values is the colorimetric way of defining metamerism. But the term metamerism is also often used when two spectrally different samples are a visual match for an individual real observer, even if the two sets of tristimulus values are not equal; this may be referred to as *perceived metamerism* when it is desirable to distinguish it from equality of tristimulus values, which may be regarded as *psychophysical metamerism*. (For a summary of terms used in connection with metamerism, see Section 6.11).

6.4 EXAMPLES OF METAMERISM IN PRACTICE

When reflection prints are made in the colour reproduction industries, the multitude of spectral power distributions of objects in original scenes is generally reproduced by colorants of cyan, magenta, and yellow colours (Hunt, 2004). In Figure 6.2 the spectral reflectance factor of a grey that absorbs uniformly throughout the spectrum (a non-selective grey) is shown (by the broken line), and that of a mixture of cyan, magenta, and yellow colorants (full line) that is metameric to it for D65; it is clear that there are considerable differences between the two curves, and that there are seven crossover points.

In Figure 6.3, the relative spectral power distributions are shown for three greys which are all metameric to one another. One grey is produced by three phosphors typical of those used in cathode-ray television or computer display devices; spectral power distributions of

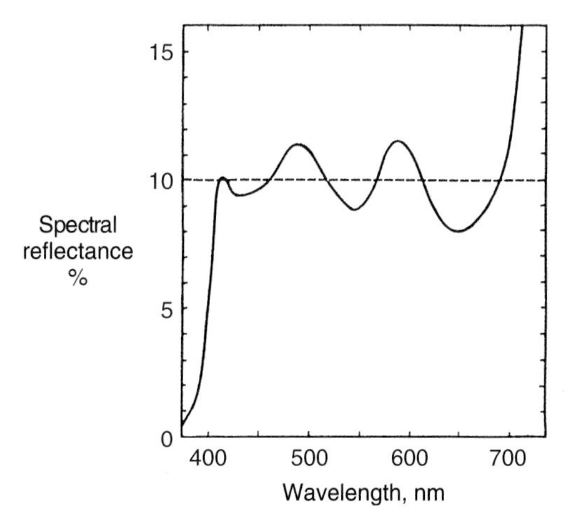

Figure 6.2 Spectral reflectance of a colour photographic grey composed of cyan, magenta, and yellow colorants (full line) that is a metameric match to a non-selective grey (broken line) in Standard Illuminant D65, for the CIE 1931 Standard Colorimetric Observer

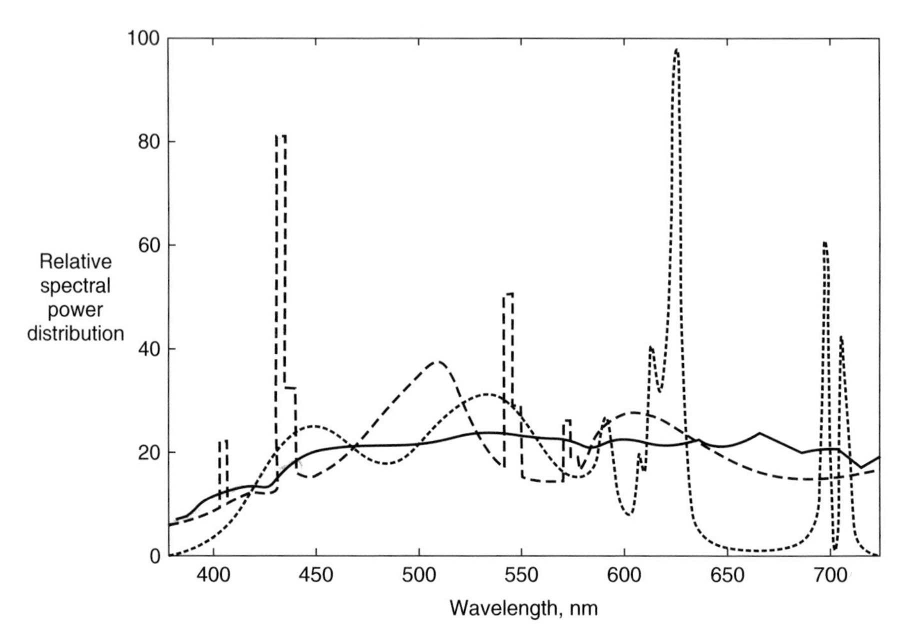

Figure 6.3 Relative spectral power distributions of three greys which are metameric to one another. Dotted line: from a mixture of light from red, green, and blue emitting phosphors typical of those used in cathode-ray television and computer displays. Broken line: from a photographic transparency illuminated by a fluorescent lamp of correlated colour temperature 5000 K. Full line: from a non-selective grey illuminated by D50

light from such phosphors undulate considerably throughout the spectrum, and especially so at the long wavelength end where the very spiky nature of the red rare-earth phosphor is much in evidence. The second grey is on a photographic transparency illuminated by a fluorescent lamp having a correlated colour temperature of 5000 K. The third grey is non-selective and is illuminated by D50. These three greys are clearly very metameric to one another. In the case of graphic arts printing, the cyan, magenta, and yellow inks absorb in similar parts of the spectrum as the dyes used in photography. But a black ink is usually used in addition, and this can reduce the degree of metamerism in the case of dark colours; and sometimes extra inks of special colours are used, and this can reduce metamerism further in some cases.

In the colorant industries, metamerism can arise when colour matches are required between different types of material. For instance, in fashion design, the same colour may be required in a dress material, in plastic buttons, and in leather shoes; or, in the automobile industry, the interior colours of the seat material, the carpet, the paint work, and the plastic trimmings may be required to match. In these applications, the use of dyes and pigments in different media precludes achieving the same spectral reflectance in all cases, but the colourists will seek to minimise the unavoidable differences by careful choice of dye or pigment mixtures.

Keeping metamerism to a minimum is thus an important consideration in the choice of colorants. Generally, of course, the smaller the differences in spectral power distribution the better. But, where differences are unavoidable, their effects should be evaluated using the appropriate indices of metamerism as discussed in Sections 6.6, 6.7, 6.8, and 6.9. Experience results in those skilled in the art knowing to some extent what differences in spectral power distribution are least harmful (Pinney and DeMarsh, 1963).

6.5 DEGREE OF METAMERISM

In Figure 6.4 are shown the spectral reflectance curves of three pairs of stimuli that are metameric in CIE Illuminant SC for the CIE 1931 Standard Colorimetric Observer. It is intuitive that the greater the difference between the spectral power distributions of a metameric pair of stimuli, or between the spectral reflectance or transmittance curves of a

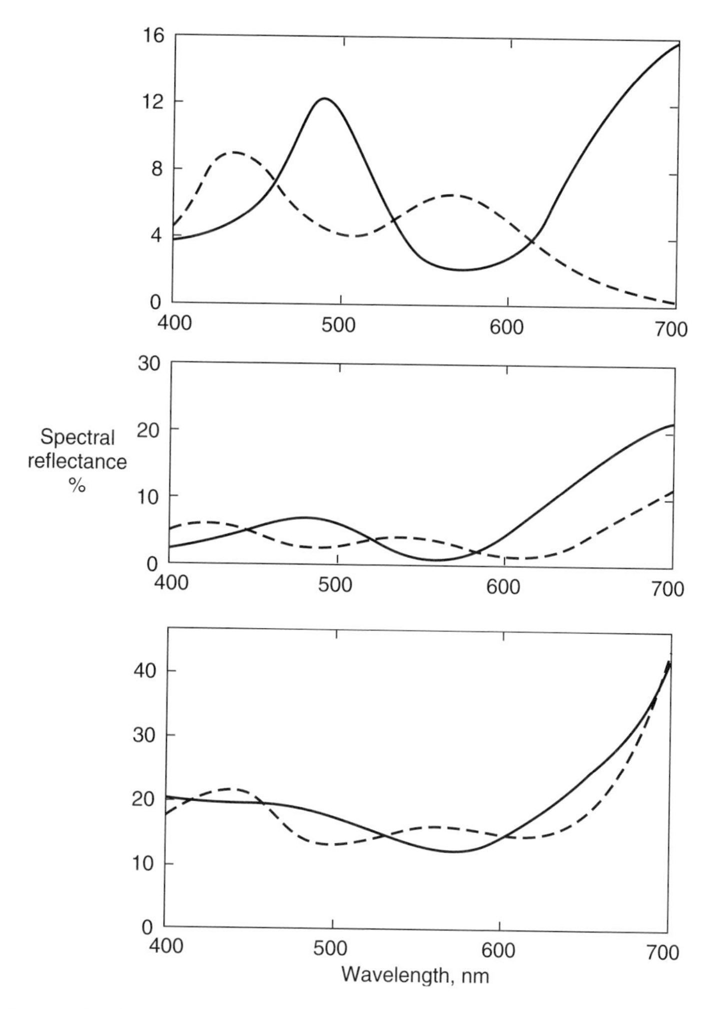

Figure 6.4 Spectral reflectances of three pairs of dyed fabric which are metameric in CIE Illuminant SC for the CIE 1931 Standard Colorimetric Observer

pair of samples that are metameric for a given illuminant, then the greater will be the colour difference between the pair when there is a change in observer, field size, or illuminant. Although this is generally the case, the concept of a general index of metamerism that would represent this phenomenon quantitatively is not soundly based. If the differences in spectral composition are averaged throughout the spectrum (Nimeroff and Yurow, 1965), then the decreasing importance of differences as the two ends of the spectrum are approached is ignored. The differences therefore need to be weighted appropriately (Nimeroff, 1968); but the correct weightings must depend on the relative importance of different parts of the spectrum, and this in turn is governed by whether the change is in observer, in field size, or in illuminant, and, in the latter case, which particular illuminants are involved. Instead of seeking a general index of metamerism, it is therefore more justifiable to use different indices according to whether the change is in observer, in field size, or in illuminant. These special indices of metamerism will now be considered.

6.6 INDEX OF METAMERISM FOR CHANGE OF ILLUMINANT

The CIE has recommended that the degree of metamerism for changes of illuminant be evaluated by calculating an Illuminant Metamerism Index, M, consisting of the size of the colour difference between a metameric pair caused by substituting, in place of a reference illuminant, a test illuminant having a different spectral composition. If the tristimulus values of the two samples, 1 and 2, under the test illuminant are $X_{1,t}$, $Y_{1,t}$, $Z_{1,t}$ and $X_{2,t}$, $Y_{2,t}$, $Z_{2,t}$, respectively, then the metamerism index is obtained by calculating the corresponding colour difference; this colour difference should preferably be calculated according to one of the recommended CIE colour-difference formulae, but whatever formula used should be stated (CIE, 1986a). It should be borne in mind that, as mentioned in Section 3.13, if a colour-difference formula is designed for one illuminant, it may not assess colour differences sufficiently accurately in another illuminant unless corrections are made for chromatic adaptation (methods for doing this will be discussed in Section 6.13) (Berns and Billmeyer, 1983).

The preferred reference illuminant is Standard Illuminant D65; if a different reference illuminant is used this should be stated. Suitable test illuminants include CIE Standard Illuminant A to represent tungsten light, and the illuminants listed in Appendix 5.3 to represent fluorescent lamps, particularly F2, F7, and F11. The most appropriate choice of test illuminant depends on the application, and it may be useful to determine the metamerism index with respect to several test illuminants. The specific test illuminant must be identified as a subscript to M, such as M_A, M_{F1}, etc.

A worked example is given at the end of the chapter.

6.7 INDEX OF METAMERISM FOR CHANGE OF OBSERVER

The CIE has recommended that the degree of metamerism for changes of observer be evaluated by calculating an Observer Metamerism Index, M_2 or M_{10}, consisting of the size of the colour difference between a metameric pair caused by substituting, in the

Table 6.1 Modifications of CIE colour-matching functions to obtain a standard deviate observer

Wavelength nm	$\Delta\bar{x}$	$\Delta\bar{y}$	$\Delta\bar{z}$	Wavelength nm	$\Delta\bar{x}$	$\Delta\bar{y}$	$\Delta\bar{z}$
380	−0.0001	0.0000	−0.0002	580	−0.0600	−0.0126	−0.0013
385	−0.0003	0.0000	−0.0010	585	−0.0637	−0.0162	−0.0011
390	−0.0009	−0.0001	−0.0036	590	−0.0656	−0.0196	−0.0009
395	−0.0026	−0.0004	−0.0110	595	−0.0638	−0.0199	−0.0008
400	−0.0069	−0.0009	−0.0294	600	−0.0595	−0.0187	−0.0006
405	−0.0134	−0.0015	−0.0558	605	−0.0530	−0.0170	−0.0005
410	−0.0197	−0.0019	−0.0820	610	−0.0448	−0.0145	−0.0004
415	−0.0248	−0.0022	−0.1030	615	−0.0346	−0.0112	0.0000
420	−0.0276	−0.0021	−0.1140	620	−0.0242	−0.0077	0.0002
425	−0.0263	−0.0017	−0.1079	625	−0.0155	−0.0048	0.0000
430	−0.0216	−0.0009	−0.0872	630	−0.0085	−0.0025	−0.0002
435	−0.0122	0.0005	−0.0455	635	−0.0044	−0.0012	−0.0002
440	−0.0021	0.0015	−0.0027	640	−0.0019	−0.0006	0.0000
445	0.0036	0.0008	0.0171	645	−0.0001	0.0000	0.0000
450	0.0092	−0.0003	0.0342	650	0.0010	0.0003	0.0000
455	0.0186	−0.0005	0.0703	655	0.0016	0.0005	0.0000
460	0.0263	−0.0011	0.0976	660	0.0019	0.0006	0.0000
465	0.0256	−0.0036	0.0859	665	0.0019	0.0006	0.0000
470	0.0225	−0.0060	0.0641	670	0.0017	0.0006	0.0000
475	0.0214	−0.0065	0.0547	675	0.0013	0.0005	0.0000
480	0.0205	−0.0060	0.0475	680	0.0009	0.0003	0.0000
485	0.0197	−0.0045	0.0397	685	0.0006	0.0002	0.0000
490	0.0187	−0.0031	0.0319	690	0.0004	0.0001	0.0000
495	0.0167	−0.0037	0.0228	695	0.0003	0.0001	0.0000
500	0.0146	−0.0047	0.0150	700	0.0002	0.0001	0.0000
505	0.0133	−0.0059	0.0117	705	0.0001	0.0000	0.0000
510	0.0118	−0.0060	0.0096	710	0.0001	0.0000	0.0000
515	0.0094	−0.0025	0.0062	715	0.0001	0.0000	0.0000
520	0.0061	0.0010	0.0029	720	0.0000	0.0000	0.0000
525	0.0017	0.0005	0.0005	725	0.0000	0.0000	0.0000
530	−0.0033	−0.0011	−0.0012	730	0.0000	0.0000	0.0000
535	−0.0085	−0.0020	−0.0020	735	0.0000	0.0000	0.0000
540	−0.0139	−0.0028	−0.0022	740	0.0000	0.0000	0.0000
545	−0.0194	−0.0039	−0.0024	745	0.0000	0.0000	0.0000
550	−0.2470	−0.0044	−0.0024	750	0.0000	0.0000	0.0000
555	−0.0286	−0.0027	−0.0021	755	0.0000	0.0000	0.0000
560	−0.0334	−0.0022	−0.0017	760	0.0000	0.0000	0.0000
565	−0.0426	−0.0073	−0.0015	765	0.0000	0.0000	0.0000
570	−0.0517	−0.0127	−0.0014	770	0.0000	0.0000	0.0000
575	−0.0566	−0.0129	−0.0013	775	0.0000	0.0000	0.0000
				780	0.0000	0.0000	0.0000

place of a reference observer, a standard deviate observer (SDO) having different spectral sensitivities (CIE, 1989). The reference observer can be either the CIE 1931 Standard Colorimetric Observer (the 2° Observer) or the CIE 1964 Standard Colorimetric Observer (the 10° Observer); the symbols M_2 and M_{10}, respectively, can be used to denote which observer is being used. The Standard Deviate Observer is obtained by changing the CIE colour-matching functions of the reference observer by applying the modifications given in Table 6.1. The colour difference should be calculated according to one of the recommended CIE colour-difference formulae, but whatever formula used this should be stated.

The use of the Standard Deviate Observer is intended to generate values of M_2 and M_{10} that are typical of the colour differences that occur when matches are made by different real observers whose colour vision is classified as normal (that is, those who are not colour deficient, see Section 1.10). (Ohta, 1985; Nayatani, Hashimoto, Takahama and Sobagaki, 1985.) Some evidence indicates that the CIE Standard Deviate Observer underestimates the range over which real observers vary (Fairchild and North, 1993).

A worked example is given at the end of the chapter.

6.8 INDEX OF METAMERISM FOR CHANGE OF FIELD SIZE

Although the CIE has not recommended an index of metamerism for a change of field size, the availability of the two CIE Standard Colorimetric Observers makes such assessments possible for changes between 2° and 10°. The 1931 Observer defines colour matches for 2° fields, and the 1964 Observer for 10° fields. Hence, for a pair of samples that is a metameric match for the 1931 Observer, their colour difference when evaluated using the 1964 Observer can be regarded as a measure of their metamerism for a change in field size. A similar assessment can be made in the reverse situation, where the samples are a metameric match for the 1964 Observer. In either case, the colour difference can be calculated using one of the recommended CIE colour-difference formulae, but whatever formula used should be stated.

A worked example is given at the end of the chapter.

6.9 COLOUR MATCHES AND GEOMETRY OF ILLUMINATION AND MEASUREMENT

If two reflecting samples match one another in one set of conditions of illuminating and measurement geometry, they will only continue to match in a different geometry if their surface characteristics are such that their gloss properties are the same. But if, for instance, one sample is glossy and the other matt, then a change in geometry must be expected to result in a breakdown of the match. For example, if the two colours match in diffuse illumination and normal viewing (d:0), then changing to 45° illumination and normal viewing (45:0), will tend to make the glossy sample look darker and (unless it is a neutral grey) more saturated. Changes of this type occur even if, in the matching geometry, the samples have identical spectral reflectances, and are therefore a spectral, rather than a metameric, match. Hence, although the term geometric metamerism has sometimes been used for this phenomenon, it is not strictly correct, because, in colorimetry, the term metamerism implies a difference in spectral composition. However, whatever term is used to denote it, the effect of geometry on the colour matching of samples of different gloss properties can be very considerable, sometimes causing a good match to become a very obvious mismatch. Once again, the resulting colour difference can be calculated using one of the recommended CIE colour-difference formulae, but whatever formula used should be stated.

These effects are of considerable importance in the paint industries, particularly, because of the widespread use of metallic and pearlescent pigments in automotive finishes; the full evaluation of such coatings requires instruments capable of measuring spectral distributions for various angles of illumination and measurement (gonio-spectrophotometers). Some multi-geometry instruments are arranged to illuminate always at 45°, but to measure at angles that differ from the illuminating direction by various amounts, such as 105°, 75°, 45° (that is, normal), 20°, and 10°.

6.10 CORRECTING FOR INEQUALITIES OF TRISTIMULUS VALUES

The two samples whose metamerism index is to be evaluated may fail to have exactly the same tristimulus values in the reference condition, a phenomenon sometimes referred to as *paramerism* (Kuehni, 1983); in this case a suitable means of correcting for this should be used and explained. Three different methods of correction have been suggested.

The first method is to multiply the X tristimulus value of sample 2 in the test condition, by X_{1r}/X_{2r}, where X_{1r} and X_{2r} are the X tristimulus values for samples 1 and 2, respectively, in the reference condition, with analogous corrections for the Y and Z tristimulus values of sample 2 in the test condition (Brockes, 1970; McLaren, 1986).

The second method is similar to the first but uses additive, instead of multiplicative, corrections; thus the X tristimulus value of sample 2 in the test condition has added to it $X_{1r} - X_{2r}$, with analogous corrections for the Y and Z tristimulus values (Brockes, 1970).

The third method involves changing the spectral power distribution of sample 2 so as to obtain an exact match in the reference condition, but making this change so as to retain, as far as possible, the differences in spectral power distribution that are attributable to metamerism. The procedure involved is as follows (Fairman, 1987):

Step 1. Regard the three colour-matching functions of the Standard Observer being used as three spectral power distributions $P_X(\lambda)$, $P_Y(\lambda)$, and $P_Z(\lambda)$.

Step 2. Find the quantities of $P_X(\lambda)$, $P_Y(\lambda)$, and $P_Z(\lambda)$ that, when added together, produce a spectral power distribution, $M_1(\lambda)$, that is an exact metameric match to the spectral power distribution, $P_1(\lambda)$, of sample 1 in the reference condition.

Step 3. Find the quantities of $P_X(\lambda)$, $P_Y(\lambda)$, and $P_Z(\lambda)$, that, when added together, produce a spectral power distribution, $M_2(\lambda)$, that is an exact metameric match to the spectral power distribution, $P_2(\lambda)$, of sample 2 in the reference condition.

Step 4. From $P_2(\lambda)$ subtract $M_2(\lambda)$ to obtain the spectral power distribution:

$$P_2(\lambda) - M_2(\lambda)$$

This spectral power distribution will have both positive and negative values, and, because $P_2(\lambda)$ and $M_2(\lambda)$ are an exact metameric match, its tristimulus values will all be zero; hence it may be thought of as a 'metameric black'.

Step 5. To the spectral power distribution, $M_1(\lambda)$, found in Step 2, add that of the metameric black obtained in Step 4, to obtain a corrected spectral power distribution, $P_{2C}(\lambda)$, for sample 2 in the reference condition:

$$P_{2C}(\lambda) = M_1(\lambda) + P_2(\lambda) - M_2(\lambda)$$

Because $M_1(\lambda)$ matches $P_1(\lambda)$ exactly, and $P_2(\lambda) - M_2(\lambda)$ has zero tristimulus values, $P_{2C}(\lambda)$ matches $P_1(\lambda)$ exactly.

Step 6. Use $P_{2C}(\lambda)$ instead of $P_2(\lambda)$ in determining the metamerism index.

6.11 TERMS USED IN CONNECTION WITH METAMERISM

The term for metamerism included in the 4th edition of the CIE International Lighting Vocabulary (CIE, 1987) is as follows:

metameric colour stimuli, metamers

Spectrally different colour stimuli that have the same tristimulus values in a specified colorimetric system. The corresponding property is called metamerism.

The following additional terms or meanings, although not endorsed by the CIE, are also found in the literature (Fairman, 1986):

(perceived) metameric colour stimuli, (perceived) metamers

Spectrally different colour stimuli that are a visual match for a particular real observer under specified viewing conditions. The corresponding property is called *perceived metamerism*. (When the meaning is clear from the context the adjective perceived can be omitted.)

parameric colour stimuli, paramers

Spectrally different colour stimuli that have nearly the same tristimulus values. The corresponding property is called *paramerism*.

indices of metamerism potential

Indices indicating the degree to which two metameric specimens may develop metamerism, derived solely from their different spectral characteristics.

special indices of metamerism

Indices of degree of metamerism associated with specific changes in illuminating or viewing conditions, such as change of illuminant, change of observer, or change of field size.

6.12 COLOUR INCONSTANCY

As mentioned in Section 1.8, the human visual system is very good at compensating for changes in the level and colour of illuminants (*adaptive colour shift*), so that there is a tendency for the appearance of colours to remain approximately constant over a wide range of conditions. Significant departures from colour constancy do occur however, as illustrated by the following situations.

First, metameric matches usually break down when the illuminant is changed. So if, in a scene, there are two colours that are spectrally different but look alike (a metameric pair), when the illuminant is changed, the two colours will then usually look different from one another; hence, even if one of them still has the same appearance, the other one must look different, and colour constancy has then not occurred for that colour. (Although Foster, Marin-Franch, Nascimento, and Amano, 2008, have shown that metamerism occurs to only a limited extent in nature, with man-made objects it occurs sufficiently often to be an important factor in the colorant industries.)

A second reason is that the spectral composition of some colours makes it impossible for them to exhibit colour constancy. This is particularly true of purple colours, which always look redder in warm lighting, such as tungsten light, than in cool lighting such as daylight. The reasons for this are illustrated in Figure 6.5.

In this figure, to make the argument uncomplicated, two simple illuminants, and four simple colours, are used. One of the illuminants (a 'daylight' type, D) has constant power throughout the spectrum, and the other (a 'tungsten' type, A) has power that increases steadily throughout the spectrum. Of the four colours, one has a uniform reflectance throughout the spectrum (a 'white'), another (a 'red') has this only in a band at the red end, another (a 'blue') only at the blue end, and the fourth (a 'purple') at both the red and blue ends;

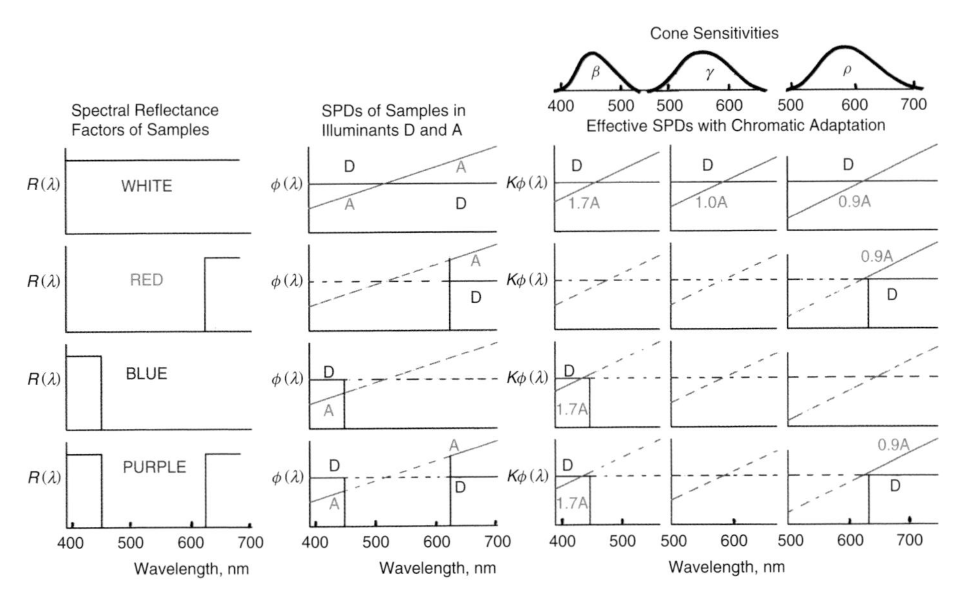

Figure 6.5 Diagrammatic explanation of why purple colours fail to show colour constancy

this is shown in the left-hand column of the figure. In the next column, the spectral power distributions of the four colours are shown for the two illuminants.

In the group of three columns at the right-hand side of the figure, the top (D and A) row shows the factor by which each of the three cone responses, ρ, γ, and β, would have to be altered for the change from illuminant D to A to result in the white exhibiting colour constancy; the β cones would have to increase their response by a factor of 1.7, and the ρ cones decrease by a factor of 0.9, while the γ cones remained unchanged. This is illustrated by the new positions of the orange lines (Effective spds with Chromatic Adaptation) in which the spd is shifted up in the case of the β cones and down in the case of the ρ cones, with no shift in the case of the γ cones.

In the next row, the effects of these shifted spds is shown on the red colour, which now gives a larger ρ response in the A illuminant than in the D illuminant (see the right-hand diagram), indicating correctly that red colours tend to look lighter and more colourful in tungsten light than in daylight. The next row shows the effect on the blue colour, which now gives a smaller β cone response in the A illuminant than in the D illuminant (see the left-hand diagram), indicating correctly that blue colours tend to look darker and less colourful in tungsten light than in daylight. Finally, the last row shows the effect on the purple colour, which now gives a larger ρ cone response and a smaller β cone response in the A illuminant than in the D illuminant (see the left-hand and the right-hand diagrams), indicating correctly that purple colours tend to look redder in tungsten light than in daylight.

The facts that metameric colours cannot always show colour constancy, and that red, blue, and purple, colours change their appearance when a bluish illuminant is changed to a reddish illuminant, illustrates that colour constancy can only be an approximate phenomenon.

These and other phenomena that can cause colour inconstancy may be summarised as follows.

As the illumination level falls:

1. There is a progressive reduction in brightness and colourfulness.

2. Below about 0.1 lux, complete loss of colour vision occurs.

As the colour of the illumination becomes increasingly different from average daylight:

3. There is progressive loss of colour compensation (for example, objects look yellower in candle light).

4. There are obvious changes in some colours (for example, in many purples).

5. There are large changes in most colours if the illuminant contains only some parts of the spectrum (for example, with low-pressure sodium street lights, which contain only yellow light).

As the field size is reduced:

6. There is progressive reduction in adaptation (for example, with images projected in dark auditoria, or with pictures printed on coloured paper).

As the stimulus size is reduced:

7. Some colours become increasingly difficult to discriminate (for example, yellow characters printed on white paper that are legible in daylight can become illegible in tungsten or candle light).

6.13 CHROMATIC ADAPTATION TRANSFORMS

There are many factors that affect the appearance of colours, and, to allow for all or most of them, a model of colour appearance is required; this will be discussed in Chapter 15. But, when the only change is in the colour of the illuminant, it is convenient to use a chromatic adaptation transform (CAT); these transforms define corresponding colour stimuli, which are pairs of colour stimuli that look alike when one is seen under one set of adaptation conditions, and the other is seen in a different set (Helson, Judd, and Warren, 1952; Breneman, 1987; Braun and Fairchild, 1996). The way in which a CAT may be used in calculating an index of metamerism for change of illuminant is as follows: if the tristimulus values of the metameric pair in the test illuminant are $X_{1,t}$, $Y_{1,t}$, $Z_{1,t}$ and $X_{2,t}$, $Y_{2,t}$, $Z_{2,t}$, then the CAT is used to transform all these values into those of the corresponding colours in a daylight type illuminant; the index is then the colour difference between these corresponding colours as evaluated by a colour-difference formula intended for use with colours in a daylight illuminant (Mori, Sobagaki, Komatsubara, and Ikeda, 1991).

Chromatic adaptation transforms can be divided into three groups: those that occur as parts of uniform colour spaces; those that depend on normalisation of cone responses; and those that form part of a colour appearance model, CAM (Luo, Clarke, Rhodes, Schappo, Scrivener, and Tait, 1991; Luo, Gao, Rhodes, Xin, Clarke, and Scrivener, 1993; Luo, Lo, Kuo, 1996; Kuo, Luo, and Bez, 1995).

Examples of the first group are the CIELUV and CIELAB systems (see Section 3.9). These systems were intended for application to conditions of adaptation to fields of chromaticity not too different from daylight. But they are sometimes used for illuminants of other chromaticities, and, because (as mentioned in Section 3.14) the formulae in the systems are normalised for the reference white, they include an allowance for chromatic adaptation that results in the reference white (and greys of the same chromaticity) always having C^* equal to zero, related adjustments being made to the values of C^* and h for other colours. These formulae therefore contain chromatic adaptation transforms, but they are only approximately valid, the approximation being much worse for CIELUV than for CIELAB.

An example of the second group is one that is often referred to as a Von Kries transformation, after the name of its originator (Von Kries, 1911). In this procedure, it is assumed that chromatic adaptation can be represented by the cone responses being multiplied (or divided) by factors that result in a reference white giving rise to the same signals in all states of adaptation. However, although Von Kries transformations are often very useful, they are not always accurate.

The third group uses parts of models of colour appearance, and usually involves more complicated formulae, such as the CAT97 transform (Hunt, 1998). Another example is given in Chapter 15; another, based on the work of Takahama, Sobagaki, and Nayatani, has been published by the CIE (Takahama, Sobagaki and Nayatani, 1984; CIE, 1986b); another is based on work by Lam and Rigg (Lam, 1985; Luo, Lo and Kuo, 1996) and is known as the Bradford transform. An elaboration of the Bradford transform, in which allowance is made for the degree of chromatic adaptation varying with the level of luminance, with the nature of the surround, and with the extent of cognitive factors (Hunt and Luo, 1998), has been found to give results in the best agreement with experimental data on corresponding colours; this transform, designated here as the CAT02 transform, is described in Section 6.15.

In these transforms all the colorimetric values can be either for the CIE 1931 Standard Colorimetric Observer or for the CIE 1964 Standard Colorimetric Observer, but it must be made clear which Observer has been used (and, if the latter, all the symbols for colorimetric measures must have a subscript 10).

In some situations the colour of the adapting illuminant is not simple to define. An example is when an observer is using a self-luminous display, as on a computer or lap-top, and comparing the appearances of its displayed colours to those of some reflecting colours illuminated by room light; the colour of the white point on the display is typically the same as that of D65, but the room may be lit, for example, by fluorescent lamps whose light has a correlated colour temperature of about 4000 K. So is the colour of the adapting illuminant D65 or 4000 K? Experiments have shown that the adaptation, under such circumstances, is usually about 60% to the display and 40% to the room light (Katoh, Nakabayashi, Ito, and Ohno, 1998; Oskoui and Pirotta, 1999).

6.14 THE VON KRIES TRANSFORM

The Von Kries transform may be expressed by saying that, if a stimulus gives rise to cone responses, ρ, γ, β, the visual signals will depend on

$$\rho/\rho_{\mathrm{w}} \quad \gamma/\gamma_{\mathrm{w}} \quad \beta/\beta_{\mathrm{w}}$$

where ρ_{w}, γ_{w}, β_{w} are the cone responses for the reference white. It is convenient to apply this procedure relative to a reference state of adaptation; in this reference state, the cone responses for the white are distinguished by the subscript wr, thus ρ_{wr}, γ_{wr}, β_{wr}. It then follows that, for a stimulus in the reference state of adaptation to have the same colour appearance as another stimulus in the state of adaptation considered, it is necessary that its cone responses ρ_{c}, γ_{c}, β_{c} be such that

$$\rho/\rho_{\mathrm{w}} = \rho_{\mathrm{c}}/\rho_{\mathrm{wr}} \quad \gamma/\gamma_{\mathrm{w}} = \gamma_{\mathrm{c}}/\gamma_{\mathrm{wr}} \quad \beta/\beta_{\mathrm{w}} = \beta_{\mathrm{c}}/\beta_{\mathrm{wr}}$$

It follows that:

$$\rho_{\mathrm{c}} = (\rho_{\mathrm{wr}}/\rho_{\mathrm{w}})\rho \quad \gamma_{\mathrm{c}} = (\gamma_{\mathrm{wr}}/\gamma_{\mathrm{w}})\gamma \quad \beta_{\mathrm{c}} = (\beta_{\mathrm{wr}}/\beta_{\mathrm{w}})\beta$$

If, then, a stimulus has tristimulus values X, Y, Z, in a state of adaptation such that the reference white has tristimulus values $X_{\mathrm{w}}, Y_{\mathrm{w}}, Z_{\mathrm{w}}$, it is possible to calculate the corresponding colour stimulus, $X_{\mathrm{c}}, Y_{\mathrm{c}}, Z_{\mathrm{c}}$, that has the same appearance in the reference state. If this reference state is a phase of daylight, then the CIELUV and CIELAB formulae can be applied to X_c, Y_c, Z_c in the usual way. To calculate X_c, Y_c, Z_c it is necessary to know the tristimulus values X_{wr}, Y_{wr}, Z_{wr} of the reference white in the reference state, and to be able to transform X, Y, Z tristimulus values to ρ, γ, β cone responses, and vice versa. These transformations are achieved by using a set of transformation equations of which the following (normalised for a D65 white) are an example:

$$\rho = 0.40024X + 0.70760Y - 0.08081Z$$

$$\gamma = -0.22630 \times +1.16532Y + 0.04570Z$$

$$\beta = 0.91822Z$$

the reverse equations being:

$$X = 1.85995\rho - 1.12939\gamma + 0.21990\beta$$

$$Y = 0.36119\rho + 0.63881\gamma$$

$$Z = 1.08906\beta.$$

The steps involved in this procedure are then as follows:

Step 1. From X_{wr}, Y_{wr}, Z_{wr} calculate ρ_{wr}, γ_{wr}, β_{wr}.

Step 2. From X_w, Y_w, Z_w calculate ρ_w, γ_w, β_w.

Step 3. Calculate ρ_{wr}/ρ_w, γ_{wr}/γ_w, β_{wr}/β_w.

Step 4. For the test colour, from X, Y, Z, calculate ρ, γ, β.

Step 5. Calculate $\rho_c = (\rho_{wr}/\rho_w)\rho$, $\gamma_c = (\gamma_{wr}/\gamma_w)\gamma$, $\beta_c = (\beta_{wr}/\beta_w)\beta$.

Step 6. From ρ_c, γ_c, β_c calculate X_c, Y_c, Z_c.

Step 7. Use X_c, Y_c, Z_c in the chosen colour-difference formula.

6.15 THE CAT02 TRANSFORM

This transform differs from the Von Kries transform in using a set of responses, R, G, B, as shown in Figure 6.6, which, unlike typical cone responses, have some negative spectral values. The CAT02 transformation forms part of the CIECAM02 colour appearance model published by the CIE (CIE, 2004), but it has not been published separately by the CIE as a chromatic adaptation transform, and it is not, therefore, designated CIECAT02.

CAT02 is defined as follows for transformations between two illuminants A and B.

Step 1. Starting data:

Sample in first illuminant A:	X_A	Y_A	Z_A
White in first illuminant A:	X_{WA}	Y_{WA}	Z_{WA}
White in second illuminant B:	X_{WB}	Y_{WB}	Z_{WB}
Luminance of adapting fields (cd m^{-2})	L_A		

L_A is required for calculating a factor, D, which allows for the degree of chromatic adaptation taking place. L_A can usually be taken as $L_W Y_b/100$ where Y_b is the Y value of the background and L_W is the luminance in $\text{cd}\,\text{m}^{-2}$ of a white in the illuminant. The luminances are assumed to be the same for both illuminants; if there is a significant difference, a Colour Appearance Model, such as CIECAM02 to be described in Chapter 15, should be used. The value adopted for L_A is not critical, and, if the value of L_A is not known, D can be set equal to 0.95 to represent typical viewing conditions for reflecting colours. The whites, normally have the same chromaticity as the illuminant. If any different white is used, its details must be clearly stated.

Transformed data to be obtained:

Test sample corresponding colour in second illuminant: X_{AC} Y_{AC} Z_{AC}
Test white corresponding colour in second illuminant: X_{WAC} Y_{WAC} Z_{WAC}

Step 2. Calculate the degree of adaptation, D:

$$D = F\left[1 - \left(\frac{1}{3.6}\right)e^{\left(\frac{-L_A - 42}{92}\right)}\right]$$

where $F = 1$ for samples seen with a surround of luminance similar to the average luminance of the sample array; and $F = 0.9$ for samples seen with a dim surround, and $F = 0.8$ for samples seen with a dark surround. But if the chromatic adaptation is complete (illuminant colour completely discounted) put $D = 1.0$; or if there is no chromatic adaptation, put $D = 0$; or if there is partial discounting put $D = 0.5F\left[1 - \left(\frac{1}{3.6}\right)e^{\left(\frac{-L_A - 42}{92}\right)}\right]$. If D is greater than 1 or less than 0, set it to 1 or 0 respectively.

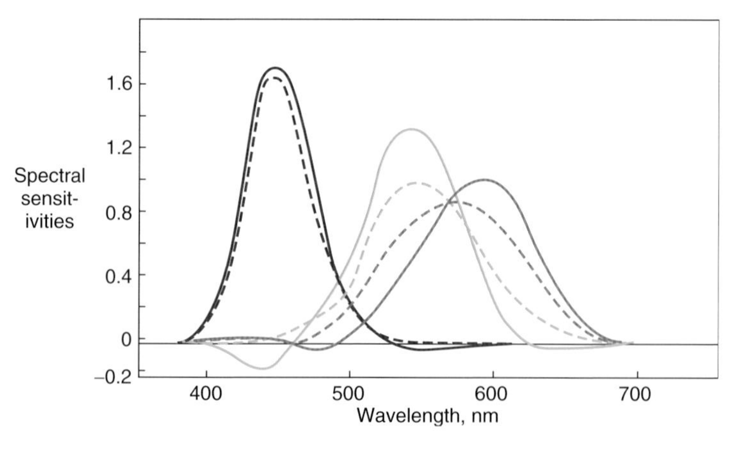

Figure 6.6 Full lines: spectral sensitivities used in the CAT02 chromatic adaptation transform (Lam, 1985, CIE, 2004). Broken lines: spectral sensitivities for cones, similar to those found by Estévez (Estévez, 1979). Both sets of curves are linear transformations of the $\bar{x}(\lambda)$, $\bar{y}(\lambda)$, and $\bar{z}(\lambda)$ functions

Step 3. For the sample and the white in the first illuminant and for the white in the second illuminant, calculate their RGB values R_A, G_A, B_A, R_{WA}, G_{WA}, B_{WA}, R_{WB}, G_{WB}, B_{WB}, using the equations:

$$R = 0.7328X + 0.4296Y - 0.1624Z$$

$$G = -0.7036X + 1.6975Y + 0.0061Z$$

$$B = 0.0030X + 0.0136Y + 0.9834Z$$

Step 4. Calculate the factors D_r, D_g, D_b:

$$D_r = (Y_{WA}/Y_{WB})D(R_{WB}/R_{WA}) + 1 - D$$

$$D_g = (Y_{WA}/Y_{WB})D(G_{WB}/G_{WA}) + 1 - D$$

$$D_b = (Y_{WA}/Y_{WB})D(B_{WB}/B_{WA}) + 1 - D$$

Step 5. Calculate the corresponding colours for the sample in the second illuminant:

$$R_{AC} = D_r.R_A$$

$$G_{AC} = D_g.G_A$$

$$B_{AC} = D_b.B_A$$

Similarly from R_{WA}, G_{WA}, B_{WA}, obtain R_{WAC}, G_{WAC}, B_{WAC}.

Step 6. Calculate the XYZ tristimulus values, X_{AC}, Y_{AC}, Z_{AC} for the corresponding colour for the sample in the second illuminant:

$$X_{AC} = 1.096124R_{AC} - 0.278869G_{AC} + 0.182745B_{AC}$$

$$Y_{AC} = 0.454369R_{AC} + 0.473533G_{AC} + 0.072098B_{AC}$$

$$Z_{AC} = -0.009628R_{AC} - 0.005698G_{AC} + 1.015326B_{AC}$$

Similarly obtain X_{WAC}, Y_{WAC}, Z_{WAC}.

The procedure for operating CAT02 in reverse, is as follows:

Step R1. From X_{AC}, Y_{AC}, Z_{AC} obtain R_{AC}, G_{AC}, B_{AC} using the same equations as in Step 3 above.

Step R2. From R_{AC}, G_{AC}, B_{AC} obtain

$$R_A = R_{AC}/D_r$$

$$G_A = G_{AC}/D_g$$

$$B_A = B_{AC}/D_b$$

Step R3. From R_A, G_A, B_A obtain X_A, Y_A, Z_A, using the same equations as in Step 6 above.

Worked examples for CAT02 are given at the end of this chapter.

6.16 A COLOUR INCONSTANCY INDEX

In various industries, particularly the colorant industries, an important issue is the degree to which colours change in appearance when the colour of the illuminant is changed. This can be checked visually by observing samples under various light sources, but this is not easy to do. If the sources are changed rapidly, the eye does not adapt properly. If enough time is allowed for full adaptation, it is difficult to remember the colour seen under the previous light source. An instrumental method is therefore desirable, and is essential in colorant-recipe prediction where many possible dye combinations might be considered, but no samples made for visual inspection. An instrumental method must be based on a chromatic adaptation transform. Such a transform allows the tristimulus values for corresponding colours to be calculated. If measurements have been made of the values of the X, Y, and Z tristimulus values, for a sample under a test illuminant, for example Standard Illuminant A, a chromatic adaptation transform can predict the values of X, Y, and Z, for a colour that looks the same under a reference illuminant, for instance D65. If the values for X, Y, and Z, for the sample under the reference illuminant are different from the values computed by the transform, then the magnitude of the colour difference, ΔE^*, can be used as a measure of the colour inconstancy of the sample when the illuminant is changed from the test to the reference.

The degree of colour inconstancy for a sample can be calculated in terms of a colour inconstancy index computed as follows.

Step 1. The tristimulus values, X, Y, Z, of a sample in a test illuminant are measured or computed, and the values X_r, Y_r, Z_r, of the sample in a reference illuminant are measured or computed similarly. The difference between these two sets of tristimulus values is the *illuminant colorimetric shift*.

Step 2. Using the CAT02 Chromatic Adaptation Transform with D set equal to 1.0, the tristimulus values, X, Y, Z, are used to compute the tristimulus values, X_c, Y_c, Z_c for the corresponding colour in the reference illuminant. The difference between these two sets of tristimulus values is the *adaptive colorimetric shift*.

Step 3. Using a suitable colour-difference formula, the colour difference, ΔE^*, defined by the difference between X_c, Y_c, Z_c, and X_r, Y_r, Z_r, is computed. This difference, ΔE^*, provides the Colour Inconstancy Index.

Notes

1. The X, Y, Z, tristimulus values used may be either for the CIE 1931 Standard Colorimetric Observer or for the CIE 1964 Standard Colorimetric Observer, but it must be made clear which Observer has been used (and, if the latter, all the symbols for colorimetric measures must have a subscript 10).
2. Both the illuminants must be specified, and, whenever possible, the reference illuminant should be D65.
3. It must be clearly stated which colour-difference formula has been used.

6.17 WORKED EXAMPLES

6.17.1 Index of metamerism for change of illuminant

The colour match below was generated by computing a three-peak stimulus that matched a non-selective grey. The two stimuli match for CIE Standard Illuminant D65, but not for CIE Standard Illuminant A. See Figure 6.7.

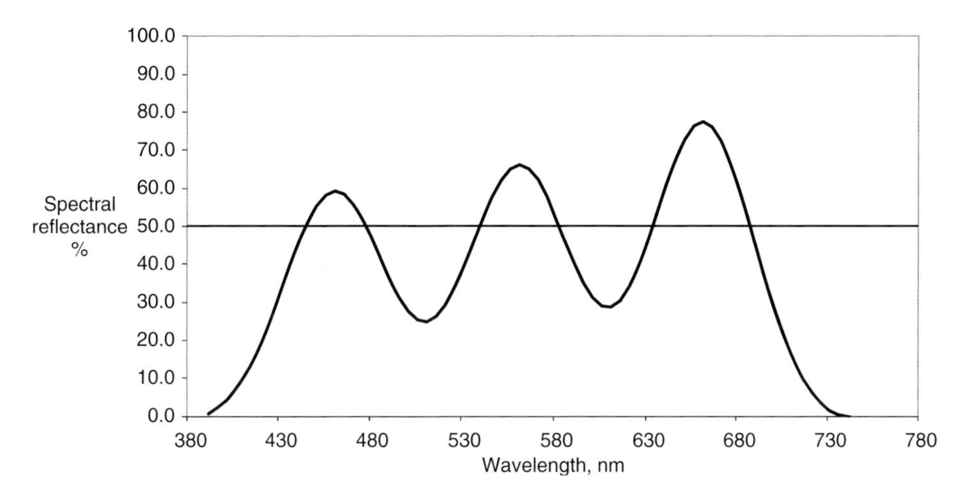

Figure 6.7 Spectral reflectances for two coloured samples that match in CIE Standard Illuminant D65 but not in CIE Standard Illuminant SA

Wave-length nm	$R1(\lambda)$	$R2(\lambda)$	Wave-length nm	$R1(\lambda)$	$R2(\lambda)$
380	2.8637	50.00	580	42.4021	50.00
385	4.3666	50.00	585	37.2372	50.00
390	6.4534	50.00	590	33.3440	50.00
395	9.2440	50.00	595	31.1338	50.00
400	12.8341	50.00	600	30.8673	50.00
405	17.2702	50.00	605	32.6356	50.00
410	22.5247	50.00	610	36.3510	50.00
415	28.4743	50.00	615	41.7475	50.00
420	34.8882	50.00	620	48.3927	50.00
425	41.4326	50.00	625	55.7151	50.00
430	47.6925	50.00	630	63.0479	50.00
435	53.2130	50.00	635	69.6902	50.00
440	57.5533	50.00	640	74.9799	50.00
445	60.3475	50.00	645	78.3690	50.00

(*Continued*)

(Continued)

Wave-length nm	$R1(\lambda)$	$R2(\lambda)$	Wave-length nm	$R1(\lambda)$	$R2(\lambda)$
450	61.3587	50.00	650	79.4893	50.00
455	60.5195	50.00	655	78.1970	50.00
460	57.9479	50.00	660	74.5854	50.00
465	53.9381	50.00	665	68.9651	50.00
470	48.9278	50.00	670	61.8126	50.00
475	43.4474	50.00	675	53.7004	50.00
480	38.0622	50.00	680	45.2187	50.00
485	33.3160	50.00	685	36.9058	50.00
490	29.6811	50.00	690	29.1946	50.00
495	27.5216	50.00	695	22.3841	50.00
500	27.0669	50.00	700	16.6344	50.00
505	28.3965	50.00	705	11.9813	50.00
510	31.4331	50.00	710	8.3643	50.00
515	35.9444	50.00	715	5.6596	50.00
520	41.5546	50.00	720	3.7116	50.00
525	47.7690	50.00	725	2.3593	50.00
530	54.0126	50.00	730	1.4535	50.00
535	59.6835	50.00	735	0.8679	50.00
540	64.2160	50.00	740	0.5023	50.00
545	67.1465	50.00	745	0.2818	50.00
550	68.1725	50.00	750	0.1532	50.00
555	67.1925	50.00	755	0.0807	50.00
560	64.3213	50.00	760	0.0412	50.00
565	59.8772	50.00	765	0.0204	50.00
570	54.3425	50.00	770	0.0098	50.00
575	48.3070	50.00	775	0.0046	50.00
			780	0.0021	50.00

	Sample 1 D65	Sample 2 D65	Sample 1 SA	Sample 2 SA
X	47.5215	47.5215	54.7442	54.9254
Y	50.0000	50.0000	50.2655	50.0000
Z	54.4395	54.4395	17.3405	17.7908
x	0.3127	0.3127	0.4474	0.4476
y	0.3290	0.3290	0.4108	0.4074
L*	76.0692	76.0692	76.2319	76.0693
a*	0.0000	0.0000	−1.1381	0.0000
b*	0.0000	0.0000	1.6312	0.0000
C*	0.0000	0.0000	1.9890	0.0000
h	224	180	125	158
Δ*E**		**0.0000**		**1.9956**

Thus the Illuminant Metamerism Index, M, has a value of 2.00 using the CIE 2 degree Standard Colorimetric Observer and the CIELAB colour-difference formula.

6.17.2 Index of metamerism for change of observer

This is the colorimetry of an NPL red tile calculated for 2 degree, 2 degree + SDO, 10 degree, and 10 degree + SDO. The illuminant is CIE Standard Illuminant D65. The values of the spectral reflectance factors are tabulated in Section 6.17.3 (See Figure 6.8).

Observer	Normal 2 degree	Deviate 2 degree	Normal 10 degree	Deviate 10 degree
X	20.8678	20.5886	19.5900	19.3227
Y	13.9896	13.9697	13.5851	13.5627
Z	7.8730	7.9950	7.7570	7.8666
X	0.4884	0.4838	0.4786	0.4742
Y	0.3274	0.3283	0.3319	0.3328
L^*	44.2181	44.1895	43.6320	43.5992
a^*	42.0793	42.5685	38.5588	38.8829
b^*	20.5017	20.4712	19.5044	19.4650
C^*	46.8080	47.2350	43.2112	43.4829
H	26	26	27	27
ΔL^*		0.0286		0.0328
Δa^*		−0.4892		−0.3241
Δb^*		0.0306		0.0394
ΔE^*		**0.4910**		**0.3281**

Thus the Observer Metamerism Index, M_2, has a value of 0.49 and the Observer Metamerism Index, M_{10}, has a value of 0.33, both using the CIELAB colour-difference formula.

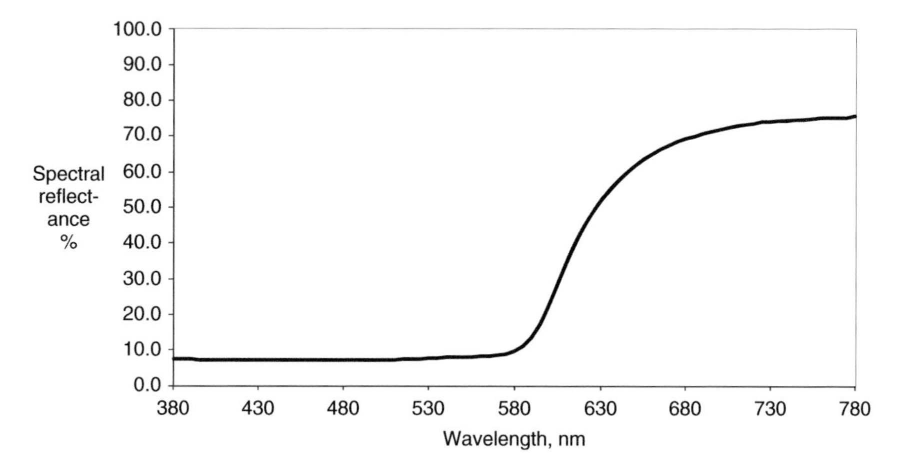

Figure 6.8 Spectral reflectance for a red coloured ceramic tile

6.17.3 Index of metamerism for change of field size

The colour differences between the 2 degree and the 10 degree observers are calculated for red, green and blue NPL tiles using CIE Standard Illuminant D65. See Figure 6.9.

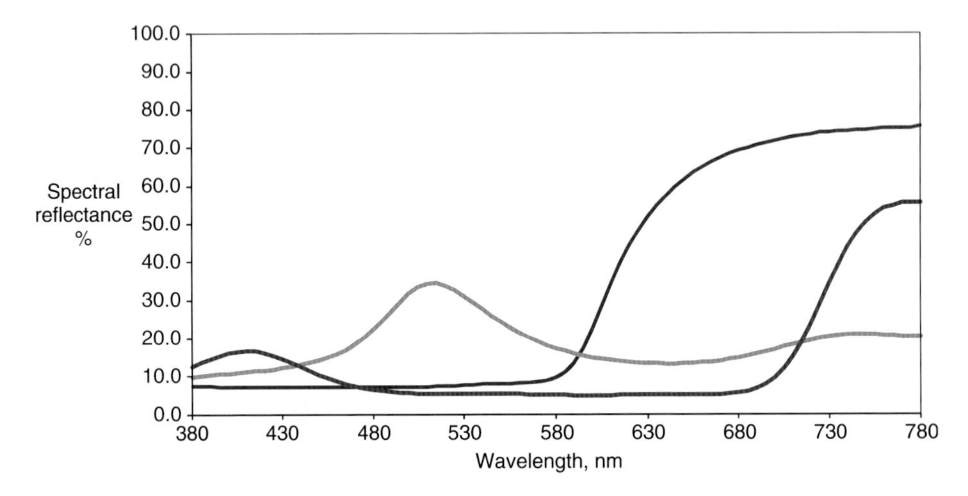

Figure 6.9 Spectral reflectances for a red, a green and a blue coloured ceramic tile

Wave-length nm	Red Tile $R(\lambda)$	Green Tile $R(\lambda)$	Blue Tile $R(\lambda)$	Wave-length nm	Red Tile $R(\lambda)$	Green Tile $R(\lambda)$	Blue Tile $R(\lambda)$
380	7.3649	9.6695	12.6639	580	9.6370	17.2861	5.0098
385	7.3553	9.9990	13.5769	585	11.0323	16.5611	4.9731
390	7.3293	10.2655	14.4832	590	13.4654	15.9001	4.9479
395	7.3048	10.4909	15.3281	595	17.3224	15.3488	4.9509
400	7.3014	10.7092	16.0367	600	22.5183	14.8760	4.9515
405	7.2942	10.9075	16.5229	605	28.3874	14.4598	4.9639
410	7.2806	11.1040	16.6880	610	34.2362	14.1174	4.9715
415	7.2798	11.3178	16.5398	615	39.6039	13.8379	4.9799
420	7.2679	11.5624	16.0935	620	44.3035	13.6168	4.9883
425	7.2596	11.8228	15.3938	625	48.2965	13.4409	4.9925
430	7.2529	12.1437	14.5237	630	51.7447	13.3159	4.9901
435	7.2454	12.5147	13.5444	635	54.7641	13.2360	4.9861
440	7.2449	12.9658	12.5478	640	57.3747	13.2109	4.9883
445	7.2475	13.5252	11.5091	645	59.6640	13.2174	4.9951
450	7.2211	14.1722	10.4648	650	61.7886	13.3040	5.0152
455	7.1931	14.9489	9.4669	655	63.5276	13.4075	5.0348
460	7.1744	15.8995	8.5712	660	65.0780	13.5691	5.0751
465	7.1736	17.0737	7.8100	665	66.3675	13.7687	5.1365
470	7.1829	18.5183	7.1829	670	67.4474	14.0235	5.2182

(Continued)

(*Continued*)

Wavelength nm	Red Tile $R(\lambda)$	Green Tile $R(\lambda)$	Blue Tile $R(\lambda)$	Wavelength nm	Red Tile $R(\lambda)$	Green Tile $R(\lambda)$	Blue Tile $R(\lambda)$
475	7.2051	20.2803	6.7148	675	68.5580	14.3596	5.3625
480	7.2293	22.3240	6.3742	680	69.3404	14.7344	5.5961
485	7.2398	24.6640	6.1104	685	70.0463	15.1958	5.9904
490	7.2311	27.1894	5.9008	690	70.7592	15.7167	6.6644
495	7.2253	29.6923	5.7466	695	71.2789	16.3027	7.7790
500	7.2317	31.8915	5.6278	700	71.7773	16.9468	9.5210
505	7.2486	33.5209	5.5215	705	72.3546	17.6390	12.0382
510	7.2795	34.4065	5.4439	710	72.9116	18.3513	15.3766
515	7.3398	34.4837	5.3959	715	73.3206	18.9916	19.5258
520	7.4252	33.8135	5.3693	720	73.6324	19.5336	24.2159
525	7.5382	32.6029	5.3635	725	73.9982	20.0103	29.3743
530	7.6668	31.0406	5.3739	730	74.1124	20.3332	34.4718
535	7.7767	29.3095	5.3967	735	74.3352	20.5800	39.3710
540	7.8761	27.5677	5.4335	740	74.4255	20.7213	43.7900
545	7.9630	25.8610	5.4654	745	74.5606	20.7566	47.4078
550	8.0357	24.2533	5.4643	750	74.7174	20.7279	50.3509
555	8.1097	22.7650	5.4113	755	74.7580	20.6937	52.3206
560	8.1891	21.4124	5.3284	760	75.1167	20.5820	53.9831
565	8.3084	20.1878	5.2292	765	75.0369	20.5291	54.7321
570	8.5140	19.1094	5.1354	770	75.2469	20.3401	55.3943
575	8.8960	18.1399	5.0653	775	75.2396	20.1738	55.3952
				780	75.6823	20.1210	55.5285

	Red Tile		Green Tile		Blue Tile	
Observer	2 degree	10 degree	2 degree	10 degree	2 degree	10 degree
X	20.8678	19.5900	15.9417	16.2292	5.8921	5.9577
Y	13.9896	13.5851	22.8142	22.8720	5.3738	5.5233
Z	7.8730	7.7570	18.1345	17.1960	10.9944	11.2851
x	0.4884	0.4786	0.2802	0.2883	0.2647	0.2617
y	0.3274	0.3319	0.4010	0.4063	0.2414	0.2426
L*	44.2181	43.6320	54.8804	54.9403	27.7743	28.1764
a*	42.0793	38.5588	−29.7736	−28.1587	9.2059	8.3657
b*	20.5017	19.5044	12.1677	13.6836	−17.6605	−18.2323
C*	46.8080	43.2112	32.1639	31.3074	19.9159	20.0599
h	26	27	158	154	298	295
Δ*L**		−0.5861		0.0598		0.4021
Δ*a**		−3.5205		1.6148		−0.8402
Δ*b**		−0.9973		1.5159		−0.5717
Δ*E**		**3.7057**		**2.2157**		**1.0929**

Thus the Field-size Metamerism Index has a value of 3.7 for the red tile, 2.2 for the green tile, and 1.1 for the blue tile.

6.17.4 CAT02 Worked Examples

Example 1

L_A	200				
F	1				
D	0.979987				
X_{WA}	Y_{WA}	Z_{WA}	98.88	90.00	32.03
X_{WB}	Y_{WB}	Z_{WB}	100.00	100.00	100.00
X_A	Y_A	Z_A	19.31	23.93	10.14
R_{WA}	G_{WA}	B_{WA}	105.92159	83.39842	33.01894
R_{WB}	G_{WB}	B_{WB}	100.00	100.00	100.00
R_A	G_A	B_A	22.78396	27.09651	10.35505
D_r	D_g	D_b	0.852693	1.077573	2.691171
R_{WAC}	G_{WAC}	B_{WAC}	90.31864	89.86788	88.85963
R_{AC}	G_{AC}	B_{AC}	19.42773	29.19847	27.86723
X_{WAC}	Y_{WAC}	Z_{WAC}	90.17772	90.00000	88.83984
X_{AC}	Y_{AC}	Z_{AC}	18.24525	24.66297	27.94090

Example 1 Reverse

X_{AC}	Y_{AC}	Z_{AC}	18.24525	24.66297	27.94090
R_{AC}	G_{AC}	B_{AC}	19.42773	29.19847	27.86723
R_A	G_A	B_A	22.78396	27.09651	10.35506
X_A	Y_A	Z_A	19.31000	23.93000	10.14000

Example 2

L_A	20	F	1	D	0.858414
X_{WA}	Y_{WA}	Z_{WA}	98.88	90.00	32.03
X_{WB}	Y_{WB}	Z_{WB}	76.03	80.00	87.11
X_A	Y_A	Z_A	19.31	23.93	10.14
R_{WA}	G_{WA}	B_{WA}	105.92159	83.39842	33.01894
R_{WB}	G_{WB}	B_{WB}	75.93612	82.83666	86.98006
R_A	G_A	B_A	22.78396	27.09651	10.35505
D_r	D_g	D_b	0.833916	1.100797	2.68552
R_{WAC}	G_{WAC}	B_{WAC}	88.32972	91.80472	88.67303
R_{AC}	G_{AC}	B_{AC}	18.99991	29.82776	27.8087
X_{WAC}	Y_{WAC}	Z_{WAC}	87.42339	90.00000	88.65849
X_{AC}	Y_{AC}	Z_{AC}	17.59012	24.76235	27.88201

Example 2 Reverse

X_{AC}	Y_{AC}	Z_{AC}	17.59012	24.76235	27.88201
R_{AC}	G_{AC}	B_{AC}	18.99991	29.82776	27.80871
R_A	G_A	B_A	22.78396	27.09651	10.35506
X_A	Y_A	Z_A	19.31000	23.93000	10.14000

REFERENCES

Berns, R.S., and Billmeyer, F.W., Proposed indices of metamerism with constant chromatic adaptation, *Color Res. Appl.*, **8**, 186–189 (1983).

Berns, R.S., and Kuehni, R.G., What determines crossover wavelengths of metameric pairs with three crossovers?, *Color Res. Appl.*, **15**, 23–28 (1990).

Breneman, E.J., Corresponding chromaticities for different states of adaptation to complex visual fields, *J. Opt. Soc. Amer. A*, **4**, 1115–1129 (1987).

Braun, K.M., and Fairchild, M.D., Psychophysical generation of matching images for cross- media color reproduction, *IS&T and SID's 4th Color Imaging Conference: Color Science and Engineering Systems, Technologies, and Applications Systems*, pp. 214-220, IS&T, Springfield, VA, USA (1996).

Brockes, A., Vergleich von berechneten Metametie-Indices mit Abmusterungsergebnissen, *Die Farbe*, **19**, 135–139 (1970).

CIE Publication 15.2, *Colorimetry*, Commission Internationale de l'Éclairage, Vienna, Austria (1986a).

CIE *CIE Journal* **5**, 16, Commission Internationale de l'Éclairage, Vienna, Austria (1986b).

CIE Publication 17.4, *International lighting vocabulary*. Commission Internationale de l'Éclairage, Vienna, Austria (1987).

CIE Publication 80, *Special metamerism index: change in observer*, Commission Internationale de l'Éclairage, Vienna (1989).

CIE Publication 159:2004, *A colour appearance model for colour management systems: CIECAM02*, Commission Internationale de l'Éclairage, Vienna, Austria (2004).

Estévez, O., *On the Fundamental Data-Base of Normal and Dichromatic Vision*, PhD Thesis. University of Amsterdam, Holland (1979).

Fairchild, M.D., and North, A.D., Measuring color-matching functions. Part I and Part II, *Color Res. Appl.*, **18**, 155–162 and 163–170 (1996).

Fairman, H.S., New terminology for metamerism revisited, *Color Res. Appl.*, **11**, 80–81 (1986).

Fairman, H.S., Metameric correction using parametric decomposition, *Color Res. Appl.*, **12**, 261–265 (1987).

Foster, D.H., Marin-Franch, I., Nascimento, M.C., and Amano, K., Coding efficiency of CIE color spaces, *IS&T and SID's 16th Color Imaging Conference: Color Science and Engineering Systems, Technologies, and Applications*, pp. 285-288. IS&T Springfield, VA, USA (2008).

Helson, H., Judd, D.B., and Warren, M.H., Object color changes from daylight to incandescent filament illumination, *Illum. Eng.*, **47**, 221–223 (1952).

Hunt, R.W.G., *Measuring Colour*, *3rd Ed.*, page 126, Fountain Press, Chichester, England (1998).

Hunt, R.W.G., *The Reproduction of Colour 6th Ed.*, Wiley, Chichester, England (2004).

Hunt, R.W.G., and Luo, M.R., Testing colour appearance models using corresponding colour and magnitude-estimation data, *Color Res. Appl.*, **23**, 147–153 (1998).

Katoh, N., Nakabayashi, K., Ito, M., and Ohno, S., Effect of ambient light on color appearance of softcopy images: mixed chromatic adaptation for self-luminous displays, *J. Elec. Imaging*, **7**, 794–806 (1998).

Kuehni, R.G., Metamerism, exact and approximate, *Color Res. Appl.*, **8**, 192 (1983).

Kuo, W.-G., Luo, M.R., and Bez, H.E., Various chromatic-adaptation transformations tested using new colour appearance data in textiles, *Color Res. Appl.*, **20**, 313–327 (1995).

Lam, K.M., *Metamerism and Colour Constancy*, PhD Thesis, University of Bradford, England (1985).

Luo, M.R., Clarke, A.A., Rhodes, P.A., Schappo, A., Scrivener, S.A.R., and Tait, C.J., Quantifying colour appearance. Part I. LUTCHI colour appearance data, *Color Res. Appl.*, **16**, 166–180 (1991). Quantifying colour appearance. Part II. Testing color models performance using LUTCHI colour appearance data. *Color Res. Appl.*, **16**, 181–197 (1991).

Luo, M.R., Gao, X.W., Rhodes, P.A., Xin, J.H., Clarke, A.A., and Scrivener, S.A.R., Quantifying colour appearance. PART III. Supplementary LUTCHI colour appearance data, *Color Res. Appl.*, **18**, 98–113 Quantifying colour appearance. PART IV. Transmissive media. *Color Res. Appl.*, **18**, 191–209 (1993).

Luo, M.R., Lo, M.-C., and Kuo, W.-G., The LLAB(l:c) colour model, *Color Res. Appl.*, **21**, 412–429 (1996).

McLaren, K. *The Colour Science of Dyes and Pigments, 2nd Ed*. Adam Hilger, Bristol, England (1986).

Mori, L., Sobagaki, H., Komatsubara, H., and Ikeda, K., Field trials on CIE chromatic adaptation formula, *Proceedings of the CIE 22nd Session*, pp. 2055–58, Commission Internationale de l'Éclairage, Vienna, Austria (1991).

Nayatani, Y., Hashimoto, K., Takahama, K., and Sobagaki, H., Comparison of methods for assessing observer metamerism, *Color Res. Appl.*, **10**, 147–155 (1985).

Nimeroff, I., A survey of papers on degree of metamerism, *Color Eng*. **6**(6), 44–46 (1968).

Nimeroff, I., and Yurow, J.A., Degree of metamerism, *J. Opt. Soc. Amer.*, **55**, 185–190 (1965).

Ohta, N. Formulation of a standard deviate observer by a non-linear optimization technique. *Color Res. Appl.*, **10**, 156–164 (1985).

Oskoui, P., and Pirotta, E., Determination of adapted white points for various viewing environments, *IS&T and SID's 7th Color Imaging Conference: Color Science and Engineering Systems, Technologies, and Applications*, pp. 101-105, IS&T Springfield, VA, USA (1999).

Pinney, J.E., and DeMarsh, L.E.J., The study of colour reproduction by computation and Experiment, *Phot. Sci.*, **11**, 249–255 (1963).

Takahama, K., Sobagaki, H., and Nayatani, Y., Formulation of a nonlinear model of chromatic adaptation for a light-gray background, *Color Res. Appl.*, **9**, 106–115 (1984).

Von Kries, J. A., *Handbuch der Physiologisches Optik*, Vol. II (W. Nagel, ed), pp. 366–369, Leopold Voss, Hamburg, Germany (1911).

7

Colour Rendering by Light Sources

7.1 INTRODUCTION

Artificial light sources have been known since antiquity. For many centuries, oil lamps and candles provided some light after dark, but the more widespread use of artificial light awaited the introduction, first, of gas mantles in the late nineteenth century, and then, most importantly, the electric filament lamp. These light sources were all of a yellower colour than average daylight, but the benefits of being able to see well indoors after dark far outweighed any disadvantages of their colour. In fact, their colour actually flatters the appearance of human skin, and it is interesting that candles are still used for social occasions, such as dinner parties. There is a tendency, in coloured illumination, for light colours to be tinged with the hue of the illuminant, and for dark colours to be tinged with the complementary hue; this is known as the *Helson-Judd effect*.

The introduction of commercially available fluorescent lamps in 1938 provided artificial light sources with a much wider range of colours, and, for any one colour, also a range of possible spectral power distributions. The selection of the most appropriate fluorescent lamp for a particular purpose required the assessment both of its *efficacy* (the amount of light produced per unit of power consumed) and its *colour rendering* (the appearance of objects illuminated by it). There was a tendency with the first fluorescent lamps for those with high efficacies to have relatively poor colour rendering, and for those with good colour rendering to have rather low efficacies, a tendency which, to some extent, still exists.

The use of gas discharge lamps in applications such as street lighting and floodlighting has introduced further examples of different colour rendering. The deep orange low-pressure sodium lamps which emit light at virtually only one wavelength, result in the appearance of colours being restricted to a range of oranges and browns. Mercury discharge lamps render bluish and greenish colours adequately, but reds appear grey or black.

Light-emitting diodes (LEDs) provide a variety of possible spectral power distributions each with its own colour rendering properties.

Measuring Colour, Fourth Edition. R.W.G. Hunt and M.R. Pointer.

7.2 THE MEANING OF COLOUR RENDERING

Colour rendering is a subjective phenomenon that depends on the spectral qualities of the light source, the object being illuminated, and the visual physiology. The CIE definition of Colour Rendering is:

> The effect of an illuminant on the colour appearance of objects by conscious or subconscious comparison with their colour appearance under a reference illuminant.

This definition cites a *reference illuminant*. It has to do so because there is no absolute colour appearance. But the choice of reference illuminant raises some interesting questions. Should the choice be daylight? Daylight can be regarded as the most basic illuminant because of its antiquity and ubiquity; but it varies with time of day, time of year, and weather conditions, so which daylight should be chosen? D65 is the preferred illuminant adopted in colorimetry, but it would be an assumption to regard it as the inevitable choice for colour rendering; in the graphic arts industry, for example, D50 has been chosen as the preferred illuminant.

Another option is to use, not just one reference illuminant, but a series having the same or nearly the same colours (chromaticities) as the test sources, but with precisely defined spectral power distributions, such as those for standard daylights for bluish sources and those for Planckian radiators for yellowish sources.

The definition also refers to a comparison of colour appearances. If a single reference illuminant is being used, the comparison could be achieved by calculating the colour differences introduced by the test source after allowance has been made for chromatic adaptation to it; this requires a choice of chromatic adaptation transform (see Chapter 6) and a colour-difference formula (see Chapter 3). If a series of reference illuminants is being used, the same procedure can be used, but in this case, the task of the chromatic adaptation transform is of a quite minor nature. But whether one or a series of reference illuminants is being used, a set of test colours has to be chosen for the colour differences to be computed. So there are three arbitrary decisions to be made: the choice of chromatic adaptation transform, the choice of colour-difference formula, and the choice of colour samples.

An elaboration of the above procedures is to use a Colour Appearance Model (see Chapter 15) instead of a chromatic adaptation transform. This could allow for a situation where a source increased the chroma of samples, but lit them at a lower level of illumination; the colour appearance model would allow for the fact that the increase in chroma could compensate for the decrease in colourfulness caused by the lower illuminance. It has been found that, if a type of lamp is used that results in an increase in the perceived saturation of most of the colours in a scene, then, for a given level of illumination, that scene appears to be brighter than with a conventional type of lamp (Thornton, 1972). The three-band type of fluorescent lamp (see Section 4.6), therefore, results in an apparently higher level of illumination than a conventional fluorescent lamp of the same light output as measured photometrically. This phenomenon has been referred to as *visual clarity*, and occurs, at least in part, because higher colourfulnesses normally occur at higher levels of illumination (Boyce and Lynes, 1976; Hunt, 1979).

In the above procedures there is a further implicit assumption being made: that good colour rendering occurs when the fewest colour differences occur between the illumination by the test and reference sources. But some changes in appearance may be beneficial. Increased colour discrimination may be of advantage in some tasks, such as the grading of diamonds; and, as already mentioned, the use of yellower sources may flatter the appearance of human skin. A flattery index, or preference index, has therefore been suggested for some applications (Judd, 1967; Jerome, 1972; Thornton, 1972).

Finally, the purpose of the entire environment may be an important factor; what is required in a dining room, for example, may be quite different from what is required in an office or in a laboratory. In this context, it has been suggested that consistency of hues may be the factor of greatest importance (Kemenade and Burgt, 1995).

In spite of the difficulties outlined above, because of the needs of industry, Colour Rendering Indices have been developed and are widely used. Some of these will now be described.

7.3 CIE COLOUR RENDERING INDICES

It was the advent of fluorescent lamps, whose spectral power distributions could be varied at will over quite a wide range, that made it desirable to have some means of expressing the degree to which any illuminant gave satisfactory colour rendering. For this purpose, the CIE recommended both a *Special Colour Rendering Index* and a *General Colour Rendering Index* (CRI) in 1965 (CIE, 1965), revised versions of which were introduced in 1974 and 1988 (CIE, 1974, 1988, and 1995). A major advantage of the CIE method is that an estimate of the colour rendering is given by a single number, but this is also the greatest disadvantage, since a lot of information is compressed into the calculation of that number.

The *CIE Special Colour Rendering Index* is given by

$$R_{\mathrm{i}} = 100 - 4.6d_{\mathrm{i}}$$

where d_{i} is the distance in the CIE U*V*W* space (see Section 3.9) between the points representing the colour concerned when illuminated by the test source and by the CIE D-illuminant closest in the CIE u,v (see Section 3.6), not the u', v', chromaticity diagram (but for sources whose correlated colour temperatures are below 5000 K the closest Planckian radiator is used instead; Robertson, 1968). U^*, V^*, W^* are given by:

$$U^* = 13W^*(u - u_{\mathrm{n}})$$

$$V^* = 13W^*(v - v_{\mathrm{n}})$$

$$W^* = 25Y^{1/3} - 17$$

The *CIE General Colour Rendering Index* (CRI) is given by

$$R_{\mathrm{a}} = 100 - 4.6\left(\frac{d_1 + d_2 + d_3 + d_4 + d_5 + d_6 + d_7 + d_8}{8}\right)$$

where d_1 to d_8 are the values of R_{i} for the Munsell colours 7.5R6/4, 5Y6/4, 5GY6/8, 2.5G6/6, 10BG6/4, 5PB6/8, 2.5P6/8, 10P6/8 (the Munsell system will be described in Section 8.4). These colours are of medium chroma. An additional six Munsell colours (Munsell 4.5R4/13, 5Y8/10, 4.5G5/8, 3PB3/11, 5YR8/4, 5GY4/4) can also be used to

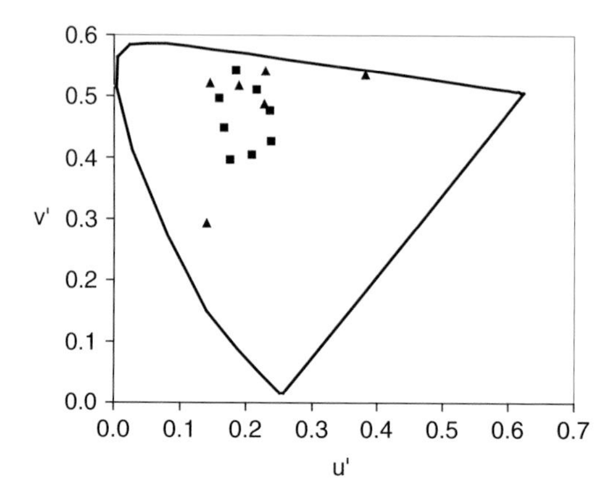

Figure 7.1 The chromaticities, in CIE Illuminant C, of the Munsell colours used in the CIE Colour Rendering Index. Squares: the eight standard colours. Triangles: the six extra colours

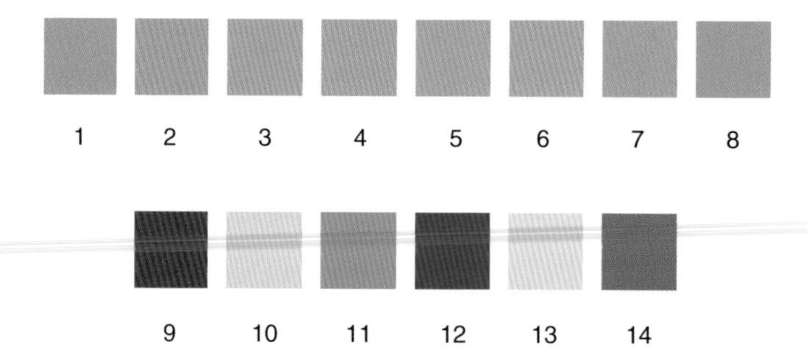

Figure 7.2 Reproductions of the colours used in the CIE Colour Rendering Index. (Because of the limitations of the printing process, the colours may have some inaccuracies)

provide further tests of the colour rendering of the source; the first four of these are of high chroma, the fifth represents Caucasian skin colour, and the sixth leaf green. The values of spectral radiance factor of each of these Munsell colours are tabulated in Appendix 7: their chromaticities for CIE Illuminant C are shown in Figure 7.1 and a reproduction of their colours in Figure 7.2. The constant 4.6 was chosen to result in a value of CRI of about 50 for a warm white fluorescent lamp. Allowance is made for any differences in chromaticity between the source considered and the nearest D-illuminant (or Planckian source) by using a Von Kries type of adjustment (Section 6.14) for chromatic adaptation. A worked example of the calculations of the CIE Colour Rendering Indices is given in Appendix 7.

A Colour Rendering Index of 100 indicates that, for the colours considered in its evaluation, the source is equivalent to the nearest D illuminant (or Planckian source). Sources having General Indices of about 90 or greater are usually considered to have very good colour rendering properties in practical applications. Some fluorescent lamps have General Indices as low as about 50; this indicates that their colour rendering is appreciably deficient in some respects, but such lamps are used in certain applications because they tend to have high efficacies, and hence are economical in power consumption.

It should also be noted that two lamps having the same value of special colour rendering index for a particular test colour do not necessarily render that colour in a similar manner. This is because the index does not indicate the direction of the colour shifts associated with the light from each lamp. Thus, one lamp may render the colour as purplish and the other as greenish but, because the colour differences are calculated to be the same, the value of the Special Colour Rendering Index will be the same. Another potential problem is that a quoted value of colour rendering index should also cite the value of the correlated colour temperature of the reference illuminant, because two lamps with different correlated colour temperatures and colour rendering can have the same value of R_a. For example, a high efficacy Warm White fluorescent lamp with a correlate colour temperature of 3000 K can have a value of CRI of 80, the same as a much lower efficacy fluorescent daylight simulator with a correlated colour temperature of 6500 K.

Development of the CIE colour rendering index based on the use of test colours, and its subsequent widespread application in the lighting industry, led to the need to establish acceptability tolerances. In one investigation (Halstead, Morley, Palmer, and Stainsby, 1971) a number of experiments were carried out using simultaneous comparison of lamps with similar chromaticities (in order to minimise the effects of chromatic adaptation between different lamps) but having different colour rendering properties. The lamps were mounted in viewing booths and used to illuminate the eight CIE test samples; observers were asked to judge the perceived differences. It was found that the just-noticeable-difference was equivalent to a difference in CRI of about 10–15 units of the CIE special index, R_i, dependent on the sample. Others, often using different viewing techniques, have found different variation in acceptability. It must be remembered however, that the concept of comparing a set of coloured samples under two similar but different light sources is somewhat artificial in that it does not represent a practical situation where such comparisons are hardly ever made.

The subject of colour rendering indices, especially as applied to LED light sources, is the subject of much on-going research (CIE, 2007).

7.4 SPECTRAL BAND METHODS

The CIE General Colour Rendering Index is based on comparing the rendering of the eight Munsell samples under the illuminant considered with the rendering of the same samples under either a CIE D illuminant or a Planckian radiator. It has been found that this index performs poorly when it is used for evaluating the rendering of white LED sources. An example is given in Figure 7.3. These two light sources have the same correlated colour temperatures (about 5000 K). LED A has a CIE Colour Rendering index, R_a, equal to 82, while for LED B it is 71; the value for the reference illuminant is 100. This implies that LED A has better colour rendering than LED B. But in fact the reverse is true; LED B renders saturated reddish colours much closer to the reference illuminant, D50, than LED A; see Figure 7.4. A modified form of the CIE colour rendering index has been proposed in which the colour space is CIELAB instead of U*V*W*, the chromatic adaptation transform is

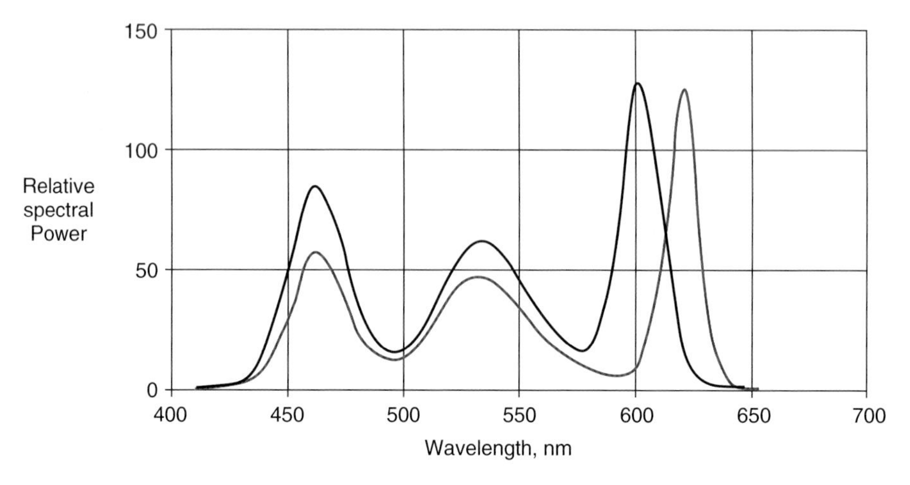

Figure 7.3 Spectral power distributions for two white LED sources: LED A, black line; LED B, magenta line

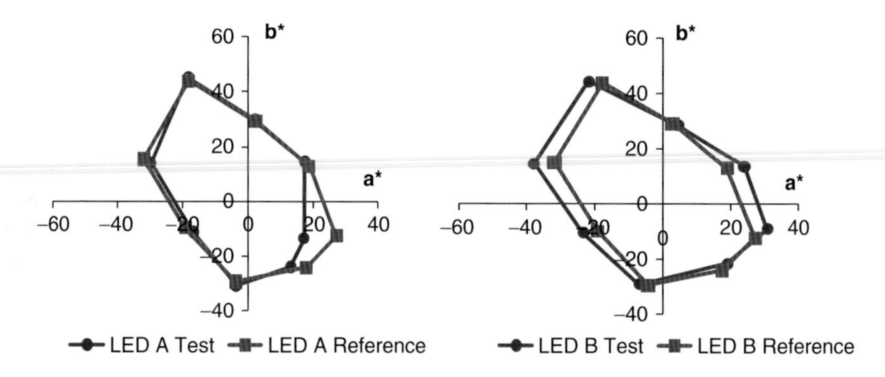

Figure 7.4 The colour rendering of two LED sources, A and B, showing the improved rendition of red colours for B, even though it has the lower of the two values of Colour Rendering Index

CAT02 (see Section 6.15) instead of Von Kries (see Section 6.14), the reference illuminant is always D65 instead of the nearest D illuminant or Planckian radiator, and the colour difference equation is CIEDE2000 instead of distance in the U*V*W* space (Park, 2010).

At the time when the CIE Colour Rendering Index was devised, there was an alternative method based on spectral bands as shown in Tables 7.1 and 7.2.

This method was first suggested by Bouma (Bouma, 1937), who recognised that the fundamental difference between two lamps was in their different spectral power distributions and that a method of comparing such distributions might be useful as an indicator of colour rendering. Bouma calculated a photometric response function by multiplying the spectral power distribution by the spectral luminous efficiency function, $V(\lambda)$, and integrating the luminance over eight broad bands. The percentage luminance in each band was

Table 7.1 Use of the spectral band method of assessing colour rendering with LED A. The tolerances are ±10% for single bands and ±5% for the average percentage difference in all pairs of contiguous bands

Spectral band (nm)	Band luminance for D50	Band luminance for LED A	Ratio of band luminances	Ratio − 1	Band deviation	Excess over tolerance
400–455	0.585	0.537	0.917	−0.083	−8%	0
400–510					−18%	13
455–510	9.52	6.80	0.715	−0.286	−29%	19
455–540					−3%	0
510–540	21.9	26.7	1.221	0.221	22%	12
510–590					−1%	0
540–590	44.3	33.3	0.752	−0.248	−25%	15
540–620					39%	34
590–620	15.8	32.1	2.032	1.032	103%	93
590–760					5%	0
620–760	7.94	0.534	0.067	−0.933	−93%	83
					Sum of excesses	269
		Figure of Merit = 1024 − 269 = 755				

Table 7.2 Same as Table 1, but for LED B

Spectral band (nm)	Band luminance for D50	Band luminance for LED B	Ratio of band luminances	Ratio −1	Band deviation	Excess over tolerance
400–455	0.585	0.519	0.887	−0.113	−11%	1
400–510					−16%	11
455–510	9.52	7.63	0.801	−0.199	−20%	10
455–540					12%	7
510–540	21.9	31.6	1.442	0.442	44%	34
510–590					7%	2
540–590	44.3	31.2	0.704	−0.296	−30%	20
540–620					−4%	0
590–620	15.8	19.2	1.218	0.218	22%	12
590–760					23%	18
620–760	7.94	9.86	1.241	0.241	24%	14
					Sum of excesses	129
		Figure of Merit = 1024 − 129 = 895				

compared with the total luminance in all bands. This reduced the spectral information to eight numbers which allowed much easier comparison. This system was adopted by the CIE in 1948 and became a British Standard in 1964. The main purpose of the index was to ensure reproducibility of a product and interchangeability between different manufacturers for a given product. To facilitate these comparisons a system of tolerances was also devised for the band values.

The eight-band system had drawbacks however. First, the band divisions selected were arranged so that they did not coincide with any of the mercury lines that are characteristic of a fluorescent lamp spectral power distribution. This resulted in the violet part of the spectrum being divided into three very narrow bands, each of very low luminance.

Crawford (1963a, 1963b) conducted experiments which led to a better band system based on the division of the visible spectrum into six bands. This system of calculating band values is included in British Standard 950 Parts I and II, *Specification for artificial daylight for the assessment of colour* (BS950, 1967a; BS950, 1967b), with a plus and minus tolerance applied to each band.

Using this band method, as shown in Tables 7.1 and 7.2, LED A has a total of 269 excesses over the tolerances, whereas the similar figure for LED B is 129. Figures of merit can be obtained by subtracting these values from a suitable number such as 1024, resulting in 755 for LED A, and 895 for LED B. The spectral power of LED B peaks more nearly at Thornton's prime wavelengths, 450 nm, 530 nm, and 610 nm (Thornton, 1971) than is the case for LED A, and this may explain its superior performance. It is not known whether this spectral band method gives satisfactory results for all sources, but there is a rather urgent need to provide a measure of the colour rendering of sources that covers LEDs and all other sources used for illumination.

7.5 OTHER METHODS FOR ASSESSING THE COLOUR RENDERING OF LIGHT SOURCES

Since the recommendation of the CIE colour rendering index was introduced, a number of other methods have been described.

The use of colour vector maps are favoured by one major lamp manufacturer (van Kemenade and van der Burgt, 1988; van der Burgt, 2009). The positions of a large number of test colours are plotted in a CIELAB a*,b* diagram and each of these points is joined by an arrow to the equivalent position of the same test colour but calculated for a reference illuminant, thus forming a vector map. The shorter the length of the arrows, the more similar is the colour rendering of the reference and test lamps, and changes in the length and direction of the arrows indicates relative changes of the test colours in terms of their hue and chroma. Clearly this method gives more information than the CIE method, and it is able to discriminate between lamps that might have the same value of CIE colour rendering index but rendering colours in a different manner.

While the use of vector maps overcomes one criticism of the CIE method, that it reduces all data to one number, there is still a desire to be able to rank lamps in order of their ability to render colours, something a vector map cannot do. A possible way to overcome this problem is to use the gamut area defined by plotting the chromaticities of the eight CIE test colours in a suitable colour space and calculating the area so defined (Thornton, 1972; Henderson and Halstead, 1975). This method has the advantage that it does not require the use of a reference illuminant; by plotting different light sources on the same chromaticity diagram it is easy to make comparisons, but it still has the disadvantage that a lamp with the same gamut area can render colours differently and only the chromaticity plot can show this.

Pointer (Pointer, 1986) described a method based on the use of a colour appearance model to calculate a series of indices that could be progressively combined according to the intended purpose. Thus, the lamp developer has available twelve indices relating to the hue, chroma, and lightness of the red, yellow, green and blue content of the test colours. Each index is a vector in that the associated change in direction (red to yellow or blue, etc. and increase or decrease in both chroma and lightness) of the appearance of the test colours under the test lamp is compared with the appearance under the reference illuminant. For marketing purposes, the indices can be combined to provide one number for use in a manner similar to the CIE index. An additional advantage of this method is that the use of a colour appearance model requires information on the absolute light level and thus an index can be defined at any level of illumination. The test colours used in this method are those of the Macbeth ColorChecker Chart® (McCamy, Marcus and Davidson, 1976).

More recently Davis and Ohno (Davis and Ohno, 2010) have described a new *Colour Quality Scale* (CQS). Like the CIE index, this is a test colour method that compares the appearance of a set of test colours when illuminated by a test lamp to their appearance under a reference illuminant. Fifteen test colours are used, all of high chroma, with the CMC-CAT2000 chromatic adaptation transform (Li, Luo, Rigg, and Hunt, 2002) and CIELAB colour space. A saturation factor is introduced that serves to negate any contribution to the colour difference that arises from an increase in the chroma of the test samples viewed under the test lamp because evidence suggests that such increases in chroma, while real, are not detrimental to the colour rendering quality of the lamp. The CQS scale is normalised to have a maximum value of 100 and in such a manner that negative values of CQS are not possible. A further factor is introduced that reduces the penalty usually associated with lamps of relatively low colour temperature, less than 3500 K.

A number of other alternative approaches to colour rendering have been suggested including a method based on categorical colour scaling (Yaguchi, Endoh, Moriyama, and Shioiri, 2005), and the concept of visual clarity, or the ability to perceive contrast between the objects in a scene (Aston and Bellchambers, 1969; Hashimoto, Yano, Shimuzi, and Nayatani, 2007); this method is based on the differences in pairs of test colours rather than single colours.

Finally, methods have been devised depending on a colour quality metric based on memory colours (Bartleson, 1960; Smet, Ryckaert, Pointer, Decononck, and Hanselaer, 2011; Hurlbert and Ling, 2005). The colour quality of a test lamp is evaluated as the degree of similarity between the colour appearance of a set of familiar objects and their associated memory colours, thus no reference illuminant is needed. It is likely that this type of memory index will be useful in addition to, rather than instead of, the present CIE recommended index.

7.6 COMPARISON OF COMMONLY USED SOURCES

Table 7.3 gives, correlated colour temperatures, General Colour Rendering Indices (CRI), and efficacies for some typical light sources.

Table 7.3 Correlated colour temperatures, general colour rendering indices, and efficacies for some typical light sources

Source	Correlated Colour Temperature	Colour Rendering Index	Efficacy
	K	CRI	lm W^{-1}
Tungsten (40W, 240V)	2650	100	10
Tungsten (40W, 110V)	2700	100	12
Tungsten (100W, 240V)	2750	100	13
Tungsten (100W, 110V)	2850	100	15
Tungsten halogen	3000	100	21
Daylight (D65)	6500	100	–
Xenon	5290	93	25
Fluorescent			
Northlight, Colour-matching	6500	93	48
Artificial Daylight	6500	92	45
Cool White	4200	58	81
Kolor-rite	4000	89	51
Natural de luxe	3500	92	45
White	3450	54	83
Warm white de luxe	3000	80	48
Warm white	3000	51	83
Three-band	4000	85	93
High pressure sodium	2000	25	100
Low pressure sodium	–	–	150
Colour corrected mercury (MBF)	3800	45	50
Colour corrected mercury (MBTF)	3800	45	20
Metal halide mercury (HMI)	6430	88	85
Metal halide mercury	4000	80	70

REFERENCES

Aston, S.M., and Bellchambers, H.E., Illumination colour rendering and visual clarity, *Light Res. Technol.*, **1**, 259–261 (1969).

Bartleson C.J., Memory colors of familiar objects, *J. Opt. Soc. Amer.*, **50**, 73–77 (1960).

BS 950:1967, *Specification for artificial daylight for the assessment of colour. Illuminant for colour matching and colour appraisal*, British Standards Institution, London (1967a).

BS 950:1967, *Specification for artificial daylight for the assessment of colour. Viewing conditions for the graphic arts industry*, British Standards Institution, London (1967b).

Bouma, P.J., Colour reproduction in the use of different sources of 'white' light, *Philips Tech. Rev.*, **2**, 1–7 (1937).

Boyce, P.R., and Lynes, J.A., Illuminance, colour rendering index and colour discrimination index, *CIE Compte Rendu, London*, P-75-35, 290–297. Commission Internationale de l'Éclairage, Vienna, Austria (1976).

CIE Publication 13, *Method of specifying and measuring colour rendering properties of light sources*, Bureau Central de la CIE, Paris, France (1965).

CIE Publication 13, *Method of specifying and measuring colour rendering properties of light sources*, 2nd *Ed.*, Bureau Central de la CIE, Paris, France (1974).

CIE Publication 13.2, *Method of specifying and measuring colour rendering properties of light sources*, *2nd Ed., Revised*, Commission Internationale de l'Éclairage, Vienna, Austria (1988).

CIE Publication 13.3, *Method of specifying and measuring colour rendering properties of light sources*, 3rd *Ed.*, Commission Internationale de l'Éclairage, Vienna, Austria (1995).

CIE Publication 177:2007, *Colour rendering of white LED light sources*, Commission Internationale de l'Éclairage, Vienna, Austria (2007).

Crawford, B.H., Colour-rendering tolerances and the colour-rendering properties of light sources, *Trans. Illum. Eng. Soc.*, **28**, 50–64 (1963a).

Crawford, B.H., The colour rendering properties of illuminants: the application of psycho-physical measurements to their evaluation, *Br. J. Appl. Phys.*, **14**, 319–328 (1963b).

Davis, W., and Ohno Y., Color quality scale, *Opt. Eng.*, **49**, 033602-1-10, (2010).

Halstead, M.B., Morley, D.I., Palmer, D., and Stainsby, A.G., Colour rendering tolerances in the CIE system, *Light. Res. Technol.*, **3**, 99–124 (1971).

Hashimoto, K., Yano, T., Shimuzi, M., and Nayatani, Y., New method for specifying color- rendering properties of light sources based on feeling of contrast, *Color Res. Appl.*, **32**, 361–371 (2007).

Henderson, S.T., and Halstead, M.B., Corrections for chromatic adaptation in the determination of colour rendering indices, *Light. Res. Technol.*, **7**, 113–124 (1975).

Hunt, R.W.G., Specifying colour appearance, *Light. Res. Technol.*, **11**, 175–183 (1979).

Hurlbert, A.C., and Ling, Y., If it's a banana, it must be yellow: the role of memory colors in color constancy, *J. of Vision*, **5**(8), 787–787 (2005).

Jerome, C.W., Flattery versus rendition, *J. Illum. Eng.*, **1**, 208–211 (1972).

Judd, D.B., A flattery index for artificial illuminants, *Illum. Engng.*, **62**, 593–598 (1967).

Kemenade, van J.T.C., and Burgt, van der P.J.M., Towards a user orientated description of color rendition of light sources, *Proc. 23rd Session of the CIE, New Delhi*, CIE 119/1, pp. 43–46, Commission Internationale de l'Éclairage, Vienna, Austria (1995).

Li, C.J., Luo, M.R., Rigg, B., and Hunt, R.W.G., CMC2000 chromatic adaptation transform, *Color Res. Appl.*, **27**, 49–58 (2002).

McCamy, C.S., Marcus, H., and Davidson, J.G., A color rendition chart, *J. Appl. Photogr. Engng.*, **2**, 95–99 (1976).

Park, Y., Approach to CRI (Colour Rendering Index) for full colour RGB LED source lighting, *IS&T and SID's 18th Color Imaging Conference: Color Science and Engineering Systems, Technologies, and Applications*, pp. 371–374, IS&T Springfield, VA, USA (2010).

Pointer, M.R., Measuring colour rendering. A new approach, *Light. Res. Technol.*, **18**(4), 175–184 (1986).

Robertson, A.R., Computation of correlated color temperature and distribution color temperature, *J. Opt. Soc. Amer.*, **58**(11), 1528–1535 (1968).

Smet, K., Ryckaert, W.R., Pointer, M.R., Deconinck, G., and Hanselaer, P., Colour appearance rating of familiar real objects, *Color Res, Appl.*, **36**, 192–200 (2011).

Thornton, W.A., Luminosity and color-rendering capability of white light, *J. Opt. Soc. Amer.*, **61**, 191–194 (1971).

Thornton, W.A., A validation of the color-preference index, *J. Illum. Eng.*, **4**, 48–52 (1972).

Van Kemenade, J.T.C., and van der Burgt, P.J.M., Light sources and colour rendering: Additional information for the R_a index, *Proceedings of the CIBSE National Lighting Conference, Cambridge*, CIBSE, London, England (1988).

Van der Burgt, P.J.M., van Kemenade, J.T.C., About color rendition of light sources: The balance between simplicity and accuracy, *Color Res, Appl.*, **35**(2), 85–93 (2009).

Yaguchi, H, Endoh, H, Moriyama, T., and Shioiri, S., Categorical color rendering of LED light sources, *Proceedings of CIE Symposium LED Light Sources*, CIE Publication x026:2005, pp. 20–23, Commission Internationale de l'Éclairage, Vienna, Austria (2005).

GENERAL REFERENCES

Cayless, M.A., and Marsden, A.M., *Lamps and Lighting*, 4th *Ed.*, Arnold, London, England (1997).

Halstead, M.B., Colour rendering: past, present and future, in *Proceedings of the Third Congress of the International Colour Association*, Adam Hilger, Bristol, England (1978).

8

Colour Order Systems

8.1 INTRODUCTION

Collections of colour samples are often used to provide examples of colour products. These may be in the form of patches of paint, swatches of cloth, pads of papers, or printings of inks, for example, according to the type of product involved. Such examples are very useful if the number of colours required is fairly limited. But, if a very large number of colours are necessary, and, if intermediate colours lying between samples are to be considered, a system is required in which interpolation can be made between samples in an unambiguous way. Such a system is referred to as a *colour order system*. (Hesselgren, 1984; Wright, 1984; Robertson, 1984; Spillmann, 1985; Hunt, 1985; McCamy, 1985).

8.2 VARIABLES

To facilitate interpolation between samples, it is usually helpful to arrange them according to major perceptual attributes of colour. If an observer is given a random collection of colour samples and asked to sort them, it is likely that the first step would be to sort them according to their hues. Thus, all the reddish colours, for example, could be separated into one group, all the yellowish into another, and so on for the greenish, bluish, and purplish colours; and another group would comprise all the colours that did not exhibit any hue at all, the white, grey, and black colours. The hue groups could then be further subdivided into intermediate hues, such as yellow-reds (that is, orange colours), green-yellows, blue-greens, purple-blues, and red-purples. Even finer subdivisions can be made, it being possible to identify constancy of hue for samples with considerable precision, particularly for saturated colours, even though different colour names may not suggest themselves for each sub-group.

Having classified the colours according to hue, the next most obvious step would probably be to classify them according to lightness. Thus, for each hue group, the observer might put all the light colours at the top of the arrangement, all the dark colours at the bottom, with the colours of medium lightness in between. The distance of a colour from

Measuring Colour, Fourth Edition. R.W.G. Hunt and M.R. Pointer.

the bottom could then be made to represent the lightness of that colour. The whites, greys, and blacks could be arranged as a separate group in a single column, in the same way, to form a grey scale with white at the top and black at the bottom.

The observer would then most likely notice that some colours were more vivid than others, and could arrange the least vivid, or pale, samples near one side of the array, say the left-hand side, and the most vivid, samples towards the right-hand side. In this case, the distance of the colour from the left-hand side could be made to represent the colourfulness, as shown in Figure 8.1. However, the colourfulness could be judged either as chroma (that is, in proportion to the brightness of a reference white for the samples), or as saturation (that is, in proportion to the brightness of the sample itself), as discussed in Section 1.9.

If now the observer were to attach the samples to suitable pages, the pages could be arranged in a three-dimensional array, with the grey scale as a central vertical axis, and the

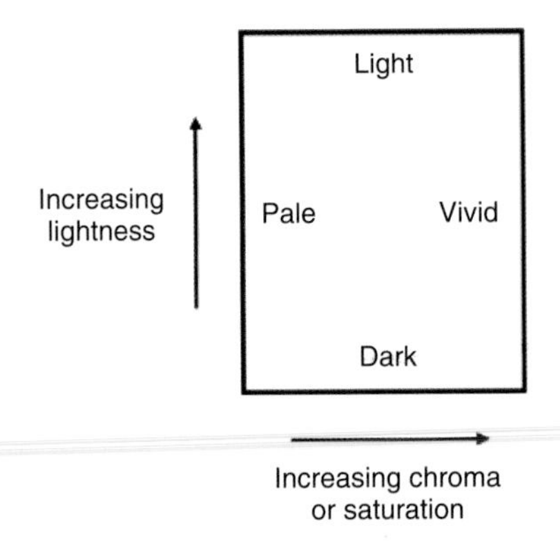

Figure 8.1 Two-dimensional array of lightness, and chroma or saturation

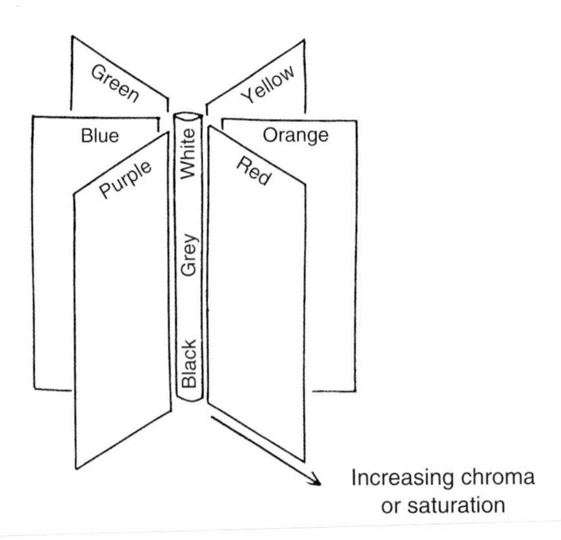

Figure 8.2 Three-dimensional array of hue, lightness, and chroma or saturation

different hue groups joining this axis at their left-hand sides, as shown in Figure 8.2. The different hue groups could be arranged in the same order as the colours of the spectrum, with the purple colours completing a circle between blue and red. This three-dimensional array of colours then provides a systematic arrangement in which interpolation is possible in any direction, so that all colours can be represented that lie within the limits imposed by the colours lying at the extreme edges of the array.

8.3 OPTIMAL COLOURS

There are theoretical limits to the extent of colour arrays of the type described in the previous section. To produce a chromaticity different from that of the illuminant, non-self-luminous samples must absorb some of the light. Assuming, for the moment, that no absorbed radiant power is reradiated by fluorescence, the absorption must result in a reduction in luminance factor; this reduction is minimal when, at each wavelength, the sample either absorbs all, or none, of the light (Rösch, 1928; MacAdam, 1935a and 1935b). Colours having this property are called *optimal colours*. In Figure 8.3, the loci of these optimal colours are shown for various luminance factors (denoted by their corresponding values of L^*) in the u′,v′ diagram for Standard Illuminants D65 and A. If the luminance factor is high, the area within the optimal colour locus is small; if the luminance factor is low, the area becomes much larger. Thus, high values of u′,v′ saturation, s_{uv}, can only be attained if the luminance factor is low. However, there is an exception in the case of yellow colours, which can have both high values of s_{uv} and high values of luminance factor; this is because the absorption of the blue light, required to produce the yellow colours, results in only a slight reduction in luminance factor, on account of the small contribution of the β cones to the achromatic signal (as discussed in Section 1.6).

Real reflecting and transmitting colours (that do not fluoresce) are even more restricted in chromaticity than indicated in Figure 8.3, because, at some wavelengths, they always have absorptions that are intermediate between total and zero. Real reflecting colours are further restricted, because light reflected from their topmost surface results in their minimum possible reflections usually being a few per cent, instead of zero, at any wavelength. This greatly reduces the range of chromaticities possible for dark colours. The colorimetric gamut of surface colours found in practice has been the subject of several studies (Pointer, 1980; ISO, 2007; Li, Li, Luo, and Pointer, 2007; Li, Li, Luo, Pointer, Cho, and Kim, 2007). The result of one such study is shown by the broken line in Figure 8.4 for chromaticity, the illuminant being D65. In Figure 8.5, in plots of L^* against C^*_{uv} for six hues, the same gamut is shown by the broken lines, the gamut for optimal colours being shown by the full lines. The hues have values of h_{uv} equal to 10° (a red), 70° (a yellow), 150° (a green), 190° (a cyan), 250° (a blue), and 330° (a magenta). It is clear that the gamuts of these real colours are substantially smaller than those of the optimal colours, except for yellows. In Figure 8.6, the same gamuts are shown in plots of L^* against C^*_{ab} for values of h_{ab} equal to 30° (a red), 90° (a yellow), 170° (a green), 190° (a cyan), 230° (a blue), and 310° (a magenta).

If colours fluoresce, they can lie outside the limits defined by the optimal colours; and, if they do lie outside, they usually have a different quality of appearance which Evans termed *fluorent* (Evans and Swenholt, 1967, 1969), a term applicable whether physical fluorescence is occurring or not.

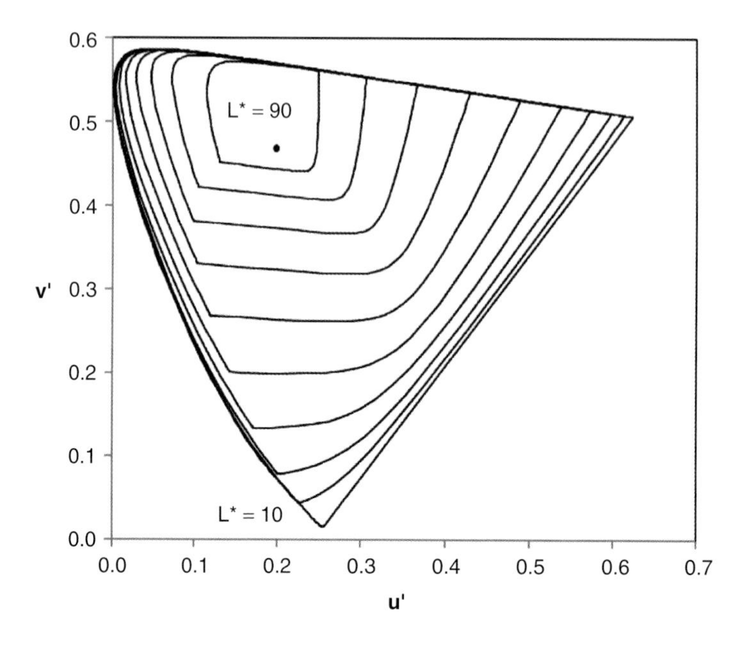

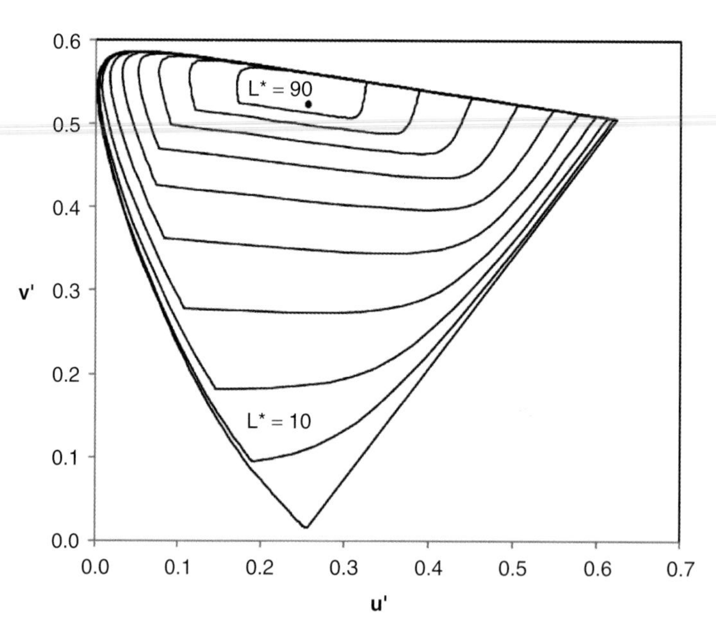

Figure 8.3 Limits of optimal colours on the u′,v′ chromaticity diagram for values of L^* equal to 10, 20, 30, 40, 50, 60, 70, 80, and 90, for a daylight illuminant (Standard Illuminant D65, upper) and for a tungsten-light illuminant (Standard Illuminant A, lower)

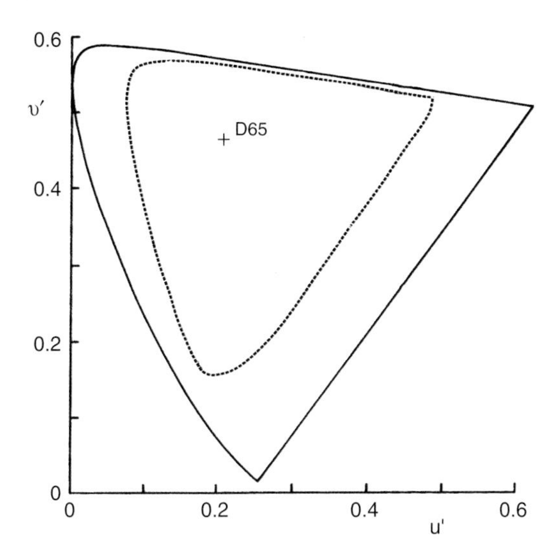

Figure 8.4 Limits of real surface colours (broken lines), for a daylight illuminant, D65 (Pointer, 1980)

8.4 THE MUNSELL SYSTEM

One of the most widely used colour order systems is the *Munsell* system, originated by the artist A.H. Munsell in 1905, and extended and refined in various ways since (Nickerson, 1976). The general arrangement follows broadly that depicted in Figures 8.1 and 8.2, and can be seen in Figure 8.7. An important feature of the Munsell system is that the colours are arranged so that, for each perceptual attribute used, the perceptual colour difference between any two neighbouring samples is, as nearly as possible, constant (Berns and Billmeyer, 1985). In Figure 8.8 shows reproductions of four Munsell hue pages; because of the limitations of colour printing, there are some deviations from the colours used in the Munsell system, but the general arrangement is shown.

In the case of the grey scale, there are ten main steps, with white designated 10, and black zero, the greys having values from 1 to 9 as they become lighter. The difference in lightness between any neighbouring pair of samples, say 3 and 4, is then intended to be perceptually as great as between any other neighbouring pair, say 7 and 8. The luminance factors that the grey samples need to have to achieve this result are affected to some extent by the background against which the samples are viewed. In the case of the Munsell system, the judgements used in its derivation were based on viewing the samples against a medium grey background having a luminance factor of about 20%. The relationship between luminance factor and number on this scale is used for all the coloured samples as well as for the greys, and is referred to as Munsell Value. The Munsell Value, V, of a sample can be intermediate between the whole numbers of the eleven basic neutral samples, and it is designated by using decimals; for example, a sample having a value of 8.5

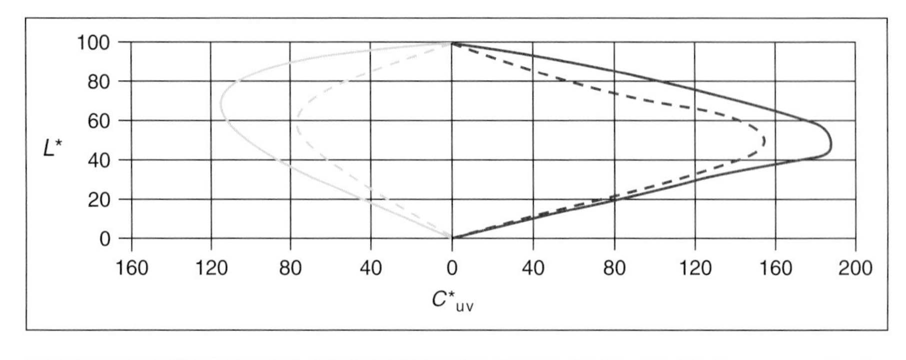

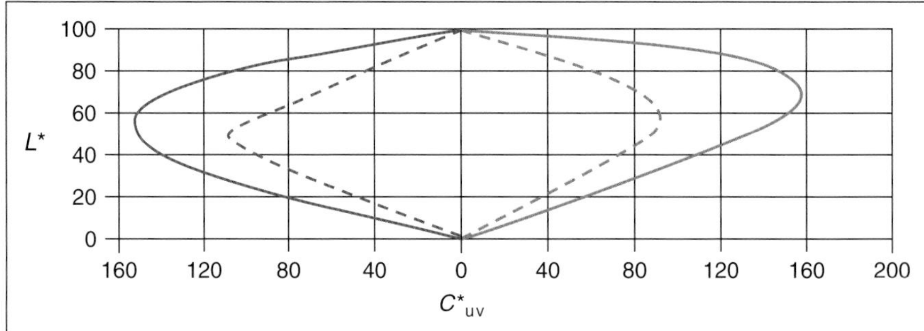

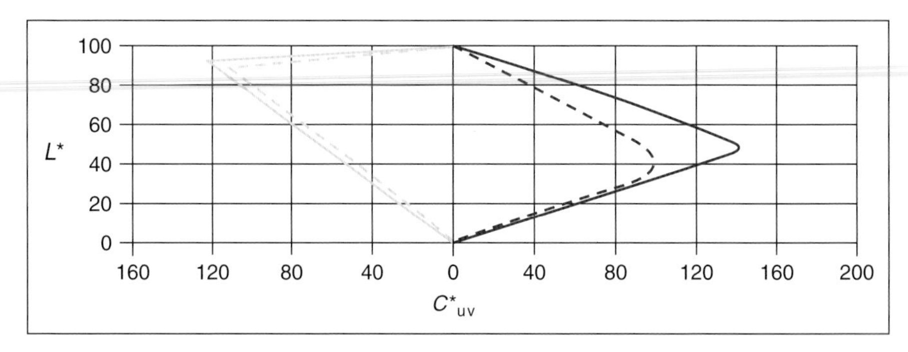

Figure 8.5 Limits of optimal colours having minimum reflectance factors of zero (full lines) and of real surface colours (broken lines), for six hues on plots of L^* against C^*_{uv} for a daylight illuminant (D65). The hues have values of h_{uv} equal to 10° (a red), 70° (a yellow), 150° (a green), 190° (a cyan), 250° (a blue), and 330° (a magenta)

would be intended to be perceptually midway in lightness between samples having values of 7 and 8.

The spacing of the hues around the grey scale axis is also intended to represent uniform differences in perceived hue between neighbouring hue pages. There are five principal hues, Red, Yellow, Green, Blue, and Purple, and they are designated: 5R, 5Y, 5G, 5B, and 5P, respectively. The intermediate hues are designated: 5YR, 5GY, 5BG, 5PB, and 5RP. Finer

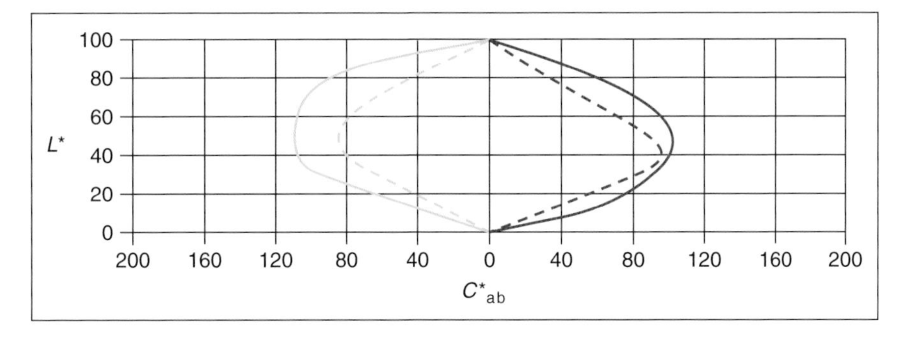

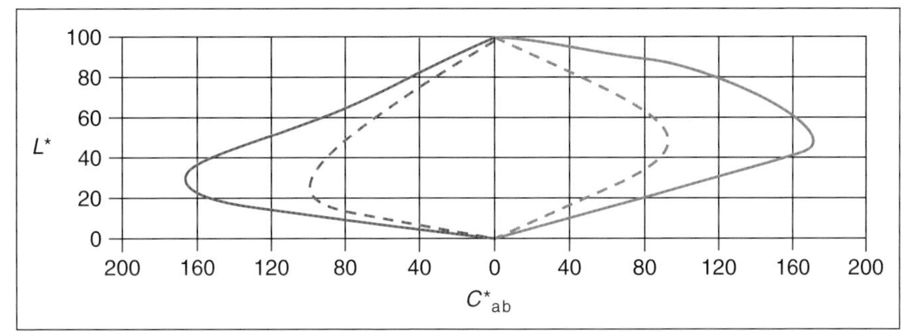

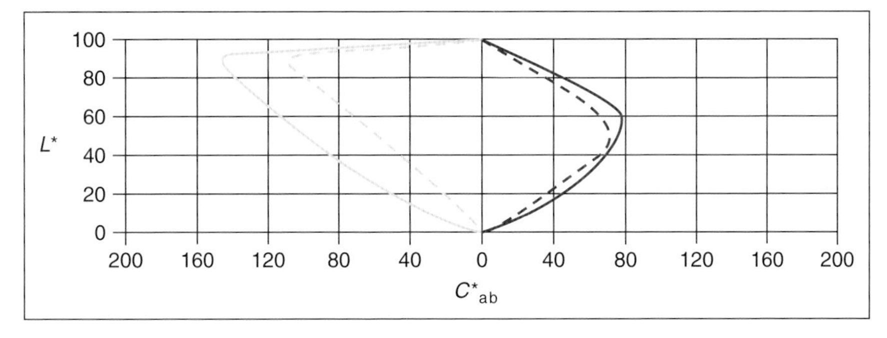

Figure 8.6 Same as Figure 8.5, but on plots of L^* against C^*_{ab}, for values of h_{ab} equal to 30° (a red), 90° (a yellow), 170° (a green), 190° (a cyan), 230° (a blue), and 310° (a magenta)

divisions between 5R and 5YR are designated: 6R, 7R, 8R, 9R, 10R, 1YR, 2YR, 3YR, and 4YR, with similar designations between other hues, as shown in Figure 8.9. Once again, finer divisions are represented by decimals; for example, a Munsell Hue of 2.5YR is intended to be perceptually midway between samples having Munsell Hues of 2YR and 3YR.

The distances of the samples from the grey scale axis are intended to represent uniform differences in perceived chroma (not saturation) and are given numbers that are typically as small as 4 or less for weak colours, and 10 or more for strong colours. Munsell Chroma

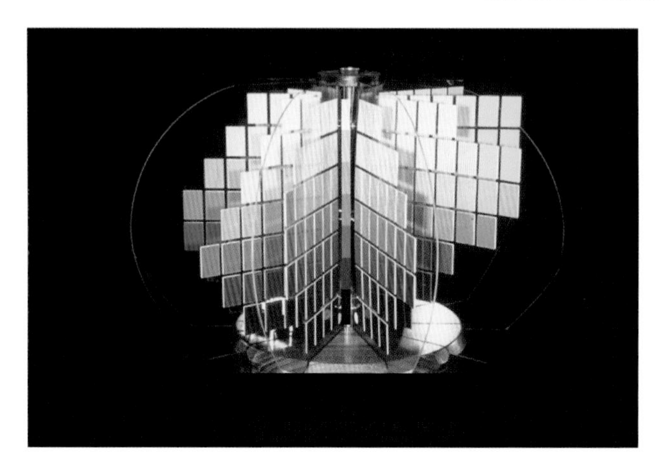

Figure 8.7 The Munsell Color solid. Because of the limitations of printing, all the colours cannot be shown accurately, but the general layout is displayed adequately. References to Munsell® in this publication are used with permission from X-Rite Inc

is indicated by an oblique line preceding the numerical value, for example /8. Once again decimals are used to indicate intermediate samples between integers; thus, for example, /8.5 indicates that the sample is intended to be perceptually midway in chroma between samples having Munsell Chromas of /8 and /9. In Figure 8.9 the dots represent samples having Munsell Chromas of /2, /4, /6, /8, and /10, for Munsell Hues at the 5 and 10 positions.

The full Munsell specification is always given in the order Hue, Value, Chroma (sometimes referred to as HVC), for example, 2.5Y6/8 indicates that the Hue is half way between 10YR and 5Y (so that it is a slightly orange yellow, 10YR being a yellowish orange and 5Y a yellow), that the lightness is slightly lighter than a medium grey (which would have a Value of 5), and that it has a fairly strong Chroma.

The maximum designation possible for Munsell Chroma depends on the hue and lightness of the colour; for dark blues it is larger than for light blues (this is because of the small contribution of the β cones to the achromatic signal); and for medium-lightness reds it is larger than for medium-lightness greens (this is because of the overlapping of the cone spectral sensitivity curves in the green part of the spectrum). The outer boundary of the Munsell Colour space is therefore not symmetrical.

If two samples of the same Munsell Hue and Chroma, but differing by 1 unit in Value, are compared with two samples of the same Hue and Value, but differing by 2 units in Chroma, it is found that the two differences are perceptually similar in magnitude; in other words, a difference of 1 unit of Value is similar in perceptual magnitude to a difference of 2 units of Chroma. This difference is also similar in perceptual magnitude to that between two samples having the same Value, and both having a Chroma of 5, but differing by 3 units of Hue. For samples of higher Chroma, the perceptual magnitude of the hue difference would be greater, and for samples of lower Chroma it would be less; this is because of the angular arrangement of the Hue parameter in the Munsell System.

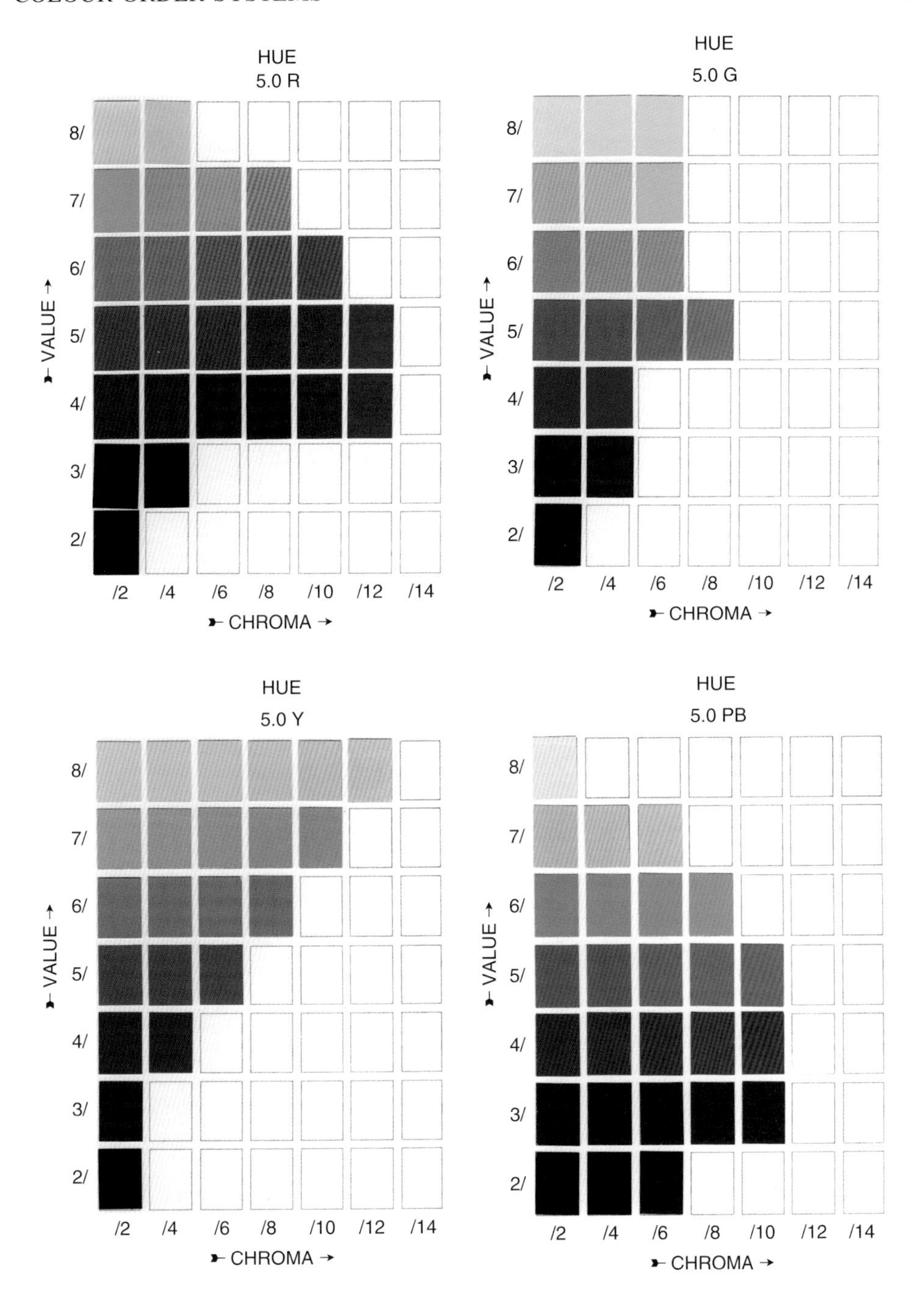

Figure 8.8 Reproductions of pages 5R, 5Y, 5G, and 5B from the Munsell Book of Color. Because of the limitations of printing, all the colours cannot be shown accurately, but the general layout is displayed adequately. References to Munsell® in this publication are used with permission from X-Rite Inc

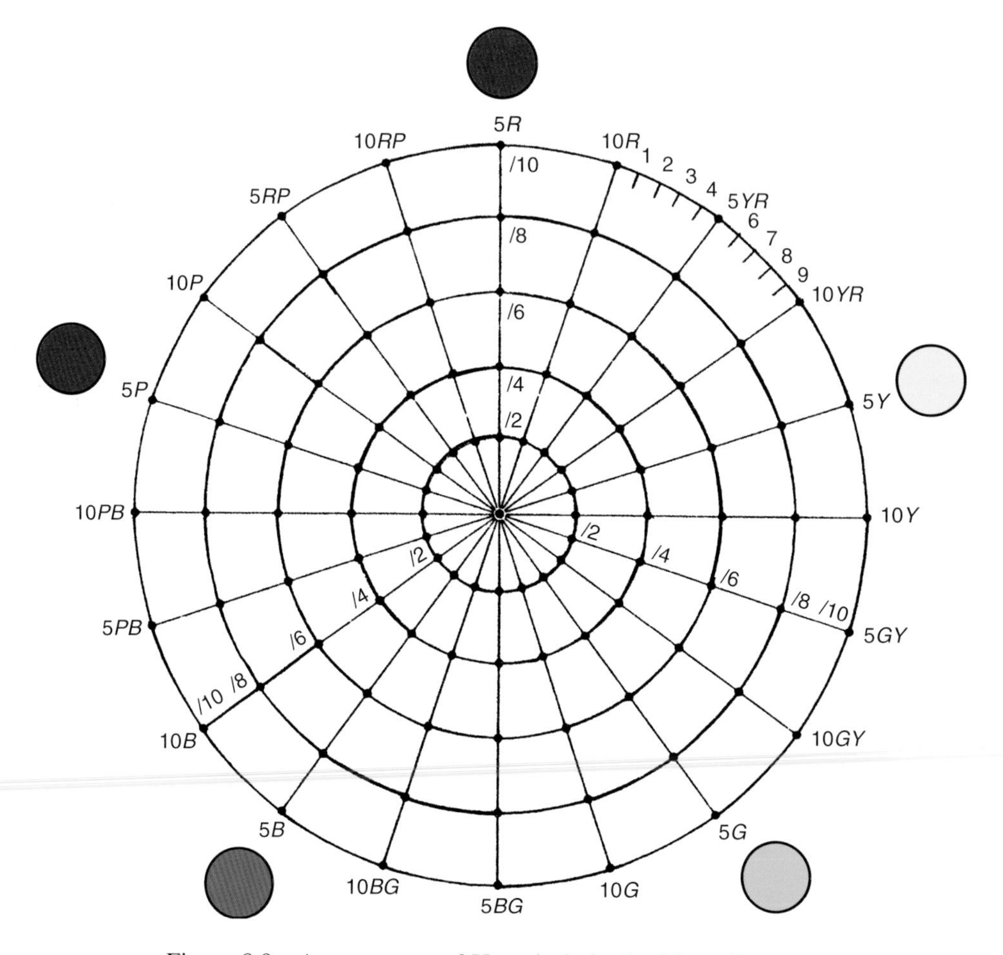

Figure 8.9 Arrangement of Hue circle in the Munsell system

8.5 THE MUNSELL BOOK OF COLOR

So far we have only considered the Munsell system, the perceptual basis on which the ordering of the colours depends. We must now consider the actual physical samples, and the way they are arranged to form an atlas, in this case the Munsell Book of Color.

The samples in the book consist of painted paper chips, typically about 17 mm × 20 mm in size; larger size pieces are also available, if required for special purposes. Most of the pages in the book are laid out in the manner shown by the examples in Figure 8.8; each page shows samples of a single Munsell Hue, there being 40 such pages, those included being for Munsell Hues 2.5YR, 5YR, 7.5YR, 10YR, 2.5Y, and so on right round the hue circle to 10R, and so back to 2.5YR. Figure 8.10 illustrates how different hue pages can have different shapes; a yellow page can have colours of high Chroma and high Value, but a blue page cannot.

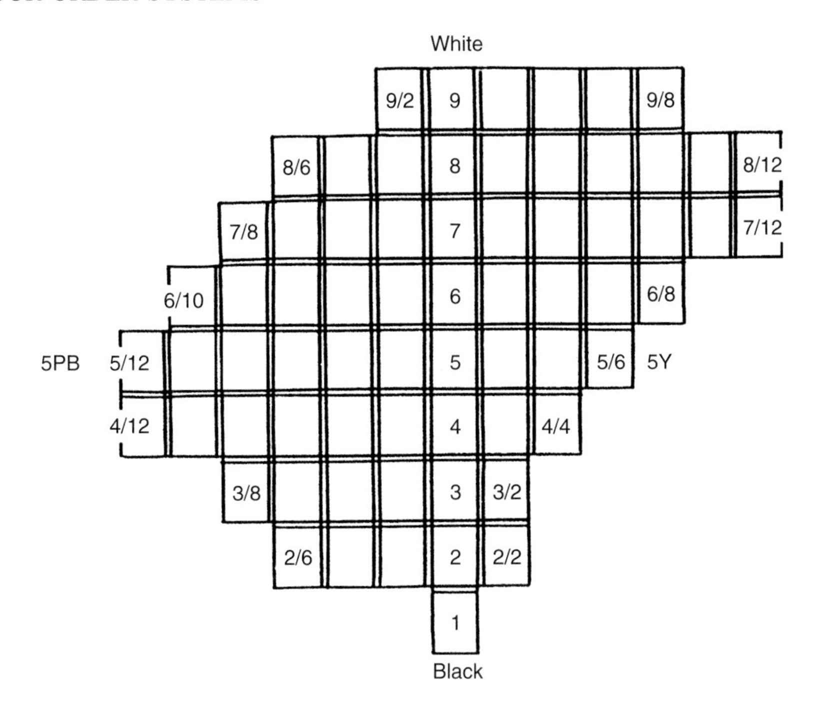

Figure 8.10 Arrangement of colours of constant Hue in the Munsell system

In 1943, a report was issued (Newhall, Nickerson, and Judd, 1943) in which, for CIE Illuminant C (SC), the X, Y, Z, tristimulus values of the samples in the Munsell Book of Color were carefully studied. It was found that, although the transitions in the tristimulus values from one sample to a neighbouring one should be smooth, there were some irregularities. A system for which the tristimulus values were smooth was then produced, known as the Munsell Renotation System. Chips made to these specifications were used for the glossy sample version of the Book of Color, introduced in 1957. For the original book of matt samples, the chips still used the original notation for some years, but since 1957 all the books produced, both matt and glossy, have been based on the renotation, which is now referred to simply as Munsell Notation (Davidson, Godlove, and Hemmendinger, 1957).

Tables of the x and y chromaticity co-ordinates and the Y tristimulus values (for CIE Illuminant C), corresponding to Munsell Notations can be found in various places (see, for instance, Wyszecki and Stiles, pages 488 to 500, 1967, or pages 840 to 852, 1982).

We must now compare the parameters of the Munsell system with some of the CIE measures that were described in Chapter 3. There is a simple approximate relationship between Munsell Value, V, and CIE 1976 lightness, L^*: L^* is approximately 10 times V. However, a Munsell Value of 10 corresponds to the perfect diffuser; hence, if L^* is evaluated using a reference white for which its reflectance factor, Y_n, is less than 100, its value for the perfect diffuser will be more than 100, and thus more than 10 times the Munsell Value.

In Figure 8.11, the u', v' chromaticities corresponding to Munsell Notation are shown for Munsell Value 5, at Munsell Hue intervals of 2.5, and at Munsell Chroma intervals

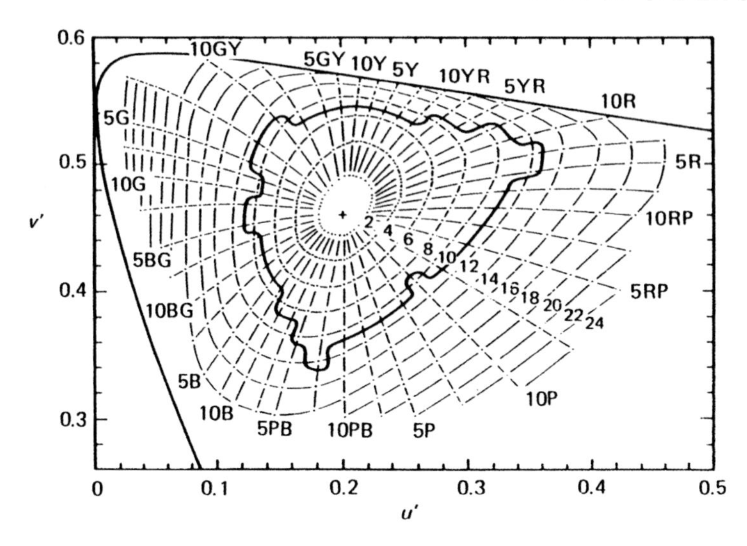

Figure 8.11 Chromaticities of colours of Munsell Value 5 shown on the u', v' diagram, for CIE Illuminant C (shown as +)

of 2. Only the colours plotting within the unbroken irregular line inside the spectral locus are included in the Munsell Book of Color (matt version); those lying outside have been extrapolated and are not based on perceptual experiments on actual chips. If the spacing of the colours in the u', v' diagram were the same as that in the Munsell System at Munsell Value 5, then, in Figure 8.11, the Munsell Hues would lie on straight lines radiating from the illuminant point, separated by equal angles all round the hue circuit, and the Munsell Chromas would lie on concentric circles, centred on the illuminant point and separated by equal distances, so as to form a grid of the same general shape as that of Figure 8.9. It is clear from Figure 8.11 that there is a tendency for this to be the case, but the conformance is only approximate. The points representing single Munsell Hues lie on lines that are slightly curved, especially for some hues; the points representing samples with constant Munsell Chroma lie on roughly circular contours which are stretched in the yellow and blue directions, and also, for the higher values of Chroma, in the red direction. As mentioned in Sections 3.4 and 3.7 loci of constant perceived hue are not exactly straight lines in a chromaticity diagram, and the Munsell hue loci represent such loci more accurately; a colorimetric system that represents hue loci more accurately will be described in Chapter 15.

In Figure 8.12 plots are shown of the loci of constant Munsell Hue and Chroma for Munsell Value 5 in u^*,v^* and a^*,b^* diagrams. Again, conformity between the Munsell system and the CIELUV and CIELAB systems would be shown by regular patterns of straight lines and concentric circles in this figure, and there is an approximate tendency for this to be so. But the circles are compressed in the purple region in the u^*,v^* diagram, and extended in the yellow region in the a^*,b^* diagram, with various irregularities of hue spacing, particularly in the purple region of the u^*,v^* diagram, and in the yellow-green region of the a^*,b^* diagram.

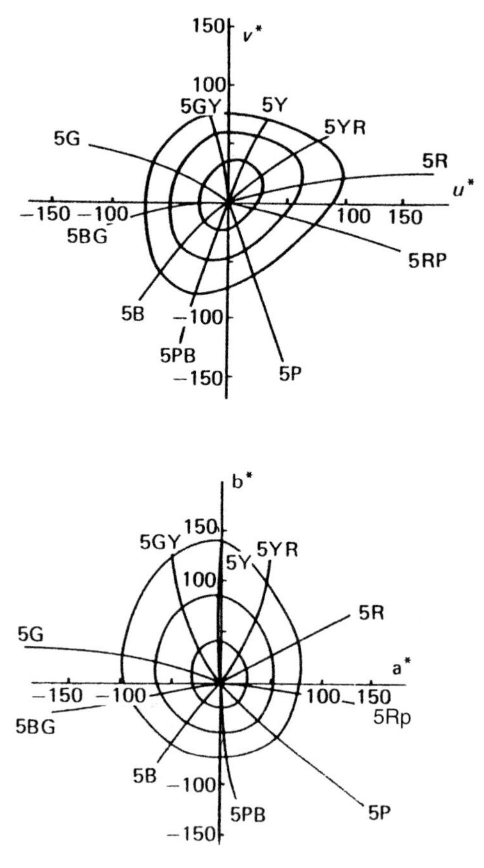

Figure 8.12 Loci of colours of constant Munsell Hues, and Munsell Chromas (of 4, 8, and 12), for Munsell Value 5, in u^*,v^* (upper) and a^*,b^* (lower) diagrams for CIE Illuminant C

In Figure 8.13, plots are shown of Munsell samples in the L^*,C^*_{uv} planes of the CIELUV system. In this case, if the spacing were the same as in the Munsell system, neighbouring points should form a series of squares (because the samples have Munsell Value differences of 1, and Munsell Chroma differences of 2). It is clear that, again, there is a tendency for this spacing to occur but the conformance is only very approximate. If C^*_{ab} of the CIELAB system is used instead of C^*_{uv} the resulting plots are broadly similar, but different in detail, as can be seen from Figure 8.14. The values of C^* are between about 5 and about 12 times those of Munsell Chroma.

The CIELUV and CIELAB systems have a general similarity to the Munsell system because they were designed with this intention. However, it is known that the Munsell system is not perfectly uniform, so that Figures 8.11, 8.12, 8.13, and 8.14 can only be used with caution to estimate the uniformity of the CIELUV and CIELAB systems.

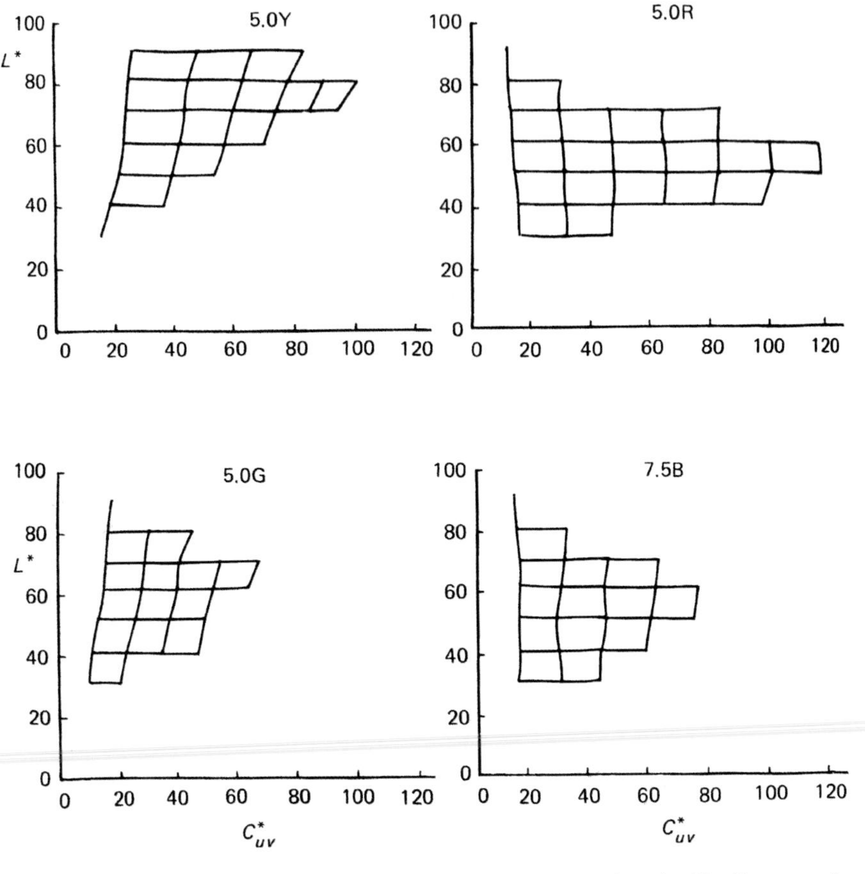

Figure 8.13 Positions of colours of four Munsell Hues plotted in L^*, C^*_{uv} diagrams for CIE Illuminant C. Adjacent samples have differences of 1 in Munsell Value and 2 in Munsell Chroma

8.6 UNIQUE HUES AND COLOUR OPPONENCY

Towards the end of the last century, the work of Young, Helmholtz, and Maxwell, amongst others, had helped to emphasise the basically trichromatic nature of colour vision. But Hering drew attention to the fact that the number of unique hues is not three, but four: red, yellow, green, and blue. Today, the trichromatic features of colour vision are associated with the three different types of cone ρ, γ, and β, and the unique hues with the colour-difference signals, as described in Sections 1.6 and 1.7. The hues red, yellow, green, and blue are said to be unique because they cannot be described in terms of any combinations of other colour names. Thus, for instance, although orange can be described as a yellowish red or a reddish yellow, red cannot be described as a yellowish blue or a bluish yellow. In fact, the four unique hues comprise two special pairs, red and green, and yellow and blue; the colours in each of these pairs are opponent, in the sense that they cannot both be perceived simultaneously as component parts of any one colour. That is, it is impossible to have a reddish

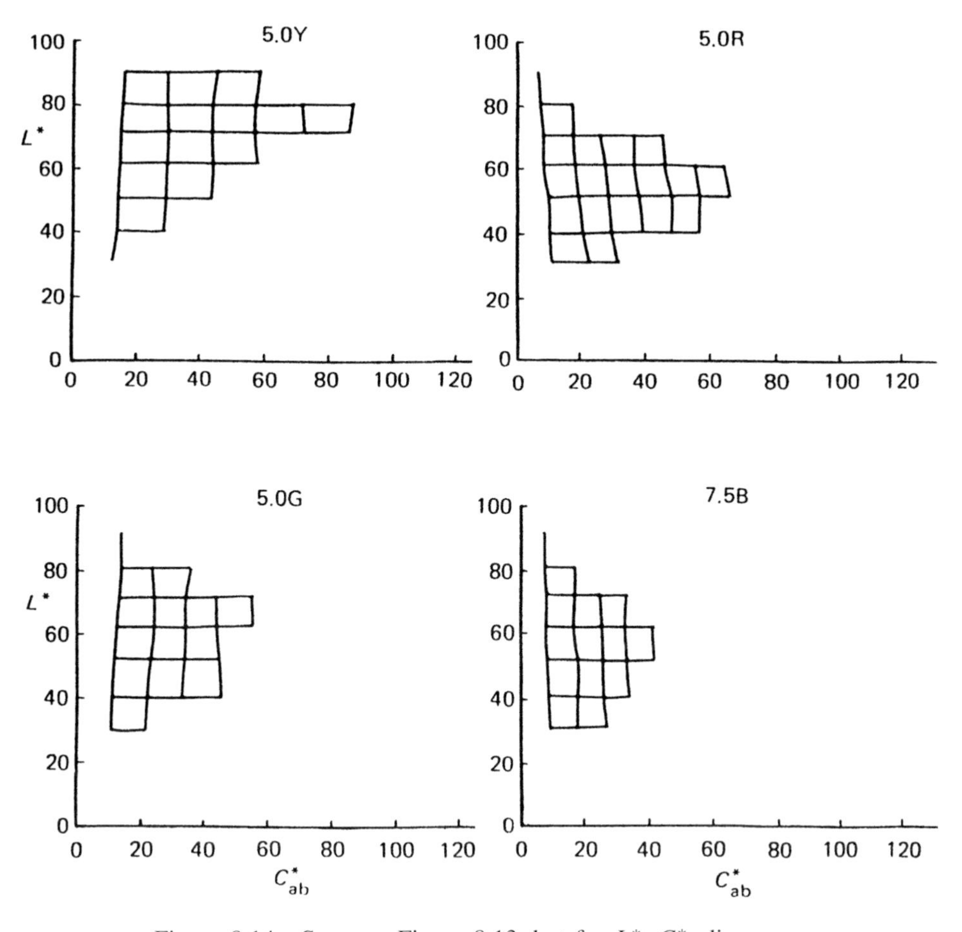

Figure 8.14 Same as Figure 8.13, but for L^*, C^*_{ab} diagrams

green, or a greenish red, or a yellowish blue, or a bluish yellow. But yellowish reds, reddish yellows, greenish yellows, yellowish greens, bluish greens, greenish blues, reddish blues, and bluish reds are all possible, and may be referred to as *binary hues*. Of course, other colour names are used, such as orange, turquoise, purple, mauve, or magenta; but, perceptually, these can all be described by using combinations of the unique hue names, yellow and red in the case of orange, green and blue in the case of turquoise, and blue and red in the case of purple, mauve, and magenta. (Some people may think in terms of green being a mixture of blue and yellow, but this comes from an association with a way of producing green colours by mixing pigments or dyes; once a green has been produced, perceptually it may be a yellowish green or a bluish green, but it cannot be both yellowish and bluish at the same time, and hence green is not perceptually a yellowish blue or a bluish yellow.)

These four unique hues, together with white and black, make six basic colours; and, as the six constitute three pairs, (red and green, blue and yellow, white and black) the trichromacy of colour vision has not disappeared, but is merely expressed in a different form. The black-white pair is different in its opponency from the unique-hue pairs, in that blackish whites and whitish blacks are possible, and in fact are so commonly experienced that the colour name grey exists to describe them.

8.7 THE NATURAL COLOUR SYSTEM (NCS)

These ideas of Hering were developed in Sweden by Tryggve Johansson and Sven Hesselgren, and, more recently, by Anders Hård and his co-workers so as to produce the *Natural Colour System* or *NCS* (Hård and Sivik, 1981). In the NCS, colours are described in terms of the relative amounts of the basic colours that are perceived to be present; these amounts are expressed as percentages. Thus, a medium grey that is perceived to have equal amounts of whiteness and blackness is described as having a whiteness of 50 and a blackness of 50. A colour that is perceived to be a pure red, with no trace of yellowness or blueness or whiteness or blackness, is described as having a redness of 100. Colours having only combinations of redness, blackness, and whiteness are represented in a triangular array, as shown in Figure 8.15. The three sides of the triangle are of equal length, with white represented by the point, W, at the top, black by the point, S, at the bottom, and red by the point, R, to the right and opposite the middle of the line WS. Colours that are perceived to have only redness and whiteness are then situated on the line WR, and their position on this line represents the proportions of redness and whiteness; for instance, a quarter along from W represents 75% whiteness and 25% redness; midway represents 50% of each; and so on. Similarly, points on the line SR represent colours having only blackness and redness; and points on WS, those having only whiteness and blackness (the greys). Points lying within the triangle represent colours having some whiteness, blackness, and redness, the relative amounts being represented by the distances, w, s, and r of the point from the lines SR, WR, and WS, respectively; these distances are expressed as percentages of the maximum

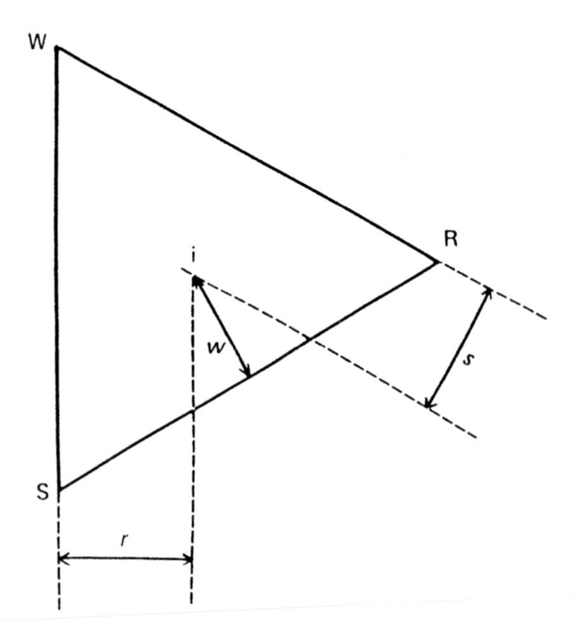

Figure 8.15 Arrangement of colours on a white-black-red plane of the NCS

possible within the triangle, and this is the distance of a corner of the triangle from the opposite side. The three lengths always add up to this maximum distance, and hence the three numbers, w, s, and r, always add up to 100. Since

$$w + s + r = 100$$

it is only necessary to quote two of the three numbers: the convention adopted is that NCS whiteness, w, is omitted; thus a colour for which, for example, $w = 20$, $s = 50$, and $r = 30$, would be specified as $s = 50$, $r = 30$.

Similar triangles exist for the other unique hues, yellow, green, and blue, in which cases the values quoted are s and y, or s and g, or s and b, respectively. Triangles also exist for hues consisting of mixtures of neighbouring pairs of unique hues, and for these intermediate hues, the right-hand point of each triangle represents colours of these hues that have no perceived whiteness or blackness content. All the triangles are then fitted into a three-dimensional array in which the white-black line is vertical and common to all triangles, and the triangles representing different hues occupy different angular positions as shown in Figure 8.16.

The specific hue of any colour is determined by its perceived contents of red and yellow, or yellow and green, or green and blue, or blue and red. Thus, if a colour is perceived to have 80% yellow, 20% red, no whiteness and no blackness, it would occupy the right-hand corner of the triangle at the Y20R position in Figure 8.16, as indicated by the point C in Figure 8.17.

If another colour had 10% whiteness, 30% blackness, 40% yellowness, and 20% redness, its hue would be determined by the ratio of its two hue contents, that is 40 to 20, which in

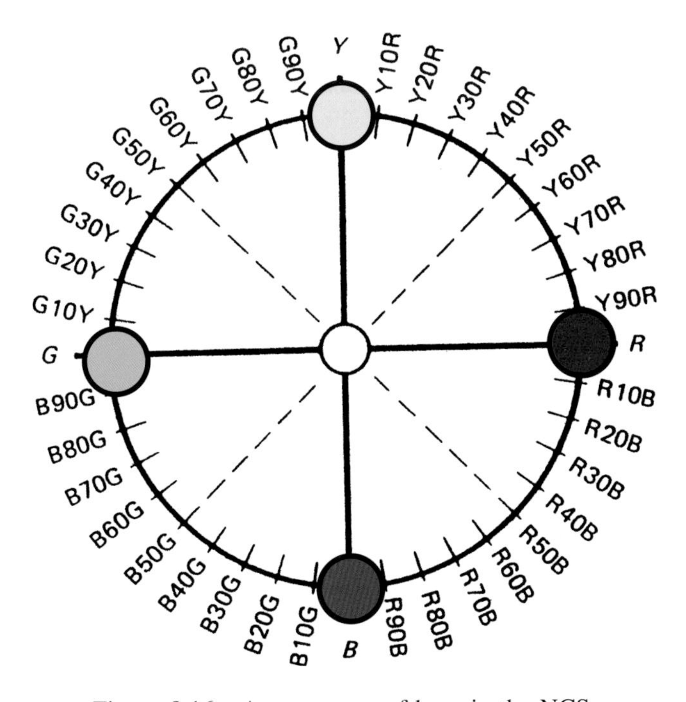

Figure 8.16 Arrangement of hues in the NCS

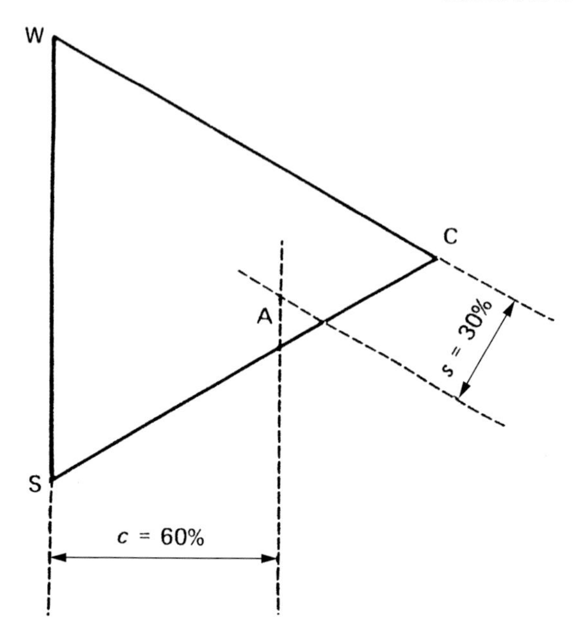

Figure 8.17 Same as Figure 8.15 but for a hue, C

percentage terms is 66% to 33%, so that it would be on the Y33R triangle; its position in this triangle would be 30% of the maximum possible distance from WC, and 60% of the maximum possible from WS as shown by the point A in Figure 8.17; this 60% is the sum of the two hue contents, and represents the total chromatic content, and this is termed the NCS chromaticness, c. The specification of the colour is then $s = 30$, $c = 60$, with a hue ratio of 66 yellow to 33 red. The NCS specification is then abbreviated to S3060-Y33R; the numbers are always given in the order blackness, chromaticness, and hue, and the letter S is used when referring to the 2nd edition of the NCS Atlas. The NCS hue is indicated by the initial letter of one unique hue, followed by the percentage of a second unique hue, and then the initial letter of the second hue, the hues always being given in the order, Y, R, B, G, Y (see Figure 8.16). Reproductions of two pages from the NCS Atlas are given in Figure 8.18; because of the limitations of colour printing, there are some deviations from the colours used in the NCS system, but the general arrangement is shown.

8.8 NATURAL COLOUR SYSTEM ATLAS

To produce an atlas based on the above concepts, over 60 000 observations were made on samples of coloured papers viewed in daylight. Based on these observations, colour samples were produced for a prototype atlas having about 1200 samples. The relative spectral power distributions of these samples were measured, for CIE Illuminant C (SC), from which their X, Y, Z tristimulus values were calculated. Comparison of the observations and the measurements was then used to obtain smoothed and adjusted X, Y, Z tristimulus values for about 16 000 NCS notations (for details, see Swedish Standard, 1982). The final atlas was intended to portray samples at every tenth percentage step in blackness and

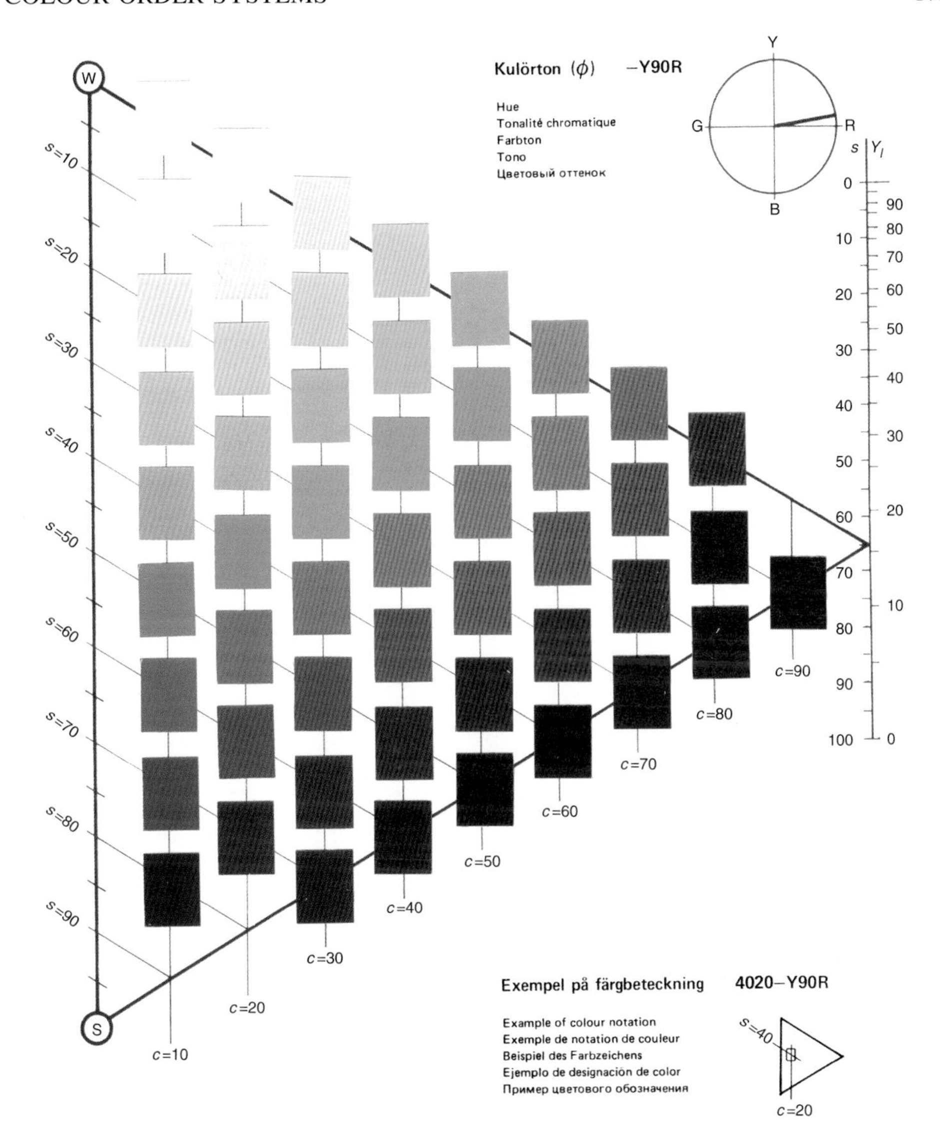

Figure 8.18 Reproduction of two pages from the NCS Atlas. Because of the limitations of printing, all the colours cannot be shown accurately, but the general layout is displayed adequately. NCS - Natural Colour System®© property of NCS Colour AB, Stockholm 2011. References to NCS®© in this publication are used with permission from NCS Colour AB

chromaticness, for 40 different hues. This would have provided over 2000 samples (55 on each hue page, plus the grey scale), but pigments do not exist to produce some of the colours, so that not all the points can be illustrated. Production atlases can have about 2000 samples, but some of these samples illustrate positions additional to the tenth percentage steps. The size of the samples in the NCS atlas is 12 mm × 15 mm; larger size pieces are also available, if required for special purposes.

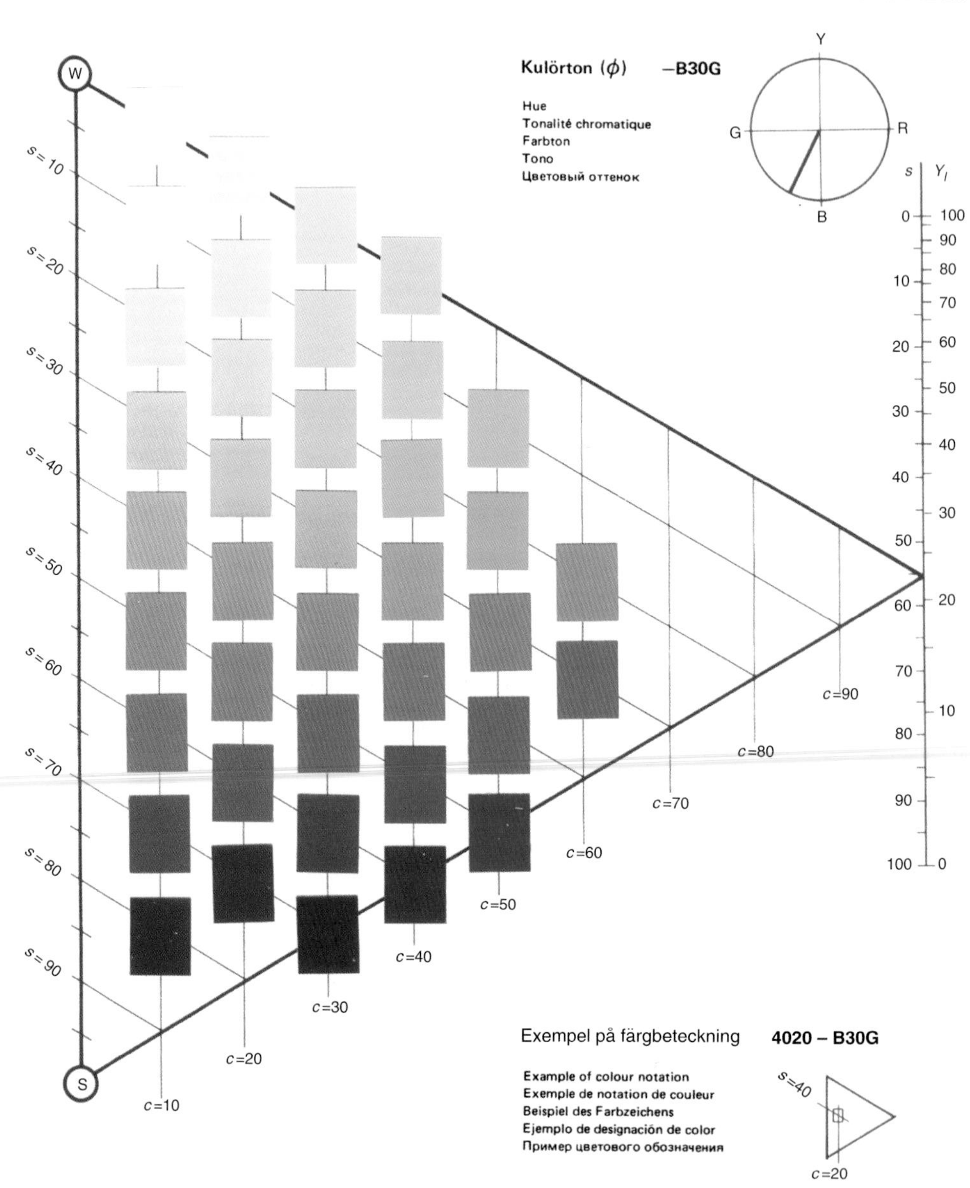

Figure 8.18 (*Continued*)

In Figure 8.19 are shown the loci of the u', v' chromaticities for the unique hues of the NCS Atlas having no blackness. The curvature of the lines is much the same as that for similar Munsell Hues, as can be seen by comparing Figure 8.19 and Figure 8.11; it can also be seen that NCS unique red corresponds approximately to Munsell 5R, unique yellow to Munsell 5Y, unique green to Munsell 5G, and unique blue to Munsell 7.5B. The Munsell system was designed with the aim of having an equal number of perceptual colour differences between its five principal hues 5R, 5Y, 5G, 5B, and 5P, and this is not

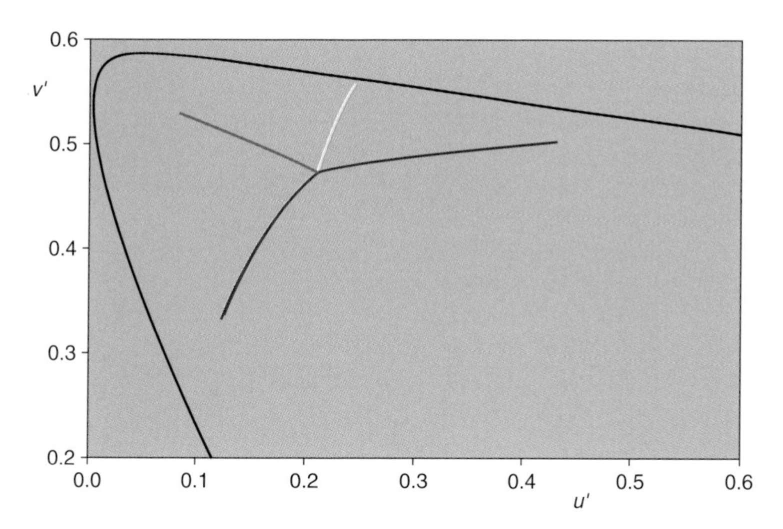

Figure 8.19 Loci of NCS unique hues on the u′,v′ diagram for Standard Illuminant D65

necessarily compatible with four of them corresponding to unique hues. It is, therefore, not surprising that 5B does not correspond to unique blue. The inclusion of the purple hue in the Munsell system was necessary because of the larger number of perceptual hue differences between blue and red than between red and yellow, or yellow and green, or green and blue. The arrangement of hues in the NCS, as shown in Figure 8.16, therefore, means that there are many more discriminable hue differences in the blue-red NCS quadrant, than in the other three NCS hue quadrants. This is illustrated in Figure 8.20. This is not necessarily to be thought of as a disadvantage; it is merely a consequence of the basis on which the NCS is formulated.

In Figure 8.21, u', v' chromaticities are shown for NCS colours of zero blackness, not only for the unique hues, but also for some intermediate hues, and for colours of constant NCS chromaticnesses of 30, 50, 70, and 90. It is clear that the general pattern is not too dissimilar from that of the Munsell system for colours of Munsell Value 5, shown in Figure 8.11. However, for purple colours, the Munsell Chroma contours tend to be nearer the illuminant point than the NCS chromaticness contours, and this can be seen more clearly in Figure 8.22, where two contours have been shown for comparison that are very similar in the red-yellow-green-blue quadrants, but appreciably different in the blue-red quadrant. The reason for this difference is not known, but the inclusion of the purple hue in the Munsell system may have had an effect on the observations in this quadrant that were absent in the NCS studies.

In Figure 8.23, the NCS network is shown for some different blackness and chromaticness levels in each of four NCS hue planes in plots of L^* against C^*_{uv}; in Figure 8.24 the same is done for L^* plotted against C^*_{ab}. The lines of constant NCS chromaticness are approximately vertical in these plots, but with a tendency to slope in towards the black point, particularly at low lightnesses; this shows that NCS chromaticness is intermediate between CIE 1976 chroma (for which the loci are vertical) and CIE 1976 saturation (for

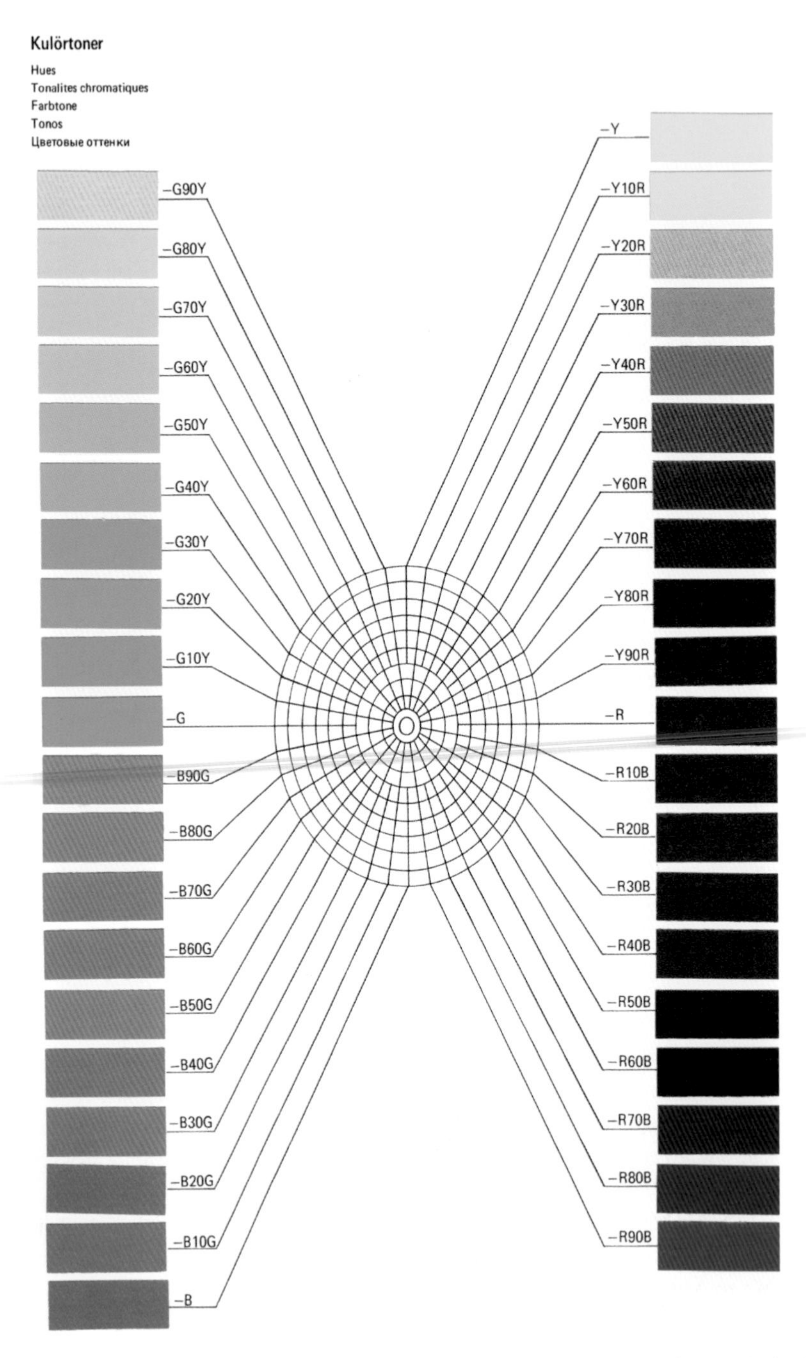

Figure 8.20 Arrangement of hues in the NCS system. Because of the limitations of printing, all the colours cannot be shown accurately, but the general layout is displayed adequately. NCS - Natural Colour System®© property of NCS Colour AB, Stockholm 2011. References to NCS®© in this publication are used with permission from NCS Colour AB

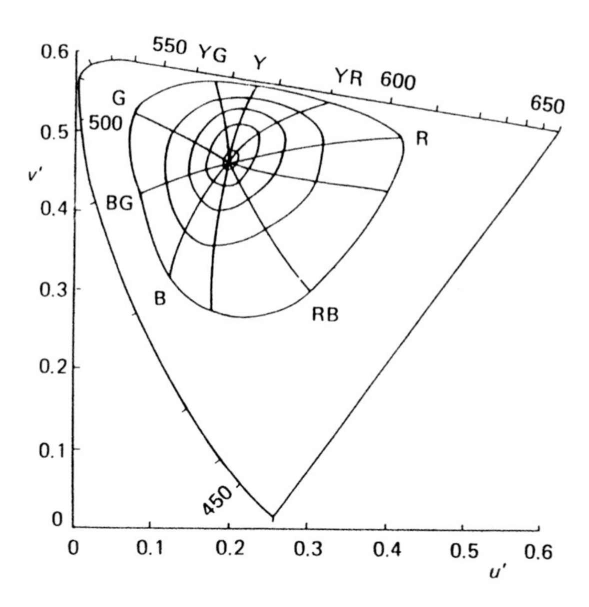

Figure 8.21 Loci of colours of zero blackness, for various hues, and for chromaticness values of 30, 50, 70, and 90, for the NCS, for CIE Illuminant C

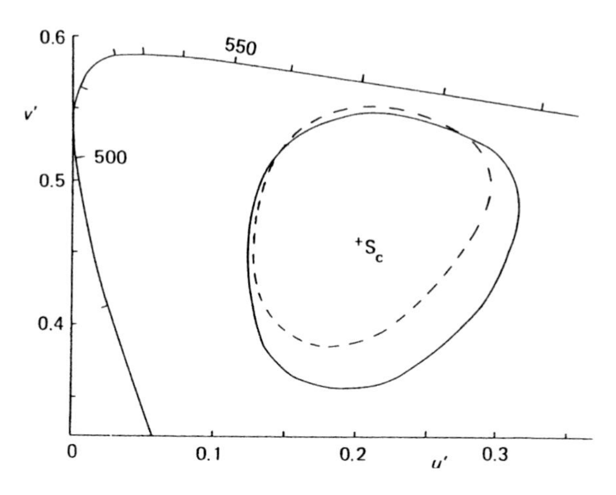

Figure 8.22 Loci of colours of a constant Munsell chroma (broken line) and a constant NCS chromaticness (full line) compared for CIE Illuminant C, both at a reflectance factor corresponding to Munsell Value 5

which the loci radiate from the black point in the L^*,C^*_{uv} plots). The lines of constant NCS blackness slope downwards as they radiate from the grey scale axis and are not horizontal, as is the case for lines of constant CIE 1976 lightness, L^*, and also for those of constant Munsell Value.

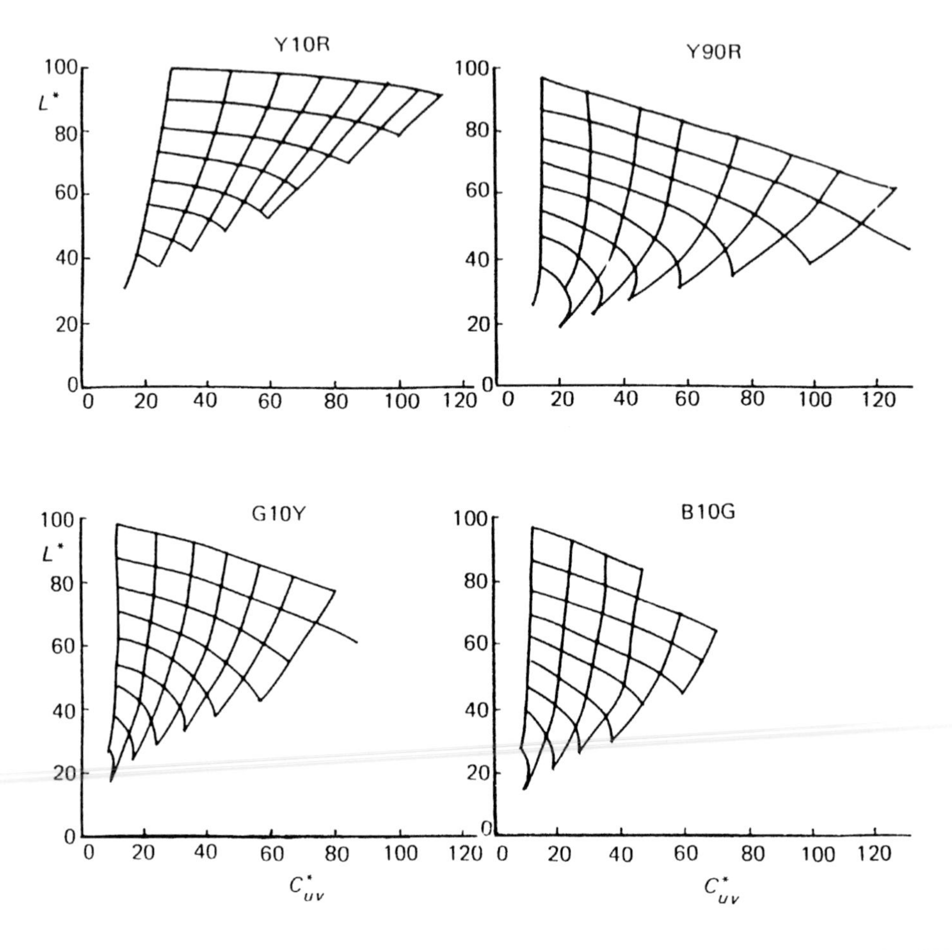

Figure 8.23 Positions of colours of four NCS hues plotted in L^*, C^*_{uv} diagrams for CIE Illuminant C

It is claimed that the attribute of NCS blackness is more readily perceived in colours than the attribute of lightness, with which Munsell Value correlates. There is also evidence that, in the use of colour in architecture, Munsell Value is not an ideal variable, and blackness might be more suitable (Gloag and Gold, 1978; Whitfield, O'Connor, and Wiltshire, 1986). It is interesting in this connection to note that colours of different hues, but having the same NCS blackness and chromaticness, are described by some designers as having a certain equivalence, sometimes referred to as equality of *nuance* or *weight*.

Variables that are more similar to NCS blackness than to Munsell Value are used in other colour systems intended as aids to colour designers; one example is the DIN system (to be described in Section 8.9).

8.9 THE DIN SYSTEM

The *DIN* (*Deutsches Institut für Normung*) system was developed by M. Richter and his co-workers in Germany (Richter and Witt, 1986). The three variables used are hue, T, saturation, S, and darkness, D, given in that order. Thus for a DIN specification of 16:6:4, for instance, $T = 16$, $S = 6$, and $D = 4$. The system uses Standard Illuminant D65, and CIE X, Y, Z tristimulus values for the samples are available (see Wyszecki and Stiles, pages 503 to 505, 1967, or pages 863 to 865, 1982).

Colours of constant dominant (or complementary) wavelength are regarded as being of constant hue. There are 24 principal hues, having values of $T = 1$ for a yellow, proceeding via reds, purples, blues, and greens, to a yellow-green of $T = 24$, and back to $T = 25 = 1$.

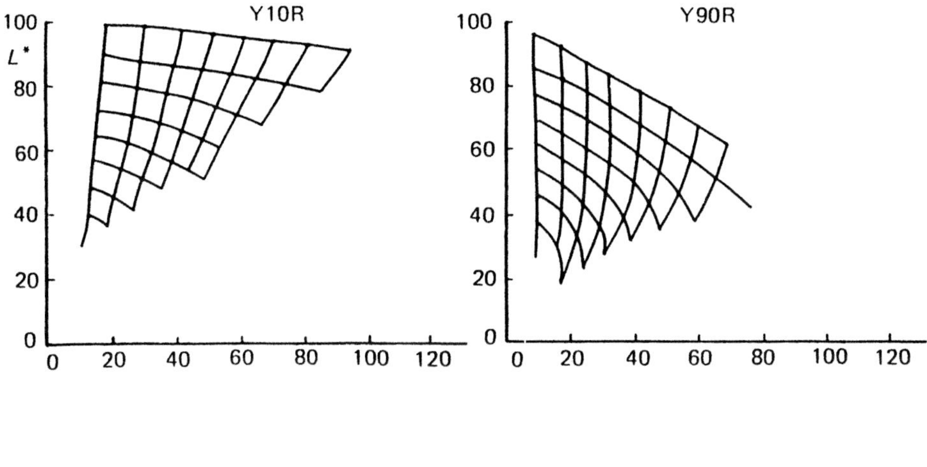

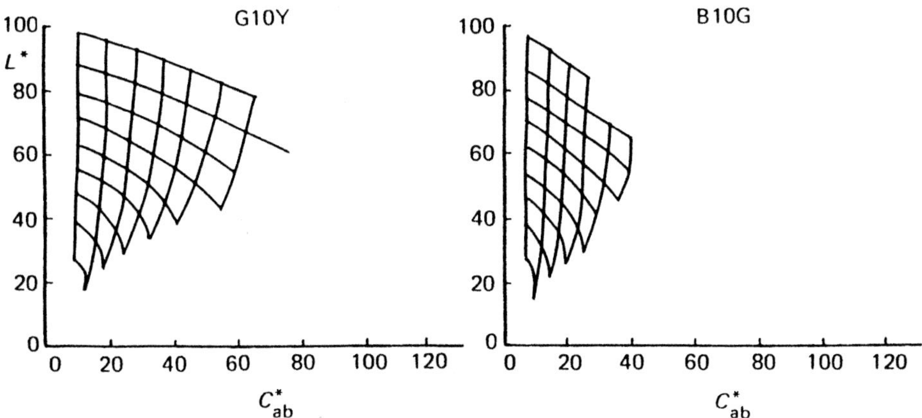

Figure 8.24 Same as Figure 8.23 but for L^*, C^*_{ab} diagrams

The 24 principal hues were chosen to represent equal hue differences between adjacent pairs, all round the hue circle.

The second of the three variables, saturation, S, is a function of distance from the point representing the reference white on a chromaticity diagram, and for colours of the same luminance factor (not the same darkness) equal saturation represents equal perceptual differences from the grey of the same luminance factor.

The third variable is related to darkness rather than to lightness, and is not related in a simple manner to luminance factor; in this respect, it is similar to blackness in the NCS system. DIN darkness, D, is evaluated as:

$$D = 10 - 6.1723\{\log[40.7(Y/Y_0) + 1]\}$$

where Y is the luminance factor of the colour, and Y_0 is the luminance factor of the optimal colour (see Section 8.3) having the same chromaticity, the suffix 0 indicating zero darkness (when $Y = Y_0$, $D = 0$).

The colour solid associated with the DIN system is formed by having the grey scale as a vertical axis, with white at the top, for which $D = 0$, and black at the bottom, for which $D = 10$. The top surface of the solid is then a portion of a sphere having the black point as its centre, as shown in Figure 8.25; this surface represents the optimal colours. Different hues, T, are then situated in 24 different planes with one edge coincident with the grey scale axis, and the same angle, 15°, between adjacent planes. DIN darkness, D, is represented by distance down from the top surface towards the black point. DIN saturation, S, is represented by angular distance out from the grey scale axis, evaluated from the black point; the angle of one saturation step is equal to 5°. Although, for simplicity, the solid is shown in Figure 8.25 as being a symmetrical cone, the maximum value of S is different for different hues, so that the edge of the top surface is not circular, and the sides of the

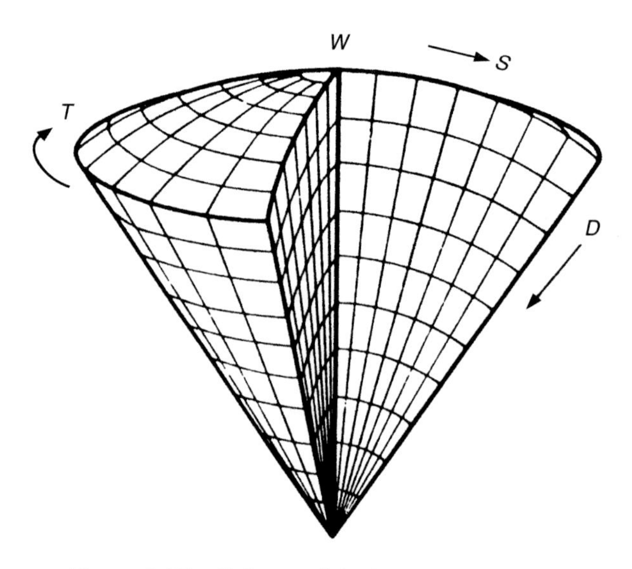

Figure 8.25 Colour solid of the DIN System

cone are at different angles for different hues. S has a maximum value of about 16, and this occurs for violet colours.

An atlas for the DIN system is available, known as the DIN Colour Chart. In the atlas, for each hue page, colours of equal DIN saturation are arranged in columns parallel to the grey scale axis, instead of in lines radiating from the black point. Samples having the same hue and darkness, but differing in DIN saturation by the same number of units, exhibit a colour difference that becomes progressively smaller as darkness increases. Hence, the dark samples on a DIN colour chart are more alike than the light samples. This result is a consequence of choosing saturation as the perceptual attribute with which one of the variables of the system correlates, instead of chroma. A column of colours of constant DIN saturation is a shadow series in which the colours all have the same chromaticity. Reproductions of two pages from the DIN system are shown in Figure 8.26; because of the limitations of colour printing, there are some deviations from the colours used in the DIN system, but the general arrangement is shown.

The size of the samples in the DIN colour Chart is 20 mm × 28 mm but larger samples of individual colours are also available.

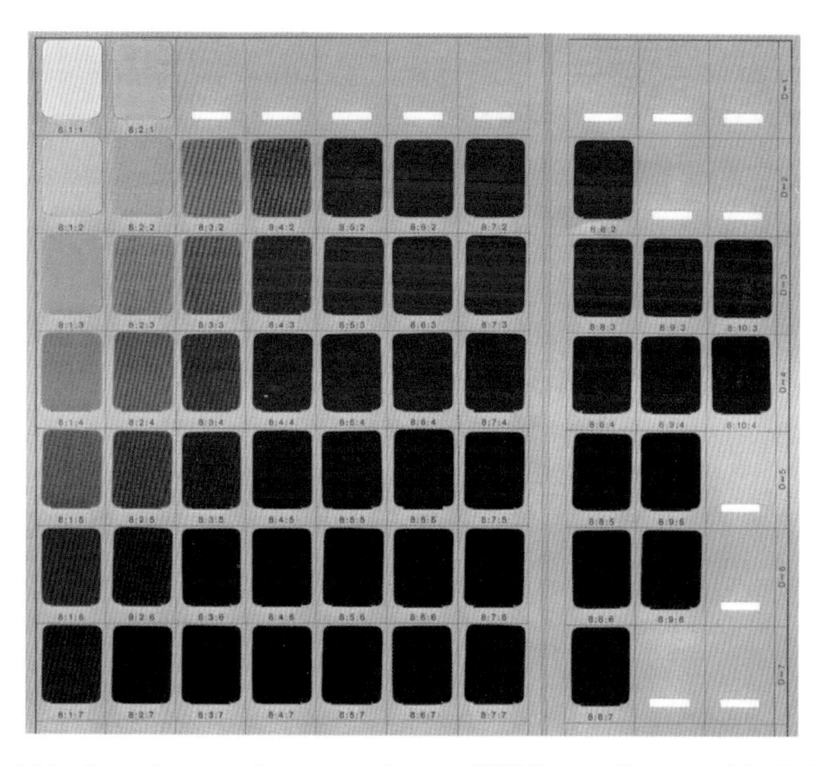

Figure 8.26 Reproductions of two pages from the DIN System. Because of the limitations of printing, all the colours cannot be shown accurately, but the general layout is displayed adequately

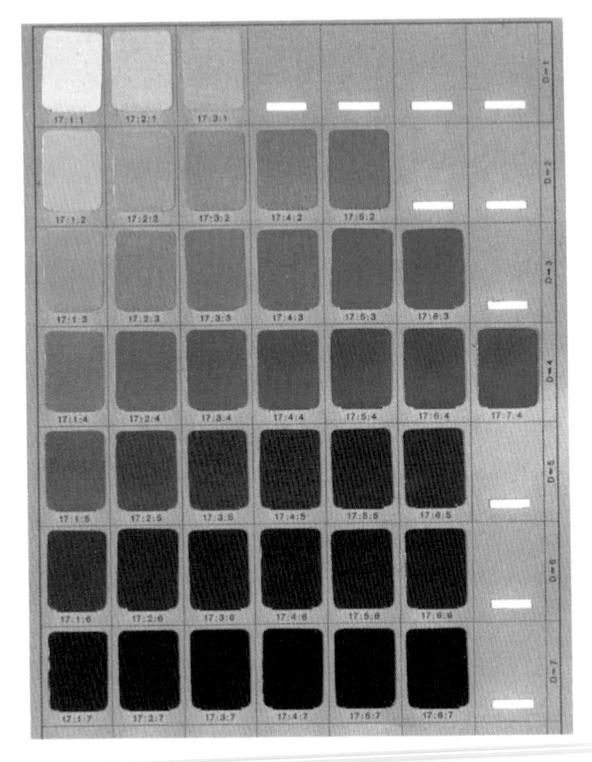

Figure 8.26 (*Continued*)

8.10 THE COLOROID SYSTEM

A colour order system designed particularly for use by architects is the *Coloroid* system, developed in Hungary by Antal Nemcsics and his co-workers (Nemcsics, 1980, 1987). The aim is to provide a system in which the colours are spaced evenly in terms of their aesthetic effects, rather than in terms of colour differences as in the Munsell system, or perceptual content as in the NCS system. One interesting feature of the Coloroid system is that notations in it can be calculated from values of CIE x, y chromaticity coordinates and luminance factors, Y, using fairly simple formulae and one look-up table. CIE Illuminant C (SC) is used (but the system is also applicable to D65).

The general arrangement is conventional in having a central vertical grey scale, with white at the top and black at the bottom, and hue planes that radiate from the grey scale.

A Coloroid specification consists of three numbers, such as 13-22-56. The first number indicates the hue. Colours of equal dominant (or complementary) wavelength are regarded as having equal hue, and the hue number, A, like 13 in the above example, is derived from a table of dominant (or complementary) wavelengths for values of A ranging from 10 at 570.83 nm to 76 at 568.92 nm; increases in A represent increases in dominant (or complementary) wavelength. The spacing of the values of A on the wavelength scale is intended to be aesthetically even. Interpolation between values is used if necessary, using values of φ, which are also tabulated against A, where:

$$\tan\varphi = (y - y_W)/(x - x_W)$$

and x_W, y_W are the values of x, y for the reference white. Values of φ are given, for Illuminants C and D65, in Table 8.1.

The second of the three numbers indicates the chromatic content, T. Colours are regarded as having the same value of T if they can be produced by additively mixing the same percentage of saturated colour of the same hue (dominant wavelength) with white and black. (By this is meant that, if the colour is produced by an additive mixture of the saturated colour and the reference white and the reference black, presented on a rapidly rotating disc, then T is constant when the saturated colour occupies a constant percentage of the area of the disc.) By saturated colour is meant a spectral colour in the range from 450 nm to 625 nm or lying on a purple line joining these two points in a chromaticity diagram. The values of T range from zero for the reference white and greys having the same chromaticity, to 100 for the saturated colours. The formula for calculating T is given in Table 8.1.

The third of the three numbers indicates the Coloroid lightness, V, which is defined as:

$$V = 10Y^{1/2}$$

where Y is the percentage luminance factor. For the perfect diffuser $Y = 100$ and hence $V = 100$. For a perfect black, $Y = 0$, and $V = 0$. The V scale thus extends from 0 to 100.

For a given value of A (hue) and V (lightness), the chromatic content T is related to Munsell Chroma, C, by the expression:

$$T = k_{AV}C^{2/3}$$

but k_{AV} varies with A and V. Hence, in the Coloroid system, while the variable A is closely correlated with hue, and V with lightness, the variable T is not related simply to either chroma or saturation; it is intended to provide an aesthetically even representation of chromatic content.

Atlases based on the Coloroid system are available for practical use by architects and other colour designers. The size of the samples is 15 mm × 28 mm.

8.11 THE OPTICAL SOCIETY OF AMERICA (OSA) SYSTEM

A colour order system that is different from the others considered so far is the OSA (Optical Society of America) system. It is dedicated primarily to producing a system in which, as nearly as possible, the distance between any two points represents the perceptual size of the colour difference between the two samples represented by them. There is, as a consequence, no attempt to include a correlate of hue, or of chroma or saturation. The OSA system was the culmination of a long series of investigations which started in 1947 and ended in 1977, in which a leading part was taken by Deane B. Judd (Nickerson, 1981).

Table 8.1 Formulae for calculations in the Coloroid system

A is derived from the values listed below for φ, where

$$\arctan \varphi = (y - y_w)/(x - x_w).$$

$$V = 10Y^{1/2}.$$

$$T = 100Ye_w(x_0 - x)/[100e_\lambda(x - x_\lambda) + Y_\lambda e_w(x_0 - x)]$$

or

$$T = [100Y(1 - ye_w)]/[100e_\lambda(y - y_\lambda) + Y_\lambda(1 - ye_w)]$$

The formula for T giving the most convenient numbers is used. The subscript w indicates that the value is for the reference white, the subscript 0 that it is for the reference black, the subscript λ that it is for the saturated colour (for which values of x_λ, y_λ, e_λ, and Y_λ are given below and in Nemscics, 1980; the values of Y_λ are the same as for the $\bar{y}(\lambda)$ function, but multiplied by 100), and

$$e_w = (1/100)(X_w + Y_w + Z_w)$$

where X_w, Y_w, and Z_w are tristimulus values for which the Y tristimulus value is the luminance factor expressed as a percentage.

The reverse equations for deriving x, y, Y from A, T, V are as follows:

$$x = [e_w x_0(V^2 - TY_\lambda) + 100Te_\lambda x_\lambda]/[e_w(V^2 - TY_\lambda) + 100Te_\lambda]$$

$$y = [V^2 + 100Te_\lambda y_\lambda - TY_\lambda]/[e_w(V^2 - TY_\lambda) + 100Te_\lambda]$$

$$Y = (V/10)^2$$

In all the evaluations, linear interpolation may be carried out between successive values (using the variable φ in the case of A; φ_C is used for CIE Illuminant C, and φ_{65} for Standard Illuminant D65).

Data for the saturated colours of the Coloroid system

A	φ_C	φ_{65}	x_λ	y_λ	e_λ	Y_λ
10	59.0	58.05	0.44987	0.54895	1.724349	94.6572
11	55.3	54.17	0.46248	0.53641	1.740845	93.3804
12	51.7	50.38	0.47451	0.52444	1.754986	92.0395
13	48.2	46.70	0.48601	0.51298	1.767088	90.6482
14	44.8	43.58	0.49578	0.50325	1.775953	89.3741
15	41.5	39.61	0.50790	0.49052	1.785074	87.6749
16	38.2	36.30	0.51874	0.43035	1.791104	86.0368
20	34.9	32.88	0.52980	0.46934	1.794831	84.2391
21	31.5	29.39	0.54137	0.45783	1.798665	82.4779
22	28.0	25.82	0.55367	0.44559	1.794822	79.9758

(*Continued*)

Table 8.1 (*Continued*)

A	φ_C	φ_{65}	x_λ	y_λ	e_λ	Y_λ
23	24.4	22.17	0.56680	0.43253	1.789610	77.4090
24	20.6	18.35	0.58128	0.41811	1.779484	77.4014
25	16.6	14.32	0.59766	0.40176	1.760984	70.7496
26	12.3	10.08	0.61653	0.38300	1.723444	66.0001
30	7.7	5.53	0.63896	0.36061	1.652892	59.6070
31	2.8	0.74	0.66619	0.33358	1.502608	50.1245
32	−2.5	−4.38	0.70061	0.29930	1.072500	32.1000
33	−8.4	−10.66	0.63925	0.26753	1.136638	30.4093
34	−19.8	−24.35	0.53962	0.22631	1.232286	27.8886
35	−31.6	−34.65	0.50340	0.19721	1.310122	25.8373
40	−43.2	−46.21	0.46041	0.17495	1.376610	24.0851
41	−54.6	−57.28	0.42386	0.15603	1.438692	22.4490
42	−65.8	−67.95	0.38991	0.13846	1.501583	20.7915
43	−76.8	−78.29	0.35586	0.12083	1.570447	18.9767
44	−86.8	−87.66	0.32195	0.10328	1.645584	16.9965
45	−95.8	−96.11	0.28657	0.08496	1.732085	14.7168
46	−108.4	−108.10	0.22202	0.05155	1.915754	9.8764
50	−117.2	−116.63	0.15664	0.01771	2.146310	3.8000
51	−124.7	−123.81	0.12736	0.05227	1.649940	8.6198
52	−131.8	−130.59	0.10813	0.09020	1.273415	11.4770
53	−138.5	−136.98	0.09414	0.12506	1.080809	13.5067
54	−145.1	−143.30	0.08249	0.15741	0.957577	15.0709
55	−152.0	−149.91	0.07206	0.18958	0.868977	16.4626
56	−163.4	−160.96	0.05787	0.24109	0.771732	18.5949
60	−177.2	−174.64	0.04353	0.30378	0.697110	21.1659
61	171.6	174.30	0.03291	0.35696	0.655804	23.4022
62	125.4	162.65	0.02240	0.41971	0.623969	26.1843
63	148.4	150.45	0.01196	0.49954	0.596037	30.1137
64	136.8	138.37	0.00425	0.60321	0.607414	36.6425
65	125.4	126.59	0.01099	0.73542	0.659924	48.5346
66	114.2	114.70	0.08050	0.83391	0.859523	71.7274
70	103.2	103.88	0.20259	0.77474	1.195684	92.6325
71	93.2	93.75	0.28807	0.70460	1.410097	99.0587
72	84.2	84.43	0.34422	0.65230	1.532830	99.9862
73	77.3	77.27	0.37838	0.61930	1.603793	99.3224
74	71.6	71.29	0.40290	0.59533	1.649449	98.1981
75	66.9	66.35	0.42141	0.57716	1.681081	97.0252
76	62.8	62.05	0.43647	0.56222	1.704981	95.8592

An OSA specification consists of three numbers, such as 3:1:5. The first represents the lightness, L. When $L = 0$, the colour has the same lightness as a medium grey of 30% reflectance factor. But, for chromatic colours having $L = 0$, the reflectance factor is usually less than 30% because of the contribution of colourfulness to brightness, and therefore to lightness (as mentioned in Sections 2.3 and 3.2). L is negative for darker colours, and positive for lighter colours. Values of L range from about -7 to about $+5$.

The second number in an OSA specification, j, represents the yellowness of the colour; j is positive for yellowish colours and negative for bluish colours. Values of j range from about -6 to about $+11$.

The third number, g, represents the greenness of the colour; g is positive for greenish colours and negative for reddish colours. Values of g range from about -10 to about $+6$.

The geometrical arrangement of the points representing the colours is such that each point is surrounded by 12 equidistant points, representing 12 nearest-neighbour colours all equally different perceptually from the colour considered. The arrangement of these points is the same as that of the centres of spheres packed as closely together as possible, each sphere touching 12 others. This geometry may be referred to as cubo-octahedral (Billmeyer, 1981), and is illustrated in Figure 8.27.

A cube has eight corners. If each of these eight corners is cut off down to the mid-point of the edges of the cube, a figure is formed having 12 corners. In Figure 8.27 these 12 corners are situated at the following positions: at the four corners of the square formed by the mid-points of the edges of the top surface of the original cube; at the four corners of the square formed by the mid-points of the vertical edges of the original cube; and at the four corners of the square formed by the mid-points of the edges of the bottom surface of the original cube. These 12 points are all equidistant from the point lying at the centre of the original cube. In the OSA system, successive values of L are located on successive horizontal planes, so that, if the five points on the centre plane of the cube of Figure 8.27

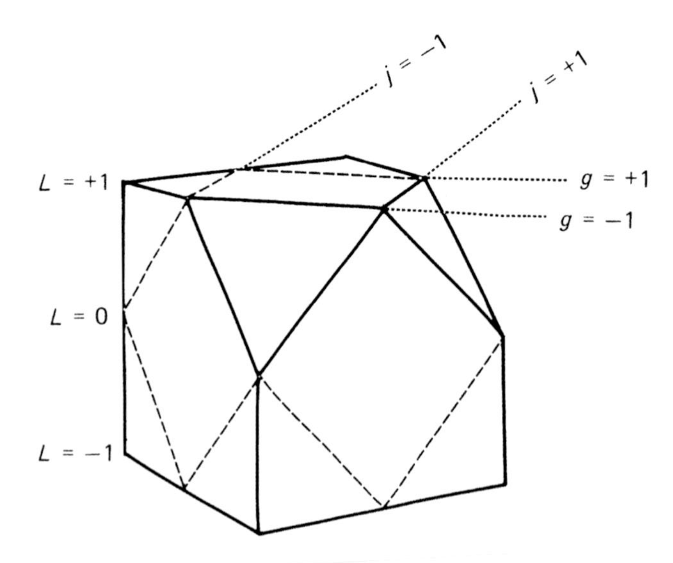

Figure 8.27 The cubo-octahedral basis of the OSA system

have $L = 0$, the four points on the plane above have $L = 1$, and the four on the plane below have $L = -1$. On the $L = 1$ and $L = -1$ planes, there is no point directly above or below the centre point, but such points occur on planes $L = 2$, 4, etc., and $L = -2$, -4, etc. On each horizontal plane, closest neighbouring points in one direction represent values of g differing by + or − 2 units, and, at right angles to that direction, values of j differing by + or − 2 units. If the centre point of Figure 8.27 represents $L = 0$, $g = 0$, $j = 0$, then the four points surrounding it on the same plane all have $L = 0$, and have $g = 2$, $j = 0$, or $g = -2$, $j = 0$, or $g = 0$, $j = 2$, or $g = 0$, $j = -2$. The four points on the plane above, all have $L = 1$, and have $g = 1$, $j = 1$, or $g = 1$, $j = -1$, or $g = -1$, $j = 1$, or $g = -1$, $j = -1$. If the distance between closest pairs of these colours is regarded as 2 units, the distance between successive horizontal planes differing in value of L by 1, is $2^{1/2}$ units. Thus, if two colours differ in their values of L, g, and j by ΔL, Δg, and Δj, respectively, then the total colour difference is given by:

$$[2(\Delta L)^2 + (\Delta g)^2 + (\Delta j)^2]^{1/2}$$

Although this is a colour-difference formula, the judgements on which the OSA system was based were not on pairs of colours exhibiting very small differences, so that it is not necessarily applicable to colours that almost match.

Samples of OSA colours, 2 inches × 2 inches (about 50 mm × 50 mm) square, have been made available, but not prearranged to represent any particular planes of the OSA system. This has the advantage that the user can select samples and arrange them on any planes passing through sample points in the OSA space. There are nine different families of parallel planes that include all of the sample points; of these, in one family all the planes are horizontal; in four families all the planes are vertical; and in the other four families all the planes are parallel to one of the four pairs of parallel cut-off corners of the original cube (see Figure 8.27). These planes represent unusual combinations of colours and are claimed to be useful in helping designers to find novel patterns for their work. Reproductions of the colours in some of these planes are shown in Figure 8.28; because of the limitations of colour printing, there are some deviations from the colours used in the OSA system, but the general arrangement is shown.

The OSA samples are all specified in terms of X, Y, Z tristimulus values for Standard Illuminant D65, for the 1964 Standard Observer (the 50 mm square samples subtending an angle of about 10° when viewed at about 300 mm) (MacAdam, 1978; Wyszecki and Stiles, pages 866 to 884, 1982), but there is no simple relationship between OSA specifications and CIE tristimulus values. Equivalent Munsell specifications have also been published for OSA specifications (Nickerson, 1981). Colours lying between the sample points are represented by decimals, for example, 3.4, −5.7.

8.12 THE HUNTER LAB SYSTEM

In the Hunter Lab system (Hunter, 1958; Billmeyer and Saltzman, 1981; Berns, 2000), the general arrangement is similar to that of the OSA system, but the emphasis is on a simple relationship to X, Y, Z tristimulus values, rather than on the closest approximation possible to uniform spacing. The variables used in the Hunter Lab system are: L as the correlate of

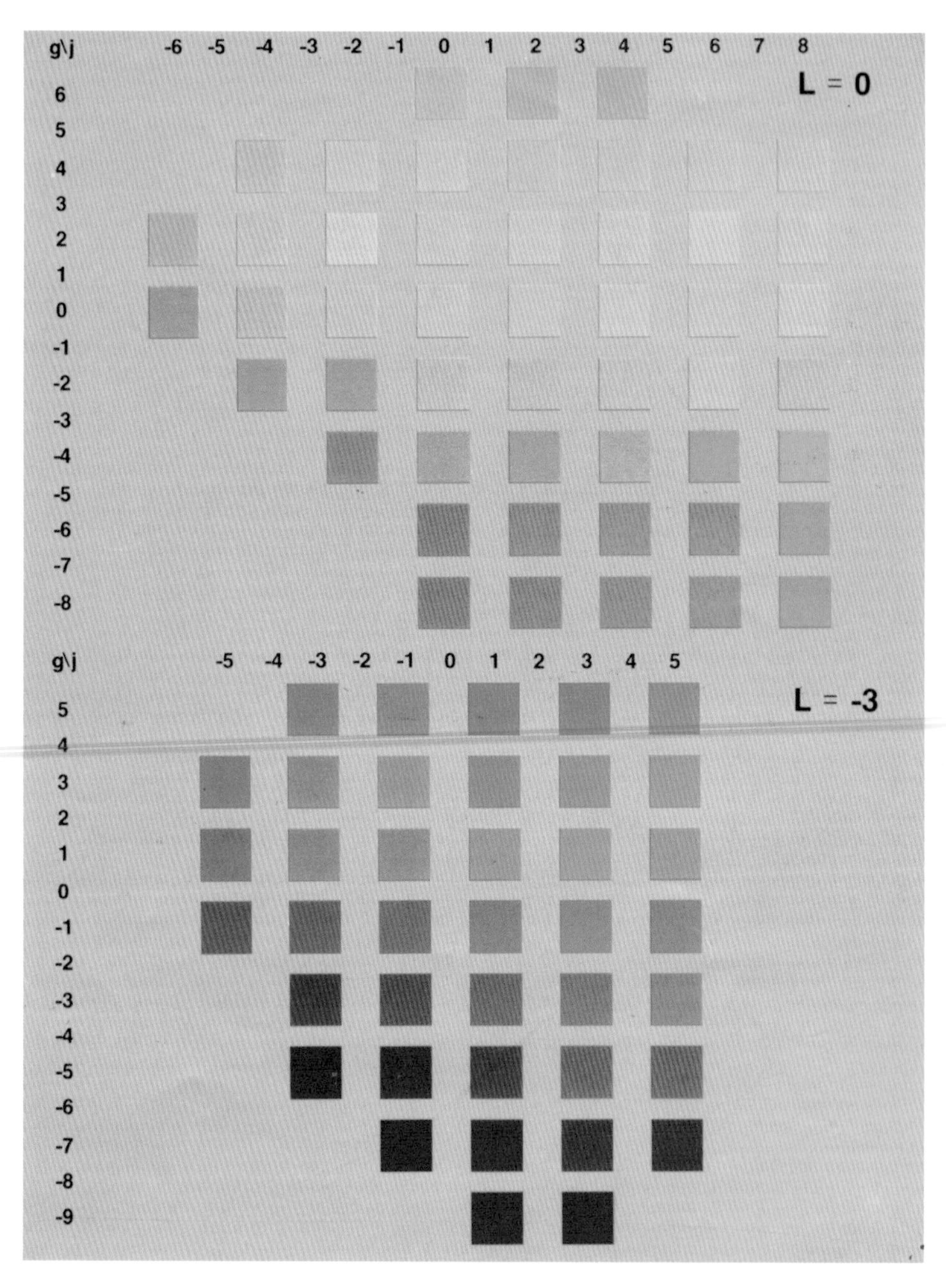

Figure 8.28 Reproductions of the colours in some of the planes of the OSA System. The planes are for: $L = 0$, $L = -3$ (upper), and for $j + g = 0$, and $L + j = 0$ (lower). Because of the limitations of printing, all the colours cannot be shown accurately, but the general layout is displayed adequately

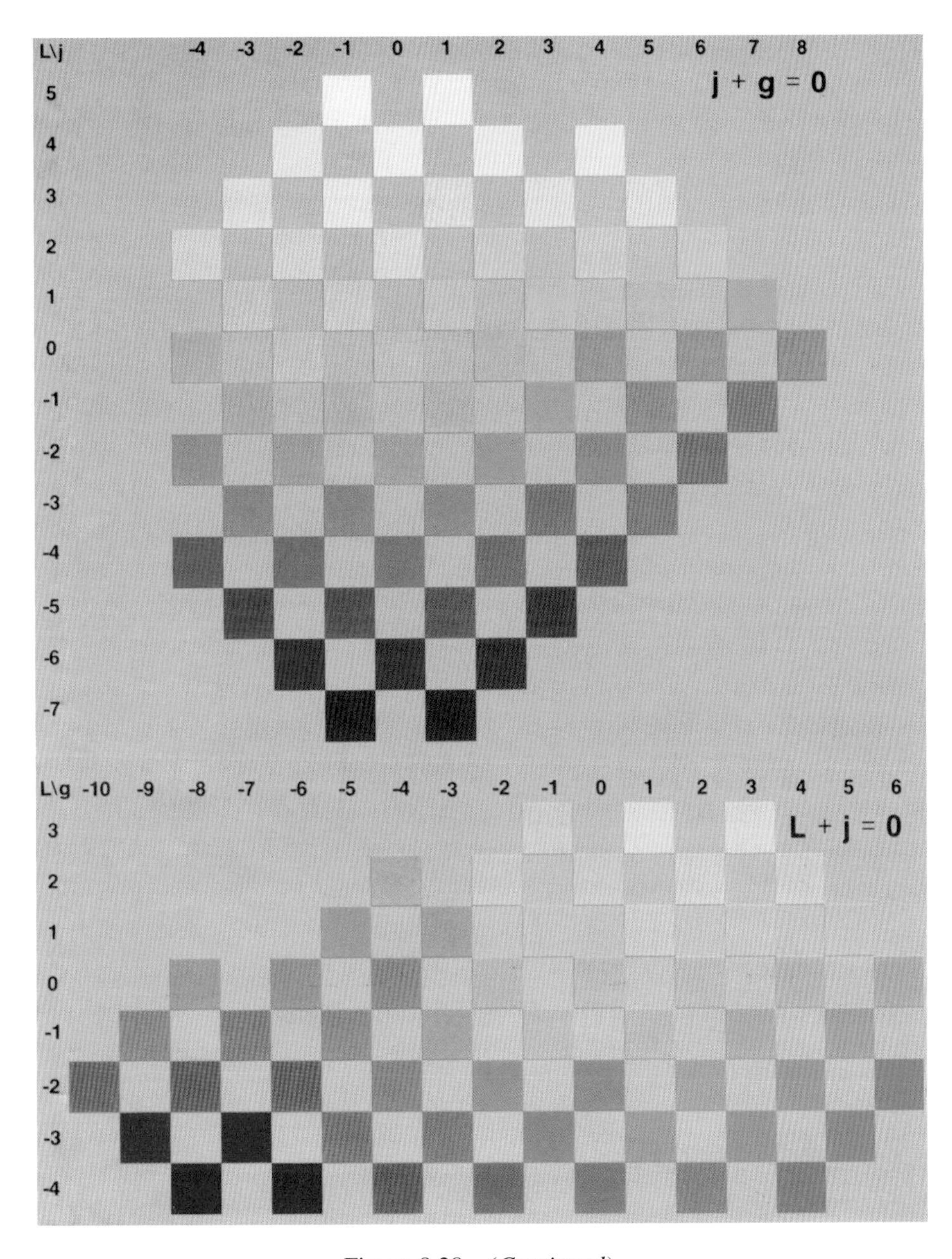

Figure 8.28 (*Continued*)

lightness, a as the correlate of redness or greenness, and b as the correlate of yellowness or blueness. These variables are defined, for CIE Illuminant C (SC), as follows:

$$L = 10Y^{1/2}$$

$$a = [17.5(1.02X - Y)]/Y^{1/2}$$

$$b = [7.0(Y - 0.847Z)]/Y^{1/2}$$

There are no samples available to represent this system, but its simple relationship to X, Y, Z tristimulus values has made it readily applicable to colour measuring instruments; as a result, many colorimeters have been marketed that produce these L, a, b values from instrumental measurements.

8.13 THE TINTOMETER

A rather different kind of colour order system is provided by an instrument called The Lovibond® *Tintometer* which is a visual colorimeter designed to facilitate the use of Lovibond glass filters, and originally developed by J.W. Lovibond in 1887 for measuring the colour of beer. The samples provided are of coloured glass, and, by superimposing them in suitable combinations, a very wide range of colours can be produced. The glasses are of three different colours, called *red* (actually a magenta to absorb green light), *yellow* (to absorb blue light), and *blue* (actually a cyan to absorb red light). They are each available in scales that provide colours ranging from that of the illuminant to highly saturated red, yellow, or blue, as the case may be. The scale divisions are arbitrary, except for two features: first, that if a red, a yellow, and a blue glass of the same value are superimposed, then the result is visually nearly neutral in CIE Illuminant C (SC); and second, that, if two glasses of the same colour are superimposed, whose numerical designations on the scale are n_1 and n_2, their combined colour is the same as that of a single glass having the designation $n_1 + n_2$ (together with a colourless glass).

The glasses are viewed against a standard white surface illuminated by a tungsten-halogen lamp. If the sample being measured is a reflecting surface, it is illuminated in a similar manner by the same lamp. In the case of transmitting samples, they are viewed against the standard white surface illuminated by the same lamp. The lamp is run at a specified voltage and a filter is situated in the eyepiece of the instrument to produce an approximation to CIE Illuminant C (SC).

The instrument is arranged such that two adjacent fields of view, seen through the viewing tube, enable the product in the sample field and a white reflective surface in the comparison field to be observed side by side, suitably illuminated. The Lovibond® glasses are introduced into the comparison field by a simple system of sliding racks, allowing the user to compare the colour of light which is either transmitted through or reflected from the sample with that transmitted through the glasses. A series of neutral glasses in racks is also supplied; these can be introduced into the sample field to dull the colour of products which are too strong to obtain a good colour match using red, yellow or blue glasses. The glasses are arranged in the racks in groups of ten, the first rack containing glasses designated from 0.1 to 0.9 in steps of 0.1, the second rack from 1.0 to 9.0 in steps of 1.0, and the third from 10.0 to the maximum in steps of 10.0. The total field size is about 2° in diameter.

The racks are positioned until a visual colour match is found for the light from the sample and its colour can then be expressed in Lovibond units; and if the match required, say, 2.5 of the red glasses, 3.7 of the yellow, and 8.6 of the blue, the Tintometer specification would be given as:

2.5 Red, 3.7 Yellow, 8.6 Blue.

Designations of this type are used in a number of industries to define colour standards for products, such as fats and edible oils. The glasses can also be made up into special combinations that give series of colours against which special tests can be evaluated. For instance, in the control of swimming pool water, indicator dyes are used in samples to determine the pH (acidity or alkalinity); the colour of the dyed water is related to the pH, and, by comparing its colour to those of special series of combination glasses, which have been previously calibrated in terms of pH, the pH is very easily determined. Such series of glasses are conveniently mounted in discs holding up to about nine glasses, and mounted in a simple comparator.

The difference in colour between two glasses whose numerical values differ by 0.1, which is the smallest interval on the scales, is very small and similar in size to a just noticeable difference in a 2° field. Altogether about 9 million different colours can be produced.

In an alternative form of the instrument, glasses of only two colours are ever used in combination, and the third variable consists of a calibrated means of reducing the amount of light passing through the glasses (Schofield 1939). With this Lovibond Schofield Tintometer it is possible to convert the readings into CIE X, Y, Z tristimulus values by means of charts provided for the purpose.

The gamut of chromaticities covered by the Tintometer instruments is very wide, but there is an area towards the green part of the spectral locus that is not covered; however, by using a supplementary cyan filter this area can be included.

8.14 THE PANTONE SYSTEM

The Pantone system provides series of coloured swatches printed on one side of thin cardboard sheets, which are bound into books or fans for convenient use. The colours on any one sheet are usually related to one another, typically being of the same hue but differing in lightness or saturation. Manufacturers in different locations can use the system to specify colours to promote consistency of the colours of products, such as the colours on flags. The system has been widely used in the graphic design and printing industries, and some of its editions include the specification of dot sizes for standard sets of inks that will reproduce the colour of each swatch. The colours are not arranged to correlate with perceptual attributes or with colorimetric measures, or to facilitate interpolation between the samples, so strictly speaking, they do not constitute a colour order system; but the system is of considerable practical value (Field, 2004).

8.15 THE RAL SYSTEM

Another system that provides samples for specifying colours in industry is the RAL system. In 1927 the German ***R**eichs**a**usschuß für **L**ieferbedingungen und Gütesicherung* (State

Commission for Delivery Terms and Quality Assurance) provided a collection of 40 colours under the name of "RAL 840". In the 1930s the specifications of the colours were changed and the collection was renamed "RAL 840 R" (R for revised). With tints constantly added to the collection, it was revised again in 1961 and changed to "RAL 840-HR", which consists of 210 colors. As "RAL 840-HR" covered only matte paint the 1980s saw the introduction of "RAL 841-GL" for glossy surfaces, limited to 193 colours. In 1993 the RAL DESIGN colour matching system was introduced, tailored to the needs of architects, designers and advertisers with 1625 colours specified in the CIELAB colour space. As in the case of the Pantone system, although not strictly speaking a colour order system, the RAL system is of considerable practical value (Ozturk, 2005).

8.16 ADVANTAGES OF COLOUR ORDER SYSTEMS

The use of colour order systems to select, measure, and check colours is very widespread. The advantages of using them include the following:

- First, they are easy to understand, because they usually have actual samples that can be seen and handled.
- Second, they are easy to use; in most circumstances side by side comparisons are made without the need for any instrumentation.
- Third, the number and spacing of the samples can be adapted for different applications, and different arrangements of the samples can be used for different purposes.
- Fourth, most colour order systems are calibrated in terms of CIE tristimulus values, and these can therefore be obtained for colours if required.

8.17 DISADVANTAGES OF COLOUR ORDER SYSTEMS

There are a number of disadvantages that occur when colour order systems are used.

- First, there is not just one colour order system in use, but several, and there is no simple means of transferring results from one system to another.
- Second, there are gaps between the samples, and this means that interpolation often has to be used both in matching a colour to the samples, and in determining the corresponding specifications.
- Third, the visual comparison between colours and the samples is only strictly valid if it is done using an illuminant having the same spectral power distribution, and geometrical arrangement, as was used for the original judgements and calibration adopted for the system (see Chapter 6). However, the errors are not likely to be large in most cases if reasonably typical indoor daylight is used; but, if any other illuminant is used, such as tungsten light, or fluorescent light, serious errors may occur. These difficulties will be more pronounced the more the colour being considered differs in spectral composition and gloss characteristics from the samples in the system. It

is therefore always advisable wherever possible to use a system having samples as similar as possible in these respects to the colours being matched. In this connection, a version of the Munsell system has been produced on textiles, known as the SCOT-Munsell system (Standard Colors of Textiles); as a further aid in obtaining similarity, the swatches in this system have one side dull and the other shiny. Keeping to a minimum the differences between the spectral compositions of the samples and those typical of the colours being measured is difficult. To produce colours of high chroma requires the use of colorants that are very spectrally selective; in other words they must absorb some wavelengths very strongly and others very weakly. But, if these colorants are mixed with grey colorants to produce samples of low chroma, these samples will be more spectrally selective than most colours of low chroma met with in practice, because these colours usually depend on colorants that are not very spectrally selective. This poses a real problem for the manufacturer of samples for systems; what is usually done is to use colorants whose spectral selectivity is low for low chroma colours, but high for high chroma colours. This means that very careful control is necessary to avoid discontinuities of colour along a series of colours of increasing chroma.

- Fourth, different observers may make slightly different matches on the same colour (this is the case even when colour-defective observers are carefully excluded). This effect (often referred to as observer metamerism, see Section 6.7) will also be more pronounced the greater the difference in spectral composition between the samples and the colours being matched.
- Fifth, there may be differences between the samples in the collection actually being used and those for which the calibration of the system applies. This can arise from manufacturing tolerances in the production of the samples, or from deterioration of the samples as a result of extensive use or fading. Naturally, manufacturers of samples for colour order systems are aware of these problems, and high consistency and good permanence of the samples is always sought.
- Sixth, some colours may lie outside the gamut of the samples available in the system. This can arise in the case of colours of very high chroma; they are often produced by pigments or dyes that are not of the highest stability, and it may not therefore be possible to make samples of these colours that are suitable for inclusion in colour order systems. Fluorescent colours are often in this category.
- Seventh, most colour order systems cannot be used for self-luminous colours, such as light sources, unless ancillary apparatus is used.

This is quite a long list of potential disadvantages, but they are of varying importance, depending on the application and the particular system being used. Thus, for example, in the Tintometer, the gaps between the samples are small enough to be insignificant in most applications; and, because the samples are made of glass, they have excellent permanence.

As has been mentioned several times, colour order systems can be calibrated in terms of their CIE tristimulus values. These calibrations can be used to determine the tristimulus values of samples that have been compared to the colours in the system, but, except in the case of the Tintometer, interpolation is usually necessary, and this is not easy to do.

REFERENCES

Berns, R.S., *Billmeyer and Saltzman's Principles of Color Technology*, 3rd *Ed*., p. 66, Wiley, New York, USA (2000).

Berns, R.S., and Billmeyer, F.W., Development of the 1929 Munsell Book of Color: A historical review, *Color Res. Appl*., **10**, 246–250 (1985).

Billmeyer, F.W., On the geometry of the OSA uniform color scales committee space, *Color Res. App*., **6**, 34–36 (1981).

Billmeyer, F.W., and Saltzman, M., *Principles of Color Technology*, *2nd. Ed*., p. 62, Wiley, New York, USA (1981).

Davidson, H.R., Godlove, M.N., and Hemmendinger, H., A Munsell book in high-gloss colours, *J. Opt. Soc. Ame* r., **47**, 336–337 (1957).

Evans, R.M., and Swenholt, B.K., Chromatic strength of colors: dominant wavelength and purity, *J. Opt. Soc. Amer*., **57**, 1319–1324 (1967).

Evans, R.M., and Swenholt, B.K., Chromatic strength of colors. III. Chromatic surround and discussion, *J. Opt. Soc. Amer*., **59**, 628–632 (1969).

Field, G.G., *Color and Its Reproduction*, 3rd Ed., p. 298, GATF, Pittsburgh, PA, USA (2004).

Gloag, H.L., and Gold, M.J., Colour Co-ordination Handbook, pp. 25–29, Building Research, Establishment Report, H.M.S.O, London, England (1978).

Hård, A., and Sivik, L., NCS – Natural Color System: a Swedish standard for color notation, *Color Res. Appl*., **6**, 129–138 (1981).

Hesselgren, S., Why color order systems?, *Color Res. Appl*., **9**, 220–228 (1984).

Hunt, R.W.G., Perceptual factors affecting colour order systems, *Color Res. Appl*., **10**, 12–19 (1985).

Hunter, R.S., Photoelectric color difference meter, *J. Opt. Soc. Amer*., **48**, 985–995 (1958).

ISO 12640-3, *Graphic technology, Prepress digital data exchange, Part 3: CIELAB standard colour image data (CIELAB/SCID)*, Annex B, International Organisation for Standardisation, Geneva, Switzerland (2007).

Li, C.J., Li, C., Luo, M.R., and Pointer, M.R., A colour gamut based on reflectance functions, *IS&T and SID's 15th Color Imaging Conference: Color Science and Engineering Systems, Technologies, and Applications*, pp. 213–217, IS&T Springfield, VA, USA (2007).

Li, X., Li, C.J., Luo, M.R., Pointer, M.R., Cho, M., and Kim, J., A New colour gamut for object colours, *IS&T and SID's 15th Color Imaging Conference: Color Science and Engineering Systems, Technologies, and Applications*, pp. 283–287, IS&T Springfield, VA, USA (2007).

MacAdam, D.L., The theory of the maximum visual efficiency of color materials, *J. Opt. Soc.Amer*., **25**, 249–252 (1935a).

MacAdam, D.L., Maximum visual efficiency of colored materials, *J. Opt. Soc. Amer*., **25**, 361–367 (1935b).

MacAdam, D.L., Colorimetric data for samples of OSA uniform color scales, *J. Opt. Soc. Amer*., **68**, 121–130 (1978).

McCamy, C.S., Physical exemplification of color order systems, *Color Res. Appl*., **10**, 20–25 (1985).

Nemcsics, A., The Coloroid color system, *Color Res. Appl*., **5**, 113–120 (1980).

Nemcsics, A., Color space of the Coloroid color system, *Color Res. Appl*., **12**, 135–146 (1987).

Newhall, S.M., Nickerson, D., and Judd, D.B., Final report of the O.S.A. subcommittee on the spacing of the Munsell colors, *J. Opt. Soc. Amer*., **33**, 385–418 (1943).

Nickerson, D., History of the Munsell Color System, Company, and Foundation, Part I. *Color Res. Appl*., **1**, 7–10, History of the Munsell Color System, Company, and Foundation, Part II. Its scientific application, *Color Res. Appl*., **1**, 69–77. History of the Munsell Color System, Company, and Foundation, Part III. *Color Res. Appl*., **1**, 121–130 (1976).

Nickerson, D., OSA uniform color scale samples: a unique set, *Color Res. Appl.*, **6**, 7–33 (1981).

Ozturk, L.D., Location of the Munsell Colors in the RAL Design System, *Color Res. Appl.*, **30**, 130–134 (2005).

Pointer, M.R., The gamut of real surface colours, *Color Res. Appl.*, **5**, 145–155 (1980).

Richter, M., and Witt, K., The story of the DIN color system, *Color Res. Appl.*, **11**, 138–145 (1986).

Robertson, A.R., Colour order systems: an introductory review, *Color Res. Appl.*, **9**, 234–240 (1984).

Rösch, S., Die Kennzeichnung der Farben, *Phys. Z.*, **29**, 83–91 (1928).

Schofield, R.K., The Lovibond Tintometer adapted by means of the Rothamsted Device to measure colours in the CIE system, *J. Sci. Instrum.*, **16**, 74–80 (1939).

Spillmann, W., Color order systems and architectural design, *Color Res. Appl.*, **10**, 5–11 (1985).

Swedish Standard, SS 01 91 03, *CIE tristimulus values and chromaticity co-ordinates for the colour samples in SS 01 91 02*, Stockholm, Sweden (1982).

Whitfield, T.W.A., O'Connor, M. and Wiltshire, T.J., The British Building-Colour Standards: a model for international application, *Color Res. Appl.*, **11**, 215–222 (1986).

Wright, W.D., The basic concepts and attributes of colour order systems, *Color Res. Appl.*, **9**, 229–233 (1984).

Wyszecki, G., and Stiles, W. S., *Color Science*, 1st *Ed.*, Wiley, New York, NY, USA (1967).

Wyszecki, G., and Stiles, W. S., *Color Science*, 2nd *Ed.*, Wiley, New York, NY, USA (1982).

GENERAL REFERENCES

Fairchild, M.D., *Color Appearance Models*, *2nd Ed.*, Chapter 5, Wiley, Chichester, England (2005).

Kuehni, R.G., *Color Space and its Divisions: Color Order from Antiquity to the Present*, Wiley-Blackwell, Hoboken, NJ, USA (2003).

9

Precision and Accuracy in Colorimetry

9.1 INTRODUCTION

Precision means the consistency with which measurements can be made of the same sample. By accuracy is meant the degree to which measurements of a sample agree with those made by a standard instrument or procedure in which all possible errors are minimised.

Precision is affected by random errors. The most common sources of random errors in photo-electric colorimeters, spectroradiometers, and spectrophotometers are electronic noise, variations in sensitivity, and sample presentation. With a modern instrument, for a given sample, the mean colour difference from the mean of a set of measurements (MCDM) can usually be expected to be about 0.1 (or less) of a CIELAB or CIELUV colour difference unit. Bearing in mind that, under average conditions, a just noticeable difference is usually regarded as corresponding to approximately one of these units, this level of precision can be considered to be very satisfactory. This performance has also been found to apply to the evaluation of colour differences of pairs of samples measured on the same instrument; it also applies both to such difference measurements made on the same day, and to series of such measurements made over a period of several weeks or more (Marcus, 1978; Billmeyer and Alessi, 1981).

Accuracy is affected by systematic errors. Common sources of systematic errors in modern instruments are wavelength calibration, detector linearity, geometry of illumination and measurement, and polarisation. Systematic errors may be associated with stray light, wavelength scale, wavelength bandwidth, reference-white calibration, thermochromism, and fluorescence (Carter and Billmeyer, 1979; Berns and Petersen, 1988). The absolute accuracy of a measurement can be equated to the total uncertainty in that measurement, which can only be determined by using a set of calibration standards, usually measured by a national standardising laboratory using the best possible instrumentation and procedures,

Measuring Colour, Fourth Edition. R.W.G. Hunt and M.R. Pointer.

and supplied with a certificate that includes an estimate of their associated measurement uncertainty (Clarke, Hanson, and Verrill, 1999; Gardner, 2000; Early and Nadel, 2004; Gardner, 2006).

A distinction has to be drawn between the accuracy of results obtained from spectroradiometers or spectrophotometers, and from colorimeters using filtered photo-cells. With spectroradiometers and spectrophotometers, the mean colour difference from the standard results (MCDS) is generally found to be at least an order of magnitude greater than the precision (MCDM) quoted above: that is, the MCDS may be about one or more CIELAB or CIELUV units (Billmeyer and Alessi, 1981). With colorimeters using filtered photocells, the accuracy can be very dependent on the spectral composition of the sample, and errors considerably in excess of one CIELAB or CIELUV unit may sometimes occur; as mentioned in Section 5.13, these errors may be reduced by determining the X, Y, Z tristimulus values from a matrix of the three filter readings (Erb, Krystek, and Budde, 1984), or from more than three filter readings (Wharmby, 1975; Kosztyán, Eppeldauer, and Schanda, 2010). The importance of accuracy depends on the application. When the same instrument is used to monitor the consistency of a product of nearly constant spectral composition, such as successive batches of nominally the same paint or dye, for instance, good precision is vital, but great accuracy is not. Accuracy becomes progressively more and more important as the measurements involve different instruments of the same type (perhaps in different locations), or instruments of different types, or samples of increasingly different spectral compositions.

When colorimetric results are compared, it is essential to ensure that like is being compared to like. Thus, the illuminant (SA, SC, or D65, etc.), the standard observer (2° or 10°), and the illuminating and measuring geometry (45°:0°, diffuse:8°, specular included or excluded, etc.) must all be the same for the two sets of data. If they are not, no meaningful comparisons can be made, unless the purpose is solely to demonstrate the effect of some difference in the measurement conditions. For example, if one measurement refers to CIE Illuminant C, and another to CIE Standard Illuminant D65, then the comparison has no meaning unless the same sample, and the same conditions (except for the illuminant) were involved, and the purpose was only to compare the colorimetry of that sample under these two illuminants.

9.2 SAMPLE PREPARATION

In many applications, colorimetric measurements are made only on selected samples from a larger population of material. It is therefore necessary that the sample be representative of the larger population. If the population involves variation in colour, then it is clearly necessary to use a number of samples sufficiently large to yield a reasonable average, if this is what is wanted, or to indicate the nature of the variability, if this is what is wanted (ASTM, 2008b). A properly designed sampling procedure, based on statistical considerations, is a necessity. It is also necessary to avoid samples that have uncharacteristic non-uniformities, such as may be caused by dirt, bubbles, streaks, or variations in flatness, for example. Visual inspection is a very sensitive method of detecting the presence of such defects, and samples should always be checked in this way; failure to do so can lead to very misleading results being obtained.

If samples are translucent, then their appearance, and the results of measurements made upon them, will be very dependent on the nature of the material behind them. It is useful, with these materials, to make measurements with both white and black backings; any difference between the two sets of readings then indicates the presence of translucency and its extent. If only one type of backing is used, this should be opaque and always the same, such as black (Wightman and Grum, 1981).

Some samples, such as woven fabrics, have directional effects, so that their colours are dependent on the directions of illumination and measurement (see Section 5.9). These differences can be explored by orientating the samples at different angles when using the 0°:45°x or 45°x:0° geometry; if the 0°:45°a, 45°a:0° or the 8°:diffuse or diffuse:8° geometries are used, the results will tend to be an average for all directions. But, in all cases, for the highest consistency of results, such samples should always be orientated in the same direction.

9.3 THERMOCHROMISM

Some materials change colour with temperature. This *thermochromism* occurs with some red, orange, and yellow glasses, and with some tiles that contain selenium (Carter and Billmeyer, 1979); it also occurs with some photographic dye images (Fairchild, Berns, Lester, and Shin, 1994). In such cases, it is necessary to standardise the temperature at which the measurements are made. For this purpose, 'room temperature' is often a sufficiently precise definition, because significant changes usually occur only when the sample is quite hot. However, although instruments are normally operated at room temperature, the samples may become quite hot under certain circumstances. If measurements are made on samples illuminated with monochromatic light, there is little chance of any appreciable thermochromism occurring. But, when samples are illuminated with 'white' light, as is necessary, for instance, when fluorescence occurs (as will be discussed in Chapter 10), then there may be a sufficient rise in sample temperature for some significant thermochromism to take place. This can be checked by taking a series of measurements with the sample left in the measuring position in the instrument; if the measurements show a steady drift with time, thermochromism may be taking place. It can be noted that many modern integrating-sphere based spectrophotometers use a xenon flash lamp to provide the illumination, and the very short duration of the flash precludes any significant rise in temperature.

9.4 GEOMETRY OF ILLUMINATION AND MEASUREMENT

For transmitting samples, unless they are either perfectly non-diffusing, or perfectly diffusing, the geometry of illumination and measurement will affect the results; but the amount of diffusion in many transparent samples is sufficiently small for the geometry to be unimportant. For reflecting samples, unless they are perfectly matt, the geometry of illumination and measurement will affect the results, and often does so very markedly (ASTM, 2006b; 2007; 2009a; 2009b). Very few reflecting samples are sufficiently matt to be unaffected by the geometry, and much of the difficulty in obtaining consistent colorimetric results from different instruments stems from this source. There are several factors that contribute to this inconsistency.

First, the standard conditions of illumination and viewing recommended by the CIE, as described in Section 5.9, are not sufficiently restrictive to avoid differences. In particular, the ±10° permissible variation of illuminating or measuring axis from the normal direction, and the permissible spread of the beam from the axis of ±8° or ±5°, can result in significant differences for some types of sample. It has been suggested that these angles should be reduced to ±2° from the nominal illuminating and measuring angles, and ±4° for beam spread, in all cases (Rich, 1988).

Second, the sizes, positions, and effective reflectances, of the gloss traps commonly present in instruments using integrating spheres can differ sufficiently to cause appreciable discrepancies (Billmeyer and Alessi, 1981).

Third, in integrating-spheres instruments, there are several other factors that can affect the results: these include the sizes, positions, and effective areas of the ports; the positions and reflectances of the baffles (see Figure 5.1) that are usually included (to prevent light travelling directly from the samples to the part of the sphere viewed by the detector, or from the part of the sphere illuminated by the input beam to the samples); and the uniformity and reflectance of the coating on the interior of the sphere (Clarke and Compton, 1986).

Fourth, if samples are on translucent supports, such as paper, some of the reflected light can be lost by diffusion in the substrate if the illuminating and measuring apertures are the same size, the effect becoming proportionally greater the smaller the sample size; errors from this *translucent blurring* (or *aperture vignetting*) can usually be made negligibly small by arranging that the measuring aperture lies either 2 mm inside or outside the edge of the illuminated area.

Finally, if the results of measurements are compared with visual inspection of samples, it is unlikely that the measuring and inspection geometry will be exactly the same, and this can have appreciable effects.

9.5 REFERENCE WHITE CALIBRATION

As mentioned in Section 5.7, although the primary standard of reflectance, established in 1969, is the perfect diffuser (CIE, 2004), this is not available as a material standard; but national standardising laboratories are able to calibrate material standards relative to it. These 'transfer' standards can then be used to calibrate working standards for general use. These working standards are often made of opal glass, or are in the form of a ceramic or porcelain-enamel tile, or a fluorinated polymer (Grum and Saltzman, 1976), or a pressed powder, such as barium sulphate (Erb and Budde, 1979). White working standards of this type are then used to normalise instruments prior to measurements being made. The normalising step in filter type colorimeters ensures that correct tristimulus values are obtained for the working standard; in spectrophotometers, the normalising step can result in corrected transmittance factors, or reflectance factors, being obtained throughout the spectrum. It has been noted that, when opal glasses are used, because they are translucent, the total amount of light they reflect in an instrument can depend on the ratio of the illuminated area to the sample port area (Billmeyer and Hemmendinger, 1981); for this reason, dense opal, or tiles, are preferable. It is possible to correct numerically for some errors in the calibration of the working standard (to be discussed in Section 9.12).

9.6 POLARISATION

In the case of non-metallic glossy materials, the fraction of the light reflected *regularly* (or *specularly*, that is, in the same direction as by a mirror) depends on its plane of

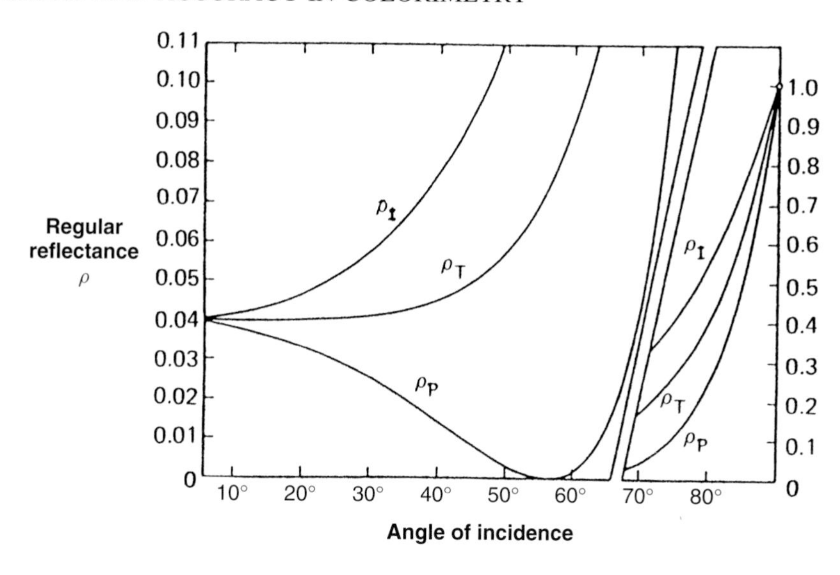

Figure 9.1 Regular reflectance from a surface as a function of angle of incidence for light polarised in the plane of the illuminating beam and the normal to the surface, ρ_I, and at right angles to it, ρ_P, and for both together, ρ_T, for a ratio of 1.5 for the refractive indices of the media at the reflecting boundary

polarisation. This is shown in Figure 9.1, where regular reflectance is plotted against angle of incidence for different planes of polarisation, for a ratio of 1.5 for the refractive indices of the media at the reflecting boundary. This can affect measurements in spectroradiometers, spectrophotometers, and colorimeters, because, in these instruments, it is possible for the light incident on the samples to become partially polarised as it passes through, or is reflected by, their optical components.

If a sample exhibits some (non-metallic) gloss, the spectral reflectance factor will depend on the polarisation of the light. Consider a non-metallic glossy sample, whose colorant is in a medium of refractive index 1.5, situated in a medium having a refractive index of 1.0 (such as air); from Figure 9.1, if the angle of incidence is 45°, the sample will reflect specularly from its topmost surface about 9% of the light, if it is fully polarised in the plane containing the beam and the normal to the surface, but about 1% if it is fully polarised at right angles to this plane. These effects can be avoided if the illumination is either normal or diffuse. In the 8:diffuse geometry, the illumination is normal to within 23° (8° plus up to 10° for the axis, plus up to 5° for the spread), and because, as can be seen from Figure 9.1, the polarising effect is still quite weak up to this angle, any effects of polarisation are likely to be quite small. Similarly, in the 0:45 geometry the illumination is normal to within 18° (up to 10° for the axis, and up to 8° for the spread), and again this is only likely to cause small effects. In the diffuse:8 geometry, the effects of polarisation should be averaged out because of the many angles of incidence involved. But if the 45:0 geometry is used, then particular attention must be paid to the possibility that polarisation may be affecting the results.

Furthermore, when the light reflected from a sample is polarised to any appreciable degree, the effective transmittance of any subsequent optical components can depend on the extent and angle of polarisation of the light; the effect of this on the measurements depends on the type of instrument used.

Liquid crystal displays (LCDs) involve passing light through crossed polarising filters, so that the emerging light is always strongly polarised. Extreme care over polarising effects is therefore necessary when making colorimetric measurements on these devices. Extruded plastic materials may also polarise the light to some extent; in this case the samples should be re-measured after rotation through a right-angle, and an average taken.

9.7 WAVELENGTH CALIBRATION

In filter-type colorimeters, the spectral sensitivity functions achieved in the instrument are only ever approximations to those theoretically required (ASTM, 2006a); the discrepancies can be regarded as errors in wavelength scale. In spectroradiometers, the wavelength scale can be checked by measuring samples of known spectral emission. In spectrophotometers, the scale can be checked by measuring samples of known spectral transmittance factor, or spectral reflectance factor; very useful samples for this purpose are didymium filters and holmium ceramic tiles, because they have rapid changes of spectral transmittance or reflectance with wavelength in several parts of the spectrum (Alman and Billmeyer, 1975; Clarke, 2006a). If the instrument has provision for adjusting the wavelength scale, then the results from the known sample can be used to determine the corrections that should be made. Some instruments make no provision for adjusting the wavelength scale, but, in this case, it may be possible to correct the data numerically (to be discussed in Section 9.12).

9.8 STRAY LIGHT

In any optical instrument some light is reflected or scattered in a non-imaging manner, and this is usually referred to as *stray light*. In spectroradiometers and spectrophotometers this can cause appreciable errors. For instance, a yellow sample may have a reflectance factor of about 95% at the long, and only about 3% at the short, wavelength end of the spectrum. If even as little as 2% of the long wavelength light is scattered in such a way as to be present when the short wavelengths are being measured, readings of 5% instead of 3% may result and this may be seriously misleading. The stray light condition of these instruments can be checked by measuring samples that have been carefully calibrated using instruments known to have very little stray light; these samples should have very low values in some parts of the spectrum. It may be possible to reduce the stray light in an instrument by cleaning the optical components, by improving the performance of baffles near the light paths, and by ensuring there are no smooth surfaces near the light path likely to cause unwanted reflections. If the amount of stray light present is appreciable and can be reliably estimated, the data can again be corrected numerically (or statistically, to be discussed in Section 9.12).

9.9 ZERO LEVEL AND LINEARITY

Noise in the detector and stray light within the optics of colorimeters, spectroradiometers, and spectrophotometers can lead to an electrical signal being present when no light is produced by a sample, and this signal may change with wavelength. Measurements on dark samples of known radiance factor, reflectance factor, or transmittance factor provide the most convenient way of checking this zero level.

It is also necessary for the electrical signals from the detector in these instruments to be proportional to the incident radiant power, or, in filter-type colorimeters, to be proportional to the integrated spectral power transmitted by each filter. Measurements on a series of neutral samples of known transmittance factor or reflectance factor provide the most convenient way of checking this feature. In some instruments the zero level can be re-adjusted, but in others it is fixed. If no adjustment is possible, the data can again be corrected numerically (to be discussed in Section 9.12).

9.10 USE OF SECONDARY STANDARDS

Reference has already been made several times to the use of known standards to help detect or correct errors (Verrill, 1987). In the case of transmission work, a set of calibrated filters should be used. One such set, provided by the National Bureau of Standards (now the National Institute for Standards and Technology, NIST, USA), is designated NBS SRM 2101–2105 (Keegan, Schleter, and Judd, 1962; Eckerle and Venable, 1977); new sets of these filters are no longer available, but similar sets can be obtained from the National Physical Laboratory (NPL) in Great Britain (Clarke, 2006b). In the case of reflection work, a set of calibrated reflecting tiles should be used (Clarke, 1969). One such set, of ceramic colour tiles, is produced by CERAM and is available with calibration from either CERAM or NPL (Malkin and Verrill, 1983; Malkin, Larkin, Verrill, and Wardman, 1997).

9.11 BANDWIDTH

Unless a sample has properties that are constant with wavelength, the width of the band of wavelengths used in spectroradiometers or spectrophotometers will generally affect the results. The bandwidths used in these instruments vary from a few tenths of a nanometre to as much as about 20 nm. The smaller the bandwidth, the more exactly the instrument is able to detect very steep or narrow variations in emission, in reflectance factor, or in transmittance factor, with wavelength; but also the smaller will be the amount of light available for detection. Conversely the larger the bandwidth, the more approximate will be the results for specimens that are very spectrally selective, but the more will be the amount of light available for detection. For best accuracy, the measurement interval and the band pass should be about equal (to be discussed in Section 9.13).

Spectroradiometers and spectrophotometers, in effect, filter the white light they use through a spectral 'window', which is equivalent to a filter that has maximum transmittance at, or near, its nominal recording wavelength, and lower transmittances at longer and shorter wavelengths. These sidebands quickly reduce to zero transmittance on either wavelength side, and this variation of transmittance with wavelength, appropriately weighted for any variation in detector spectral sensitivity, is referred to as the *bandwidth function*. If the bandwidth function varies with wavelength this can introduce further errors (Fairman, 2010). Numerical corrections can also be made for some effects of bandwidth function (to be discussed in Section 9.12).

9.12 CORRECTING FOR ERRORS IN THE SPECTRAL DATA

As mentioned in Sections 9.5, 9.7, 9.8, 9.9, and 9.11, numerical corrections for known errors in spectral data can be provided, and these can improve accuracy and agreement between instruments. Systematic methods for deriving such corrections have been suggested by Robertson (Robertson, 1987), by Berns and Petersen (Berns and Petersen, 1988) and by Clarke (Clarke, 2006a). These methods are applicable to spectral measurements of either reflectance factor or transmittance factor; but, for simplicity, their description will be limited here to reflectance factor data. The methods depend on devising equations that are likely to represent the form of certain types of error present in typical instruments. The technique works well for correcting errors in photometric scale and in wavelength (Berns and Reniff, 1997). It does not address differences in measurement that are attributable to differences in geometry between instruments, such as in aperture size, illuminating and measuring angles, or sphere design.

Consider, for instance, an instrument with an error in its zero setting, perhaps caused by stray light, by an imperfect black trap, or by an improperly adjusted analogue-to-digital converter. This type of error is likely to produce a constant offset throughout the spectrum. Hence, if $R_m(\lambda)$ is the measured value of the reflectance factor at the various wavelengths, λ, the true value $R_t(\lambda)$ is likely to be given by:

$$R_t(\lambda) = R_m(\lambda) + B_0$$

where B_0 is a constant (which, as also in the cases of the constants B_1, B_2, B_3, and B_4, to be introduced later, could be positive or negative).

An error in the assumed value of the reflectance factor of the white working standard of the instrument, caused, for instance, by it being miscalibrated or improperly handled or maintained, is likely to produce a constant percentage error throughout the spectrum. Hence, in this case:

$$R_t(\lambda) = R_m(\lambda) + B_1 R_m(\lambda)$$

where B_1 is another constant. By combining the effects represented by both the above equations, photometric scale errors in $R_m(\lambda)$ can be related to $R_t(\lambda)$ by a straight line where B_1 is the slope of the line and B_0 is the intercept.

Non-linearity in the response of the photo-detector, or in its associated electronic circuits, would result in the photometric scale errors being non-linear. Such errors can be represented approximately by:

$$R_t(\lambda) = R_m(\lambda) + B_2[R_m(\lambda)]^2$$

where B_2 is another constant.

An error in the wavelength scale consisting of a simple shift of all wavelengths by the same increment in the same direction is likely to produce errors that are proportional to the slope of the curve of $R_m(\lambda)$ plotted against λ: in regions where the curve is flat, there would be no error; small errors would occur where the curve had a low slope; large errors would occur where the curve had a high slope. Hence, in this case:

$$R_t(\lambda) = R_m(\lambda) + B_3 \mathrm{d}R_m/\mathrm{d}\lambda$$

where $dR_m/d\lambda$ is the slope of R_m plotted against λ, and where B_3 is another constant. (This is the simplest form of wavelength error, corresponding, for instance, to a misalignment of a diffraction grating in a monochromator, and it may not represent the true form of the error of the instrument; but formulae representing more complicated forms of error can be devised; Berns and Petersen, 1988.)

An instrument possessing too large a bandwidth function is likely to produce errors that are proportional to the rate of change with wavelength of the slope of the curve of $R_m(\lambda)$ plotted against λ: in regions where this curve is a straight line, there would be no error, because the extra long wavelengths would be balanced by the extra short wavelengths; but, where this curve is not straight, this balance is upset, and errors occur to a degree that is proportional to the rate of change of the slope with wavelength. Hence, in this case:

$$R_t(\lambda) = R_m(\lambda) + B_4 d^2 R_m/d\lambda^2$$

where $d^2 R_m/d\lambda^2$ is the rate at which the slope of the curve of R_m plotted against λ changes with wavelength, and where B_4 is another constant.

If an instrument had all these systematic errors, the true reflectance factor could then be calculated by combining all five of the above equations, thus:

$$R_t(\lambda) = R_m(\lambda) + B_0 + B_1 R_m(\lambda) + B_2[R_m(\lambda)]^2 + B_3 dR_m/d\lambda + B_4 d^2 R_m/d\lambda^2$$

To make such calculations, it is necessary to determine the values of B_0, B_1, B_2, B_3, and B_4. But, with real instruments, the errors will not usually correspond exactly to those assumed above. Hence, those values of B_0, B_1, B_2, B_3, and B_4 that give the best representation of the real errors are required.

A method of correction can then be formulated that requires the following steps to be carried out.

Step 1. A set of stable calibrated chromatic and neutral reference materials is obtained, together with their certified spectral reflectance factor data. The set of Ceramic Colour Standards from CERAM or NPL in Great Britain is a good example (Malkin and Verrill, 1983; Malkin, Larkin, Verrill, and Wardman, 1997). The calibration must be for the same geometry of illumination and viewing as in the instrument being used for measurement. The data must be available at whatever wavelength range and wavelength interval is adopted.

Step 2. Each reference material is measured on the measuring instrument. The reported results should be the average of at least three measurements. When making these measurements, care should be taken to avoid raising the temperature of the materials to a level different from the reported temperature applying to the calibration.

Step 3. A chromatic material is selected that has a wide range of reflectance factor values and steep slopes as a function of wavelength. This material will be used to characterise and correct the systematic errors of the instrument. Careful selection of this material is important; because of thermochromism, some materials will change their measured reflectance factor dramatically with temperature changes. Berns and Petersen found that the cyan tile of the CERAM set was a good compromise between spectral properties and little thermochromic change in measured reflectance factor (Berns and Petersen, 1988). For the selected material, the differences between the true result, $R_t(\lambda)$, and its measured result, $R_m(\lambda)$, obtained in Step 2, are then calculated:

$$R_t(\lambda) - R_m(\lambda)$$

If the sampling is at an interval of 10 nm from 400 nm to 700 nm, for example, this yields 31 values.

Step 4. Using the measurements, $R_m(\lambda)$, from the selected material, a table of values, F_3, is calculated that approximates the slope of the curve of $R_m(\lambda)$ plotted against wavelength at the same 31 wavelengths; and a table of values, F_4, is calculated that approximates the rate of change of this slope with wavelength, at the same 31 wavelengths. The values F_3 and F_4 are calculated as follows:

$$F_{3i} = [R(\lambda_{i+1}) - R(\lambda_{i-1})]/[\lambda_{i+1} - \lambda_{i-1}]$$

$$F_{4i} = [R(\lambda_{i+1}) - R(\lambda_{i-1}) - 2R_i(\lambda)]/[(\lambda_{i+1} - \lambda_{i-1})/2]^2$$

The subscript, i, indicates the wavelength considered, and the subscripts i+1 and i−1 indicate neighbouring sampling wavelengths. (At the ends of the spectrum, the values of F_3 and F_4 for the first and last measured wavelengths can be set equal to those of the second, and second to last, wavelengths, respectively.)

Step 5. If B_0, B_1, B_2, B_3, and B_4 are set equal to some trial values, a set of trial corrected reflectance factors $R'_c(\lambda)$, is obtained for the 31 wavelengths:

$$R'_c(\lambda) = R_m + B_0 + B_1 R_m(\lambda) + B_2[R_m(\lambda)]^2 + B_3 F_3(\lambda) + B_4 F_4(\lambda)$$

These values can be compared with the true values $R_t(\lambda)$ for this material to obtain the value of the error $e(\lambda)$ at each of the 31 wavelengths:

$$e(\lambda) = R_t(\lambda) - R'_c(\lambda)$$

Those values of B_0, B_1, B_2, B_3, and B_4 are now obtained that result in the sum of the squares of the values of $e(\lambda)$ at the 31 wavelengths being the minimum possible. (This multiple linear regression procedure is available in the form of readily applicable computer programmes.)

Step 6. Optimum corrected values, $R_c(\lambda)$, of the data for the selected material are then obtained as:

$$R_c(\lambda) = R_m(\lambda) + B_0 + B_1 R_m(\lambda) + B_2[R_m(\lambda)]^2 + B_3 F_3(\lambda) + B_4 F_4(\lambda)$$

where B_0, B_1, B_2, B_3, and B_4 have the values that minimise $e(\lambda)$.

Step 7. Step 6 is repeated for the other materials for which the true values are known, and it should be found that the errors have been reduced. (If there are still important errors, different formulae, or a larger number of formulae, for representing the errors may be needed; see Berns and Petersen, 1988.)

Step 8. Step 6 is used to correct all measured data obtained from the instrument using the same geometry.

9.13 CALCULATIONS

As described in Section 2.6, CIE tristimulus values, X, Y, Z, are obtained by weighting spectral power (or other radiant quantity) distributions with the CIE colour-matching functions, $\bar{x}(\lambda)$, $\bar{y}(\lambda)$, $\bar{z}(\lambda)$. This is expressed mathematically by the summations:

$$X = K[P_1\bar{x}_1 + P_2\bar{x}_2 + \ldots\ldots + P_n\bar{x}_n]$$

$$Y = K[P_1\bar{y}_1 + P_2\bar{y}_2 + \ldots\ldots + P_n\bar{y}_n]$$

$$Z = K[P_1\bar{z}_1 + P_2\bar{z}_2 + \ldots\ldots + P_n\bar{z}_n]$$

where the subscripts 1, 2......... n indicate the values of the quantities denoted, at a series of equally spaced wavelength intervals throughout the spectrum from wavelength 1 to wavelength n; the values, P, are appropriate measures of spectral power, such as the colour stimulus function (the absolute measure of a radiant quantity per small constant-width wavelength interval throughout the spectrum), $\varphi_\lambda(\lambda)$, or the equivalent relative measure, the relative colour stimulus function, $\varphi(\lambda)$, which is often derived as the product of a relative spectral power distribution, $S(\lambda)$, for a source, and the spectral transmittance factor or the spectral reflectance factor, $R(\lambda)$, for a sample; and K is a suitably chosen constant. For transmitting or reflecting samples, K is given a value that results in Y being equal to 100 for the perfect transmitter or the perfect reflecting diffuser; hence, in this case:

$$K = 100/[S_1\bar{y}_1 + S_2\bar{y}_2 + \ldots\ldots + S_n\bar{y}_n]$$

The results obtained by such summations differ according to the size of the wavelength interval used, and the CIE regards the correct result as that obtained when the interval is infinitesimally small. This is expressed mathematically by replacing the summations by the equivalent integrals:

$$X = K\int P(\lambda)\bar{x}(\lambda)\mathrm{d}\lambda$$

$$Y = K\int P(\lambda)\bar{y}(\lambda)\mathrm{d}\lambda$$

$$Z = K\int P(\lambda)\bar{z}(\lambda)\mathrm{d}\lambda$$

$$K = 100/\int S(\lambda)\bar{y}(\lambda)\mathrm{d}\lambda$$

The process of integration requires that the quantities being integrated have values that are defined continuously throughout the spectrum. For quantities defined by the CIE at 1 nm intervals (this includes $\bar{x}(\lambda)$, $\bar{y}(\lambda)$, and $\bar{z}(\lambda)$, and $S(\lambda)$ for Standard Illuminants D65 and A), this is achieved by linear interpolation of the values within each 1 nm interval. For quantities only available at larger intervals, a more elaborate method of interpolation, such as one of those given in Table 9.1, is preferable (CIE, 2005).

A colorimeter that uses filtered photo-cells performs this type of integration by virtue of the fact that the response of the photo-cell is the result of the additive effects of the filtered radiation throughout the part of the spectrum transmitted by the filter. The difficulty with

Table 9.1 Interpolation formulae

The known values of the spectral quantity are designated:
F_1, F_2, F_3, F_4 at wavelengths
$\lambda_1, \lambda_2, \lambda_3, \lambda_4$.
It is required to estimate the value of F at a wavelength λ between the wavelengths λ_2 and λ_3.

Third degree polynomial formula
This formula can only be used when the known values of F are at regular wavelength intervals. The required value of F is given by:

$$F = F_2 + p\Delta F + p(p-1)\Delta_2 F/2 + p(p-1)(p-2)\Delta_3 F/6$$

where
ΔF is the first difference, $F_2 - F_3$,
$\Delta_2 F$is the second difference, the average of $(F_1 - F_2) - (F_2 - F_3)$ and $(F_2 - F_3) - (F_3 - F_4)$
$\Delta_3 F$ is the third difference, $(F_1 - F_2) + (F_3 - F_4)$
and $p = (\lambda - \lambda_2)/(\lambda_3 - \lambda_2)$

Lagrange formula
This formula can be used when the known values of F occur at any wavelengths. The required value of F is given by:

$$\begin{aligned} F = {} & F_1(\lambda-\lambda_2)(\lambda-\lambda_3)(\lambda-\lambda_4)/(\lambda_1-\lambda_2)(\lambda_1-\lambda_3)(\lambda_1-\lambda_4) \\ & +F_2(\lambda-\lambda_1)(\lambda-\lambda_3)(\lambda-\lambda_4)/(\lambda_2-\lambda_1)(\lambda_2-\lambda_3)(\lambda_2-\lambda_4) \\ & +F_3(\lambda-\lambda_1)(\lambda-\lambda_2)(\lambda-\lambda_4)/(\lambda_3-\lambda_1)(\lambda_3-\lambda_2)(\lambda_3-\lambda_4) \\ & +F_4(\lambda-\lambda_1)(\lambda-\lambda_2)(\lambda-\lambda_3)/(\lambda_4-\lambda_1)(\lambda_4-\lambda_2)(\lambda_4-\lambda_3) \end{aligned}$$

this type of instrument is that it is usually impossible to find filters that, when combined with the spectral sensitivity of the unfiltered photo-cell, result in a perfect match to the $\bar{x}(\lambda)$, $\bar{y}(\lambda)$, $\bar{z}(\lambda)$ functions. Thus, although integration is carried out, errors are introduced on account of the incorrect spectral sensitivities achieved.

The alternative method of obtaining tristimulus values, computation from spectral data, has the advantage that correct spectral sensitivities can be simulated in the computation, but true integration is not carried out because the computations always consist of summations at discrete wavelength intervals (Billmeyer and Fairman, 1987).

The rate at which the $\bar{x}(\lambda)$, $\bar{y}(\lambda)$, $\bar{z}(\lambda)$ functions, and the vast majority of spectral power distributions, change with wavelength is sufficiently slow that summations of data at 1 nm intervals are usually considered to be equivalent to integration (Fairman, 1983). The only exceptions are light sources emitting very sharp spectral lines, as, for example, in the case of fluorescent lamps; in such cases, allocation of the spectral power in the line to the two closest 1 nm wavelengths in proportion to their proximity to the line, is usually considered to be equivalent to integration.

However, even with summations at 1 m intervals, there is the possibility of error. If the data for the spectral power distribution are obtained by sampling the spectrum through a spectral 'window' that is wider than 1 nm, then the data do not usually correctly represent

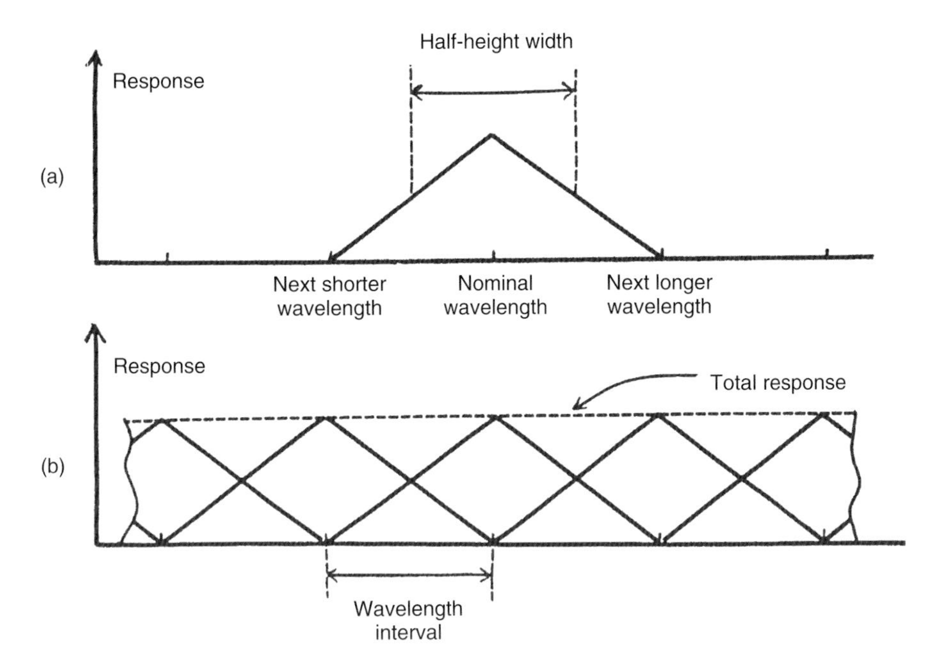

Figure 9.2 Spectral bandwidth functions. (a) Triangular. (b) Effective response for the summation of triangular bandwidth functions, separated by a wavelength interval equal to their half-height width, for a spectral power distribution that is constant with wavelength

the value of the spectral power distribution at each 1 nm wavelength (Stearns, 1981a). Ideally, the sampling of the spectrum should be through a 'triangular window' in which, as illustrated in Figure 9.2(a), the response is at a maximum at the nominal wavelength, is zero at the two neighbouring wavelengths, and falls linearly from maximum to zero over the intermediate wavelengths; such a bandwidth function has a width, at its half height, equal to the wavelength interval. (As mentioned in Section 9.11, the bandwidth function represents the combined effects of both the spectral transmittance of the measuring instrument and the spectral sensitivity of the detector.) For a spectral power distribution that is constant with wavelength, summation of data obtained with a 1 nm triangular bandwidth function, at successive 1 nm wavelengths, results in perfect integration (see Figure 9.2(b)). For spectral power distributions that are not constant with wavelength, perfect integration does not occur; but, because, in general, spectral power distributions are as likely to increase as to decrease with wavelength, the triangular window is at least impartial in the way in which the data are treated.

If the illuminant spectral power data are available at 1 nm intervals, it is appropriate to use these with sample data obtained with a triangular window having a half-height width of 1 nm. To obtain such data, an instrument with a slit-width corresponding to about 1 nm would have to be used. But such instruments tend to have poor signal-to-noise ratios (especially with dark samples), because their monochromators transmit so little light. It is therefore much more common to use instruments having wider slits. These wider slits usually still result in approximately triangular windows of bandwidth, but of course with greater half-height width. Another problem with using data at 1 nm intervals is the very large number of values that have to be included in the summations. For these reasons, most determinations of tristimulus values by the summation of spectral data use wavelength intervals of either 5 nm, 10 nm, or 20 nm instead of 1 nm.

When the bandwidth function is triangular with a half-height equal to the wavelength interval, and the summations are obtained using as weights the entries for the 1 nm $\bar{x}(\lambda)$, $\bar{y}(\lambda)$, $\bar{z}(\lambda)$ functions at each nominal wavelength, then, for a 5 nm interval, the errors are usually less than about 0.2 ΔE^*_{ab}; but, if a 10 nm bandwidth and interval is used, errors of up to about 0.7 ΔE^*_{ab} may be encountered; and with a 20 nm bandwidth and interval, errors of up to about 3 ΔE^*_{ab} may occur (Venable, 1989). Even larger errors are possible if data corresponding to the wrong bandwidth function are used. Attempts to reduce these errors have been made in various ways.

First, the measured data can be interpolated (using one of the methods given in Table 9.1) to provide values at every 5 nm; but this procedure does not usually improve the accuracy of the results (Venable, 1989).

Second, deconvolution procedures can be used that are designed to make the data more like those that would have been obtained with a 1 nm window (to which the entries in the tables of data refer). Instead of using the measured data, P, deconvoluted data, P_c, are used, obtained as:

$$P_{c\lambda} = 1.166P_{\lambda} - 0.083P_{\lambda-w} - 0.083P_{\lambda+w}$$

where w is the half-height bandwidth and interval (Stearns and Stearns, 1988). Thus each measurement of P is affected by two neighbouring measurements separated along the wavelength scale by $\pm$w. What this does is to change the values of $P(\lambda)$ from those of the broad window to those that approximate equivalent values for a 1 nm window at the same nominal wavelengths. With 1 nm window values for the nominal wavelengths available, 1 nm window values for all the intermediate wavelengths can be obtained by interpolation (see Table 9.1). These 1 nm values can then be used with the 1 nm $\bar{x}(\lambda)$, $\bar{y}(\lambda)$, $\bar{z}(\lambda)$ data to approximate integration (Stearns, 1981b; Stearns, 1981c; Stearns and Stearns, 1981). But it has been found (Venable, 1989) that, if the estimated 1 nm values are used just at the nominal wavelengths (at 5 nm, 10 nm, or 20 nm intervals as the case may be), then the errors are not great, so that the interpolation procedure can be omitted, without incurring too much inaccuracy. Thus, data deconvoluted to a 1 nm bandpass can be used with weights at the nominal wavelengths that are correct for 1 nm data, such as those given in Appendix 3. However, in ASTM document E308 - 08 (ASTM, 2008a; Fairman, 1985; Fairman, 1995), tables for 10 and 20 nm intervals are provided that have been based on interpolating to 1 nm values (very occasionally the interpolation procedure leads to very small negative values for some entries). For ease of computation, weighting values for products of illuminants and colour-matching functions can be produced.

Third, optimised weights can be used (Venable, 1989). The principle of optimised weights can be illustrated by considering, first, a filter type colorimeter that uses two different filters for approximating the major and minor lobes of the $\bar{x}(\lambda)$ function. Pairs of readings through the two filters have to be added together, and the best weights to use for the two contributions to the final result are those that tend to minimise the errors from the correct result, represented by integration with the true $\bar{x}(\lambda)$ function. Because, in general, spectral data of samples are as likely to increase as to decrease with wavelength, a reasonable approach is to optimise the weights for a non-selective neutral sample with the illuminant being used. This amounts to selecting the weights so that the difference between the computed observer-illuminant product and the true observer-illuminant product, D_{nm}, evaluated at successive 1 nm wavelengths, sums to zero over the entire spectrum. Data obtained from a spectroradiometer or spectrophotometer can be thought of as having been obtained through a number of filters, N, where N is the number of nominal wavelengths used, each filter having a transmittance equal to the bandwidth function of the instrument; N is usually not less than 16 (20 nm intervals from 400 nm to 700 nm). The best weights to use for combining these N contributions to the final result will be those that tend to minimise the errors from the true result. Again, it is reasonable to optimise the weights for a non-selective neutral sample with the illuminant being used. Optimization of the N weights is carried out by considering D_{nm} in wavelength intervals equal to the wavelength separation of adjacent spectrophotometric readings. Each interval is used with its centre at successive 1 nm positions along its wavelength range. The sum of the values of D_{nm} in each interval, ΣD_{nm}, is then calculated. The weights are then chosen, by an iterative procedure so that ΣD_{nm} is as near zero as possible in all the intervals at all the positions (minimum root-mean-square deviations). This ensures that there is no appreciable wavelength range where the differences, D_{nm}, are all of the same sign.

The advantage of using optimised weights can be illustrated by considering the case where the equi-energy stimulus, S_E, is the illuminant, and where the bandwidth function, instead of being triangular, is rectangular, and covers the bandwidth bounded by adjacent mid-points between neighbouring nominal wavelengths, as shown in Figure 9.3(a). Again, we consider the $\bar{x}(\lambda)$ function as an example. In this case, because the spectral power for S_E is constant with wavelength, and the bandwidth is rectangular, the computed observer-illuminant product is constant over each bandwidth, and has the shape of a histogram, as shown in Figure 9.3(b), with the heights of the steps being equal to the weights used; and the true observer-illuminant product is the same as the colour-matching function. For a non-selective neutral, the difference between the computed and true tristimulus values will then be equal to the difference between the area under the histogram and the area under the true colour-matching function. In a single bandwidth, the difference between the computed and true observer-illuminant product will be equal to the difference in area under the step considered and under the colour-matching function in the same bandwidth. If weights are used that are simply the 1 nm entries, the height of each step will always be that of the true colour-matching function at the nominal wavelength at the centre of the bandwidth; in this case, the differences between the areas will only be zero if the colour-matching function is a straight line over the bandwidth, as shown in Figure 9.3(c). When the colour-matching function has curvature, the areas will no longer be equal, as shown in Figure 9.3(d). But, if the height of the step is altered, the areas can be made equal, as shown in Figure 9.3(e). These altered heights are the optimised weights; they are obtained by an iterative procedure that minimises the differences in area for bandwidths having their centres at successive 1 nm positions throughout the spectrum.

It is clear from the above discussion that the optimum weights will be different for each combination of observer (1931 or 1964), wavelength interval, bandwidth function, and illuminant, and that they are not easy to calculate; but they can be very effective in reducing errors. Table 6 of the ASTM E308 document provides optimised weights for both the CIE 1931 and 1964 Observers, for 10 and 20 nm intervals, and for Standard Illuminants

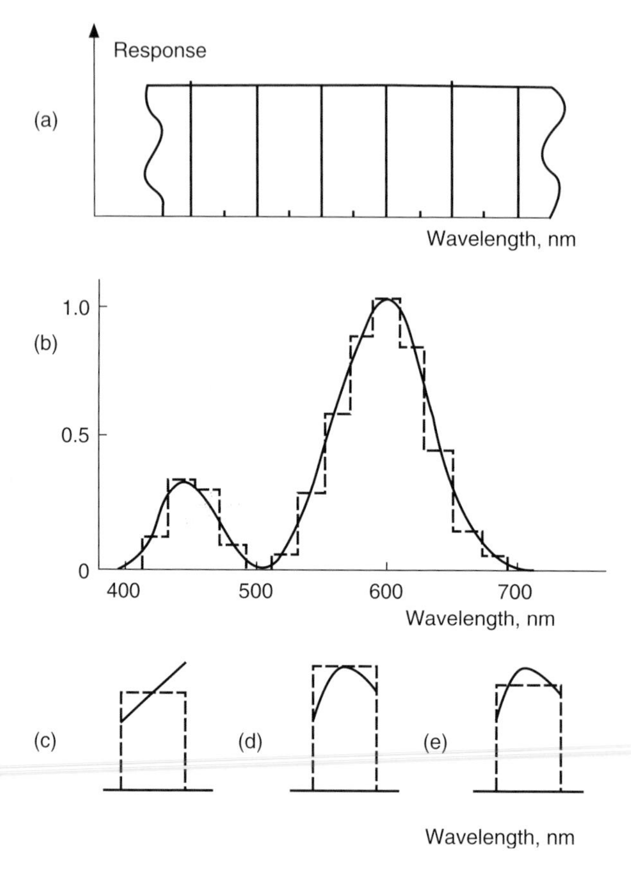

Figure 9.3 (a) Rectangular bandwidth functions that cover the whole of the spectrum without any overlapping. (b) An observer-illuminant function approximated by a histogram. (c) Element of a histogram in a single bandwidth, using a step height equal to the value of a function at the mid-wavelength, the function being straight throughout the bandwidth considered. (d) Same as (c) but with the function being curved. (e) Same as (d) but using a step height adjusted to equalise the areas under the step and under the curve

A, C, D50, D55, D65, and D75, and for fluorescent illuminants F2, F7, and F11 (ASTM, 2008b). These values are reproduced in Appendix 8, and are intended for use with data that have not been corrected for band-pass errors. (The values for F2, F7, and F11 in Table 6 were obtained from those of Table 5 of ASTM E308 with deconvolution; Fairman, 1995). Optimum weights for any other observer and illuminant combination can be obtained using the method described in ASTM E2022.

In Table 9.2 are shown typical maximum errors for three cases as found in the study by Venable (Venable, 1989): using measured data with table entries (MT); using deconvoluted data with table entries (DT); and using measured data with optimised weights (MO); in all cases the data is assumed to have been measured with triangular windows having the same half-height bandwidth as the wavelength interval, and the table entries are selected from the CIE 1 nm tables. For the MT case, the errors range from small (0.2 ΔE^*_{ab}) at

Table 9.2 Typical maximum errors for different types of summation

Half height bandwidth	Wavelength interval	Data used	Weights used	Case	Typical maximum errors (ΔE^*_{ab})
5 nm	5 nm	Measured	Table entries	MT	0.2
5 nm	5 nm	Deconvoluted	Table entries	DT	0.02
5 nm	5 nm	Measured	Optimised	MO	0.002
10 nm	10 nm	Measured	Table entries	MT	0.7
10 nm	10 nm	Deconvoluted	Table entries	DT	0.06
10 nm	10 nm	Measured	Optimised	MO	0.01
20 nm	20 nm	Measured	Table entries	MT	3.0
20 nm	20 nm	Deconvoluted	Table entries	DT	0.7
20 nm	20 nm	Measured	Optimised	MO	0.2

5 nm to considerable (3 ΔE^*_{ab}) at 20 nm. For the DT case, the errors range from very small (0.02 ΔE^*_{ab}) at 5 nm to significant (0.7 ΔE^*_{ab}) at 20 nm. For the MO case, the errors range from negligible (0.002 ΔE^*_{ab}) at 5 mn to small (0.2 ΔE^*_{ab}) at 20 nm. The samples used in this study were the NBS SRM 2101–2104 red, yellow, green, and blue glass filters (see Section 9.10) together with a didymium glass filter; and, for each filter the curve of spectral transmittance factor against wavelength was shifted in the long wavelength direction by nineteen 1 nm increments to generate nineteen extra test samples. Some other samples might give larger errors.

The data given in Appendix 3 are the 1 nm entries selected at 5 nm intervals, rounded to four places of decimals. The results of Table 9.2 indicate that, if weights corresponding to these table entries are used with data obtained with a half-height triangular bandwidth of 5 nm, the typical maximum error is only 0.2 ΔE^*_{ab}. This indicates that the 5 nm table entries in Appendix 3 can be used with reasonable confidence in conjunction with spectral data obtained with half-height triangular bandwidths of 5 nm (or less than 5 nm, because the deconvoluted data used with the 5 nm entries have a typical maximum error of only 0.02 ΔE^*_{ab}, provided that no line emissions from lamps are missed). But, if these table entries are used at only 10 nm or 20 nm intervals, or with data obtained with half-height bandwidths of more than 5 nm, appreciable errors may occur.

As mentioned in Section 5.12, if data at the two ends of the spectrum are not available, they should be provided by setting the missing values equal to the best available estimate or to the nearest measured values of the appropriate quantities. To ignore these wavelengths is equivalent to assuming zero values for these missing quantities and this is likely to lead to greater errors (Erb and Krystek, 1984). The full range of wavelengths for which the 1 nm values are available for the $\bar{x}(\lambda)$, $\bar{y}(\lambda)$, $\bar{z}(\lambda)$ functions is from 360 nm to 830 nm; but, when, as is often the case in practice, the computations are only made to 4 places of decimals, it is sufficient to use a range of wavelengths from 360 nm or 380 nm to 780 nm, because the values above 780 nm are then all zero, and those below 380 nm are then nearly zero and are often combined with low values of the spectral power distribution.

In view of these inaccuracies that can arise from the computation of tristimulus values, it is very important that any results that are to undergo critical comparison should have been derived using the same bandwidth function, wavelength interval, wavelength range, and method of computation including the choice of weights; this applies both to the evaluation of colour differences, and also to the evaluation of the variables in the CIELAB and CIELUV systems, which depend on the relationships between samples and reference whites. This was illustrated in Table 5.3, where the table entries for the $\bar{x}(\lambda)$, $\bar{y}(\lambda)$, $\bar{z}(\lambda)$ functions, and for $S(\lambda)$ for Standard Illuminants A and D65, were taken at 1 nm intervals from 360 nm to 830 nm, at 5 nm intervals from 380 nm to 780 am, at 10 nm intervals from 380 nm to 780 nm and from 400 nm to 700 nm, and at 20 nm intervals from 400 nm to 700 nm; both the wavelength interval and the wavelength range affected the results obtained, with differences from the 1 nm results varying from about 0.01 to 1.4 ΔE^*_{uv} or ΔE^*_{ab}.

9.14 PRECAUTIONS TO BE TAKEN IN PRACTICE

It is clear from the topics discussed in this chapter that there are many sources of potential imprecision and error in colorimetry, spectroradiometry, and spectrophotometry. Useful precautions that can be taken in practice include the following.

1. Ensure by visual inspection that all samples are properly selected, and are clean, uniform, and not misshapen.
2. Store all working standards in protective containers when not in use, and ensure that they are clean and in good condition whenever used.
3. Ensure that equipment is maintained and operated according to the manufacturer's instructions.
4. Mount all reflecting samples on standard backings, and measure all textured samples at the same orientation.
5. When samples are illuminated with white light in instruments, cheek for thermochromism by making successive measurements over a suitable period of time.
6. Select the appropriate geometry of illumination and measurement, and use it without change through each set of measurements.
7. If polarisation is suspected, do not use 45° illumination.
8. When computing tristimulus values:
   a. If possible use a 5 nm wavelength interval with a bandwidth half-height of 5 nm or less for the spectral measurements, and use the 5 nm entries in the 1 nm tables of $\bar{x}(\lambda)$, $\bar{y}(\lambda)$, $\bar{z}(\lambda)$, such as the table entries given in Appendix 3 or in similar sets of tables (for instance, those in ASTM E308 (ASTM, 2008a)). If 5 nm data are being used that have not been corrected for band-pass, it is preferable to correct them by using the deconvolution formula before carrying out the summations.
   b. If a wavelength interval greater than 5 nm has to be used, make the spectral measurements with a bandwidth function half-height equal to the wavelength interval, and use the measured data with entries at the greater interval in the 1 nm tables of $\bar{x}(\lambda)$, $\bar{y}(\lambda)$, $\bar{z}(\lambda)$. If the data have not been corrected for bandpass errors, it is preferable to correct them by using the deconvolution formula before carrying out the summations.

c. If the data are for a wavelength interval, observer, and illuminant, that are included in the ASTM E308, and the data have not been bandpass corrected use the Table 6 values of that document (reproduced in Appendix 8).

d. When using data with a range of wavelengths less than from 380 nm to 780 nm, do not set the missing spectral powers equal to zero, but equal to the best available estimate or to the nearest measured value.

9. When comparing tristimulus values, or any measures derived from them, use the same measuring and computing procedures for all the data; this is particularly important when parameters in colour spaces or colour differences are being calculated.

10. When measuring fluorescent samples (to be discussed in Chapter 10), always illuminate them with white light.

11. Make repeated measurements on working standards to check precision.

12. Measure calibrated samples to check accuracy.

13. If the highest accuracy is required, use the correction procedure described in Section 9.12.

14. When quoting results list the following conditions under which they were obtained:

Geometry of illumination and measurement.

Measuring source (used for fluorescent samples).

Type of spectrophotometer or spectroradiometer used (or bandwidth) for any spectral measurements, or type of photo-electric colorimeter used.

Illuminant and Standard Observer used for tristimulus values.

Method of computation used for deriving tristimulus values from spectral data (including wavelength range and interval, and the use of any correction procedures or optimised weights).

REFERENCES

Alman, D.H. and Billmeyer, F.W., A review of wavelength calibration methods for visible-range photoelectric spectrophotometers, *J. Chem. Educ.*, **52**, A281–A290 (1975).

ASTM Document E308, *Standard practice for computing the colors of objects by using the CIE system*, ASTM International, West Conshohocken, PA, USA (2008a).

ASTM Document E1164, *Standard practice for obtaining spectrophotometric data for object-color evaluation*, ASTM International, West Conshohocken, PA, USA (2009a).

ASTM Document E1331, *Standard test method for reflectance factor and color by spectrophotometry using hemispherical geometry*, ASTM International, West Conshohocken, PA, USA (2009b).

ASTM Document E1345, *Standard practice for reducing the effect of variability of color measurement by use of multiple measurements*, ASTM International, West Conshohocken, PA, USA (2008b).

ASTM Document E1347, *Standard test method for color and color-difference measurement by tristimulus (filter) colorimetry*, ASTM International, West Conshohocken, PA, USA (2006a).

ASTM Document E1348, *Standard test method for transmittance and color by spectrophotometry using hemispherical geometry*, ASTM International, West Conshohocken, PA, USA (2007).

ASTM Document E1349, *Standard test method for reflectance factor and color by spectrophotometry using bidirectional geometry*, ASTM International, West Conshohocken, PA, USA (2006b).

ASTM Document E2022, *Standard practice for calculation of weighting factors for tristimulus integration*, ASTM International, West Conshohocken, PA, USA (2008).

Berns, R. S. and Petersen, K.H., Empirical modelling of systematic spectrophotometric errors, *Color Res. Appl.*, **13**, 243–256 (1988).

Berns, R. S., and Reniff, L., An abridged technique to diagnose spectrophotometric errors, *Color Res. Appl.*, **22**, 51–60 (1997).

Billmeyer, F.W., and Alessi, P.J., Assessment of color-measuring instruments, *Color Res. Appl.*, **6**, 195–202 (1981).

Billmeyer, F.W., and Fairman, H.S., CIE method of calculating tristimulus values, *Color Res. Appl.*, **12**, 27–36 (1987).

Billmeyer, F.W., and Hemmendinger, H., Instrumentation for colour measurement and its performance, in *Golden Jubilee of Colour in the CIE*, pp. 98–112, Society of Dyers and Colourists, Bradford, England (1981).

Carter, E.C., and Billmeyer, F.W., Material standards and their use in color measurement, *Color Res. Appl.*, **4**, 96–100 (1979).

CIE Publication 15:2004, *Colorimetry*, *3rd ed.*, Commission Internationale de l'Éclairage, Vienna, Austria (2004).

CIE Publication 167:2005, *Recommended practice for tabulating spectral data for use in colour computations*, Commission Internationale de l'Éclairage, Vienna, Austria (2005).

Clarke, F.J.J., Ceramic colour standards – an aid for industrial colour control, *Printing Technology Magazine*, **13**(3), 101–113 (1969).

Clarke, F.J.J., and Compton, J.A., Correction methods for integrating-sphere measurement of hemispherical reflectance, *Color Res. Appl.*, **11**, 253–262 (1986).

Clarke, P.J., Hanson, H.R., and Verrill, J.F., Determination of colorimetric uncertainties in the spectrophotometric measurement of colour, *Analytica Chimica Acta*, **380**, 277–284 (1999).

Clarke, P.J., *Surface Colour Measurements*, National Physical Laboratory Good Practice Guide No. 96, National Physical Laboratory, Teddington, England (2006a).

Clarke, P.J., *Regular Transmittance Measurements*, National Physical Laboratory Good Practice Guide No. 97, National Physical Laboratory, Teddington, England (2006b).

Early, E.A., and Nadel, M.E., Uncertainty analysis for reflectance colorimetry, *Color Res. Appl.*, **29**, 205–216 (2004).

Eckerle, K.L., and Venable, W.H., Jr., 1976 remeasurement of NBS spectrophotometer integrator filters, *Color Res. Appl.*, **2**, 137–141 (1977).

Erb, W., and Krystek, M., Truncation error in colorimetric computations, *Color Res. Appl.*, **8**, 17–30 (1983).

Erb, W., Krystek, M., and Budde, W., A method for improving the accuracy of tristimulus colorimeters, *Color Res. Appl.*, **9**, 84–88 (1984).

Erb, W., and Budde, W., Properties of standard materials for reflection, *Color Res. Appl.*, **4**, 113–118 (1979).

Fairchild, M.D., Berns, R.S., Lester, A.A., and Shin, H.K., Accurate color reproduction of CRT displayed images as projected 35mm slides, *IS&T and SID's 2nd Color Imaging Conference: Color Science and Engineering Systems, Technologies, and Applications*, pp. 69–73, IS&T Springfield, VA, USA (1994).

Fairman, H.S., On analytical versus numerical integration in tristimulus calculations, *Color Res. Appl.*, **8**, 245–246 (1983).

Fairman, H.S., The calculation of weight factors tristimulus integration, *Color Res. Appl.*, **10**, 199–203 (1985).

Fairman, H.S., Results of the ASTM field test of tristimulus weighting functions, *Color Res. Appl.*, **20**, 44–49 (1995).

Fairman, H.S., An improved method for correcting radiance data for bandpass error, *Color Res. Appl.*, **35**, 328–333 (2010).

Gardner, J.L., Uncertainty estimation in colour measurement, *Color Res. Appl.*, **25**, 349–355 (2000).

Gardner, J.L., *Uncertainties in Surface Colour Measurement*, National Physical Laboratory Good Practice Guide No. 95, National Physical Laboratory, Teddington, England (2006).

Grum, F., and Saltzman, M., New white standard of reflectance, in *Proceedings of the 18th Session CIE, London, 1975*, pp. 91–97, CIE Publication 36, Bureau Central de la CIE, Paris, France (1976).

Keegan, H.J., Schleter, J.B., and Judd, D.B., Glass filters for checking performance of spectrophotometer-integrator systems of color measurement, *J. Res. Natl. Bur. Stand.*, **66A**, 203 (1962).

Kosztyán, Z.T., Eppeldauer, G.P., and Schanda, J.D., Matrix-based color measurement corrections of tristimulus colorimeters, *Appl. Optics*, **49**, 2288–2301 (2010).

Malkin, F., and Verrill, J.F., Advances in standards and methodology in spectrophotometry, In CIE Publication 56, *Proc. CIE 20th Session, Amsterdam*, pp. 49–63, Commission Internationale de l'Éclairage, Vienna, Austria (1983).

Malkin, F., Larkin, J.A., Verrill, J.F., and Wardman, R.H., The BCRA-NPL Ceramic Color Standards, Series II - master spectral reflectance and thermochromism data, *J. Soc. Dyers Colorists*, **113**, 84–94 (1997).

Marcus, R.T., Long-term repeatability of color-measuring instrumentation, *Color Res. Appl.*, **3**, 29 (1978).

Rich, D.C., The effect of measuring geometry on color matching, *Color Res. Appl.*, **13**, 113–118 (1988).

Robertson, A.R., Diagnostic performance evaluation of spectrophotometers, In Burgess, C., and Mielenz, K.D., Eds., *Advances in Standards and Methodology in Spectrophotometry*, pp. 287–286, Elsevier, New York, NY, USA (1987).

Stearns, E.I., Influence of spectrophotometer slits on tristimulus calculations, *Color Res. Appl.*, **6**, 78–84 (1981a).

Stearns, E.I., A new look at the calculation of tristimulus values, *Color Res. Appl.*, **6**, 203–206 (1981b).

Stearns, E.I., The determination of weights for use in calculating tristimulus values, *Color Res. Appl.*, **6**, 210–212 (1981c).

Stearns, E.I. and Stearns, R.E., Calculation of tristimulus weights by integration, *Color Res. Appl.*, **6**, 207 (1981).

Stearns, E.I., and Stearns, R.E., An example of a method for correcting radiance data for bandpass error, *Color Res. Appl.*, **13**, 257–259 (1988).

Venable, W.H., Accurate tristimulus values from spectral data, *Color Res. Appl.*, **14**, 260–267 (1989).

Verrill, J., Survey of reference materials for testing the performance of spectrophotometers and colorimeters, *CIE Journal*, **6**(1), 23–31, Commission Internationale de l'Éclairage, Vienna, Austria (1987).

Wharmby, D.O., Improvements in the use of filter colorimeters, *J. Phys. E: Scientific Instrum.*, **8**, 41–43 (1975).

Wightman, T.E., and Grum, F., Low-reflectance backing materials for use in optical radiation measurements, *Color Res. Appl.*, **6**, 139–142 (1981).

10

Fluorescent Colours

10.1 INTRODUCTION

When a sample fluoresces, some of the power incident on it is re-emitted with a change of wavelength. Therefore, at each wavelength, the total light re-emitted consists of the sum of that caused by reflection or transmission (where there has been no change of wavelength) and that caused by fluorescence (where there has been a change of wavelength).

Fluorescence often occurs in paper, as a result of *optical brightening agents* being added in order to increase their apparent whiteness; this is achieved by these agents absorbing ultra-violet radiation and re-emitting it in the blue part of the spectrum to counteract the tendency for paper to have lower bluish reflectances and thus to appear yellowish. Similar agents are also sometimes used to provide brilliant coloured textiles.

For the sake of simplifying what is inevitably a complex matter, we will consider only opaque samples; transmitting samples can be dealt with in a similar manner.

10.2 TERMINOLOGY

The following terms are used to discuss the effects of fluorescence:

Luminescence: emission of radiation in excess of that caused by thermal radiation.

Photoluminescence: radiation being absorbed by a substance that causes light to be emitted with a change of wavelength; if the emission is immediate it is termed *fluorescence*; if the emission continues appreciably after the absorbing radiation is removed, it is termed *phosphorescence*.

Spectral reflected radiance factor, $\beta_S(\lambda)$: ratio, at a given wavelength, of the radiance produced by reflection by a sample to that produced by reflection by the perfect reflecting diffuser identically irradiated.

Measuring Colour, Fourth Edition. R.W.G. Hunt and M.R. Pointer.

Spectral luminescent radiance factor, $\beta_L(\lambda)$: ratio, at a given wavelength, of the radiance produced by luminescence by a sample to that produced by reflection by the perfect reflecting diffuser identically irradiated. (The more general term *luminescent* includes *fluorescent* and *phosphorescent* phenomena.)

Spectral total radiance factor, $\beta_T(\lambda)$: the sum, at a given wavelength, of the spectral reflected radiance factor and the spectral luminescent radiance factor.

$$\beta_T(\lambda) = \beta_S(\lambda) + \beta_L(\lambda)$$

Spectral conventional reflectometer value, $\rho_C(\lambda)$: the apparent reflectance factor obtained when a fluorescent sample is measured relative to a non-fluorescent standard white sample, using monochromatic illumination and polychromatic detection.

The term *radiance factor* has been used in the above definitions, because *reflectance factor* applies only to reflected light and not to fluorescent light, which is produced by photoluminescence. The CIE recommends *spectral radiance factor*, β_e, instead of *spectral total radiance factor*, β_T, but the latter term is used in this chapter to emphasise its composite nature when fluorescence occurs. To measure the radiance factor, the cone of light collected should be negligibly small, but with real instruments this may not be so; however, the CIE has not defined terms for luminescent and total radiation collected within a defined cone, hence, in the following discussions, such instrumental measurements are regarded as approximations to the radiance factor.

10.3 USE OF DOUBLE MONOCHROMATORS

The most fundamental way of evaluating the colour of fluorescent samples is to irradiate them with monochromatic light and then, for each incident wavelength of the spectrum, to record the amount of light returned at every wavelength. For an illuminant of known spectral power distribution (such as Illuminant D65) these data can then be used to calculate

Table 10.1 Steps for the double monochromator method

Step 1.	With the sample irradiated by a stable, readily available, source, and the first monochromator set at wavelength 1, the spectral total radiance factor for that wavelength of irradiation, $\beta_T(\lambda)_1$, is measured throughout the spectrum by setting the second monochromator at a series of wavelengths throughout the spectrum.
Step 2.	The procedure in Step 1 is repeated with the first monochromator set at each of the wavelengths 2, 3, 4, etc., up to the last wavelength, n, throughout the spectrum.
Step 3.	The spectral total radiance factor, $\beta_T(\lambda)$, is calculated for the desired illuminant as: $\beta_T(\lambda) = S_1\beta_T(\lambda)_1 + S_2\beta_T(\lambda)_2 + S_3\beta_T(\lambda)_3 + \ldots\ldots\ldots + S_n\beta_T(\lambda)_n$ where $S_1, S_2, S_3, \ldots\ldots\ldots\ldots\ldots S_n$ are the values of spectral power of the desired illuminant at the wavelengths, 1, 2, 3, etc., to n.

the total spectral radiance factor throughout the spectrum, from which tristimulus values can be evaluated in the usual manner (Donaldson, 1954; Grum, 1972; Minato, Nanjo, and Nayatani, 1985). The steps necessary for this method are given in Table 10.1.

Although this procedure is correct, it entails many calculations; the required instruments have been constructed, but they require costly components, and hence they are usually only found in research and standardising laboratories. More simple methods are therefore often used in practice.

10.4 ILLUMINATION WITH WHITE LIGHT

The most useful practical approach is to use an instrument in which the sample is irradiated with a suitably chosen 'white' light, after which it is dispersed into the spectrum for the usual evaluation against the spectrum produced similarly by the perfect diffuser. It is essential that the sample is irradiated with white light (polychromatic irradiation), the dispersion of the light into the spectrum taking place subsequently (monochromatic detection). If this is not done, and the light is first dispersed into the spectrum (monochromatic irradiation) and then used to irradiate the sample, the light produced by the fluorescence, although of various wavelengths, will all be treated as if it were of the same wavelength as the irradiating light. Errors typical of those that are consequently produced (the *spectral conventional reflectometer values*, $\rho_C(\lambda)$) are shown in Figure 10.1 by the curve marked 'conventional'.

If the sample is illuminated with tungsten light, then Standard Illuminant A can be used in the form of Standard Source A, a tungsten filament lamp having a colour temperature of 2856 K. But tungsten light contains only a small proportion of the ultraviolet radiation that usually plays a very important part in the excitation of fluorescence in samples; so the main interest is in the evaluation of fluorescent materials under daylight illuminants. This requirement, however, poses a difficult problem regarding the choice of white light to

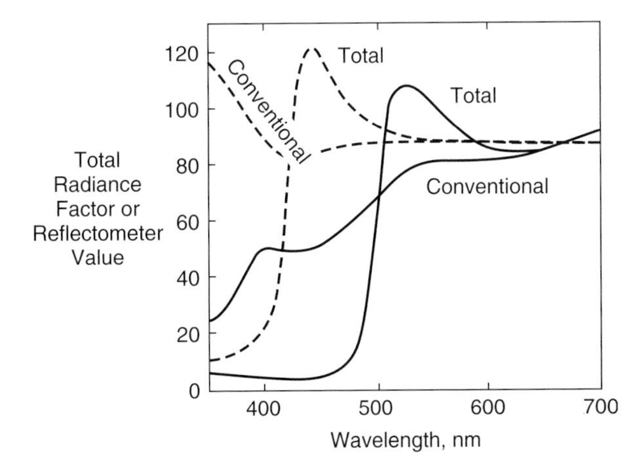

Figure 10.1 An example of the difference between total radiance factor ('total'), and conventional reflectometer value ('conventional'), when the sample fluoresces. Full lines: yellow sample. Broken lines: white sample. (After Grum and Bartleson, 1980, page 246)

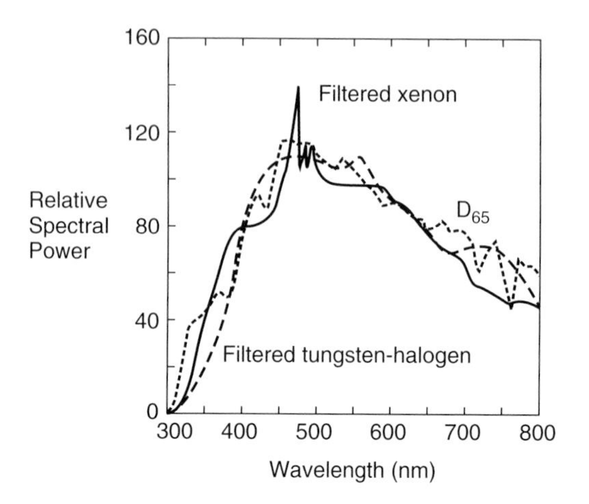

Figure 10.2 Relative spectral power distributions of a filtered xenon source (full line), and of a filtered tungsten-halogen source (broken line), intended to approximate illuminant D65, whose relative spectral power distribution is shown by the dotted line. (After Grum and Bartleson, 1980, page 268)

use. CIE Illuminant C is realisable as a source, but is deficient in the amount of ultraviolet radiation that is included. On the other hand, CIE Standard Illuminant D65, which has a more correct ultraviolet content, is not realisable as a source.

However, it is possible to use an illuminant that approximates a daylight distribution (such as that of D65), and then to make corrections that give results that are satisfactory for many practical purposes (Grum and Costa, 1977). Examples of sources that approximate D65 are a xenon lamp and a tungsten-halogen lamp, each with suitably chosen filters; in Figure 10.2 relative spectral power distributions are given for such sources, together with that for D65 for comparison. To correct for the effects of such differences in a source, it is necessary to know how the measured spectral total radiance factor is composed of the contributions from the spectral reflected radiance factor and the spectral luminescent radiance factor, and the range of wavelengths over which the fluorescing agent is excited. Four methods of obtaining this information will be described later. Assuming that it has been obtained satisfactorily, the procedure for making the corrections is then as follows (Billmeyer and Chong, 1980; Grum and Bartleson, 1980; Billmeyer, 1988).

10.5 CORRECTING FOR DIFFERENCES BETWEEN AN ACTUAL AND THE DESIRED SOURCE

The use of an actual source instead of the desired source does not affect the values obtained for the spectral reflected radiance factor. By its definition, it is given by:

$$\beta_S(\lambda) = [\beta_R(\lambda).k_a S(\lambda)]/[\beta_p(\lambda).k_a S(\lambda)]$$

where $\beta_R(\lambda)$ is the fraction of the radiance produced by reflection by the sample relative to that produced by the perfect diffuser identically irradiated, $S(\lambda)$ is the relative spectral power distribution of the source, k_a is a constant that converts $S(\lambda)$ to absolute radiance values, and $\beta_P(\lambda)$ is the fraction of the radiance reflected by the perfect diffuser relative to itself. Because the $k_a S(\lambda)$ terms cancel out, $\beta_S(\lambda)$ is independent of the spectral power distribution of the source; and because $\beta_P(\lambda)$ is equal to unity at all wavelengths, $\beta_S(\lambda)$ reduces to $\beta_R(\lambda)$, as is to be expected.

However, the spectral luminescent radiance factor, $\beta_L(\lambda)$, certainly is affected by the spectral power distribution of the source. By its definition, it is given by:

$$\beta_L(\lambda) = F(\lambda)/[\beta_P(\lambda).k_a S(\lambda)]$$

where $F(\lambda)$ is the spectral distribution of the radiance produced by fluorescence. Because $\beta_P(\lambda) = 1$, we have:

$$\beta_L(\lambda) = F(\lambda)/[k_a S(\lambda)]$$

and thus $\beta_L(\lambda)$ is dependent on $S(\lambda)$. If the measuring source contains too great a proportion of ultraviolet radiation, $\beta_L(\lambda)$ will tend to be too high, and if this proportion is too small, $\beta_L(\lambda)$ will tend to be too low.

The spectral distribution of the radiance produced by fluorescence, $F(\lambda)$, is usually at longer wavelengths than those of the irradiating source, and is proportional to the number of quanta absorbed from it (Stokes' Law). Hence:

$$F(\lambda) = k_b Q$$

where k_b is a constant, and Q represents the number of quanta from the source absorbed per second by the fluorescing agent; Q can be calculated as follows:

$$Q = k_c \int \alpha(\lambda') k_d S(\lambda') \lambda'.d\lambda'$$

where k_c is another constant, $\alpha(\lambda')$ is the spectral absorptance of the fluorescing agent, $S(\lambda')$ is the relative spectral power distribution of the irradiating source over the range of wavelengths, λ', that excite the fluorescing agent, and k_d is a constant that converts $S(\lambda')$ to absolute power values. $S(\lambda')$ is multiplied by λ' to allow for the fact that, for a given amount of power of a monochromatic radiation, the number of quanta per second is proportional to the wavelength.

To simplify the following steps, it is now assumed that $\alpha(\lambda')$ is constant with wavelength; although this is unlikely to be strictly true, if, as is often the case, the fluorescing agent is present in high concentration, $\alpha(\lambda')$ will be nearly equal to unity at most wavelengths and hence will not vary much. The above equation can then be rewritten as:

$$Q = k_c k_d \alpha \int S(\lambda') \lambda'.d\lambda'$$

Approximate numerical values of Q can be obtained from:

$$Q = k_c k_d \alpha [S_1(\lambda'_1) + S_2(\lambda'_2) + S_3(\lambda'_3) + \ldots\ldots\ldots\ldots S_n(\lambda'_n)]$$

where the subscript 1 indicates the shortest wavelength of the fluorescent excitation region, the subscripts 2 etc. indicate wavelengths at regular wavelength intervals within that region, and the subscript n indicates the longest wavelength of that region.

Because $\beta_L(\lambda) = F(\lambda)/[k_a S(\lambda)]$ and $F(\lambda) = k_b Q$, we have:

$$\beta_L(\lambda) = k_b Q/[k_a S(\lambda)]$$

It follows from the above that, for two sources, indicated by the subscripts d denoting the desired source, and m denoting the actual source used in the measuring instrument:

$$\frac{\beta_{Ld}(\lambda)}{\beta_{Lm}(\lambda)} = \frac{[S_1(\lambda'_1) + S_2(\lambda'_2) + S_3(\lambda'_3) + \ldots\ldots\ldots\ldots S_n(\lambda'_n)]_d / S_d(\lambda)}{[S_1(\lambda'_1) + S_2(\lambda'_2) + S_3(\lambda'_3) + \ldots\ldots\ldots\ldots S_n(\lambda'_n)]_m / S_m(\lambda)}$$

the constants, k_a, k_b, k_c, k_d, and α all cancelling out. This expression is then used to obtain the desired spectral luminescent radiance factor $\beta_{Ld}(\lambda)$ from the measured spectral luminescent radiance factor $\beta_{Lm}(\lambda)$.

The spectral total radiance factor for the sample illuminated by the desired illuminant, $\beta_{Td}(\lambda)$, is then calculated as $\beta_{Ld}(\lambda)$ plus the spectral reflected radiance factor, $\beta_S(\lambda)$.

$$\beta_{Td}(\lambda) = \beta_{Ld}(\lambda) + \beta_S(\lambda).$$

The four methods that will now be described for separating the contributions of the spectral reflected radiance factor, β_S, and the spectral luminescent radiance factor, β_L, to the spectral total radiance factor, β_T, and for determining the range of wavelengths exciting fluorescence, may be referred to as the *two-monochromator* method, the *two-mode method, the filter-reduction method*, and the *luminescence-weakening method*.

10.6 TWO-MONOCHROMATOR METHOD

In the two-monochromator method, the sample is illuminated with monochromatic light by means of a first monochromator, and then a second monochromator is used to transmit light

Table 10.2 Steps for the two-monochromator method

Step 1.	The spectral total radiance factor, $\beta_T(\lambda)$, is measured with the sample irradiated by a stable, readily available, source that approximates D65 as closely as possible.
Step 2.	The spectral reflected radiance factor, $\beta_S(\lambda)$, is measured by irradiating the sample with monochromatic light and using the second monochromator to isolate light only of the same wavelength.
Step 3.	The spectral luminescent radiance factor, $\beta_L(\lambda)$, is evaluated as the difference between the spectral total radiance factor and the spectral reflected radiance factor.
Step 4.	The range of wavelengths over which the fluorescing agent is excited is estimated as follows. The longest wavelength is taken as that at which the 'Total' curve is at its greatest height above the 'Reflected' curve. The shortest wavelength is taken as the shortest wavelength (on the short wavelength side of the fluorescent emission band) at which the spectral reflected radiance curve is similar to that of a version of the material being studied that contains no fluorescing agent (this indicating that the fluorescing agent is not absorbing any radiation).

only of the same wavelength; in this way, any fluorescent light generated by irradiance will, because of its wavelength shift, not pass through the second monochromator, and β_S alone is determined. By scanning the wavelength range of interest with the two monochromators in step for wavelength and by using a narrow bandpass for each, β_S can be measured with high accuracy. This is a simpler procedure than measuring the complete spectrum of the light emitted by the sample for each wavelength of irradiation, and is therefore a useful technique. The steps necessary for this method are given in Table 10.2.

However, if, as is often the case, only a single monochromator is available, then one of the other methods has to be used, although they tend to be less accurate. In one study of the three remaining methods (Alman, Billmeyer, and Phillips, 1976), it was found that the filter-reduction method was preferable for samples with small amounts of fluorescence, and the luminescence-weakening method for samples with large amounts of fluorescence; the two-mode method was found to be less accurate, but it is a useful diagnostic tool, and is therefore now described first.

10.7 TWO-MODE METHOD

In the two-mode method (Simon, 1972), an instrument is required that can be used in both the polychromatic-irradiation and the monochromatic-irradiation modes; the latter mode is referred to as a *conventional reflectometer*. Such an instrument provides a good means of showing whether any fluorescence is occurring: this is illustrated by the differences between the curves for spectral total radiance factor and for spectral conventional reflectometer value shown in Figures 10.1 and 10.3. The steps necessary for this method are given in Table 10.3.

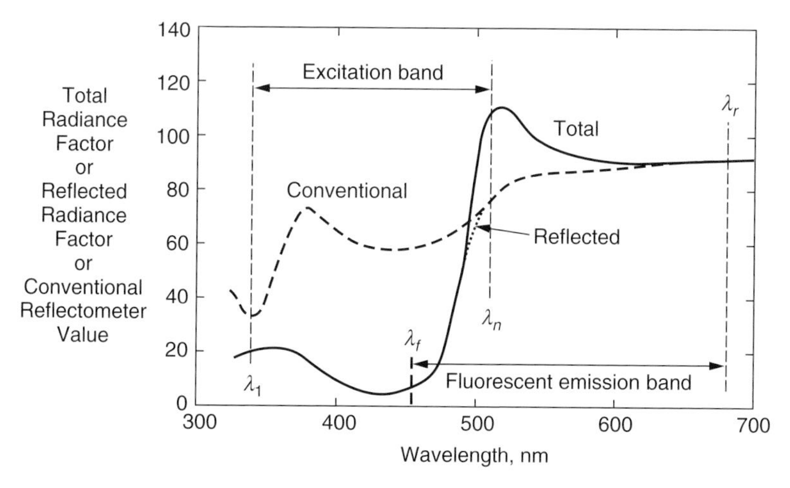

Figure 10.3 Total radiance factor ('Total'), reflected radiance factor ('Reflected'), and conventional reflectometer value ('Conventional') for a green fluorescent sample. The excitation band of wavelengths is from λ_l to λ_n. The fluorescent emission band of wavelengths is from λ_f to λ_r. (After Grum and Bartleson, 1980, page 276)

Table 10.3 Steps for the two-mode method

Step 1.	The spectral total radiance factor, $\beta_T(\lambda)$, is measured with the sample irradiated by a stable readily available source that approximates D65 as closely as possible.
Step 2.	The spectral conventional reflectometer value, $\rho_C(\lambda)$, is measured, preferably using the same instrument, but operating it in a monochromatic-irradiating mode.
Step 3.	The spectral reflected radiance factor, $\beta_S(\lambda)$, is deduced. This is given by the spectral total radiance factor at wavelengths shorter than those of the fluorescent emission band (the shortest wavelength of which is usually about 50 nm below that at which the 'Total' and 'Conventional' curves cross), and by the conventional spectral reflectometer value at wavelengths longer than those of the region of excitation for fluorescence (the longest wavelength of which is usually about 15 nm above that at which the 'Total' and 'Conventional' curves cross). For wavelengths within both these regions the results have to be estimated by interpolation. (See Figure 10.3.)
Step 4.	The spectral luminescent radiance factor, $\beta_L(\lambda)$, is evaluated as the difference between the spectral total radiance factor and the spectral reflected radiance factor.
Step 5.	The range of wavelengths over which the fluorescent agent is excited is estimated as follows. The longest wavelength is the point of intercept of the values for the spectral reflected radiance factor and the spectral conventional reflectometer value; the shortest wavelength is that at which the total spectral radiance factor and the spectral conventional reflectometer value are closest. This is illustrated in Figure 10.3.

10.8 FILTER-REDUCTION METHOD

In the filter-reduction method (Eitle and Ganz, 1968), a two-mode instrument is not necessary, and the measurements can be carried out using only a polychromatic-irradiating monochromatic-detecting instrument. The white light irradiating the sample is used both unfiltered and with a series of filters each of which sharply cuts out all radiation of wavelengths shorter than a chosen value. These chosen wavelengths cover the region where both excitation and fluorescent emission occur (see Figure 10.3). The steps necessary for this method are given in Table 10.4.

10.9 LUMINESCENCE-WEAKENING METHOD

In the luminescence-weakening method (Allen, 1973), as was the case for the filter-reduction method, a two-mode instrument is not necessary, and the work can be carried out using only a polychromatic-irradiating monochromatic-detecting instrument. The white light irradiating the sample is used both with and without two filters, one of which cuts out all the radiation that causes fluorescence, and the other that reduces that radiation appreciably. The steps necessary for this method are given in Table 10.5.

Table 10.4 Steps for the filter-reduction method

Step 1.	The spectral total radiance factor, $\beta_T(\lambda)$, is measured with the sample irradiated by a stable readily available source that approximates D65 as closely as possible. At wavelengths shorter than the short-wavelength boundary, λ_f, of the fluorescent emission band, the curve obtained provides the spectral reflected radiance factor. The wavelength, λ_f, is usually about 80 nm below the wavelength, λ_T, at which $\beta_T(\lambda)$ is maximum in the fluorescent emission band.
Step 2.	A sharp cut-off filter is introduced into the incident beam that completely eliminates all wavelengths below λ_T. At wavelengths above the cut-off wavelength of the filter, this curve provides the spectral reflected radiance factor. It is now necessary to determine the spectral reflected radiance factor at the intermediate wavelengths, between λ_f, and λ_T.
Step 3.	A cut-off filter is introduced into the incident beam that eliminates all wavelengths below a wavelength in the region between λ_f and λ_T. This reduces the fluorescent emission to give an approximation to the spectral reflected radiance factor.
Step 4.	Step 3 is repeated using filters with different chosen cut-off wavelengths (between λ_f and λ_T) to obtain further approximations to the spectral reflected radiance factor. These approximations are then used to obtain a best estimate of the true spectral reflected radiance factor, the most accurate estimate at each wavelength always being the lowest of those available.
Step 5.	The spectral luminescent radiance factor, $\beta_L(\lambda)$, is evaluated as the difference between the spectral total radiance factor and the spectral reflected radiance factor.
Step 6.	The range of wavelengths over which the fluorescing agent is excited is estimated as follows. The longest wavelength is taken as that at which the 'Total' curve is at its greatest height above the 'Reflected' curve. The shortest wavelength is taken as the shortest (on the short wavelength side of the fluorescent emission band) at which the spectral reflected radiance curve is similar to that of a version of the material being studied that contains no fluorescing agent (this indicating that the fluorescing agent is not absorbing any radiation).

10.10 PRACTICAL CONSIDERATIONS

If one of these four methods, or a similar procedure, is not carried out, the best that can then be done is to use the polychromatic irradiation mode with a source that approximates D65 as closely as possible, and to record the exact nature of the source used. Various sources have been suggested for this purpose, and some are much more appropriate than others (Terstiege, 1989; Imura, 2007). One method is to use a spectrophotometer with an ultraviolet absorbing filter which can be inserted to a varying extent into the white-light irradiating beam, so as to adjust its ultraviolet content. A setting for the position of the filter is then chosen that most nearly produces spectral radiance measurements that are the

Table 10.5 Steps for the luminescent-weakening method

Step 1. The spectral total radiance factor, $\beta_T(\lambda)$, is measured with the sample irradiated by a stable readily available source that approximates D65 as closely as possible. At wavelengths shorter than the short-wavelength boundary, λ_f, of the fluorescent emission band, the curve obtained provides the spectral reflected radiance factor. The wavelength, λ_f, is usually about 80 nm below the wavelength, λ_T, at which $\beta_T(\lambda)$ is maximum in the fluorescent emission band.

Step 2. A sharp cut-off filter is introduced into the incident beam that completely eliminates all wavelengths below λ_T, and a 'Completely filtered (CF)' version of $\beta_T(\lambda)$ is obtained, $\beta_{TCF}(\lambda)$. At wavelengths above the cut-off wavelength of the filter, this curve provides the spectral reflected radiance factor. It is now necessary to determine the spectral reflected radiance factor at the intermediate wavelengths between λ_f and λ_T.

Step 3. Instead of the sharp cut-off filter, a partial cut-off filter is introduced into the incident beam that eliminates all wavelengths below a wavelength just above the minimum of the unfiltered $\beta_T(\lambda)$ curve, and a 'Partly filtered (PF)' version of $\beta_T(\lambda)$ is obtained, $\beta_{TPF}(\lambda)$. The spectral transmittance curve, $T(\lambda)$, of this filter has to be known, and can be measured on the spectrophotometer being used.

Step 4. The curves obtained with and without the partial cut-off filter are now used to determine the spectral reflected radiance factor, $\beta_S(\lambda)$, from the equation:

$$\beta_S(\lambda) = [\beta_{TPF}(\lambda)T(\lambda) - \beta_T(\lambda)k]/[T(\lambda) - k]$$

the value of the constant k being given by:

$$k = T(\mathrm{w})[\beta_{TPF}(\mathrm{w}) - \beta_S(\mathrm{w})]/[\beta_T(\mathrm{w}) - \beta_S(\mathrm{w})]$$

where w is the lowest-wavelength at which readings were obtained with the sharp cut-off filter; the values of $\beta_{TPF}(\mathrm{w})$, $\beta_T(\mathrm{w})$, and $T(\mathrm{w})$ are read off their curves at the wavelength w, and the value of β_S at this wavelength is the same as that of β_{TCF} and can thus be obtained from the curve of β_{TCF}.

Step 5. The spectral luminescent radiance factor, $\beta_L(\lambda)$, is evaluated as the difference between the spectral total radiance factor and the spectral reflected radiance factor.

Step 6. The range of wavelengths over which the fluorescing agent is excited is estimated as follows. The longest wavelength is taken as that at which the 'Total' curve is at its greatest height above the 'Reflected' curve. The shortest wavelength is taken as the shortest (on the short wavelength side of the fluorescent emission band) at which the spectral reflected radiance curve is similar to that of a version of the material being studied that contains no fluorescing agent (this indicating that the fluorescing agent is not absorbing any radiation).

same as known true values for a set of standard samples for the desired illuminant, such as D65. A set of white fluorescent samples is sometimes used for the standard samples.

In evaluating fluorescent samples, the following are important areas to keep under control:

1. The illuminating system in the measuring instrument must be fully defined and standardised.
2. The samples must be stable under the measurement conditions. In the white-light irradiating mode, the amount of radiant power falling on the samples must not affect their colour by heating or fading.
3. The appropriate illuminating and measuring geometry must be used. If an integrating sphere is used, then the specimen modifies the spectral power distribution of the illumination in the white-light irradiating mode (Alman and Billmeyer, 1976), but this effect can be negligible if a large sphere with small ports is used.
4. The coating on the interior surface of any integrating sphere that is used must be non-fluorescent.

Justification for the procedure in Step 4. That the procedure in step 4 is correct can be shown as follows. Using the subscripts PF to indicate 'Partly Filtered' and CF to indicate 'Completely Filtered', we can write:

$$\beta_T(\lambda) = \beta_S(\lambda) + F(\lambda)/[k_a S(\lambda)]$$

$$\beta_{TPF}(\lambda) = \beta_S(\lambda) + f_{PF} F(\lambda)/[k_{aPF} S_{PF}(\lambda)]$$

where f_{PF} is a constant that allows for the different level of fluorescent excitation produced by the partly filtered source. If the spectral transmittance of the partial cut-off filter is $T(\lambda)$, then

$$T(\lambda)S(\lambda) = S_{PF}(\lambda)$$

Hence:

$$\beta_{TPF}(\lambda) = \beta_S(\lambda) + f_{PF} F(\lambda)/[k_{aPF} T(\lambda) S(\lambda)]$$

Eliminating $F(\lambda)/S(\lambda)$ we obtain:

$$\beta_S(\lambda) = [\beta_{TPF}(\lambda)T(\lambda)k_{aPF} - \beta_T(\lambda)k_a f_{PF}]/[k_{aPF}T(\lambda) - k_a f_{PF}]$$

Changing the power of the partial cut-off light source would change f_{PF} and k_{aPF} in the same proportion; and f_{PF} is inversely proportional to k_a, because, if k_a were increased, $F(\lambda)$ would increase in proportion and f_{PF} would have to decrease in the same proportion to keep $f_{PF}F(\lambda)$ in the same relationship to the power of the partial cut-off light source $k_{aPF}S_{PF}(\lambda)$. Therefore, where k is a new constant:

$$f_{PF} = k k_{aPF}/k_a$$

Substituting this expression for f_{PF} we obtain:

$$\beta_S(\lambda) = [\beta_{TPF}(\lambda)T(\lambda) - \beta_T(\lambda)k]/[T(\lambda) - k]$$

If w represents a particular wavelength, we can write the above equation for w, solve for k, and obtain:

$$k = T(\mathrm{w})[\beta_{\mathrm{TPF}}(\mathrm{w}) - \beta_{\mathrm{S}}(\mathrm{w})]/[\beta_{\mathrm{T}}(\mathrm{w}) - \beta_{\mathrm{S}}(\mathrm{w})]$$

REFERENCES

Allen, E., Separation of the spectral radiance factor curve of fluorescent substances into reflected and fluoresced components, *Appl. Optics*, **12**, 289–293 (1973).

Alman, D.H., and Billmeyer, F.W., Integrating-sphere errors in the colorimetry of fluorescent materials, *Color Res. Appl.*, **1**, 141–145 (1976).

Alman, D.H., Billmeyer, F.W., and Phillips, D.G., *A comparison of one-monochromator methods for determining the reflectance of opaque fluorescent samples*, Proc. 18th Session CIE, London 1975, pp. 237–244, Bureau de la CIE, Paris, France (1976).

Billmeyer, F.W., Intercomparison on measurement of (total) spectral radiance factor of luminescent specimens, *Color Res. Appl.*, **13**, 318–326 (1988).

Billmeyer, F.W., and Chong, T.-F., Calculation of the spectral radiance factors of luminescent samples, *Color Res. Appl.*, **5**, 156–168 (1980).

Donaldson, R., Spectrophotometry of fluorescent pigments, *Brit. J. Appl. Phys.*, **5**, 210–214 (1954).

Eitle, D., and Ganz, E., Eine Methode zur Bestimmung von Normfarbwerten Für fluoreszierende Proben, *Textilveredlung*, **3**, 389–392 (1968).

Grum, F., Use of true reflectance and fluorescence for color evaluation of achromatic and chromatic fluorescent materials, In *CIE Proceedings 1971 (Barcelona)*, Paper 71.22. Bureau Central de la CIE, Paris, France (1972).

Grum, F., and Bartleson, C.J., *Color Measurement*, Academic Press, New York, NY, USA (1980).

Grum, F., and Costa, L.F., Color evaluation by fluorescence measurement without the need for multiple illumination sources, *Tappi*, **60**, 119–121 (1977).

Imura, K., New method for measuring an optical property of a sample treated by FWA, *Color Res. Appl.*, **32**, 195–200 (2007).

Minato, H., Nanjo, M., and Nayatani, Y., Colorimetry and its accuracy in the measurement of fluorescence by the two-monochromator method, *Color Res. Appl.*, **10**, 84–91 (1985).

Simon, F.T., The two-mode method for measurement and formulation with fluorescent colorants, *J. Color Appearance*, **1**(4), 5–11 (1972).

Terstiege, H., Artificial daylight for measurement of optical properties of materials, *Color Res. Appl.*, **14**, 131–138 (1989).

11

RGB Colorimetry

11.1 INTRODUCTION

In the years immediately following the introduction of the CIE system of colorimetry in 1931, visual instruments providing additive mixtures of controllable beams of red, green, and blue light were used to obtain colour matches on samples, from which CIE tristimulus values were calculated. The skill required to make such matches, and particularly the length of time that it took even an experienced observer to match large numbers of samples, have resulted in such instruments now being obsolete. Visual colour matches are still made when colour atlases are used (see Chapter 8); approximate CIE specifications can be obtained from such matches, and, if matches are made on the Lovibond Schofield Tintometer, CIE specifications that are accurate can be derived (see Section 8.13). But, today, most colorimetry depends on responses from photo-detectors, either in colorimeters employing filtered photo-detectors, or in spectroradiometers or spectrophotometers to provide spectral data for calculation.

However, there remain two areas where the production and specification of colours by additive mixtures of beams of red, green, and blue light are still carried out. In visual research work, this is still a very useful way of providing colour stimuli in a continuously variable form; and in television and computer displays, the colours are also produced by such additive mixtures. In this chapter, we shall therefore discuss the colorimetric procedures involved.

11.2 CHOICE AND SPECIFICATION OF MATCHING STIMULI

The gamut of chromaticities that can be matched with a set of three colour matching stimuli is defined by the triangle formed by the three points representing them in a chromaticity diagram. In Figure 11.1, such a triangle is shown, in the x,y and in the u′,v′ diagrams, for the three phosphors chosen as standard for European colour television (BREMA, 1969;

Measuring Colour, Fourth Edition. R.W.G. Hunt and M.R. Pointer.

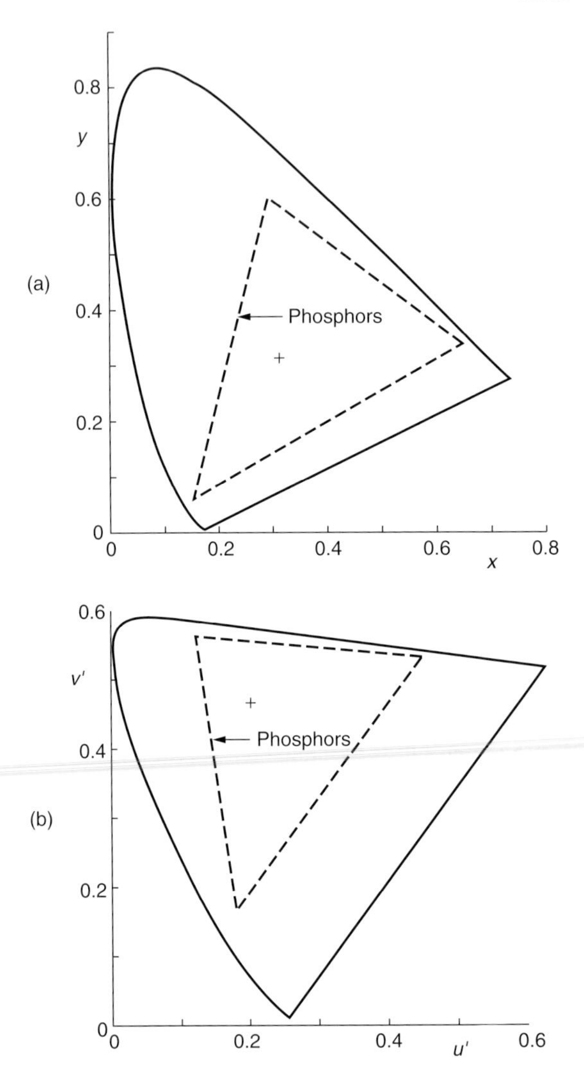

Figure 11.1 Gamut of (E.B.U) CCIR 709 broadcast television phosphors in (a) the x,y chromaticity diagram, and (b) the u′,v′ chromaticity diagram

Sproson, 1978; Hunt, 2004); the chromaticities of these *primaries*, as they are usually called, are:

Red	$x = 0.64$	$y = 0.33$	$u' = 0.451$	$v' = 0.523$
Green	$x = 0.29$	$y = 0.60$	$u' = 0.125$	$v' = 0.562$
Blue	$x = 0.15$	$y = 0.06$	$u' = 0.175$	$v' = 0.158$

It can be noted that these primaries are the same as those specified for sRGB, a standard colour space for use on displays, printers, and the internet (IEC 61966-2-1:1999).

It is clear that the triangles in Figure 11.1 cover less than half the domain of chromaticities lying within the spectrum locus and the purple boundary. If more saturated matching stimuli, lying on the spectrum locus, had been chosen, then more of the domain of chromaticities would have been included in the triangles; but, even if primaries for television and computer display devices were available having such chromaticities, their luminous efficacies would be very low, so that the luminance levels attainable with such primaries would also be very low. Low levels of luminance result in poor precision in colour matching, and in low brightness and colourfulness in displayed colours. The choice of primaries nearly always requires a compromise between covering as much of the chromaticity domain as possible and having adequate luminance; the primaries whose chromaticities are shown in Figure 11.1 were the result of such a compromise for television.

The triangle for the primaries appears to be particularly lacking in its coverage of greenish colours in the x,y diagram, whereas in the u′,v′ diagram a greater lack is apparent for purplish colours. This provides a good illustration of the way in which the x,y diagram can be misleading; as already shown in Figure 3.5, colour differences of a given size are represented by much larger distances in the greenish than in the purplish regions of this diagram. Such distortions are much reduced in the u′,v′ diagram, so that its use is more appropriate, although it is not without some distortions itself, as can be seen in Figure 3.6.

11.3 CHOICE OF UNITS

As already explained in Section 2.4, if the amounts of red, green, and blue are measured in photometric units, such as units of luminance, a match on a white is represented by three very unequal numbers, the value for the blue being particularly low. It is, therefore, common practice to use units defined such that the amounts of the three primaries required to match a suitably chosen white are equal to one another. In television displays this white is often set to approximate to the chromaticity of D65 but bluer whites are sometimes used. The luminances of these units are denoted by L_R, L_G, and L_B.

11.4 CHROMATICITY DIAGRAMS USING *r* AND *g*

Having defined the chromaticities of the three matching primaries, and the units in which they are to be measured, the colorimetric system is complete, and the amounts of red, green, and blue light needed to match any colour are the *tristimulus values*, R, G, B, in the system. From these can be obtained corresponding *chromaticity coordinates*:

$$r = R/(R + G + B)$$

$$g = G/(R + G + B)$$

$$b = B/(R + G + B)$$

Since $r + g + b = 1$, if two of these values are known the third can be deduced by subtracting their sum from 1. A chromaticity diagram in which one of the coordinates is plotted against one of the others can therefore be used, and it is customary to use r and g for this purpose, as shown in Figure 11.2.

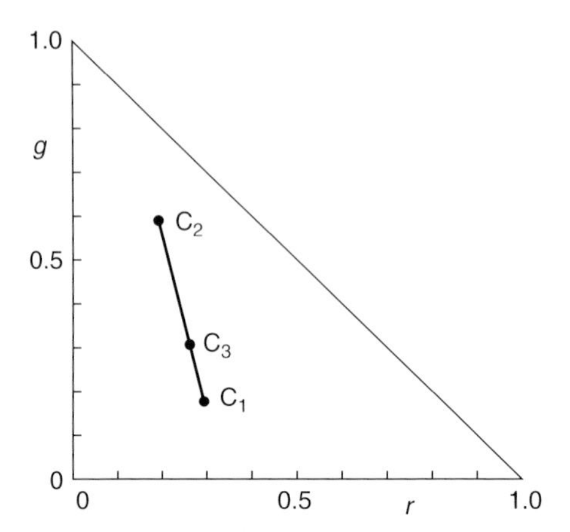

Figure 11.2 r,g chromaticity diagram. The colour C_3 can be matched by an additive mixture of the colours C_1 and C_2

Additive mixtures of colours on r,g chromaticity diagrams, such as that shown in Figure 11.2, lie on the straight line joining the two points representing the constituent colours in the mixture. If the calculation given in Section 3.5 for mixtures on the x,y diagram is reworked for mixtures on an r,g diagram, it is found that the chromaticity, r_3, g_3 of the colour C_3, produced by the additive mixture of m_1 luminance units of colour C_1, and m_2 luminance units of colour C_2, and having chromaticities r_1, g_1 and r_2, g_2, respectively, is the same as the centre of gravity of weights, W_1 and W_2 placed at the points representing C_1 and C_2, respectively, where:

$$W_1 = m_1/(L_R r_1 + L_G g_1 + L_B b_1)$$

$$W_2 = m_2/(L_R r_2 + L_G g_2 + L_B b_2)$$

This means that the point C_3 divides the line C_1C_2 in the ratio

$$(C_1C_3)/(C_2C_3) = W_2/W_1$$

The *centre of gravity law of colour mixture* thus applies to mixtures in the r,g diagram, but is a little more complicated than in the x,y diagram where the weights used are m_1/y_1 and m_2/y_2 or in the u′,v′ diagram where the weights used are m_1/v'_1 and m_2/v'_2 (see Section 3.5).

11.5 COLOUR-MATCHING FUNCTIONS IN RGB SYSTEMS

If the red, green, and blue primaries are used to match equal amounts of power per small constant-width wavelength interval throughout the spectrum, the results constitute a set of colour-matching functions. If the results are plotted against wavelength, a set of curves similar to those shown in Figure 11.3 is obtained. If these curves are compared with those shown in Figure 2.4, it can be seen that they are broadly of the same type but differ in detail. These differences result from the differences in chromaticities between the television

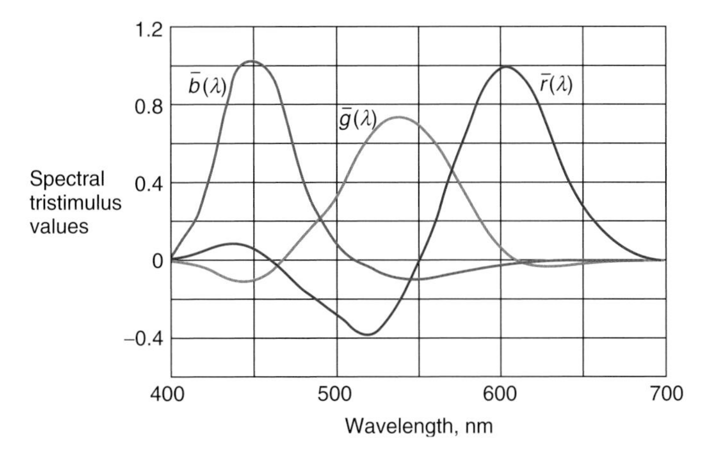

Figure 11.3 Colour-matching functions for E.B.U. phosphors

phosphors used as the primaries in Figure 11.3, and the monochromatic wavelengths of 700 nm, 546.1 nm, and 435.8 nm used as the matching stimuli in Figure 2.4.

11.6 DERIVATION OF XYZ FROM RGB TRISTIMULUS VALUES

If CIE X,Y,Z tristimulus values from the R,G,B tristimulus values of a red, green, and blue system are required, then a set of transformation equations is generated. These are usually in the form:

$$X = A_1R + A_2G + A_3B$$

$$Y = A_4R + A_5G + A_6B$$

$$Z = A_7R + A_8G + A_9B$$

with the corresponding reverse equations in the form:

$$R = B_1X + B_2Y + B_3Z$$

$$G = B_4X + B_5Y + B_6Z$$

$$B = B_7X + B_8Y + B_9Z$$

where A_1 to A_9 and B_1 to B_9 are constants. The reverse equations can be used to transform the CIE colour-matching functions $\bar{x}(\lambda)$, $\bar{y}(\lambda)$, $\bar{z}(\lambda)$ into $\bar{r}(\lambda)$, $\bar{g}(\lambda)$, $\bar{b}(\lambda)$, the corresponding functions for the CIE Standard Observer in the red, green, and blue system being used. The curves of Figure 11.3 were obtained in this way, and therefore represent the colour-matching properties of the CIE 1931 Standard Colorimetric Observer in this system.

The calculation of the constants A_1 to A_9 and B_1 to B_9 in the transformation equations proceeds as follows. The initial data is usually in the form of the CIE chromaticities of the primaries, and of the white used for equalising their units. These chromaticities are best derived by computation in the usual way from spectral power data. We can denote them as follows:

	x	y	z
Red	a_1	a_2	a_3
Green	a_4	a_5	a_6
Blue	a_7	a_8	a_9
White	j_1	j_2	j_3

It is convenient now to use equations to represent colour matches. When this is done, the equations are written in the form:

$$C(\mathrm{C}) \equiv R(\mathrm{R}) + G(\mathrm{G}) + B(\mathrm{B})$$

The sign $\equiv$ in these equations means 'matches', the italicised symbols C, R, G, B represent the amounts of the colours in the match, and the bracketed symbols (C), (R), (G), (B) indicate the colours to which these amounts refer; these bracketed symbols are only labels, and there is no sense in which C(C) is to be regarded as C being multiplied by (C), for instance. Although these labels are unnecessary in the above equations because the symbols themselves indicate the colour concerned, this is not generally so, particularly when actual numbers are used for the amounts. The same convention is used for the X,Y,Z system, in equations of the type:

$$C(\mathrm{C}) \equiv X(\mathrm{X}) + Y(\mathrm{Y}) + Z(\mathrm{Z})$$

The chromaticity co-ordinates can be regarded as tristimulus values for a certain amount, k_1 of (R), k_2 of (G), and k_3 of (B). We can therefore represent these amounts in colour-matching equations, as follows:

$$k_1(\mathrm{R}) \equiv a_1(\mathrm{X}) + a_2(\mathrm{Y}) + a_3(\mathrm{Z})$$

$$k_2(\mathrm{G}) \equiv a_4(\mathrm{X}) + a_5(\mathrm{Y}) + a_6(\mathrm{Z})$$

$$k_3(\mathrm{B}) \equiv a_7(\mathrm{X}) + a_8(\mathrm{Y}) + a_9(\mathrm{Z})$$

The matches on the white in the two systems can be represented by the colour-matching equations:

$$k_4(\mathrm{W}) \equiv H_1(\mathrm{R}) + H_2(\mathrm{G}) + H_3(\mathrm{B})$$

$$k_4(\mathrm{W}) \equiv J_1(\mathrm{X}) + J_2(\mathrm{Y}) + J_3(\mathrm{Z})$$

where H_1, H_2, H_3 are the amounts of (R), (G), (B), respectively, required to match the white, and J_1, J_2, J_3 are proportional to the x, y, z chromaticity co-ordinates, j_1, j_2, j_3, of the white, but such that J_2 is equal to the luminance factor of the white. It is now necessary to evaluate k_1, k_2, k_3, and in order to do this it is required to solve the three simultaneous equations for k_1(R), k_2(G), k_3(B) to obtain the equivalent equations for:

$$1.0(\mathrm{X}) \equiv c_1 k_1(\mathrm{R}) + c_2 k_2(\mathrm{G}) + c_3 k_3(\mathrm{B})$$

$$1.0(\mathrm{Y}) \equiv c_4 k_1(\mathrm{R}) + c_5 k_2(\mathrm{G}) + c_6 k_3(\mathrm{B})$$

$$1.0(\mathrm{Z}) \equiv c_7 k_1(\mathrm{R}) + c_8 k_2(\mathrm{G}) + c_9 k_3(\mathrm{B})$$

Substituting (X), (Y), (Z) in the equation

$$k_4(\mathrm{W}) \equiv J_1(\mathrm{X}) + J_2(\mathrm{Y}) + J_3(\mathrm{Z})$$

we obtain

$$\begin{aligned} k_4(\mathrm{W}) \equiv\ & J_1[c_1k_1(\mathrm{R}) + c_2k_2(\mathrm{G}) + c_3k_3(\mathrm{B})] \\ & + J_2[c_4k_1(\mathrm{R}) + c_5k_2(\mathrm{G}) + c_6k_3(\mathrm{B})] \\ & + J_3[c_7k_1(\mathrm{R}) + c_8k_2(\mathrm{G}) + c_9k_3(\mathrm{B})] \end{aligned}$$

and comparing the result with the equation

$$k_4(\mathrm{W}) \equiv H_1(\mathrm{R}) + H_2(\mathrm{G}) + H_3(\mathrm{B})$$

we obtain:

$$\begin{aligned} k_1 &= H_1/(J_1c_1 + J_2c_4 + J_3c_7) \\ k_2 &= H_2/(J_1c_2 + J_2c_5 + J_3c_8) \\ k_3 &= H_3/(J_1c_3 + J_2c_6 + J_3c_9) \end{aligned}$$

Hence k_1, k_2, k_3 are evaluated, and can be used to obtain:

$$\begin{aligned} 1.0(\mathrm{R}) &\equiv (a_1/k_l)(\mathrm{X}) + (a_2/k_1)(\mathrm{Y}) + (a_3/k_1)(\mathrm{Z}) \\ 1.0(\mathrm{G}) &\equiv (a_4/k_2)(\mathrm{X}) + (a_5/k_2)(\mathrm{Y}) + (a_6/k_2)(\mathrm{Z}) \\ 1.0(\mathrm{B}) &\equiv (a_7/k_3)(\mathrm{X}) + (a_8/k_3)(\mathrm{Y}) + (a_9/k_3)(\mathrm{Z}) \end{aligned}$$

If then a colour has been produced or measured in the (R), (G), (B) system by amounts R of (R), G of (G), and B of (B), this can be expressed as:

$$C(\mathrm{C}) \equiv R(\mathrm{R}) + G(\mathrm{G}) + B(\mathrm{B})$$

and by substituting the expressions for 1.0(R), 1.0(G), 1.0(B) in this equation we obtain:

$$\begin{aligned} C(\mathrm{C}) \equiv\ & (Ra_1/k_1 + Ga_4/k_2 + Ba_7/k_3)(\mathrm{X}) + \\ & (Ra_2/k_1 + Ga_5/k_2 + Ba_8/k_3)(\mathrm{Y}) + \\ & (Ra_3/k_1 + Ga_6/k_2 + Ba_9/k_3)(\mathrm{Z}) \end{aligned}$$

So the X,Y,Z tristimulus values are calculated from the R,G,B tristimulus values by the equations:

$$\begin{aligned} X &= (a_1/k_1)R + (a_4/k_2)G + (a_7/k_3)B \\ Y &= (a_2/k_1)R + (a_5/k_2)G + (a_8/k_3)B \\ Z &= (a_3/k_1)R + (a_6/k_2)G + (a_9/k_3)B \end{aligned}$$

Table 11.1 Computation of transformation equations between RGB and XYZ systems

XYZ system

White(D_{65}) $x = 0.3127 = w_1$ $y = 0.3290 = w_2$ $z = 0.3583 = w_3$

$X = 100w_1/w_2 = J_1$ $Y = 100w_2/w_2 = J_2$ $Z = 100w_3/w_2 = J_3$

RGB system

White $R = 100 = H_1$ $G = 100 = H_2$ $B = 100 = H_3$

	x	y	z
Red	$0.64 = a_1$	$0.33 = a_2$	$0.03 = a_3$
Green	$0.30 = a_4$	$0.60 = a_5$	$0.10 = a_6$
Blue	$0.15 = a_7$	$0.06 = a_8$	$0.79 = a_9$

Computation

$b_1 = a_5a_9 - a_6a_8 = 0.4680$ $b_4 = a_3a_8 - a_2a_9 = -0.2589$ $b_7 = a_2a_6 - a_3a_5 = 0.0150$

$b_2 = a_6a_7 - a_4a_9 = -0.2220$ $b_5 = a_1a_9 - a_3a_7 = 0.5011$ $b_8 = a_3a_4 - a_1a_6 = -0.0550$

$b_3 = a_4a_8 - a_5a_7 = -0.0720$ $b_6 = a_2a_7 - a_1a_8 = 0.0111$ $b_9 = a_1a_5 - a_2a_4 = 0.2850$

$|A| = b_1 + b_4 + b_7 = 0.2241$ $|A| = b_2 + b_5 + b_8 = 0.2241$ $|A| = b_3 + b_6 + b_9 = 0.2241$

The equality of these three values for $|A|$ checks that the values of b are correct.

$c_1 = b_1/|A| = 2.0884$ $c_2 = b_4/|A| = -1.1553$ $c_3 = b_7/|A| = 0.0669$

$c_4 = b_2/|A| = -0.9906$ $c_5 = b_5/|A| = 2.2361$ $c_6 = b_8/|A| = -0.2454$

$c_7 = b_3/|A| = -0.3213$ $c_8 = b_6/|A| = 0.0495$ $c_9 = b_9/|A| = 1.2718$

$c_1 + c_4 + c_7 = 1.0000$ $c_2 + c_5 + c_8 = 1.0000$ $c_3 + c_6 + c_9 = 1.0000$

The equality of these three summations checks that the values of c are correct.

$k_1 = H_1/(J_1c_1 + J_2c_4 + J_3c_7) = 1.5519$ $k_2 = H_1/(J_1c_2 + J_2c_5 + J_3c_8) = 0.8390$ $k_3 = H_1/(J_1c_3 + J_2c_6 + J_3c_9) = 0.8311$

$$X = (a_1/k_1)R + (a_4/k_2)G + (a_7/k_3)B = 0.4124R + 0.3576G + 0.1805B$$
$$Y = (a_2/k_1)R + (a_5/k_2)G + (a_8/k_3)B = 0.2126R + 0.7152G + 0.0722B$$
$$Z = (a_3/k_1)R + (a_6/k_2)G + (a_9/k_3)B = 0.0193R + 0.1192G + 0.9505B$$

$$R = c_1k_1X + c_4k_iY + c_7k_1Z = 3.2410X - 1.5374Y - 0.4986Z$$
$$G = c_2k_2X + c_5k_2Y + c_8k_2Z = -0.9692X + 1.8760Y + 0.0416Z$$
$$B = c_3k_3X + c_6k_3Y + c_9k_3Z = 0.0556X - 0.2040Y + 1.0570Z$$

The equation for Y gives the relative luminances, L_R, L_G, L_B, as 0.2126, 0.7125, 0.0722.

The set of reverse transformation equations enabling R, G, B to be obtained from X, Y, Z are given by the expressions:

$$R = c_1k_1X + c_4k_1Y + c_7k_1Z$$

$$G = c_2k_2X + c_5k_2Y + c_8k_2Z$$

$$B = c_3k_3X + c_6k_3Y + c_9k_3Z$$

These coefficients were obtained in the colour matching equations for 1.0(X), 1.0(Y), and 1.0(Z); because they are now used in tristimulus value equations, their positions are transposed diagonally.

In Table 11.1, a worked example is given of the above type of computation.

11.7 USING TELEVISION AND COMPUTER DISPLAYS

When television and computer displays are being used, it is necessary, for critical work, to check the characteristics of the display. The chromaticities of the primaries define the three primaries involved (Hunt, 2004); the chromaticity of the reference white defines the units in which they are expressed; and the relationship between the voltages applied to the three channels and the amount of light produced by the corresponding emitter enables tristimulus values to be calculated from the three applied voltages. In an ideal display, the displayed colour would have the same tristimulus values as those calculated. However, in a given actual display, there may be differences between the displayed and calculated colours; these can arise from at least four sources (Cowan, 1983).

First, the display output may vary with time. A convenient way of checking this is to compare a displayed grey scale with a standard grey scale produced by means that are known to be reasonably stable. A very convenient way of doing this is to have a small fluorescent lamp emitting light of chromaticity the same as that of the reference white, with a series of non-selective neutral grey filters wrapped round it at convenient intervals. By comparing the displayed and standard grey scales on a daily, or if necessary more frequent, basis, the stability of the monitor is checked in quite a comprehensive fashion. The relationship between the applied voltages, and the amounts and colours of the light produced by the phosphors, is checked; although small identical changes in all three channels may pass unnoticed, these are not usually very important, but different changes in the channels are readily detected.

Second, the relationship between the applied voltages and the colour produced may vary over the area of the display. This can be checked by displaying a constant signal all over the display. However, because the eye is not very sensitive to changes in colour that are spatially very gradual it is advisable to check the colour at different positions on the display using some form of comparator that enables the colours of separated areas to be seen or measured side by side.

Third, the chromaticities of the primaries may not be the same as those given in the display specification. This can arise in several different ways. The primaries may not be quite the same chemically or physically as those intended by the manufacturer. They may alter with extensive usage. In cathode-ray tubes there may be some minor excitation of a phosphor by electrons from the wrong guns; a phenomenon known as *beam-landing errors*. To check these effects requires spectroradiometry, and this should be done at different positions, because beam-landing errors are usually very position-dependent (King and Marshall, 1984).

Fourth, there may be some interaction between the strengths of the signals; for instance, the amount of red light produced by the signal in the red channel, may be affected by changes in the strengths of the green and blue signals. This can only be checked by producing a variety of colours, in which the strengths of all three signals are altered, and comparing the predicted with the displayed colours (Cowan and Rowell, 1986).

A convenient way of relating displayed colour to digital counts in a computer controlled monitor has been described by Berns, Motta, and Gorzinski (Berns, Motta, and Gorzinski, 1993). By carrying out radiometric measurements relative to a display's maximum exitance, its performance can be characterised by a two-stage model. The first stage is a non-linear transformation relating normalised digital-to-analogue converter values to monitor RGB tristimulus values, using model parameters of gain, offset, and gamma (the slope of the relationship between the logarithm of the input signal and the logarithm of the amount of light produced). The second stage is a linear transformation from the R,G,B tristimulus values to CIE XYZ tristimulus values. The model is established using eight test colours: the maximum signals in each of the red, green, and blue channels, on their own, are used to establish the linear transformation to X,Y,Z; and five white or grey colours are used to establish the nine gain, offset, and gamma coefficients of the three channels. This procedure, although only using the eight carefully selected colours, has been found to result in errors that are usually less than one CIELAB ΔE^* unit over the gamut of displayed colours.

REFERENCES

Berns, R. S., Motta, R.J., and Gorzynski, M.E., CRT colorimetry. Part I: theory and practice, *Color Res. Appl.*, **18**, 299–314 (1993) and CRT colorimetry. Part II: metrology, *Color Res. Appl.*, **18**, 315–325 (1993).

B.R.E.M.A., Technical performance targets for a PAL colour television broadcasting chain, *The Radio and Electronic Engineer*, **38**, 201–216 (1969).

Cowan, W.B., An inexpensive scheme for calibration of a colour monitor in terms of CIE standard coordinates, *Computer Graphics*, **17**, 315–321 (1983).

Cowan, W.B., and Rowell, N., On the gun independence and phosphor constancy of colour video monitors, *Color Res. Appl.*, **11**, S34–S38 (1986).

Hunt. R.W.G., *The Reproduction of Colour*, 6th *Ed.*, page 388, Wiley, Chichester, England (2004).

King, P.A., and Marshall, P.J., *I.B.A. Technical Review No. 22, Light and Colour Principles*, page 46, Independent Broadcasting Authority, London, England (1984).

IEC 61966-2-1:1999: *Colour measurement and management in multimedia systems and equipment – Part 2-1: Default RGB colour space – sRGB*, International Electrotechnical Commission, Geneva, Switzerland (1999).

Sproson, W.N., PAL System I phosphor primaries: the present position, *Proc. I.E.E.*, **125**, 603–605 (1978).

GENERAL REFERENCE

Sproson, W.N., *Colour Science in Television and Display Systems*, Hilger, Bristol, England (1983).

12

Colorimetry with Digital Cameras

12.1 INTRODUCTION

The availability of digital cameras has stimulated interest in their use in a wide range of commercial and industrial applications, from the archiving of art objects and other artefacts, to medical and forensic imaging, quality control in manufacturing, and aerial surveying.

Some of these applications are essentially instrumental in that the final requirement is not necessarily an image on a computer display but an array of CIE colorimetric measures in a data file. Thus, the components of the imaging chain, Figure 12.1, can be used to gather quantitative data concerning the original subject and may be interrupted for assessment at some point before a final image is produced. In many applications there may be no conventional 'image' beyond the optics of the camera, and the remainder of the imaging chain involves the manipulation and analysis of an electronic virtual image, which possesses no immediate visual significance. What is required is a fully characterised relation between the subject, the electronic virtual image, and the visual properties of any displayed image.

The use of digital cameras makes it possible to carry out colorimetric measurements on objects with complicated shapes or patterns, a task which would be very tedious with a tele-spectroradiometer. For example, the colours in a carpet or a garment can be measured in context, and image processing used to show a customer what the carpet might look like with a different selection of colours. Further analysis enables close approximations of the spectral reflectances of the selected colours to be found for subsequent input into dye-match prediction software to enable corresponding dye recipes to be formulated. Another example involves the detection of the quality of food products on a production line, for example baked products such as cakes, bread, rolls and buns. A tolerance can be set for the required colour of the finished product and the colour monitored and used to control the temperature of the oven to prevent burning or under-cooking. Vegetables, for example peas, can be artificiality coloured to produce uniformity of colour through the packet: measurement of the actual product colour enables calculation of the amount of artificial colour required. Tomatoes have to be harvested when green so that, after what might be a long sea voyage, they have ripened to red: digital images, and associated colorimetry, can be used to ascertain the precise time to pick. All of these potential applications require the

Measuring Colour, Fourth Edition. R.W.G. Hunt and M.R. Pointer.

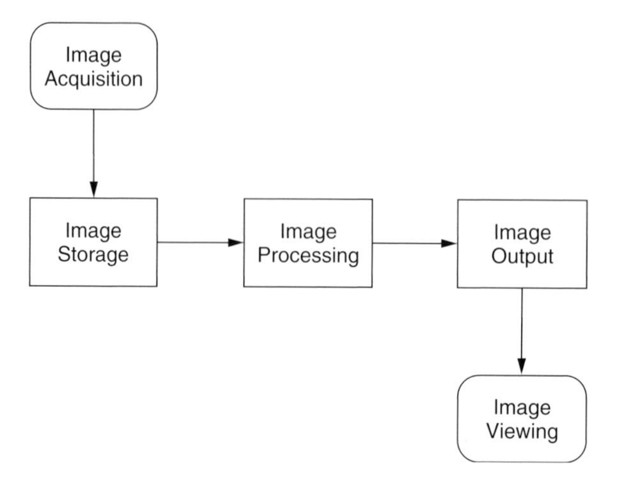

Figure 12.1 The imaging chain

colour measurement of objects that are not flat and are inherently non-uniform in that their surfaces exhibit a range of colours. This range can easily be captured by a digital camera to give an array of information that can be transformed to appropriate CIE colorimetric values.

Colour fastness is a measure of how permanent a colour is on a fabric and the colour can be adversely affected by a number of factors including exposure to light, exposure to water, as in washing, and to normal wear and tear, such as rubbing against a surface. Various test methods have been created to assess how the colour is affected by these different parameters and a numerical value is then established to indicate the degree of colour change. Traditionally the tests have been performed by visual evaluation of a test fabric against a control fabric, which is usually white or un-dyed. This test procedure has been automated by the use of a digital camera-based system that takes an image of the appropriate test and control fabrics which are laid out in a defined manner, finds the various patches in the image and calculates the colorimetry of each area; an index is then calculated that correlates with the traditional, visually derived value (Cui, Luo, Rhodes, Rigg, and Dakin, 2003a, 2003b; Cui, Luo, Rigg, Butterworth, and Dakin, 2004; Cui, Luo, Rigg, Butterworth, Maplesden, and Dakin, 2004). Another example is the use of a digital camera to measure the uniformity of coloured objects, such as textiles (Günay, 2009).

12.2 CAMERA CHARACTERISATION

To use a digital camera as a tristimulus colorimeter requires signal processing to obtain *device-independent coordinates*, usually CIE XYZ tristimulus values, from the RGB output data and it is the form and efficiency of this *characterisation process* that dictates the overall accuracy of the camera as a colour measuring device.

The basic RGB 'values' of a specific pixel in an image can be calculated from a knowledge of the spectral power distribution of the light coming from the equivalent location in the original scene, $E(\lambda)$, integrated with the respective spectral responsivities of the three colour channels of the camera detector and its associated filters, $S_i(\lambda)$, i = r, g, b (Holst 1996):

$$R = \int E(\lambda) S_r(\lambda) d\lambda$$

$$G = \int E(\lambda) S_g(\lambda) d\lambda$$

$$B = \int E(\lambda) S_b(\lambda) d\lambda$$

Assuming the light from the scene is present by reflection, then the spectral power distribution of the scene element can be substituted by the product of the spectral power distribution of the light source, $P(\lambda)$, and the reflectance factor of the object in the scene element, $R_f(\lambda)$. In the real situation, this substitution is not straightforward because some points in the scene might be self-luminous, i.e. emitters of light. These may be obvious light sources, or not so obvious specular highlights caused by reflection of the light source. In general, however, $E(\lambda)$ in the above equations can be replaced to give:

$$R = \int P(\lambda) R_f(\lambda) S_r(\lambda) d\lambda$$

$$G = \int P(\lambda) R_f(\lambda) S_g(\lambda) d\lambda$$

$$B = \int P(\lambda) R_f(\lambda) S_b(\lambda) d\lambda$$

where the integration is taken over a suitable wavelength range in the visible part of the spectrum, for example, from 380 nm to 780 nm.

The values of R, G and B are usually calculated by summation and not integration, i.e.

$$R = \Sigma P(\lambda) R_f(\lambda) S_r(\lambda) d\lambda$$

$$G = \Sigma P(\lambda) R_f(\lambda) S_g(\lambda) d\lambda$$

$$B = \Sigma P(\lambda) R_f(\lambda) S_b(\lambda) d\lambda$$

where the summations are taken over a suitable wavelength range, such as 380 nm to 780 nm, and use data defined within equal finite wavelength intervals, for example, every 5 nm or 10 nm.

The values of R, G and B calculated using the above equations are *device-dependent* in that they will be different, for example, for a camera having a different spectral responsivity. Thus, the problem to be solved is how to transform these device-dependent coordinates into device-independent coordinates. The device-independent coordinates usually chosen are the CIE tristimulus values, X, Y, Z, which are calculated in a similar manner to the R, G, B values above.

$$X = \Sigma P(\lambda) R_f(\lambda) \bar{x}(\lambda) d\lambda$$

$$Y = \Sigma P(\lambda) R_f(\lambda) \bar{y}(\lambda) d\lambda$$

$$Z = \Sigma P(\lambda) R_f(\lambda) \bar{z}(\lambda) d\lambda$$

The difference is that the $\bar{x}(\lambda)$, $\bar{y}(\lambda)$, $\bar{z}(\lambda)$ functions are unique and represent the colour matching functions of a CIE standard colorimetric observer. (As described in Section 2.6

there are, in fact, two such CIE standard observers with slightly different colour matching functions, whose differences are attributable to differences in field size: the ensuing discussion can be applied to either set.)

Thus, the problem becomes one of finding a mathematical relationship between the R, G, B and corresponding X, Y, Z values.

12.3 METAMERISM

Because the spectral responsivities of the red, green, and blue sensors of a camera are different from CIE colour matching functions, the camera will 'see' colours differently compared to the human eye. Thus, two colours having different spectral power distributions that appear the same to the eye (a metameric pair) may not be perceived as the same by the camera, an effect that may be called *eye-camera metamerism* (Hong, Luo and Rhodes, 2001).

Consider two colours defined by spectral power distributions E_1 and E_2. If they are metameric to the human eye then the following holds:

$$\int E_1(\lambda)\bar{x}(\lambda)\mathrm{d}\lambda = \int E_2(\lambda)\bar{x}(\lambda)\mathrm{d}\lambda$$

$$\int E_1(\lambda)\bar{y}(\lambda)\mathrm{d}\lambda = \int E_2(\lambda)\bar{y}(\lambda)\mathrm{d}\lambda$$

$$\int E_1(\lambda)\bar{z}(\lambda)\mathrm{d}\lambda = \int E_2(\lambda)\bar{z}(\lambda)\mathrm{d}\lambda$$

If the colour matching functions $\bar{x}(\lambda)$, $\bar{y}(\lambda)$, $\bar{z}(\lambda)$ are changed to the camera responsivities $S_r(\lambda)$, $S_g(\lambda)$, $S_b(\lambda)$ then the following will generally be true:

$$\int E_1(\lambda)S_r(\lambda)\mathrm{d}\lambda \neq \int E_2(\lambda)S_r(\lambda)\mathrm{d}\lambda$$

$$\int E_1(\lambda)S_g(\lambda)\mathrm{d}\lambda \neq \int E_2(\lambda)S_g(\lambda)\mathrm{d}\lambda$$

$$\int E_1(\lambda)S_b(\lambda)\mathrm{d}\lambda \neq \int E_2(\lambda)S_b(\lambda)\mathrm{d}\lambda$$

Thus, a camera may generate two different sets of R, G, B values for two colour samples that look the same to the human eye. The major consequence of this is that the characterisation method derived to perform best for one set of samples may not perform best for another set, even if that set is identical in a colorimetric sense but not in a spectral sense (that is, there is metamerism present). This can occur, for example, when the two sets use different dyes. Thus the characterisation method is not only camera dependent, but also dependent on the actual samples used for the characterisation procedure. The magnitude of this effect can be demonstrated by using a specific set of colours to derive a characterisation and then using it to predict the colorimetry of both the original set of colours and another set of colours whose CIE colorimetry is also known by independent measurement. The colour differences between the two sets of data can then be combined to give a measure of the degree of metamerism between them.

12.4 CHARACTERISATION METHODS

There are several characterisation methods, based on:

- the measurement of the spectral responsivity of the camera;

- the colorimetry of a number of coloured samples;
- the use of neural networks;
- the derivation of basis functions that represent the spectral reflectance of the individual colours in the original scene;
- the use of multi-spectral analysis.

12.4.1 Spectral responsivity

Investigation of the above equations shows a certain similarity between the equations for deriving the camera responses R, G, B, and the equations for deriving the CIE X, Y, Z tristimulus values, the difference being the spectral responsivities. Thus if a relationship can be found between the camera spectral responsivities and the CIE colour matching functions then this same relationship can be used to transform R, G, B values to X, Y, Z values (Thomson and Westland, 1999). The usual place to start when looking for such a relationship is to assume that the camera responses are a linear combination of the CIE colour matching functions, thus they can be related by equations of the form:

$$X = m_{11}R + m_{12}G + m_{13}B + k_1$$

$$Y = m_{21}R + m_{22}G + m_{23}B + k_2$$

$$Z = m_{31}R + m_{32}G + m_{33}B + k_3$$

Written in matrix form this becomes:

$$\begin{pmatrix} X \\ Y \\ Z \end{pmatrix} = M \begin{pmatrix} R \\ G \\ B \end{pmatrix} + \begin{pmatrix} k_1 \\ k_2 \\ k_3 \end{pmatrix}$$

where k_1, k_2, k_3 are constants that may or may not be statistically significant. This reduces to finding the solution to three applications of a linear regression procedure.

Because it is an assumption, rather than a fact, that the camera responses are a linear combination of the CIE colour matching functions, it is reasonable to investigate the use of more complex forms of equation to model the relationship. For example, the functions in Table 12.1 are suggested (Hong, Luo, and Rhodes, 2001).

From a mathematical perspective there is, of course, no limit to the order and number of terms in the polynomial; in practice the complexity can be constrained by the accuracy required and the number of data points available. It is also reasonable to remove the need for a constant term, designated by the k in the above equations; its inclusion, however, can be statistically significant.

While linear regression is a useful and reasonable starting point for this type of analysis, the extension to polynomial regression does not always represent a sensible, physical, model between two sets of data such as those considered above. Polynomials are essentially oscillating functions and, while they can predict data satisfactorily by interpolation, they are often unstable when required to provide extrapolation to points outside the basic data set. It is also not easy to ascribe a physical meaning to the equations obtained, a fact that

Table 12.1 Functions to relate camera R, G, B values to CIE X, Y, Z tristimulus values

	Function
1	$M = \mathrm{f}(R, G, B, k)$
2	$M = \mathrm{f}(R, G, B, RGB, k)$
3	$M = \mathrm{f}(R, G, B, RG, RB, GB, k)$
4	$M = \mathrm{f}(R, G, B, RG, RB, GB, RGB, k)$
5	$M = \mathrm{f}(R, G, B, RG, RB, GB, R^2, G^2, B^2, k)$
6	$M = \mathrm{f}(R, G, B, RG, RB, GB, R^2, G^2, B^2, RGB, k)$
7	$M = \mathrm{f}(R, G, B, RG, RB, GB, R^2, G^2, B^2, RGB, R^3, G^3, B^3, k)$

is not essential but, if it were possible, would provide a useful insight into the physics of the relationship between eye and camera responses.

It should also be noted that the above characterisation method requires measurement of the spectral responsivity of the digital camera which is not an easy measurement to make in the sense that it requires a set of images of the light output of a suitable calibrated monochromator be available.

The spectral responsivity of a digital camera can be determined directly using either one of two methods. The most direct method is to determine the responsivity by taking a series of images of a set of narrow band interference filters of known spectral transmittance, illuminated using light of known spectral power distribution. Alternatively, a set of images can be made of the output slit of a scanning monochromator. In this case, the power output of the monochromator must be measured separately using a spectroradiometer (Park, Kim, Park, and Eem, 1995; Martinez-Verdú, Pujol, and Capilla, 2002).

12.4.2 Colorimetry

This may be regarded as an extension of the above solution in that the regression procedures are applied, not to the data comprising the spectral responsivities, but to the RGB and XYZ data of a suitable number of test colours (Mullikan, Van-Viet, Boddeke, Van Der Feltz, and Young, 1994; Wu, Allebach, and Analoui, 2000). Thus, a number of test colours are illuminated by a suitable light source and images captured using the camera to be characterised. Average RGB values are obtained from repeated exposures of each coloured sample. It is also necessary to be able to calculate the XYZ tristimulus values of the coloured samples, and this can be done, either by measuring the spectral power distribution directly using a tele-spectroradiometer, or by measuring the spectral power distribution of the light source using a spectroradiometer and also the spectral reflectance of each of the physical coloured samples using a spectrophotometer. The XYZ values are then calculated using the equations described above using the required CIE observer and illuminant. Regression analysis can be applied in a manner similar to that applied to the values of spectral responsivity. Some regression techniques offer the further advantage that they are constrained to permit, for example, a white sample to be correctly reproduced with defined colorimetric values (Finlayson and Drew, 1997; Hong, Luo and Rhodes, 2001).

It should be noted that there is no inherent reason why XYZ values must be used to represent the device-independent colour space. CIELAB coordinates L^*, a^*, b^* can provide reasonable alternatives representing as they do a more uniform distribution of colours.

12.4.3 Neural networks

The application of the principles of regression to this problem is perhaps intuitively obvious. Other techniques, however, can be utilised. Neural networks are computer models that attempt to imitate some of the functions of the human brain using certain basic structures (Westland, 1998). The generalised unit of such a network receives input from other units, computes a weighted sum of these inputs, and produces an output that is itself some function of the weighted inputs.

Networks based on the multi-layer principle account for the majority of practical applications of neural networks. They can be conceived as being constructed of multiple layers of input/output 'cells', each layer receiving a weighted input from the previous layer. The middle layers are often referred to as 'hidden' because their functionality is hidden from the user in the sense that they do not provide any opportunity for input or output. Use of a multiple layer system has two distinct phases, learning and recall. The learning phase requires the network to be presented with a known set of input-output pairs that constitutes a training set. The recall phase involves the actual use of the result of the learning phase to produce useful output data.

In reality a set of data may be divided into two, the first sub-set used to train the network and the second used to calculate the precision of its prediction capability. The application to camera characterisation is obvious in that a set of test colours with derived values of R, G, B and calculated values of X, Y, Z provides suitable data for training such a system (Kang and Anderson, 1992).

A potential disadvantage of such a software application is that the weights derived during the training exercise are not always available to the end-user. Thus, the actual network application must be included in the characterisation software.

12.4.4 Basis functions

A characteristic vector analysis of a set of spectral reflectance data that vary only slowly in the visible part of the spectrum can lead to an algorithm that gives nearly colorimetrically accurate values of spectral reflectance from the RGB output of a digital camera (Lee, 1988). Together with prior knowledge of the illuminant spectral power distribution, this can thus lead to luminance factor and chromaticity information.

As noted above, each R, G, B pixel in a digitised image has a value proportional to a weighted integral over the visible spectrum that depends on the spectral power distribution of the incident radiation and the spectral response of the respective channels of the CCD camera. Further, the incident radiation can be considered as the multiple of the spectral power distribution of the light source and the spectral reflectance of the discrete samples. This can be written in the matrix form:

$$a = \boldsymbol{R}\,\boldsymbol{P}\,\boldsymbol{D}$$

where:

$\boldsymbol{a}$ is a 1×3 row vector - the R, G, B pixel values.

$\boldsymbol{R}$ is a $1 \times m$ row vector whose elements are the surface reflectances at m equally spaced wavelength intervals across the spectrum.

$\boldsymbol{P}$ is an $m \times m$ diagonal matrix whose non-zero entries are the spectral power distribution of the illuminant.

$\boldsymbol{D}$ is an $m \times 3$ matrix representing the spectral response of the digitising system.

This equation can be thought of in two ways. First, it may consist of a known $1 \times m$ vector $\boldsymbol{RP}$ and an unknown vector $\boldsymbol{D}$. Alternatively, if the spectral illumination and the device spectral response are both known then the $m \times 3$ matrix $\boldsymbol{PD}$ is a known and the unknown quantity is the surface reflectance, $\boldsymbol{R}$.

In principle, if the elements of two of the arrays on the right side of the equation are known, together with the corresponding R, G, B pixel values on the left side, then the equation can be solved for the unknown array.

One obvious difficulty with this approach is that for all $m > 3$ instances the equation is undetermined. From a practical viewpoint, this implies that the equation can only be solved for three values of m through the visible spectrum. If it is assumed however, that the spectral reflectance of many pigments, and naturally occurring materials, varies both slowly and smoothly with wavelength, then the actual spectral reflectances of a set of colour samples can be replaced by the first few components of a characteristic vector analysis. This in effect reduces the dimensionality of the $\boldsymbol{R}$ matrix and makes solution possible. Knowledge of the scene reflectances, derived via the camera responsivities and the R, G, B values of the image of that scene, enables the colorimetry of the scene to be calculated.

A number of practical methods of characterising digital cameras using this technique have been proposed, many relying on the inherent smoothness of the reflectance spectra (Cheung, Westland, Li, Hardeberg, and Connah, 2005). Thus a reflectance $P(\lambda)$, sampled at equal intervals of wavelength λ may be approximated by a weighted sum of m basis functions $B_i(\lambda)$ with $i = 1 \ldots n$ such that, in matrix notation:

$$\boldsymbol{P} = \boldsymbol{BW}$$

where, if the spectral properties are sampled at w wavelength intervals, $\boldsymbol{P}$ is a $w \times n$ matrix of values of reflectance representing n samples, $\boldsymbol{B}$ is $w \times m$ matrix of basis functions, $B_i(\lambda)$ and $\boldsymbol{W}$ is an $m \times n$ matrix of weights.

Rearranging the above equation:

$$\boldsymbol{W} = \boldsymbol{B}^{-1}\boldsymbol{P}$$

where $\boldsymbol{B}^{-1}$ is the inverse of matrix $\boldsymbol{B}$. The problem then reduces to finding a solution to this equation because the matrix $\boldsymbol{B}$ is not square and thus not easily inverted.

Solutions suggested include a simple linear approach that assumes three basis functions (Maloney and Wandell, 1986), a higher order method (Shi and Healey, 2002), the calculation of a set of all possible surface reflectances for a given camera response followed by the selection of specific examples (Finlayson and Morovic, 2005; Morovic and Finlayson, 2006).

12.4.5 Multi-spectral analysis

Much of the above discussion has assumed that the spectral analysis in the camera is based on broad-band sampling into three channels. There is, however, no inherent reason why more than three channels should not be incorporated into a characterisation process and improved colorimetric accuracy is claimed by the use of such methods (Chiba University, 1999). For example, a seven channel device is reported that uses a set of approximately 50 nm bandwidth interference filters to sample the spectrum. To characterise such a device requires the solution to the following series of equations:

$$X = m_{11}F_1 + m_{12}F_2 + m_{13}F_3 + m_{14}F_4 + m_{15}F_5 + m_{16}F_6 + m_{17}F_7 + k_1$$

$$Y = m_{21}F_1 + m_{22}F_2 + m_{23}F_3 + m_{24}F_4 + m_{25}F_5 + m_{26}F_6 + m_{27}F_7 + k_2$$

$$Z = m_{31}F_1 + m_{32}F_2 + m_{33}F_3 + m_{34}F_4 + m_{35}F_5 + m_{36}F_6 + m_{37}F_7 + k_3$$

where F_i (i = 1, 7), are the responses obtained using the seven filters respectively. Thus, the method should lead to greater accuracy, because more information that is independent is being provided at the input stage, but at the expense of greater processing complexity compared with the standard 3×3, linear, form.

12.5 PRACTICAL CONSIDERATIONS IN DIGITAL CAMERA CHARACTERISATION

12.5.1 Choice of test colours

The initial step is often the careful selection of a set of test colours. These should span the range of colours of interest and should be characterised, preferably by spectral data. Arrays of useful colours include the IT8 target (CGATS 1993), comprising 264 coloured patches and used, for example, in creating device profiles for digital scanners; the samples of a colour atlas, for example, the Munsell Book of Color in its various guises, with 225 samples in its simplest form; the X-Rite (Macbeth) ColorChecker® Classic with 24 patches (McCamy, Marcus, and Davidson 1976), Figure 12.2; and the X-Rite ColorChecker® Digital SG with 140 patches. It is arguable that the visual importance of the grey scale is reflected in the use of all six neutral samples when the ColorChecker Classic is used, together with its complement of colours representative of real scenes and subjects.

Because of the effect of metamerism, it must be recognised that two targets could be colorimetrically identical, that is each sample would have the same XYZ values, but give different RGB values when imaged by a camera. This would lead to different characterisation matrices, M.

12.5.2 Choice of illuminant

The issue of metamerism also indicates the importance of the illuminant. Ideally, it should have a defined spectral power distribution but, in practice, a variety of distributions may be encountered, even when supplied as allegedly matching CIE specifications such as Standard Illuminant D65. This power distribution may be approximated for example, by using filtered tungsten halogen lamps, or specified fluorescent lamps, as supplied in viewing

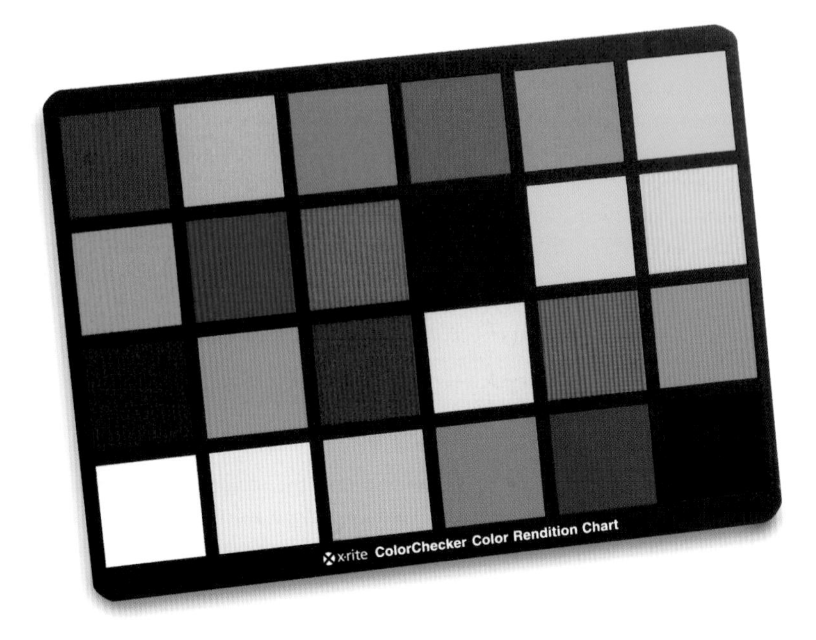

Figure 12.2 The GretagMacbeth ColorChecker® Chart – now the X-Rite ColorChecker® Classic. Because of the limitations of printing, all the colours cannot be shown accurately, but the general layout is displayed adequately. References to Munsell® in this publication are used with permission from X-Rite Inc

booths designed for daylight viewing. The spectral power distributions of such illuminants are usually significantly different from that of D65: for example, mercury emission lines are never completely excluded from the output of the fluorescent lamps.

Thus, the choice of illuminant comes down to one of practicality. If a viewing booth is to be used to provide a practical source of illumination, then the illuminant is established and its spectral power distribution can be measured. There is no inherent reason however, why the illuminant used to light the objects to be measured should be the same as that used to calculate their colorimetry in the characterisation process: this is especially so if the required colorimetry is to use one of the CIE designated illuminants, for example Illuminant D65. The fact that the illuminant is different is essentially compensated for in the characterisation process.

12.5.3 Uniformity of illumination

Uniformity of illumination of both the sample and the sensor plane is important if objectively correct colour reproduction is required. The sample may appear to be uniformly illuminated but measurement can reveal considerable variation. Completely even illumination is seldom achieved but variation should be reduced as much as possible. If the unevenness of illumination is determined as the spatial non-uniformity of the sensor output when the camera records a uniform white or grey card, then it is possible to factor out the non-uniformity as part of the characterisation process. It is most rigorous when carried out at exposure because it then includes any off-axis fall-off in photometric efficiency of the camera objective, as well as deficiencies in the illumination system and in the sensor device itself. A practical set-up using a viewing booth might use a number of fluorescent lamps at a suitable distance to provide reasonably even diffuse illumination. An alternative might use two directional light sources at 45° to the subject plane to achieve near-uniform illumination.

12.5.4 Camera white balance

The issue of white balance is also important. Digital cameras usually possess a mechanism for achieving neutral balance given suitable information about the white point. The mechanism employed for white balance is customarily provided in the controlling firmware, resident in the camera, and varies from camera to camera. It may or may not be disabled if required, and may involve any of the following non-exclusive list:

- a diffusing white lens cap to allow the camera to be calibrated to neutral balance;
- the use of a standard white or, better, neutral grey in the scene to assist setting up by the operator;
- a grey world assumption: the camera is programmed to assume that the integrated scene is neutral at some mid-grey;
- an assumption that the lightest part of the subject is either white or a specular reflection of the light source, so that the camera firmware is programmed to adjust the colour balance accordingly.

All these methods may cause colour and tone reproduction errors if the subject or its lighting is atypical. A fully automatic camera may present insoluble problems if it is completely automatic, i.e. designed entirely to eliminate operator over-ride.

12.5.5 Display characteristic compensation

An imaging system may carry out some image processing within the image acquisition device or in the host computer. Such manipulations are designed to optimise images according to criteria that may be concealed from the operator. Examples include manipulation by a $1/\gamma$ function to compensate for the power law transfer characteristic of a typical display.

12.6 PRACTICAL EXAMPLE

As an example of the digital camera characterisation process, results are presented below using a typical camera, two colour charts, and a suitable illumination system (Pointer, Attridge, and Jacobsen, 2001).

12.6.1 Input data

Data were obtained by use of the following components:

Camera:	a typical digital single lens reflex camera fitted with a 50 mm lens
Test samples:	i. the X-Rite Macbeth ColorChecker® Classic chart – 24 coloured patches ii. a second colour chart with 170 samples based on Munsell Color samples
Light source:	a viewing booth fitted with fluorescent lamps that simulated CIE Standard Illuminant D65.

Colorimetric data were obtained for the two separate sample sets by measurement of the spectral reflectance of each sample using a suitable spectrophotometer and calculating CIE tristimulus values using the equations described above. The CIE 2° Standard Observer and CIE Standard Illuminant D65, were specified. It should be noted that this was not the exact illuminant used physically to illuminate the test charts; it represents the illuminant for which the output colorimetry is required after the camera characterisation process.

RGB values were obtained from images of the test charts using suitable software. As part of this process an average RGB value was assigned to a suitable area of each of the colour patches in each chart.

12.6.2 Output Data

It was necessary to optimise the matrix, M. The initial, linear, approach was replaced by polynomial regression to optimise the matrix relating the measurements of the sample and reproduction sets of colours. Both CIE XYZ values and CIELAB L^*, a^*, b^* coordinates were used as input to the regression software. The CIELAB colour difference formula was used to quantify differences, ΔE^*_{ab}, between actual chart values, calculated from the measured spectral reflectance data, and those derived from the digital images.

Inclusion of the median value of colour difference as a measure of central tendency was made having regard to the non-Gaussian distribution of the ΔE^*_{ab} values, which cannot be negative. The median alone was judged inadequate for the investigation which invariably detected outliers in the distributions found, so the maximum ΔE^*_{ab} was also evaluated.

Table 12.2, shows values of median and maximum CIELAB colour difference, ΔE^*_{ab} for two charts with different numbers of test patches, linear and quadratic regression and two colour spaces. It is seen that, for these particular camera/chart/space/regression combinations the 24 patch chart, quadratic regression and CIELAB space gives the lowest median value of colour difference, as well as the lowest maximum value. Figure 12.3 shows the distribution of the values for CIELAB colour difference in terms of both frequency of occurrence and cumulative frequency of occurrence.

It should be noted that the rather high values of colour difference, ΔE^*_{ab}, obtained for the maximum values are usually attributable to high chroma colours that are very dark (a relatively high value of C^* and a low value of L^*) but also to very dark neutral colours, for example the black of the Macbeth ColorChecker. Both these groups of samples are near the bottom of the colour solid where the shape of a colour gamut tends to a point that represents black (L^*, C^* equal to zero).

Table 12.2 Values of median and maximum colour difference for combinations of sample number, regression type and colour space

Samples	Regression	Space	Median	Maximum
24	Linear	X, Y, Z	11.58	14.21
170	Linear	X, Y, Z	3.98	15.70
24	Linear	L^*, a^*, b^*	4.49	10.82
24	Quadratic	X, Y, Z	3.10	10.62
170	Quadratic	X, Y, Z	2.91	15.47
24	Quadratic	L^*, a^*, b^*	1.89	6.30

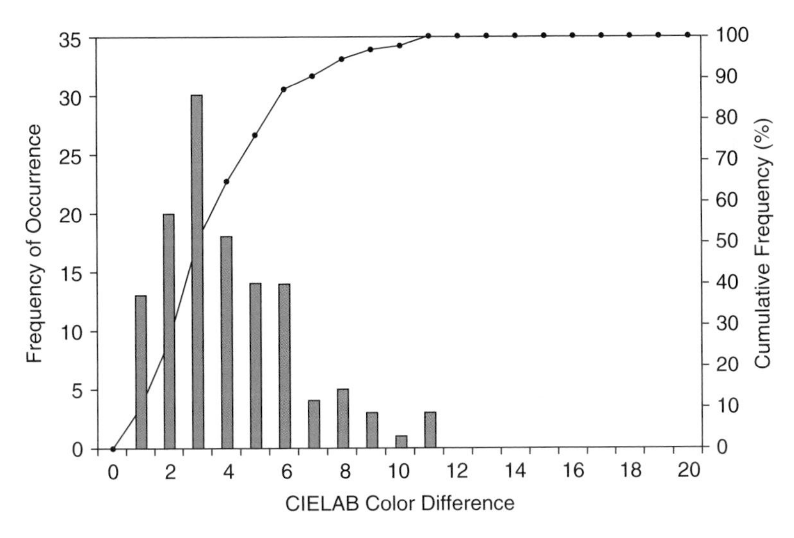

Figure 12.3 Typical data from a camera characterisation procedure. The frequency of occurrence of a particular value of colour difference is plotted as a distribution, left, and a cumulative distribution, right

12.6.3 Variation between regression

Table 12.3 shows the results, in terms of median and maximum value of CIELAB ΔE^*_{ab}, of applying the various functions listed in Table 12.1. It is shown that the size of the colour median difference decreases with increasing complexity of the regression.

12.6.4 Variation between cameras

Table 12.4 shows the results, in terms of median and maximum value of CIELAB ΔE^*_{ab} of applying the characterisation process to ten different cameras. It is shown that the size of the colour median difference varies between 0.7 and 2.8, and that the camera that gives the highest median value does not necessarily give the highest maximum value.

Table 12.3 The results of applying a series of functions to two different charts

	Macbeth Chart		Chart 2	
Function	Median ΔE^*_{ab}	Maximum ΔE^*_{ab}	Median ΔE^*_{ab}	Maximum ΔE^*_{ab}
1	5.1	20.9	5.3	27.4
2	5.6	18.1	3.8	20.8
3	4.7	15.6	4.0	19.9
4	5.1	15.6	3.8	20.8
5	3.2	10.6	2.9	13.1
6	3.2	10.2	2.9	13.7
7	2.9	9.5	2.9	15.0

Table 12.4 The results of applying a quadratic characterisation model to the Macbeth Chart using ten different cameras

Camera	Median ΔE^*_{ab}	Maximum ΔE^*_{ab}
1	4.1	10.7
2	2.9	4.4
3	0.7	2.6
4	2.4	20.5
5	1.1	4.1
6	1.1	3.0
7	2.8	8.4
8	1.8	10.1
9	2.8	8.8
10	2.6	7.7

12.6.5 Effect of non-uniformity

Table 12.5 shows the results of applying different degrees of non-uniformity to an image of the Macbeth Chart using two different cameras. The uniformity refers to the difference between the illuminance at the centre of the chart and the average of that at each of the four corners.

12.7 DISCUSSION

Reasonable characterisation of a typical digital camera is possible using the 24 patch Macbeth ColorChecker and a suitable quadratic regression routine assuming a correction is made for the non-uniformity of the illumination. The choice of colour chart, the colour space for application of the analysis, and the degree of regression applied, are a matter of choice based on a suitable parameter, for example CIELAB colour difference, ΔE^*_{ab}. Little advantage is to be gained by using a very large set of colour patches for characterisation.

Table 12.5 The results of applying a linear characterisation model to the Macbeth Chart using two different cameras and different levels of subject illumination uniformity

	Camera 3		Camera 7	
Uniformity	Median ΔE^*_{ab}	Maximum ΔE^*_{ab}	Median ΔE^*_{ab}	Maximum ΔE^*_{ab}
0%	1.0	4.3	3.7	9.0
5%	1.3	3.2	3.8	9.6
10%	1.8	3.8	3.9	10.2
15%	1.9	5.6	4.0	11.1
20%	2.4	7.4	4.2	13.1
25%	3.4	9.2	4.7	15.2
30%	4.1	11.0	5.2	16.9

The uniformity of the subject illumination is a strong contributor to the characterisation process. The non-uniformity that is almost inevitably present in the camera-exposing situation must also be factored out as part of the characterisation procedure. Failure to do this can give rise to a 2 – 3 ΔE^*_{ab} increase in median colour difference (see Table 12.5).

The issue concerning the choice of test target and the number of colours in that target is unresolved. The main criterion that should be adopted is to choose test colours that are similar to those likely to be measured. If a general colour-measuring instrument is required then this may not be a simple choice.

REFERENCES

CGATS/ANSI IT8.7/2 *Graphic Technology - Color reflection target for input scanner calibration*. ANSI, USA (1993).

Chiba University, *Proceedings of the International Symposium on Multispectral Imaging and Color Reproduction for Digital Archives*, Society of Multispectral Imaging of Japan, Chiba University, Japan (1999).

Cheung, V., Westland, S., Li, C., Hardeberg, J., and Connah, D., Characterisation of trichromatic color cameras by using a new multispectral imaging technique, *J. Opt. Soc. Amer. A*, **22**, 1231–1239 (2005).

Cui, G., Luo, M.R., Rhodes, P.A., Rigg, B., and Dakin, J., Grading textile fastness. Part 1: Using a digital camera system, *Color. Technol.*, **119**(4) 212–218 (2003).

Cui, G., Luo, M.R., Rhodes, P.A., Rigg, B., and Dakin, J., Grading textile fastness. Part 2: Development of a new staining fastness formula, *Color. Technol.*, **119**(4) 219–224 (2003).

Cui, G., Luo, M.R., Rigg, B., Butterworth, M., and Dakin, J., Grading textile fastness. Part 3: Development of a new fastness formula for assessing change in colour, *Color. Technol.*, **120**(5) 226–230 (2004).

Cui, G., Luo, M.R., Rigg, B., Butterworth, M, Maplesden, N., and Dakin, J., Grading textile fastness. Part 4: An inter-laboratory trial using DigiEye systems, *Color. Technol.*, **120**(5) 231–235 (2004).

Finlayson, G.D., and Drew, M.S. Constrained least-squares regression in color spaces, *J. Electronic Imaging*, **6**, 484–493 (1997).

Finlayson, G.A., and Morovic, P., Metamer sets, *J. Opt. Soc. Amer. A*, **22**, 1–11 (2005).

Günay, M., Determination of dyeing levelness using surface irregularity function. *Color Res. Appl.*, **34**, 285–290 (2009).

Holst, G.C. *CCD Arrays, Cameras and Displays*, SPIE Optical Engineering Press, Bellingham, WA, USA (1996).

Hong, G., Luo, M.R., and Rhodes, P.A., A study of digital camera colorimetric characterisation based on polynomial modelling, *Color Res. Appl.*, **26**, 76–84 (2001).

Kang, H.R., and Anderson, P.G. Neural network applications to the color scanner and printer calibrations, *J. Electronic Imaging*, **1**, 125–135 (1992).

Lee, R.L., Colorimetric calibration of a video digitizing system: Algorithm and applications, *Color Res. Appl.*, **13**, 180–186 (1988).

Maloney, L.T., and Wandell, B.A., Color constancy: a method for recovering surface spectral reflectance, *J. Opt. Soc. Amer. A*, **9**, 1905–1913 (1986).

Morovic, P., and Finlayson, G.D., Metamer-set based approach to estimating surface reflectance from camera RGB, *J. Opt. Soc. Amer. A*, **23**, 1814–1822 (2006).

Park, S.O., Kim, H.S., Park, J.M., and Eem, J.K., Developments of spectral sensitivity measurement system of image sensor devices, *Proc. IS&T/SID 3rd Color Imaging Conference*, IS&T, Springfield, VA, USA, (1995).

McCamy, C.S., Marcus, H., and Davidson, J.G., A color rendition chart, *J. Appl. Phot. Eng*,. **2**, 95–99 (1976).

Martinez-Verdú, F., Pujol, J., and Capilla, P., Calculation of the color-matching functions of digital cameras from their complete spectral responsitivities, *J. Imaging Sci. Technol.*, **46**, 15–25 (2002).

Martinez-Verdú, F., Pujol, J., and Capilla, P., Characterization of a digital camera as an absolute tristimulus colorimeter, *J. Imaging Sci. Technol.*, **47**, 279–295 (2003).

Mullikan, J.C., Van-Viet, L.J., Boddeke, F.R., Van Der Feltz, G., and Young, I.T. Methods for CCD camera characterisation, *Proc. SPIE*, 2173, 73–84 (1994).

Pointer, M.R., Attridge, G.G., and Jacobson, R.E., Practical camera characterisation for colour measurement, *Imaging Science Journal*, **49**, 63–80 (2001).

Shi, M., and Healey, G., Using reflectance models for color scanner calibration, *J. Opt. Soc. Amer. A*, 645–656 (2002).

Thomson, M.G.A., and Westland, S. Colour camera calibration by parametric fitting of sensor responses, *Color Res. Appl.*, **26**, 442–449 (2001).

Westland, S. Artificial neural networks explained. Part 1. *J. Soc. Dyers & Colourists*, **114**, 274. Part 2, *J. Soc. Dyers & Colourists*, **114**, 312 (1998).

Wu, W., Allebach, J.P., and Analoui, M. Imaging colorimetry using a digital camera, *J. Imaging Sci. Technol.*, **44**, 267–279 (2000).

GENERAL REFERENCES

Kang, H.R., *Computational Color Technology*, SPIE Press, Bellingham, WA, USA (2006).

Sharma, G. *Digital Color Imaging Handbook*, CRC Press, Boca Raton, FL, USA (2002).

Westland, S., and Ripamonti, C., *Computational Colour Science using MATLAB*, Wiley, Chichester, England (2004).

13

Colorant Mixtures

13.1 INTRODUCTION

In the textile and paint industries, individual dyes and pigments are mixed to provide the great variety of colours that are needed; and in the photographic and printing industries mixtures of dyes and inks are also used to obtain the required colours. These colours are produced by combinations of subtractive colorants, each of which absorbs light in some parts of the spectrum. In this chapter, consideration will be given to the ways in which the colour produced is related to the concentrations of the components in a colorant mixture. The simplest situation is when the colorants are non-diffusing and are situated in a transparent layer. Rather more complicated is the situation where non-diffusing dyes are in a layer which is in optical contact with a reflecting diffuser. More complicated still is the situation where the colorants not only absorb the light, but also diffuse it.

13.2 NON-DIFFUSING COLORANTS IN A TRANSMITTING LAYER

The colouring properties of a non-diffusing colorant can conveniently be characterised by its spectral transmission properties. In Figure 13.1 (top) curves of spectral transmittance factor, $T(\lambda)$, are shown for a magenta dye at three concentrations, in the ratios 1 to 2 to 4 times; these curves change shape in a rather complicated way as the concentration changes. If, instead of spectral transmittance factor, $T(\lambda)$, use is made of the corresponding spectral transmission density, $D(\lambda)$, which is given by $\log_{10}(1/T(\lambda))$, then, as can be seen from Figure 13.1 (centre), the curves for the different concentrations are all the same shape apart from being multiplied by a factor that is constant with wavelength. Finally, if instead of spectral transmission density, $D(\lambda)$, its logarithm, $\log_{10}D(\lambda)$, is plotted, then, as shown in Figure 13.1 (bottom), three curves of the same shape are obtained, whose separation on the $\log_{10}D(\lambda)$ axis is related to the concentration of the dye; the shape of the $\log_{10}D(\lambda)$ curve can then be regarded as characterising the colouring properties of the dye.

Measuring Colour, Fourth Edition. R.W.G. Hunt and M.R. Pointer.

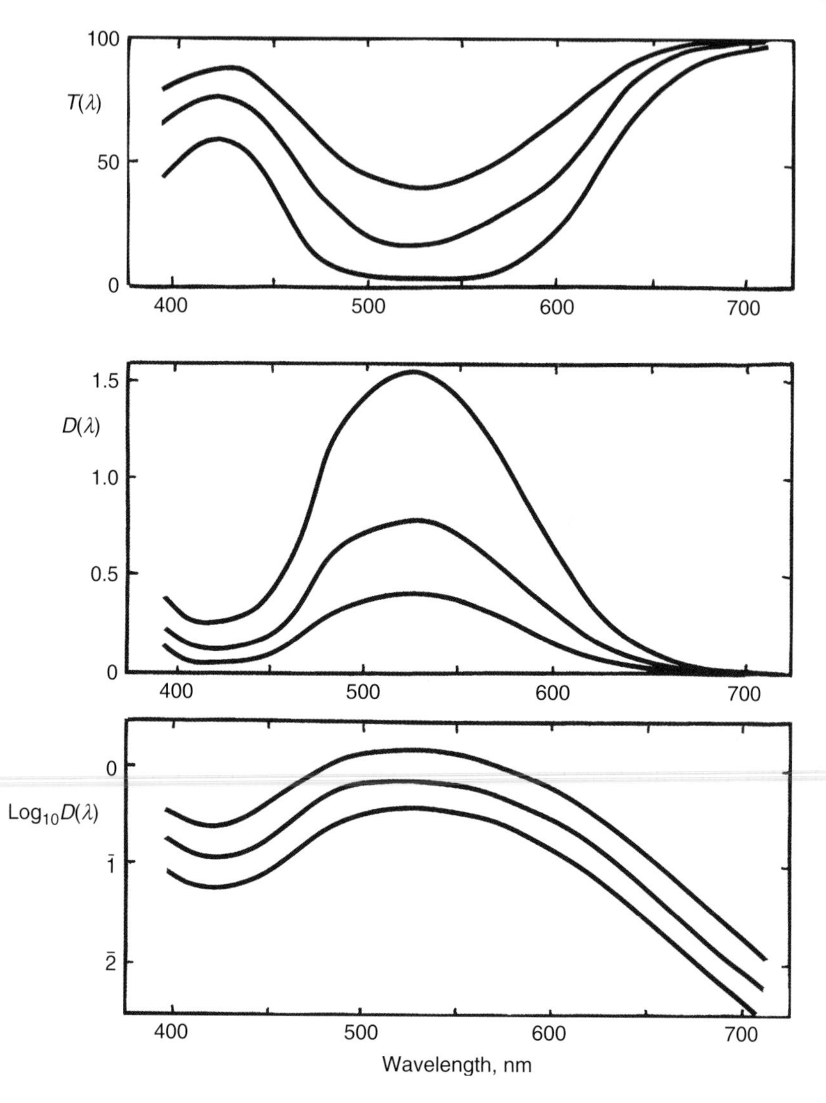

Figure 13.1 The spectral absorption properties of a non-diffusing magenta dye at three concentrations, in the ratios 1 to 2 to 4 times, plotted: (top) in terms of spectral transmittance factor, $T(\lambda)$; (centre), the corresponding spectral transmission density, $D(\lambda)$; and (bottom) $\log_{10} D(\lambda)$

When non-diffusing colorants of different colours are combined in a single transmitting layer, their densities are additive, so that the total spectral transmission density is given by:

$$D(\lambda) = D_1(\lambda) + D_2(\lambda) + D_3(\lambda) + D_0(\lambda)$$

where $D_1(\lambda)$ is the spectral transmission density of colorant 1 measured internally, that is without the losses occurring at the interface of the layer with the outside air; $D_2(\lambda)$ is the similar spectral transmission density of colorant 2, and similarly for colorant 3, and $D_0(\lambda)$ is the spectral transmission density attributable to the interface of the layer with the outside air. The colorimetric measures of such mixtures of colorants are then obtained by converting $D(\lambda)$ to $T(\lambda)$ and using the appropriate illuminant data and standard observer weighting functions (see Section 2.6).

13.3 NON-DIFFUSING COLORANTS IN A LAYER IN OPTICAL CONTACT WITH A DIFFUSING SURFACE

If a non-diffusing colorant layer is placed on top of a diffusely reflecting surface, with an air gap between them, then, for light at right-angles to the layer, the resultant spectral reflection density is equal to twice that of the spectral transmission density of the colorant layer plus the spectral reflection density of the reflecting surface (apart from any differences caused by inter-reflections between the reflecting surface and the colorant layer); the doubling of the density of the colorant layer occurs because the light has to pass through it twice, once on the way to the reflecting surface, and a second time on the way back. However, if the colorant layer is placed on the reflecting layer without an air gap between them, that is in *optical contact*, then the situation is more complicated.

If a colorant layer has a refractive index appreciably higher than air (such as about 1.5, for instance, which is the approximate value for many transparent solids, such as plastics, glass, and gelatin), and if it is in optical contact with a diffusing support, such as paper, then only light reflected within a limited angular cone of 40° semi-angle can escape, and the rest is totally internally reflected and has to make a second attempt to escape, as shown in Figure 13.2. Multiple passes of the light through the colorant layer therefore take place. As the number of these multiple passes increases, their effects quickly become insignificant if

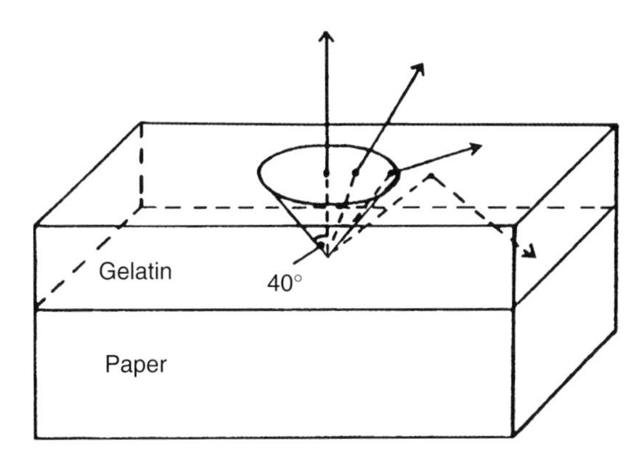

Figure 13.2 Many transparent solids (such as gelatin) have a refractive index of about 1.5. If a colorant layer using such a solid is placed in optical contact with a diffusely reflecting white surface, only light reaching its top surface within a cone of semi-angle approximately 40° is able to emerge from the colorant layer; the rest of the light is totally internally reflected back to the diffusing surface from where it makes further attempts to emerge

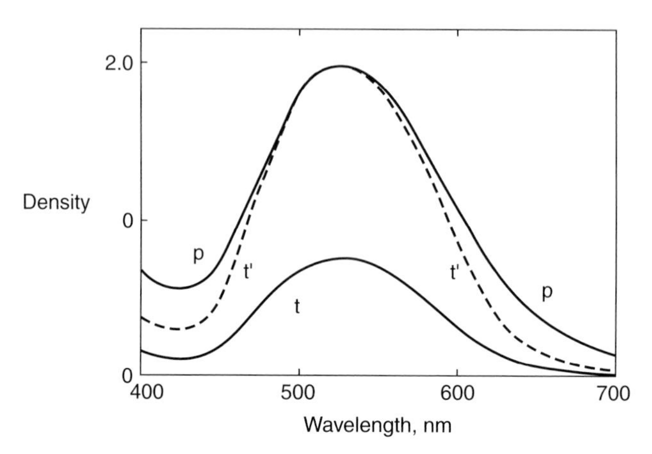

Figure 13.3 Spectral density curves for the same magenta dye: t, in a transparent layer; p, in the same layer placed in optical contact with a diffusely reflecting white surface; t′, curve t multiplied by 2.61 to make the peak density equal to that of curve p. Curve p is broader and has higher secondary absorptions

the absorption of the layer is high; but, if the absorption is low, the effects of the multiple passes decrease to an insignificant level only after many passes. The result is that the multiple passes amplify low absorptions more than high absorptions, thus broadening the absorption band of colorants. In Figure 13.3, curve t shows the spectral transmission density of a magenta dye; curve p shows the spectral reflection density of the same dye when in a gelatin layer in optical contact with paper; curve t′ shows curve t multiplied by a constant to make its maximum value the same as the maximum for curve p. If curves p and t are compared, it is clear that, when the dye is in the layer in optical contact with the paper, the maximum density is increased by more than a factor of 2; and if curves p and t′ are compared, it is clear that the density increases by even greater factors for lower densities, so that the effect is to broaden the green absorption of the magenta dye and to increase its red and blue absorptions. This means that, when a given dye is used in optical contact on a diffusing reflector, its colour becomes of lower purity and darker (so that its chroma is reduced) as compared to the same dye at the higher concentration that gives the same maximum density in a layer not in optical contact.

The relationship between the spectral densities produced by colorants for reflection and transmission can be established for a given reflecting surface and a given geometry of illumination and measurement. A calibration curve of this type is shown in Figure 13.4 (in which the reflection density scale is twice that of the transmission density scale). Curve t shows the relationship when the colorant layer is not in optical contact with the reflecting surface; the reflection density is simply twice the transmission density (represented in the figure by a line of slope 45°). When the colorant layer is in optical contact with the reflecting surface, at low transmission densities, the reflection density can be as much as five times the transmission density; the slope of the curve then decreases from this high value, as the transmission density increases and the multiple passes through the colorant layer become less and less important. In the curve p, of Figure 13.4, the slope approaches

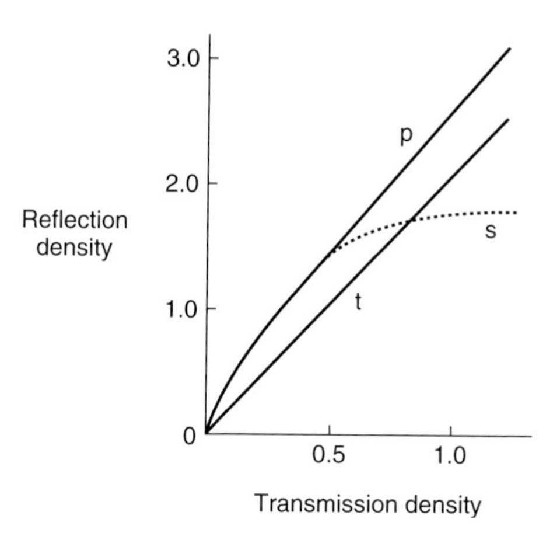

Figure 13.4 Reflection density plotted as a function of transmission density for a colorant: p, when the colorant layer is in optical contact with a diffusely reflecting white surface; t, when the optical contact is broken, and the reflection density is simply twice the transmission density; s, the way in which the top surface reflections modify curve p in typical room viewing conditions

a value of about 2; but the very high reflection densities shown by the curve p are not normally reached in practice because of the light reflected from the top-most surface of the colorant layer, which typically limits the maximum reflection density to about 1.7, as shown by the line s in the figure. The actual maximum value reached by the curve s depends on the glossiness of the surface of the colorant layer and on the geometry of illumination and measurement used (see Section 5.8). The highest densities occur with very glossy samples and 45:0 or 0:45 geometry; the lowest densities occur with matt samples and diffuse:8 or 8:diffuse geometries.

Colorimetric measures for layers of colorants in optical contact with diffusing surfaces can be obtained as follows: first, the spectral transmission density curve of the colorant mixture being used is obtained; second, a calibration curve of the type shown in Figure 13.4 is used at each wavelength to obtain the corresponding spectral reflection density curve; third, the corresponding spectral reflectance factor values, $R(\lambda)$, can be found, for use with the appropriate illuminant data and standard observer weighting functions (see Section 2.6).

As a colorant is increased in concentration, if it has several unequal absorptions, the larger absorptions progressively have greater effects than the smaller. Thus cyan dyes become bluer at higher concentrations because their secondary green absorptions are larger than their secondary blue absorptions. When inks are used in half-tone images, the hue is similar to that of the solid ink for all dot sizes, because, as the dots become smaller, white light is added to the colour of the solid ink; thus, with cyan inks, half-tone (or dot) scales look bluish throughout, but continuous tone scales look blue-green in the lighter tones but become bluer in the darker tones. Similarly, magenta continuous scales look redder as the ink is increased in concentration but in half-tone scales the hue remains constant. The greater effects of the secondary absorptions of the inks at high concentrations also result in half-tone scales being greyer than continuous tone scales.

13.4 LAYERS CONTAINING COLORANTS WHICH DIFFUSE AND ABSORB LIGHT

When pigments are used as colorants, as in many paints and inks, or when dyes are used on fibres as in textiles, some of the light, in addition to being absorbed, is usually scattered. This makes the relationship between the amounts of colorants and the colorimetry very much more complicated than is the case with non-scattering colorants (McDonald, 1997; Nobbs, 1997).

If we consider the situation within a horizontal colorant layer, then any light that is travelling parallel to the boundaries of the layer need not be considered; this is because for every amount of light travelling in one direction, there will be an equal amount travelling in the opposite direction so that the total amount can be regarded as zero (although this will not be true near an edge of the layer, the area of the layer is usually regarded as being sufficiently large compared to its thickness that edge effects can be ignored). A flux of light travelling in any direction can be divided into its vertical and horizontal components, and, since the horizontal components can be ignored, consideration can be confined to the vertical components only.

In Figure 13.5 a thin section inside a colorant layer is considered. The total thickness of the layer is denoted by D, and the thin layer is of thickness $\mathrm{d}x$, and is situated a distance x from the top of the layer. Two vertical fluxes of light have to be considered, that travelling downwards, the i flux, and that travelling upwards, the j flux. The intensities of these two fluxes at the distance x from the top surface are denoted as i_x and j_x. If $K\mathrm{d}x$ is the fractional amount of light lost from a flux by absorption as it passes through the thickness $\mathrm{d}x$, and $S\mathrm{d}x$ is the fractional amount of light lost by scattering as it passes through the thickness $\mathrm{d}x$, then K is termed the absorption coefficient and S is termed the scattering coefficient. Light absorbed in the thin layer is lost, but light scattered from the i direction is added to the j direction and vice-versa.

The net change in each flux as a result of passing through the thin layer is obtained by adding the fluxes lost and gained thus:

$$
\begin{array}{cccc}
\mathrm{d_i}x = & -Ki_\mathrm{x}\mathrm{d}x & -Si_\mathrm{x}\mathrm{d}x & +Sj_\mathrm{x}\mathrm{d}x \\
-\mathrm{d_j}x = & -Kj_\mathrm{x}\mathrm{d}x & -Sj_\mathrm{x}\mathrm{d}x & +Si_\mathrm{x}\mathrm{d}x \\
\uparrow & \uparrow & \uparrow & \uparrow \\
\text{net} & \text{loss by} & \text{loss by} & \text{gain} \\
\text{change} & \text{absorption} & \text{scattering} & \text{by scattering}
\end{array}
$$

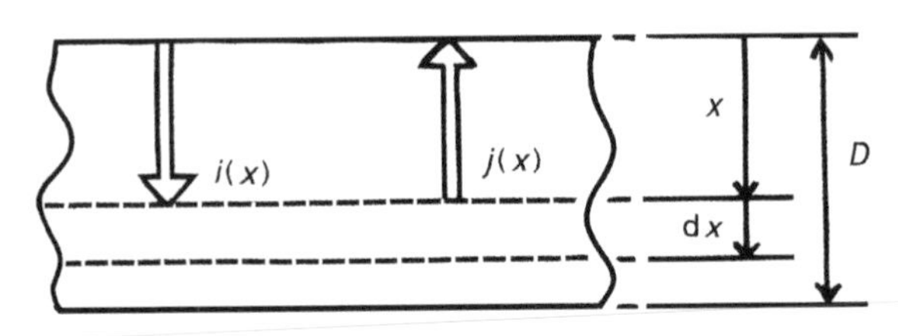

Figure 13.5 Situation in a horizontal layer of scattering colorants

The negative sign in the j_x flux arises because the direction of the j flux is opposite to that of x. Dividing these equations by dx, we obtain:

$$\mathrm{d}i_x/\mathrm{d}x = -(K + S)i_x + Sj_x$$

$$\mathrm{d}j_x/\mathrm{d}x = -Si_x + (K + S)j_x$$

The ratio of the upward to the downward fluxes at a point x from the top of the layer is called the reflectance ratio, r_x:

$$r_x = j_x/i_x$$

The way in which r_x changes with position in the layer is found by differentiating the equation for r_x with respect to x to obtain

$$\mathrm{d}r_x/\mathrm{d}x = [i_x(\mathrm{d}j_x/\mathrm{d}x) - j_x(\mathrm{d}i_x/\mathrm{d}x)]i_x^2$$

from which it follows that:

$$\mathrm{d}r_x/\mathrm{d}x = -S{r_x}^2 + 2(K + S)r_x - S$$

If we now consider that the layer is so thick that further increases in thickness, however great, do not significantly change its reflectance, then the reflectance ratio, r_x, will not change with x, so that:

$$\mathrm{d}r_x/\mathrm{d}x = 0 = -S{r_x}^2 + 2(\mathrm{K} + S)r_x - S$$

The value of r_x in this situation is denoted by R_∞ and hence:

$$0 = -S{R_\infty}^2 + 2(K + S)R_\infty - S$$

This is a quadratic equation which can be solved to give:

$$R_\infty = 1 + K/S - [(1 + K/S)^2 - 1]^{1/2}$$

from which

$$K/S = (1 - R_\infty)^2/2R_\infty$$

This is known as the Kubelka-Munk function (Kubelka and Munk, 1931; Kubelka, 1948). It shows that the reflectance of an opaque layer depends only on the ratio of K to S and not on their individual values. The K and S coefficients are related to ε, the linear absorption extinction coefficient, and σ, the linear scattering extinction coefficient, of the layer; these relationships are approximately:

$$K = 2\varepsilon \qquad S = \sigma$$

When several scattering colorants are used in the same layer, the resultant values, K_T and S_T, of K and S are approximately given by the sum of the individual values. Thus for a mixture of three colorants, 1, 2, and 3:

$$K_T = K_1 + K_2 + K_3$$

$$S_T = S_1 + S_2 + S_3$$

For matt samples measured with an integrating sphere, the measured reflectance factor usually approximates R_∞, but for glossy samples, or for 45:0 or 0:45 measurements, the measured reflectance factor may differ significantly from R_∞.

So far, no allowance has been made for the effects of the interface between the outside air and the colorant layer at its top boundary. To allow for this we denote the flux travelling in air down towards the boundary as I, the flux travelling in the colorant layer up towards the boundary as J, the fractions of the incident flux, I, that are reflected and transmitted by the interface as r_e and t_e, respectively; the fractions of the incident flux, J, that are reflected and transmitted by the interface as r_i and t_i, respectively; and the measured reflectance factor as ρ. Then:

$$\rho = r_e + t_e t_i R_\infty/(1 - r_i R_\infty)$$

This equation is often referred to as the Saunderson equation (Saunderson, 1942). When the refractive index of the vehicle used in the colorant layer is 1.5 (a figure typical of many of the polymeric materials used in paints and inks), and flux I is a collimated beam and flux J is a diffuse beam, $r_e = 0.040$, $r_i = 0.600$, $t_e = 0.960$, and $t_i = 0.400$. The Saunderson equation can be rearranged as:

$$R_\infty = [\rho - r_e]/[t_e t_i + r_i(r - r_e)]$$

In the textile and paint industries, the prediction of recipes of dye and pigment mixtures to produce particular colours is a very important activity. Most of the procedures that have been worked out use the Kubelka-Munk analysis as a starting point, but many elaborations have been developed to give better accuracy of prediction in particular circumstances (Berns and Mohammadi, 2007). Very useful discussions of these topics are given in *Colour Physics for Industry*, edited by Roderick McDonald and published by the Society of Dyers and Colourists (McDonald, 1997).

13.5 THE USE OF MULTI-SPECTRAL ANALYSIS TO REDUCE METAMERISM IN ART RESTORATION

When colorants are added to areas of a painting to restore its appearance, it is desirable to obtain, not just a colour match which might be metameric, but, as nearly as possible, a spectral match; in this case the match will not be upset by changes in illuminant or observer. A spectral match would be achieved if the colorants used are the same as those in the original, but these might be toxic, unavailable, have poor working properties, or have poor light-fastness. This means that it is desirable to find usable colorants that result in an approximate spectral match to the original.

Each area to be restored could be measured using a spectrophotometer, but this is a lengthy operation, and may require the difficult task of identifying and locating many different areas. It has therefore been found useful to use abridged spectrophotometry of the whole art work. This is done by using a high resolution digital camera with about six different coloured filters, a technique known as *multi-spectral imaging* (Mohammadi, Nezamabadi, Berns, and Taplin, 2004; Zhao, Berns, Okumura, and Taplin, 2005; Zhao and Berns, 2007). A training target is used to convert the camera outputs to approximate spectral reflectances (see Section 12.4). An optical database of possible retouching materials is then used to derive a paint that minimises metamerism; this paint may consist of one colorant, or a combination of several colorants (Taplin and Berns, 2005).

Another useful application of this technique is to use the estimated spectral reflectances of the original to determine which colorants were used by the artist, and at what concentrations, and with what working method (Berns, Taplin, Imai, Day, and Day, 2006).

REFERENCES

Berns, R.S., and Mohammadi, M., Single-constant simplification of Kubelka-Monk turbid- media theory for paint systems – a review, *Color Res. Appl.*, **32**, 201–207 (2007).

Berns, R. S., Taplin, L.A., Imai, F.H., Day, E.A., and Day, D.C., A comparison of small- aperture and image-based spectrophotometry of paintings, *Studies in Conservation*, **50**, 253–266 (2006).

Kubelka, P. J., New contributions to the optics of intensely light-scattering materials, Part I, *J. Opt. Soc. Amer*. **38**, 448–457 (1948).

Kubelka, P. and Munk, F., Ein Bewitrag zur Optik der Farbenstriche, *Z Tech. Physiki*. **12**, 593–601 (1931).

McDonald, R., In *Colour Physics for Industry*, *2nd Ed*., pp. 209–291, ed. Roderick McDonald, Society of Dyers and Colourists, Bradford, England (1997).

Mohammadi, M., Nezamabadi, M., Berns, R.S., and Taplin, L.A., Spectral imaging target development based on hierarchical cluster analysis, *IS&T and SID's 12th Color Imaging Conference: Color Science and Engineering Systems, Technologies, and Applications*, pp. 59–64, IS&T Springfield, VA, USA (2004).

Nobbs, J.H., In *Colour Physics for Industry*, *2nd Ed*., pp. 292–372, ed. Roderick McDonald, Society of Dyers and Colourists, Bradford, England (1997).

Saunderson, J.L., Calculation of the colour of pigmented plastics, *J. Opt. Soc. Amer*., **32**, 727–736 (1942).

Taplin, L.A., and Berns, R.S., Practical spectral capture systems for museum imaging, *Proc. 10th Congress of the International Colour Association*, pp. 1287–1290, Granada, Spain (2005).

Zhao, Y., Berns, R.S., Okumura, Y., and Taplin, L.A., Improvement of spectral imaging by pigment mapping, *IS&T and SID's 13th Color Imaging Conference: Color Science and Engineering Systems, Technologies, and Applications*, pp. 40–45, IS&T Springfield, VA, USA (2005).

Zhao, Y., and Berns, R.S., Image-based spectral reflectance reconstruction using the Matrix R method, *Color Res. Appl*., **32**, 343–351 (2007).

GENERAL REFERENCES

Hunt, R.W.G., *The Reproduction of Colour*, *6th Ed*., Wiley, Chichester, England (2004).

McDonald, R., *Colour Physics for Industry*, *2nd Ed*., ed. Roderick McDonald, Society of Dyers and Colourists, Bradford, England (1997).

14

Factors Affecting the Appearance of Coloured Objects

14.1 INTRODUCTION

The overall appearance of an object is a combination of different attributes, which are produced by the interaction of the object with the light falling upon it. Spectral absorption and diffuse reflection of the light by the pigments or other colorants within the object give it the attribute of colour. Most objects reflect some light from their surface, which we then perceive as gloss or haze. The amount of scattering of the light as it is transmitted through parts of an object, prompts a judgement on translucency. Finally, spatial irregularities on or near a surface may be seen as texture. Although, at the moment, it is not possible to make a single measurement called *appearance* – and this will probably never be achievable – it is possible to design instruments to 'measure' at least some of the various components of appearance and to make some kind of correlation to visual perception. Thus, a goal of making measurements that ensures appropriate quality control in a manufacturing process is achievable, but the measurement process will be multi-dimensional, product specific, and probably application specific.

The relationship between the physical structure of a material and its optical radiation properties (scattering, reflection, etc.) is complex. For example, selective absorption, which is largely responsible for the colour of a material, takes place during the passage of light through that material. Scattering occurs where light encounters interfaces between particles and resin, fibre, air, etc. Normally, when particle sizes are made smaller, less light is absorbed during the passage through each particle resulting in less colour being apparent; at the same time, the total particle surface becomes greater, leading to an increase in light scattering, or diffusion, since reflection occurs at the particle surfaces.

14.2 MEASURING OPTICAL PROPERTIES

It is possible to divide the characterisation of the optical properties of materials into at least four groups, colour, gloss, translucency and surface texture (CIE, 2006), as shown in

Measuring Colour, Fourth Edition. R.W.G. Hunt and M.R. Pointer.

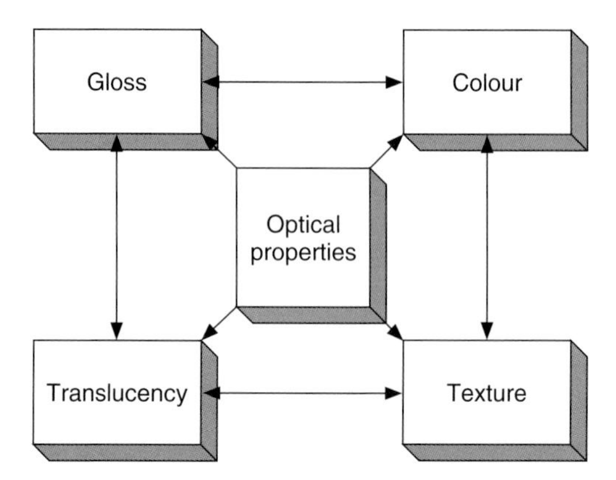

Figure 14.1 The suggested sub-division of the optical properties of materials into measurement groups – with interactions between those groups

Figure 14.1. These groups are perhaps not definitive but they are intuitive and represent useful categories for measurement, especially in view of the fact that measurement techniques already exist for some of them; they are not independent in that there must be links between colour and gloss, and colour and translucency, etc.

14.3 COLOUR

The measurement of the colour property of objects is considered in other chapters of this book, and in Chapters 15, 16, and 17 models for predicting colour appearance are described. But there are some phenomena that can greatly affect the appearance of coloured objects for which there are as yet no generally accepted quantitative measures; these include *successive contrast, simultaneous contrast* (sometimes termed *chromatic induction*), and *assimilation* (sometimes termed *spreading effect*). Some of these phenomena can be of great practical importance in some industries, such as those involved with signage, woven fabrics, and tapestries. Examples of these phenomena will now be presented to illustrate their potential to change colour appearance.

14.3.1 Successive contrast

As has already been mentioned in Sections 1.8 and 6.13, when light falls on the retina of the eye it adjusts its sensitivity to compensate for the general intensity and colour of the stimulation. This can occur in a few seconds or minutes and the result can last for a similar time; this is termed *successive contrast*. A practical example of this is when an observer watches a television screen with a white point of colour similar to that of D65 and then looks at objects in a room lit with light of significantly lower correlated colour temperature, such as tungsten light or compact fluorescent lamps; in such cases, when the images of the objects fall in the area that had been used for viewing the television screen, they look yellower than normal. A demonstration of successive contrast is provided in Figure 14.2.

Figure 14.2 A demonstration of successive contrast. The two lower colours are the same. Stare at the upper black dot for about 20 seconds, and then look at the lower black dot; the lower left-hand colour now appears greenish and the lower right-hand colour appears reddish

14.3.2 Simultaneous contrast

The appearance of a colour can be greatly affected by the presence of other colours around it; this is termed *simultaneous contrast* (or *chromatic induction*). Simultaneous contrast can result in large changes in the appearance of colours in items such as woven fabrics and tapestries. The French chemist Michel Eugène Chevreul, as director of Gobelin, the famous carpet manufacturer, was one of the first to investigate the phenomenon; in 1839 he introduced his law: 'two adjacent colours, when seen by the eye, will appear as dissimilar as possible' (Chevreul, 1839).

In Figure 14.3 the effect of simultaneous contrast on lightness is demonstrated. It is well known that a dark surround makes a colour look lighter, and a light surround makes it look darker. But it is also true that a dark surround lowers apparent contrast, and this is evident in the right-hand part of the figure. This contrast-lowering effect occurs quite strongly when pictures are projected in cinemas, and has to be countered by increasing the contrast of the picture being projected.

In Figure 14.4 a further example of simultaneous contrast affecting lightness is shown. Areas C and D reflect the same amount of light, but C, with its lighter surround, looks darker than D, with its darker surround; but the effect is much greater with areas A and B, which also reflect the same amount of light; this is because, in addition to simultaneous

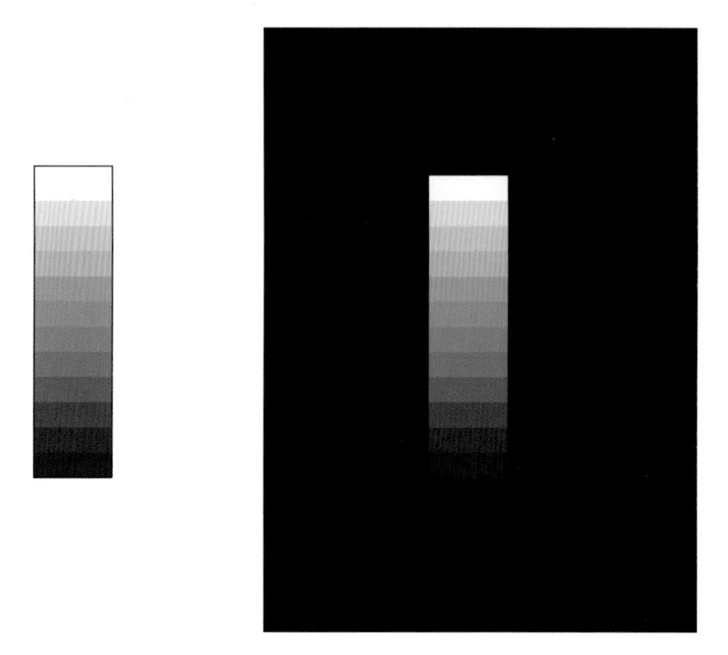

Figure 14.3 The two grey wedges are identical, but simultaneous contrast makes the right-hand wedge look lighter and of lower contrast

Figure 14.4 The two areas labelled A and B reflect the same amount of light. But simultaneous contrast and cognitive effects make A look much darker than B. The effect of simultaneous contrast alone is shown by areas C and D, which also reflect the same amounts of light, but whose difference is not enhanced by a cognitive effect. (The chequerboard figure produced by Edward H. Adelston, 1993.)

contrast, cognitive effects are present as the brain interprets the picture in terms of the recognition of objects.

In Figures 14.5 and 14.6 the effect of simultaneous contrast on colour is demonstrated. In Figure 14.5, although, in each of the six diagrams, the same ink is used for each pair of smaller rectangles, their backgrounds make them appear of different colours, because of simultaneous contrast. In Figure 14.6, in each of the six diagrams, the same ink is used to print both the crosses, but their backgrounds make them appear of different colours; the effect is much greater than in Figure 14.5, because of the greater contact of the backgrounds with the crosses. This demonstrates the important fact that that simultaneous contrast depends not only on the colours involved but also on their geometrical pattern.

14.3.3 Assimilation

When stimuli are seen at small angular subtenses, the opposite of simultaneous contrast can occur, when colours become *more*, instead of *less*, like their surroundings, an effect termed *assimilation* (or *spreading effect*).

In Figure 14.7 examples of assimilation are demonstrated. Each rectangle has the same background colour as the rectangle beneath it. The white patterns on the upper rectangles make their background colours appear lighter; the dark patterns on the lower rectangles make their background colours appear darker. This is the opposite of the effect of simultaneous contrast, which would make the backgrounds of the upper rectangles appear darker, and those of the lower rectangles appear lighter. The magnitude of the effect may be enhanced by viewing the patterns at a distance. Demonstrations of assimilation have also been published by Evans (Evans, 1948), and by Wright (Wright, 1969). The likely causes of the effect include scattering of light in the eye, and the fact that the colour difference signals in the visual system have lower resolution than that of the achromatic signal (Hunt, 1967).

14.4 GLOSS

Gloss perception is associated with the way that an object reflects light, particularly depending on the way that light is reflected from the surface of the object at and near the specular, mirror angle, direction. Gloss is normally perceived independent of colour; it may, however, be affected by the underlying colour of the object or itself affect the perceived colour of the object (Willmouth, 1976). It is usual, however, for the perception of gloss to be abstracted from the total visual experience as separate from colour. Hunter and Judd (Hunter and Judd, 1939) first defined specular gloss as the ratio of the light reflected from a surface at a specified angle to that incident on the surface at the same angle on the other side of the surface normal (see Figure 14.8). Hunter recognised, however, that the perception of gloss involves more than just the specular reflection (Hunter and Harold, 1987).

Where I is the amount of incident light, S is the amount of specularly reflected light, D is the diffuse reflectance normal to the surface, and B is the off-specular light, Hunter defined:

- specular gloss as proportional to S/I;
- sheen as proportional to S/I at grazing angles of incidence and viewing;

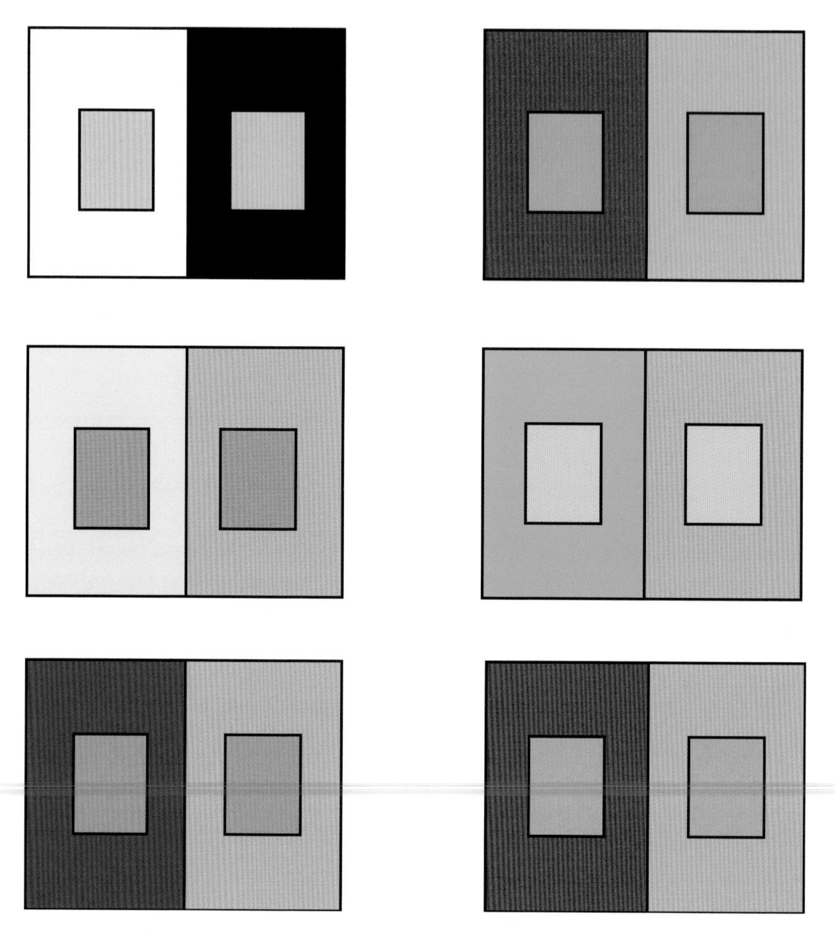

Figure 14.5 In the top left-hand diagram the two smaller rectangles are printed with ink of the same colour, but their backgrounds make them appear different, because of simultaneous contrast. Similarly, in the other five diagrams the colours in the smaller rectangles appear different although the ink used is the same for each pair

- contrast gloss or lustre as proportional to D/S;
- absence-of-bloom gloss (a measure of the haze or a milky appearance adjacent to the specularly reflected light) as proportional to $(B - D)/I$;
- and distinctness-of-image gloss as the sharpness of the specularly reflected light.

A surface can appear very shiny if it has a well-defined specular reflectance at the specular angle. The perception of an image reflected in the surface can be degraded by appearing unsharp, or by appearing to be of low contrast. The former is characterised by the distinctness-of-image gloss (see Figure 14.9) and the latter by the haze or contrast gloss (haze is the inverse of absence-of-bloom) (Tingle and Potter, 1961; Smith, 1997). An added

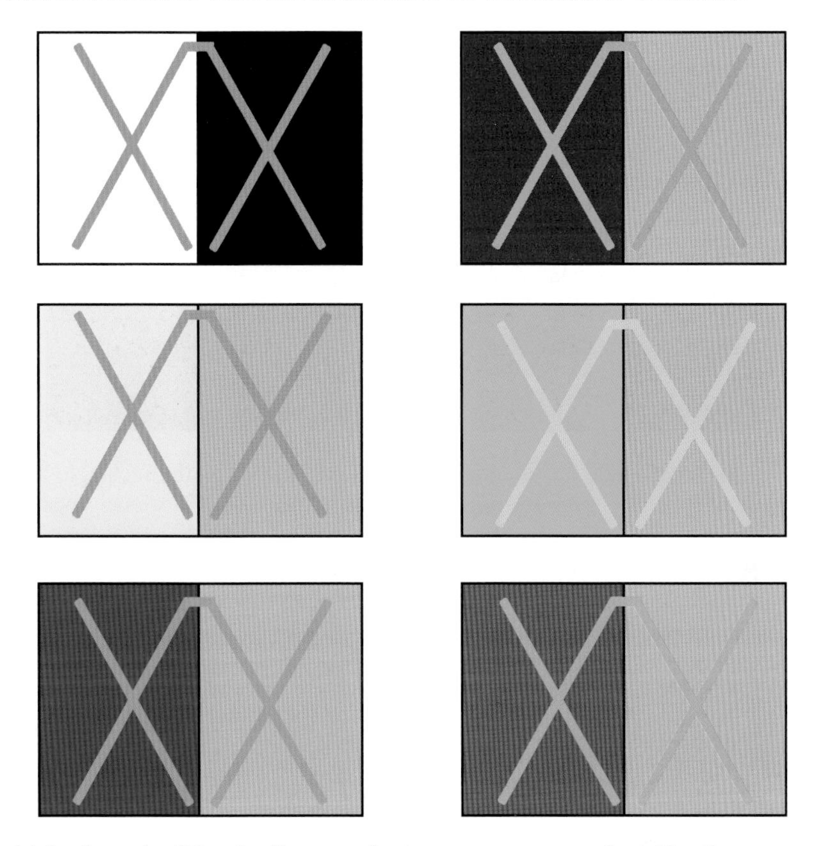

Figure 14.6 In each of the six diagrams the two crosses are produced by the same coloured ink (as can be seen at the top of each diagram), but simultaneous contrast makes the left-hand one look different from the right-hand one (after a similar figure in Albers, 1975). The colours used for the crosses and the backgrounds are the same as in Figure 14.5, but the differences in apparent colour are much larger; this is because of the greater contact of the background with the crosses than with the rectangles. This demonstrates that simultaneous contrast is not only affected by the colours involved, but can also be strongly dependent on their geometric pattern

complexity is caused by surface non-uniformity leading to an effect known as orange peel. This effect can be caused, for example, by uneven coating of the acrylic overcoat on an automobile finish leading to a relatively low frequency 'ripple'. It should be noted that it is not always the 'top' surface of a material that contributes to the gloss. The quality of the colour image produced by inkjet printing technology varies, for example, depending on the type and quality of the substrate, the raw stock paper (Konda and Harold, 1999; Béland and Bennet, 2000). To get a high laydown of ink requires a relatively rough surface to give a high surface area on which to print; this, however, tends to be a relatively low gloss surface and so the print will not look 'photographic' or have a high gloss unless other measures are taken to achieve this.

Figure 14.7 Each rectangle has the same background colour as the rectangle beneath it. The white patterns on the upper rectangles make their background colours appear lighter; the dark patterns on the lower rectangles make their background colours appear darker. This is the opposite of the effect of simultaneous contrast, which would make the backgrounds of the upper rectangles appear darker, and those of then lower rectangles appear lighter. The magnitude of the effect may be enhanced by viewing the patterns at a distance

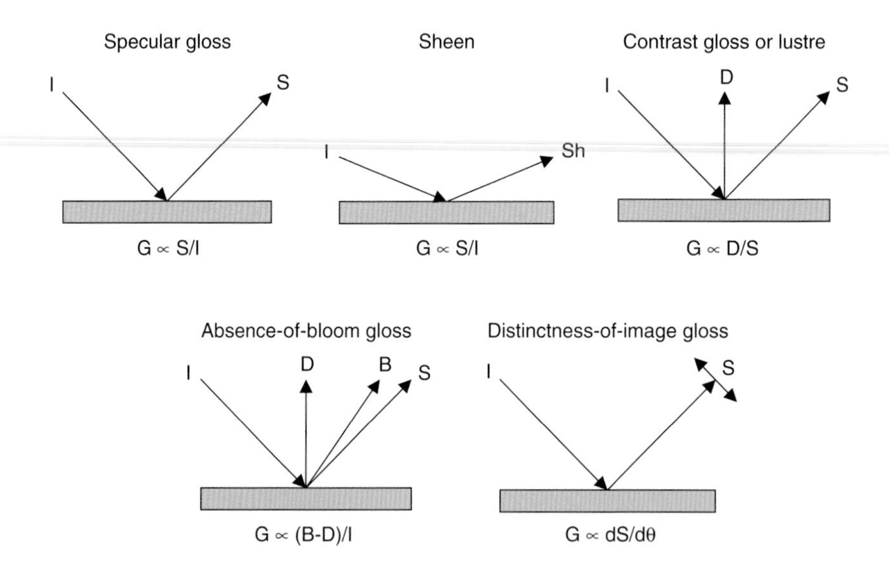

Figure 14.8 Hunter's five types of gloss, G, associated with the amounts of incident light, I, specularly reflected light, S, diffuse reflectance, D, and off-specular light, B

14.4.1 Measuring gloss – gloss meters

Many developments of gloss measurement have been carried out as part of the technical work of the American Society for Testing and Materials (ASTM International) commencing in 1925 with the instrument constructed by Pfund (Pfund, 1930). This used parallel light to illuminate the sample at 20° with a detector placed at 20° on the other side of the normal.

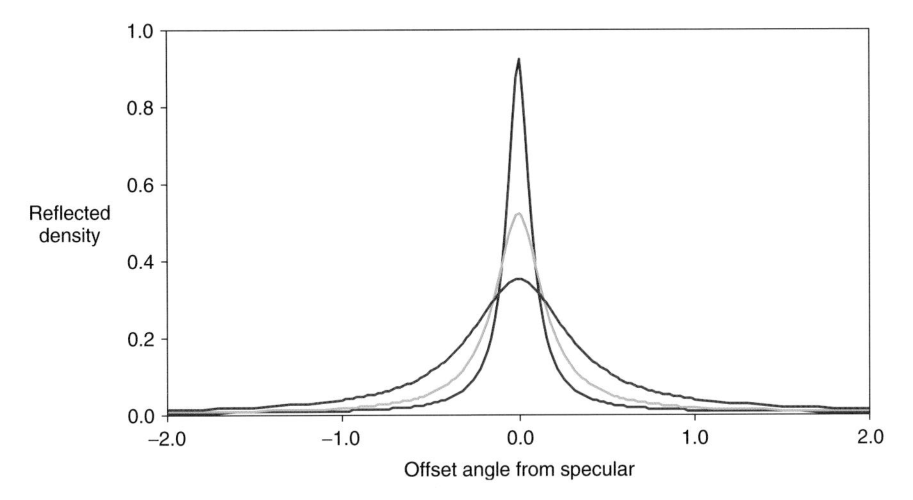

Figure 14.9 Examples of distinctness-of-image measurements at three levels of gloss

Hunter and Judd later incorporated this design into an ASTM Method (ASTM 2008; BS EN ISO 2000) which designates three angles (20°, 60°and 85°) for measurement, depending on the relative gloss of the surface. Measurements are made relative to a highly polished black glass standard with a refractive index of 1.567, using a source/detector combination whose implied spectral power distribution approximates that of CIE illuminant C and whose spectral response approximates that of the CIE luminous efficiency function $V(\lambda)$. The gloss of the standard is assigned a value of 100 for each geometry. In order to differentiate the gloss of different samples it is necessary to select the appropriate measuring geometry. The sample is first measured with 60° geometry. If the gloss value is higher than 70 (high gloss), then it is re-measured at 20° and if less than 10 (low gloss), it is re-measured at 85°. These values are defined following some experimental work described by Byk-Gardner (Byk-Gardner, 2000) in which 13 black glass tiles were visually ranked from matt to high-gloss and the results compared with measurements at the three angles, as shown in Figure 14.10. It was found that the largest difference could be detected by using 85° for low gloss samples, 60° for semi-gloss samples and 20° for high gloss samples.

Some argue that gloss is very much a 'second-order' visual attribute in that it results from an interpretation by the brain of first-order signals (Obein, Leroux, and Viénot, 2001; Obein, Knoblauch, Chrisment, and Viénot, 2002; Obein, Leroux, Knoblauch, and Viénot, 2003; Obein, Knoblauch, and Viénot, 2004). This implies that an observer must look at the surface of an object from two or three different angles to receive enough information to be able to attribute a value to the gloss of that surface. Their work showed that the scaled visual gloss of a set of custom designed black samples, obtained using a pair-comparison technique, was not linearly related to the corresponding values obtained from a gloss meter (see Figure 14.11): for matt samples the visual scale undergoes compression, and for very high gloss samples the gain of the visual responses rises steeply. In the intermediate range, the two scales are almost linearly related. Analysis of the data also shows that the observers exhibit a form of gloss constancy. When data obtained using two different observing angles (60° and 20°) are plotted versus a unique abscissa, for example the sample number in the

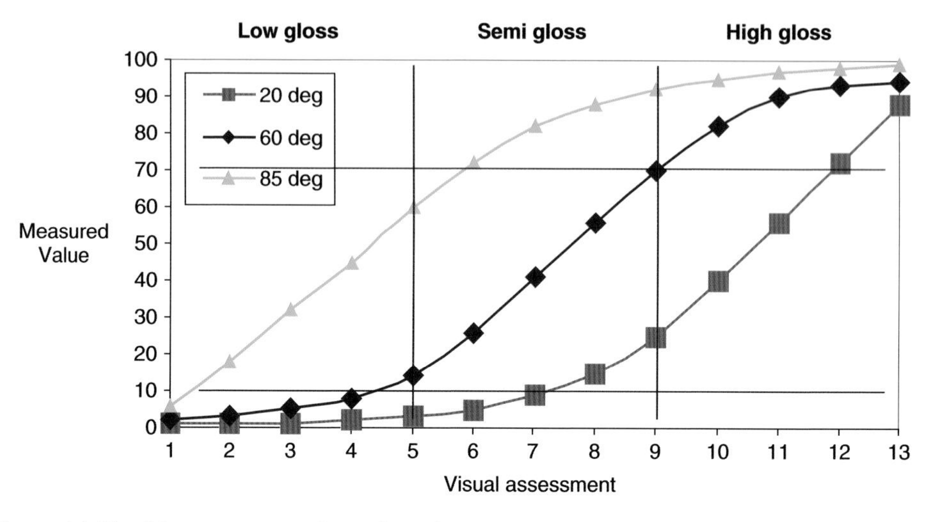

Figure 14.10 Measurements of a series of samples, at three angles, as a function of visual assessment of their gloss

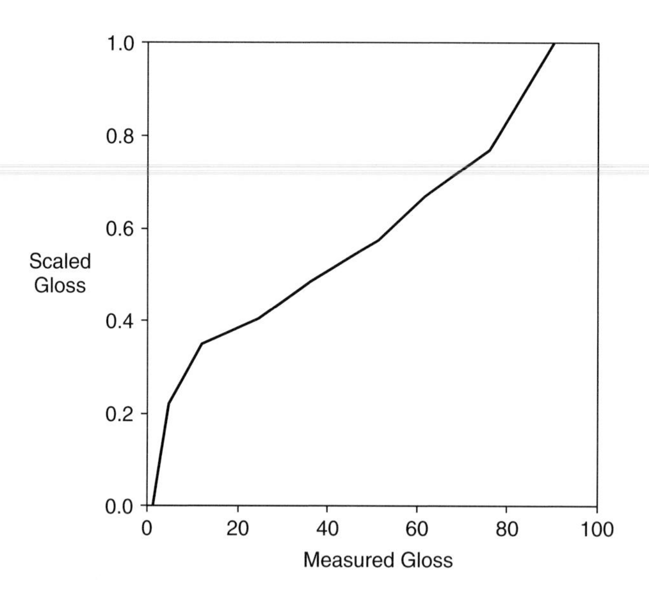

Figure 14.11 Gloss as estimated by one observer (Scaled Gloss) plotted against gloss measured at 60° (Normalised Specular Gloss) for a series of black samples

series, the two plots superimpose. This would indicate that, although the flux that is collected by the eye varies according to the angle of view, an observer is able to recover a visual gloss index that is inherent to the surface. Thus, just as observers can assign a colour to a sample under lights of different spectral power distribution, they can also assign a gloss value to a surface despite changes in the geometry of illumination and viewing.

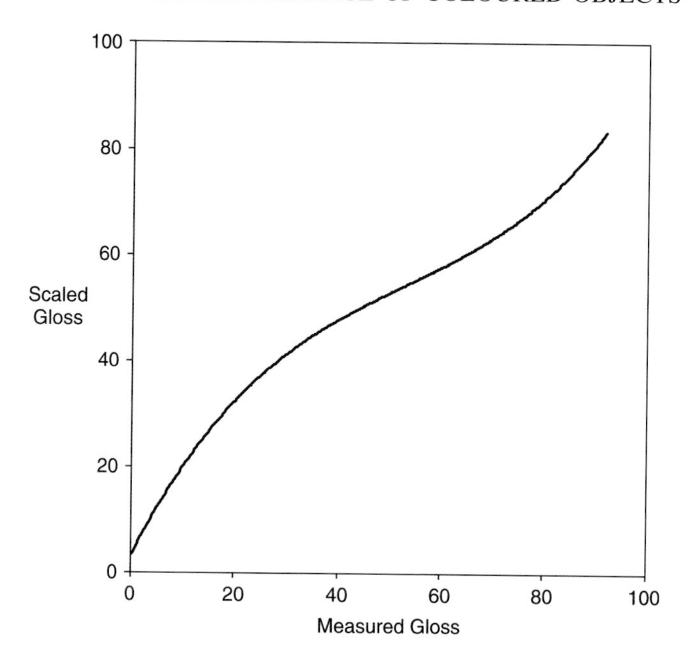

Figure 14.12 Visually scaled gloss as a function of measured gloss at 60°

Work by Ji, Luo, Li, Cui, Pointer, and Dakin (Ji, Luo, Li, Cui, and Pointer, 2004; Ji, Pointer, Luo, and Dakin, 2006) supports the findings of Obein, Leroux, Viénot, Knoblauch, and Chrisment. Eighty-four samples with varying gloss, including both neutral and coloured samples, were scaled 20 times by 14 observers. Mean observer data are plotted in Figure 14.12. The fitted curve is a third order polynomial but this gives a similar shape to that discussed by Obein, Leroux, Viénot, Knoblauch, and Chrisment.

14.4.2 Measuring gloss – goniophotometers

The most informative and precise technique for the measurement of the gloss of surfaces is undoubtedly goniophotometry – the measurement of the intensity of reflected light as a function of viewing angle (Barkas, 1939; Tighe, 1978). It is often the case that the angle of illumination of a goniophotometer is also variable. Typical goniophotometric data obtained using this type of instrument are shown in Figure 14.13, which shows the relative amount of light reflected from a white gloss tile and a white matt tile when illuminated with a parallel beam of light incident at 8° to the surface normal. The gap in the data, to the left of the figure, indicates the position of the illuminating beam; the gloss peak is shown to the right of the figure. The measurements made using the matt tile show only a small gloss peak – for a perfectly matt sample, there would be no gloss peak.

14.4.3 Discussion – gloss

The measurement of correlates of the phenomenon known as gloss is not without its problems. Sève, in a review article in 1993 (Sève, 1993), noted that 'the CIE had been grappling

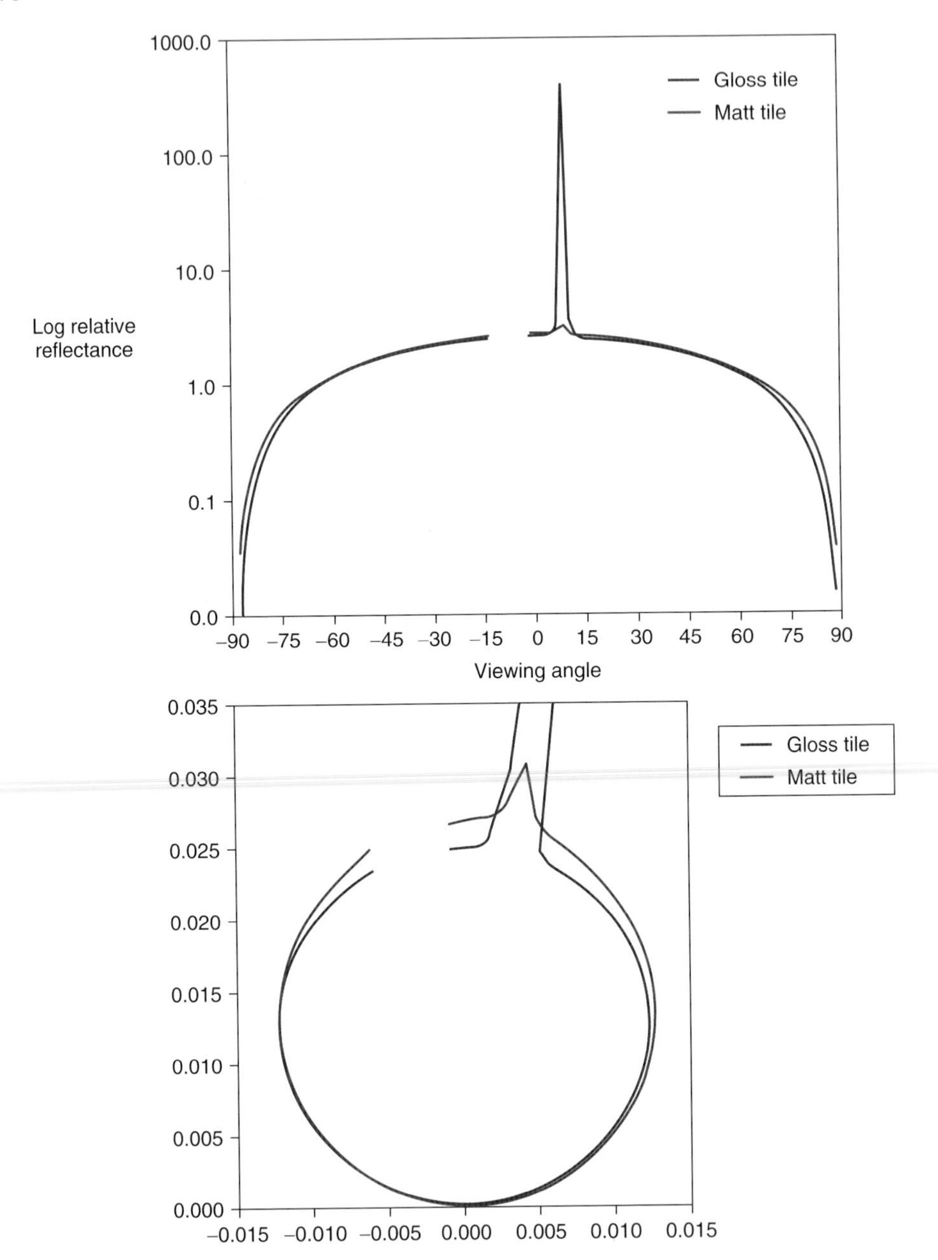

Figure 14.13 *Upp*er. Goniophotometric data plot showing the relative amount of light reflected from a white gloss tile and a white matt tile over the angular range ±85°, when illuminated with a parallel beam of light incident at 8° to the surface normal. *Lower*. The same data plotted as the loci of the upper ends of lines whose lengths are proportional to the relative amounts of reflected light and whose lower ends are at the point 0.000 on the abscissa, and whose angles relative to a perpendicular at that point are the angles at which those amounts of light are reflected. (The numbers on the axes are the reflectance x the sine of the angle vs, the reflectance x the cosine of the angle)

with the subject for over 20 years and, while they had produced a state-of-the-art report in 1986 (CIE, 1986), little had changed in the interim period'. Going back further Sève noted 'Only a small number of research papers have been published on gloss in its history of about fifty years'. Sève goes on to try to elucidate what the problems might be and to suggest solutions. The biggest problem seems to be that the instruments currently available use an arbitrary choice of gloss scale. This is compounded by the fact that instruments from different manufacturers, while making measurements at nominally the same angles, show wide variation in the aperture size and beam geometry. Sève proposes that the measurement of gloss be related to the measurement of luminance factor in the specular direction; he also discusses the influence of polarisation on the measurements obtained using instruments.

This would seem to move the gloss scale from being based on physics (with respect to a glass of defined refractive index) to being based on visual data. It would also serve to enable the gloss values, measured at different angles of illumination, to be compared: with the present system, the scale at each measurement angle is separately normalised to have a maximum value of 100.

Another view suggests that gloss should be considered under a more general heading of goniometric reflectance factor with a scale defined by a white Lambertian diffuser and using a set of defined illumination angles (Lindstrand, 2004). This provides generality but may not best represent industrial requirements. It can be argued that this concept forms a subset of the more general measurement of the Bidirectional Reflectance Distribution Function (BRDF) which gives the reflectance of a target as a function of illumination geometry and viewing geometry. The BRDF depends on wavelength and is determined by both the structural and optical properties of the surface (Nicodemus, 1965).

14.5 TRANSLUCENCY

Translucency occurs between the extremes of complete transparency and complete opacity. Within the concept of total appearance, translucency has an important part to play because an object may 'appear' different depending not just on its colour, but also on the appearance of that colour affected by the relationship between the light transmitted, the light reflected, and the light scattered by the body of the object (Hutchings and Luo, 2005).

In order to understand the relationship between, for example, the structure of many food products and the consumer perception of quality, it has been found necessary to understand the part played by translucency (Hutchings, 1999).

If it is possible to see an object or scene through a material then that material is said to be transparent. If it is possible to see only a 'blurred' image through the material then it has a degree of transparency the extent of which is a property of the particular material (Willmouth, 1976; Willmouth, 1986). This blurring, or loss of information, is caused by the diffusion of light as it passes through the material. If it is obvious that no light is being transmitted by the material then it is said to be opaque, and opacity can be considered a property of the material whereby the passage of light is inhibited, i.e. the opposite of transparency.

Other terms used to describe similar visual effects include:

- clarity, defined in terms of the ability to perceive the fine detail of images through the material;

- haze, defined as a property of the material whereby objects viewed through it appear to be reduced in contrast;
- and translucency, a property of the material by which it transmits light diffusely without permitting a clear view of objects beyond it.

These terms all imply a scattering or diffusing mechanism within the material, but there is an important distinction between clarity and haze. Consider a target that consists of a series of sets of black and white bars and each set is of a different spatial frequency (Webber, 1957). For a material with a high value of clarity and a low value of haze, it will be possible to discern a high spatial frequency pattern irrespective of the contrast between the black and the white bars at the highest discernible frequency. For a material with a high value of haze but low value of clarity it will be possible to distinguish only a blurred image at the higher frequencies, because the contrast between the black and white bars appears much reduced. Thus, the concept of translucency can perhaps be regarded as a descriptor of the combined effects defined above as clarity and haze. This implies that it is a more generic term and, perhaps, should be limited to use as a subjective term, keeping clarity and haze as descriptors of objective, or measurable, correlates. Turbidity, which is defined as the reduction in transparency because of the presence of particulate matter in the material is also important.

In general terms, the transparency of the material is determined by the balance between the directly transmitted light and the scattered light; it is the former that conveys information from which the visual perception of the image is derived, and the latter which degrades that image. Indeed, it might be thought that the use of the terms translucency, clarity, haze, opacity, etc., is superfluous and that *transparency* is both totally applicable and sufficient. It has been demonstrated above, however, that there is a clear distinction between the perceived contrast of an image and the detection of fine detail in that image: thus the terms haze and clarity are required. Indeed, it can be demonstrated that a series of materials that lead to a decrease in haze might show an increase in clarity.

Hunter and Harold, in their book *The Measurement of Appearance* (Hunter and Harold, 1987), attempt to bring order to this apparently complicated situation by defining four classes of object (surfaces):

Class of Object	*Dominant Light Distribution*
Diffusing surfaces (opaque, non-metals)	Diffuse reflection
Metallic surfaces	Specular reflection
Translucent materials	Diffuse transmission
Transparent materials	Specular transmission

This division is based on some simplifying assumptions. Not all objects fit clearly into one of the four categories because, for example, the specular and diffuse components of reflected and transmitted light are seldom completely separable.

14.5.1 Haze of transparent plastics

The definition of haze adopted by ASTM Test Method D1003 is in terms of the properties of the material (ASTM, 2007), rather than the effect those properties have on objects viewed through the material. An immediate advantage of this definition is that it leads to an obvious measurement technique and the ASTM Method describes a suitable device (Billmeyer and Chen, 1985; Weidner and Hsia, 1979). Four measurements are needed, made by mounting the sample at the input port of an integrating sphere with items on the opposite sphere wall:

i. the incident light, T_1, with no sample and a white standard on the sphere wall;

ii. the total light transmitted by the specimen, T_2, with the sample in place and a white standard on the sphere wall;

iii. the light scattered by the instrument, T_3, with no specimen and a light trap on the sphere wall;

iv. and the light scattered by the instrument and the specimen T_4, with the specimen in place and a light trap on the sphere wall.

Although not strictly necessary, it is usual to calculate the total luminous transmittance of the specimen, T_t, as well as the haze, H. The equations used are:

$$T_t = (T_2/T_1) \times 100\%$$

$$H = [(T_4/T_2) - (T_3/T_1)] \times 100\%$$

Note that if the diffuse transmittance, T_d, is defined by:

$$T_d = [T_4 - T_3(T_2/T_1)]/T_1$$

then the haze, H, can be defined by:

$$H = T_d/T_t \times 100\%$$

The spectral and geometric requirements detailed in the Standard are precise and must be adhered to in order that comparative measurements can be made.

14.5.2 Discussion – translucency

It is clear that the concept of translucency is a useful one, but it has become apparent that it means different things to those in different industries. Thus a single, simple, definition of translucency is unlikely to be achieved. The property referred to as translucency however, can be linked to other, objective, properties, such as opacity or transparency, and further research into the relationship between these is needed.

14.6 SURFACE TEXTURE

The fourth aspect of appearance assessment to be considered is that of surface texture. The concept of texture is intuitively obvious but is difficult to define. ASTM (ASTM, 2010a) gives the following definition:

> The visible surface structure depending on the size and organisation of small constituent parts of a material; typically, the surface of a woven fabric.

While this definition is satisfactory, it is by no means complete and is, perhaps, spoilt by the example. If this is a definition of surface texture, it is necessary to differentiate between texture associated with physical, topological, variability in a surface, and subsurface texture, texture associated with spatial variation in appearance caused by non-uniformity of colorant. Examples of the former are woven fabric and the surface of a metal, and of the latter is a metallic automobile finish where the variation caused by the spatial distribution of the discrete aluminium flakes is visible through an acrylic varnish overcoat which itself gives a smooth, high gloss, finish. It might also be necessary to differentiate between two-dimensional and three-dimensional variation in a surface. An example of the first might be a printed picture of a piece of wood and of the second, the surface of a piece of real wood. Another consideration is the difference between a pattern and surface texture. It might be held that a pattern is an inherent part of a surface whereas surface texture is concerned with the perception of that pattern and is thus a function of other variables including the direction of illumination and the viewing distance.

Hutchings suggests that, in Standard English, the word 'texture' refers to the character of the 'bumpiness' of a surface, i.e. to a mechanical discontinuity as distinct from a discontinuity of the apparent colour (Hutchings, 2003). Hence, he suggests that, relating to the deconstruction of appearance, we should term this bumpiness character as *surface texture*. Any discontinuity of colour can then be regarded as belonging to *colour uniformity* or *colour pattern*. The definition of whether something is bumpy (rough) or flat is based on visual perception alone. Therefore, surface texture, for its existence, depends on whether we can 'see' that the surface has a mechanical discontinuity. For example, in the region of the continuous circumference of a high gloss tomato, the surface may seem to have a degree of colour non-uniformity while having no element of surface texture.

In the past, surface texture has been assessed by the judgement of an inspector, either by eye or even by fingernail. In order to put a number to the surface texture, a more accurate means of measurement is required. A typical surface-measuring instrument will consist of a stylus with a small tip (the fingernail), a gauge or transducer, a traverse datum and a processor. The surface is measured by moving the stylus across the surface. As the stylus moves up and down along the surface, the transducer converts this movement into a signal which is then exported to a processor which converts it into a number and usually a visual profile. There are also systems that measure the profile of a surface without contact, such as confocal systems.

For correct data collection, the gauge needs to pass over the surface in a straight line such that only the stylus tip follows the surface under test. This is done using a straightness datum. This can consist of some form of datum bar that is usually lapped or precision ground to a high straightness tolerance. Two attributes are distinguished: roughness, and waviness. Roughness is a measure of the process marks produced during the creation of the surface and other factors such as the structure of the material, and waviness is a longer wavelength variation in surface away from its basic form (BS1134, 1998; BS1134, 1990; ISO 4287, 1970).

The human response to texture can be characterised using terms like fine, coarse, grained, smooth. Alternatively, texture can be described as a variation in tone intensity or lightness (Singh, Markou, and Singh, 2002) and structure. Other responses to a physical surface may be in terms of its roughness, smoothness, ripple, or orange peel, or its apparent mottle or speckle. Whichever scheme is used the words are attempting to describe

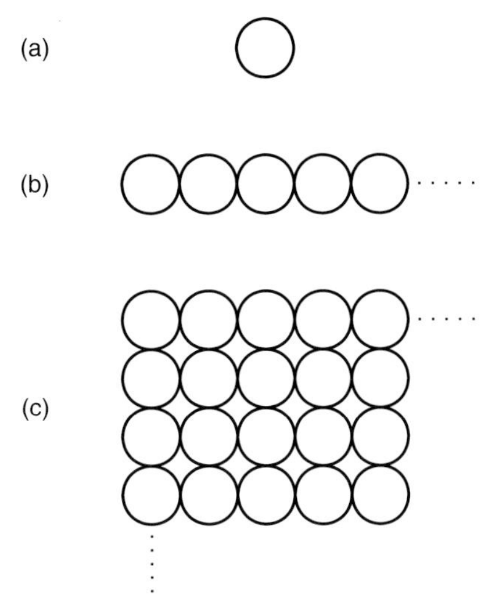

Figure 14.14 (a) Texture primitive; (b) Pattern generated by a line of primitives; (c) Two-dimensional texture pattern generated by an array of lines of primitives

a variation and it is pertinent to try to establish what it is that is actually varying. The building blocks of texture can be considered as texture elements – those elements of the physical surface that are perceived to be different (Gonzalez and Woods, 1993). These have variously been called texture primitives, texture elements, textons, and texels. Texture can then be described by the number and types of primitives, and by their relative spatial relationships, a process called texture classification. It should also be noted that the perception of physical texture is scale dependent; the viewing distance from the physical surface will influence the perception: it is also dependent on the angles of illumination and viewing (Dana, van Ginneken, Nayar, and Koenderink, 1999).

Consider a texture primitive, a circle, Figure 14.14 (a), that builds up into a line of circles, (b), and then an array of circles (c). This simple example shows the formation of a (regular) textured pattern based on the simple texture primitive, a circle. If this formation process could be reversed, and applied to more complex patterns (textures), to define the primitive and its spatial repeat, then a suitable analysis technique, texture segmentation, has been devised.

If texture is considered as a variation along a linear dimension of a material, then there are a number of techniques that can be used to provide numbers that relate to this variation. By including the analysis of several, parallel, lines then a spatial map can be built up. The advent of digital image processing techniques, based on the output from digital cameras, for example, has led to the easy availability of data to enable this process to be carried out and has, indeed, led to techniques that analyse the complete image data in a direct spatial sense. This, in turn, has led to the published literature concerned with texture analysis being prolific. There are applications, based on digital imaging, in the military (target recognition in camouflage situations, and terrain classification from aerial images), in machine vision (product detection during manufacture to find inadequate mixing in biscuits or other defects in the product), and in the medical world (for remote diagnosis and radiographic image interpretation). In these applications, it is sometimes possible to compare a sample image with a 'standard' image but in many circumstances, it is necessary to analyse the spatial distributions in the actual image to detect the required perturbation.

It should be noted that, in some applications, it is the detection of objects in the image that is important (pattern recognition); in other applications, it is an analysis of the noise pattern that requires an understanding of the texture (texture segmentation).

14.6.1 The psychophysics of texture

One reason it is important to study the psychophysics of texture perception is that the performance of various texture algorithms is evaluated against the performance of the human visual system doing the same task (Tuceryan and Jain, 1998; Landy and Graham, 2004). Julesz has studied texture perception extensively in the context of texture discrimination (Julesz, 1965; Julesz, 1975; Julesz, 1981). The question posed was 'When is a texture pair discriminable, given that they have the same brightness, contrast and colour?' Julesz concentrates on the spatial statistics of the image grey levels that are inherent in the definition of texture by keeping other illumination-related properties the same. The results of the experiments suggested that two textures are not discriminable if their second-order statistics are identical. (First-order statistics are to do with the likelihood of observing a specific pixel value at a random position in an image, second-order statistics are defined as the likelihood of observing a pair of pixel values at two defined locations in the image.) Thus, Julesz favours an explanation of texture discrimination in terms of spatial distribution. Beck, Prazdny, and Rosenfeld (Beck, Prazdny, and Rosenfeld, 1983), however, argue that the perception of texture is primarily a function of spatial frequency and not the result of grouping of spatial information.

14.6.2 Illumination and texture

Chantler argues, with some justification, that the appearance of surface texture is dependent on the direction and angle of the illumination (Chantler, 1995; Chantler and McGunnigle, 2000; Gonzalez, 2002). Thus, many researchers recognise that the texture appears different when viewed from different directions but do not consider that the texture will also appear different when viewed from a fixed direction, but under variable direction and angle of illumination. Chantler shows that the use of directed illumination during image capture, can act as a directional filter of texture, and that the directional characteristics of the texture are not just a function of surface relief but are affected by the angle of the illumination. Chantler further shows that the use of the analysis of stereo pairs, with their associated differences in illumination and viewpoint, gives advantages in surface texture classification (McGunnigle and Chantler, 2000).

14.6.3 Analysis techniques for texture

Overall, texture analysis can be considered under two headings: syntactic and analytical, as shown in Figure 14.15. The former is essentially subjective and, as discussed above, there are, as yet, no common perceptual scales that can be assigned to such an analysis (Lu and Fu, 1978). On the other hand, digital images can be thought of as arrays of data

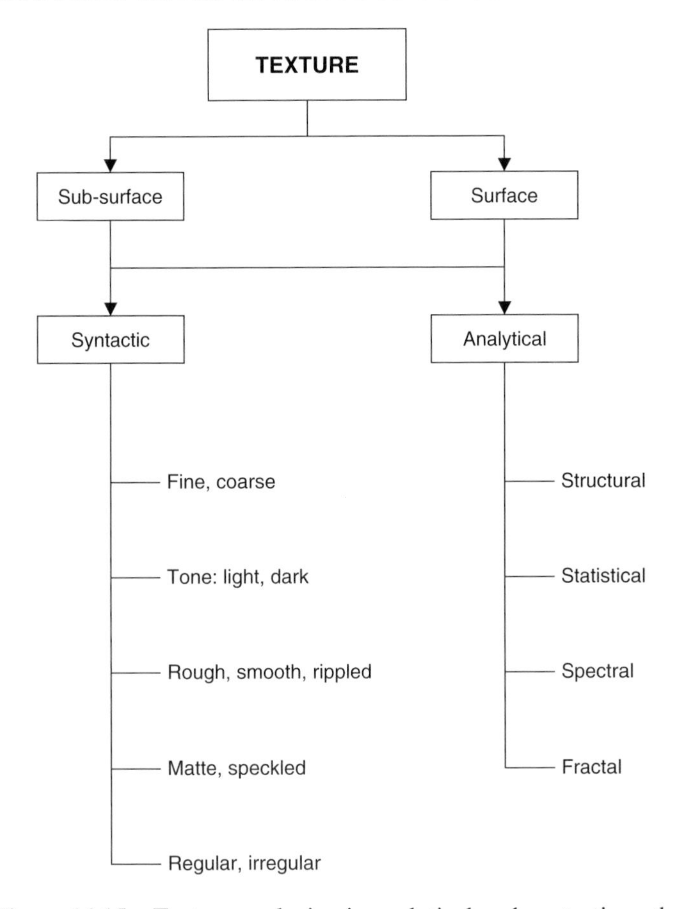

Figure 14.15 Texture analysis via analytical and syntactic paths

representing the spatial world and, as such, they lend themselves to computational analysis and a number of techniques have been proved useful.

Those methods described in the literature are called by different names by different authors but they can loosely be discussed under the two headings:

- structural – where the texture is considered to be defined by sub-patterns, called *primitives* (examples are those methods based on autocorrelation and spatial frequency (Fourier) analysis);
- statistical – where the texture is defined by a set of statistics extracted from the entire textural region (examples are methods based on co-occurrence matrices or on run length).

The different methods capture the inherent coarseness or fineness of texture (Haralick, 1979; Sharma, Markou and Singh, 2001; Singh, and Singh, 2002). Fractal based texture classification is another approach that attempts to find correlation between texture coarseness and fractal dimension.

14.6.4 Spatial texture: uniformity and colour difference

The CIELAB colour-difference metric, ΔE^*_{ab}, is suitable for measuring colour difference of uniform colour targets that subtend at least 2° visual angle in size (CIE, 2004). Most real images, however, are not made up of large uniform areas and many psychological studies have shown that discrimination and appearance of small-field, fine-patterned colours, differ from similar measurements made using large uniform fields (Noorlander and Koenerdick, 1983; Poirson and Wandell, 1993; Bäuml and Wandell, 1996). Therefore, applying CIELAB ΔE^*_{ab} to predict local colour reproduction errors in patterned images does not give satisfactory results. For example, when a continuous-tone coloured image is compared with a half-tone version of the image, a point-by-point computation of the CIELAB difference produces large colour differences at most image points. Because the half-tone dots vary rapidly in a spatial sense, these differences are blurred by the eye and the reproduction may still preserve the appearance of the original. In general, as the spatial frequencies present in the target increase, to give finer variation in a spatial sense, colour differences become harder to see, especially those differences that align in the blue-yellow colour direction.

The S-CIELAB metric has been proposed as a measure of perceptual colour fidelity to include this perceived blurring and, when applied to an image containing structure or texture, may give a more realistic measure of spatial colour variation (Zhang and Wandell, 1996; Zhang, Farrell, and Wandell, 1997; Zhang and Wandell, 1998). The S-CIELAB metric adds a spatial pre-processing step to the standard CIELAB ΔE^*_{ab} metric to account for the spatial-colour sensitivity of the human eye. The S-CIELAB calculation is illustrated in Figure 14.16 (Zhang and Wandell, 1996). The key components are the colour transformation from a three colour-channel system to a three opponent-channel system comprising luminance, L, red-green, C_1, and blue-yellow, C_2, channels; this is followed by two-dimensional spatial filtering algorithms based on the contrast sensitivity functions

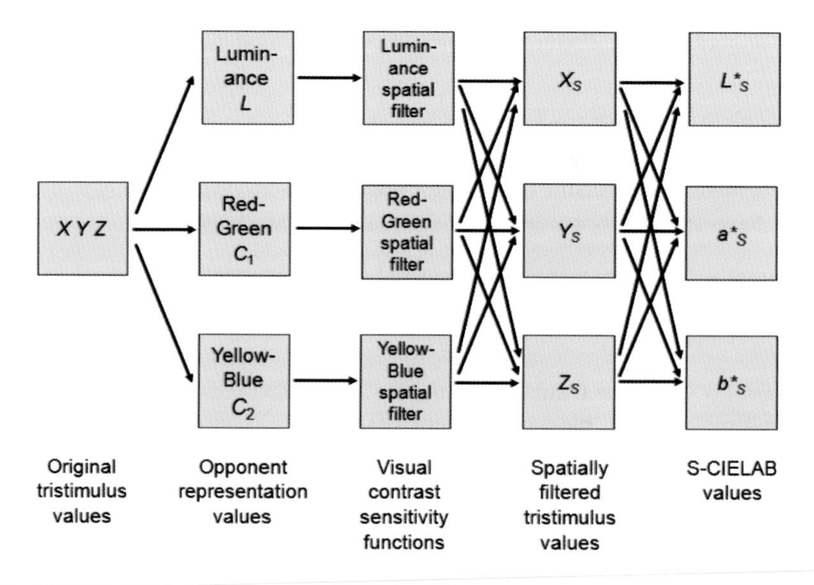

Figure 14.16 A flow chart showing the steps to compute S-CIELAB

for each channel, before inverse transformation back to CIE X, Y, Z, for subsequent calculation of CIELAB ΔE^*_{ab}. It is also possible to use a colour difference formula other than the CIELAB formula, such as CIEDE2000 described in Section 3.11 (Johnson and Fairchild, 2002).

14.6.5 Texture and colour

A further weakness in most discussions of texture is that it has been assumed that the perception of texture is a tonal effect and that chromaticity is of secondary importance. When dealing with colour and colour-difference however, the chromaticity of the texture may have a great impact on the perceived difference in colour between two samples. Xin, Shen, and Lam showed that increasing the texture of pairs of samples increased the acceptability of colour differences (Xin, Shen, and Lam, 2005). Thus, texture is an additional parametric effect that might need to be incorporated in future colour-difference metrics. The assessment of colour and colour difference in the presence of varying 'coarseness' of metallic automobile coatings has been investigated (Ji, Luo, and Kirchner, 2005; Kitaguchi, Westland, and Luo, 2005; Han, Cui, and Luo, 2003; Shao, Xin, and Chung Fu-Lai, 2003). Suzuki and Baba adopted a different approach in that they considered the colour appearance of textile samples by relating their perception with measurements made using a goniospectrophotometer (Suziki and Baba, 2005). They also considered the local gloss or micro-brilliance of textured paint surfaces (Arai and Baba, 2005).

14.6.6 Discussion – surface texture

The above descriptions serve to show that surface texture is a complex subject that has no unique mathematical interpretation and certainly no defined method of measurement. The advent of high-speed computers with suitable digital imaging software has opened up the possibility of analysing images of surface and sub-surface texture and obtaining suitable measures based on the statistics of those images. It has been argued that first-order statistics, those based on the likelihood of observing a particular pixel value at a specified location, can be correlated with the perception of texture. Second-order statistics, based on the likelihood of perceiving a difference in pixel value at two specified locations, however, do not correlate well with visually perceived texture (Julesz, 1981). That this is so is unfortunate, because the more powerful techniques, for example, the autocorrelation function derived by Fourier analysis, have the ability to reduce greatly the amount of data to a more manageable form. One rider to the above conclusion needs to be noted. Much of the early work on 'texture analysis' was carried out using patterns of information produced by a computer, and these were often patterns of regular arrays of characters. This is not related, except in exceptional circumstances, to the real world.

An example of a successful application of image analysis to a surface texture problem is that discussed by Xu (Xu, 1996) and by Lee and Sato (Lee and Sato, 2001) who discuss fabric appearance. This is routinely tested during textile and apparel manufacture in both a quality-control and a product-improvement environment. Traditionally it is evaluated by observers following a number of standardised test methods (ASTM, 2010b). These methods are however, characterised by poor accuracy because of their subjective nature. Using a digital camera-based system, wrinkle (an undesirable crease of short and irregular deformation), stain (a discoloured area), pills (bunches or balls of tangled fibres which are held on the surface of the fabric by one or more fibres), crease (a break or line in a fabric, generally caused by a sharp fold inserted intentionally by the application of pressure, heat or moisture), smoothness (similar to wrinkle but measured over larger areas), and colour, were all successfully assessed to give an overall fabric grading.

Another successful application of image analysis involves the characterisation of the topography of the surface of paper for smoothness and porosity (Chinga, Gregersen, and Dougherty, 2003). Suitable filtering of images enabled measures to be obtained that correlated with conventional laser profilometry measurements. A further example involves the prototyping and transforming of facial textures for perception research. It is found that the apparent morphology of the perception of a face with age is compounded by the presence of features that relate to age, notably stubble and wrinkles. Using wavelet analysis, it has proved possible to model the changes and improve the ability to transform facial images (Tiddeman, Burt, and Perret, 2001).

A weakness of many systems for the analysis of texture is that the human visual system is not considered. For a complete and effective system, it can be imagined that the viewing distance and the size of the image (or sample) should be required knowledge. This is because texture that is visible on a sample viewed at one distance may not be visible if the viewing distance is increased. The property of the visual system that is pertinent to this phenomenon is the contrast-sensitivity function.

The contrast-sensitivity function (CSF) provides a theoretical framework whose aim is to account for the visual detection and spatial discrimination of luminance and chromaticity in a defined form. The basic idea is that spatial information from any small area of the retinal image passes through an array of parallel, linear, spatial filters. The current consensus is that any given filter responds most strongly to a target (such as a bar) of specific width and orientation, and that different filters prefer different widths and orientations. Because the spatial characteristics of a bar can be described by its width and height or by spatial frequency, the filters are alternatively described as being either tuned to size and orientation, or as being tuned to spatial frequency and orientation (Campbell and Robson, 1968; Blakemore and Campbell, 1969). It should be noted that much of the evidence for spatial-frequency channels is based on psychophysical measurements carried out at contrast levels near detection threshold. The extent to which the theoretical position has to be modified to account for vision in the everyday environment is unclear: in everyday vision, contrast levels are commonly well above threshold, thus giving rise to the possibility of interactions between the filters and a consequent breakdown of their independence. Indeed, there is psychophysical evidence that, at suprathreshold levels of contrast, nonlinear interactions between differently tuned spatial filters are by no means negligible in human observers (DeValois, Morgan, and Snodderly, 1974). It has been found however, that the CSF provides a simple and robust method when used in image-difference metrics and machine vision applications, and, until a more complete model of human-visual response is available, the CSF will play a fundamental role in image analysis (Kitaguchi, Westland, Owens, Luo, and Pointer, 2004). An interesting application of texture (spatial pattern) analysis is found in work to assess the impact of dental fluorosis which occurs during the development of teeth and is caused by over-exposure to fluoride. The visual effect is of yellowing of the teeth, white spot and pitting or mottling of the enamel. A discrimination experiment was conducted whereby a number of images of the same face were digitally altered to simulate the effects of fluorosis at four levels viewed at five different distances. These images were viewed by over 200 observers and the acceptability correlated with the simulated viewing distance and the degree of mottling (Edwards, MacPherson, Simmons, Harper, and Stephan, 2005).

14.7 CONCLUSIONS

Starting from a definition of appearance, this chapter has attempted to describe the components of a framework on which a set of measurements can be made to provide important correlates of visual appearance. It is recognised that the interactions between the various components of the framework are complex, that physical parameters relating to objects are influenced, at the perception stage, by the physiological response of the human visual system and, in turn, by the additional psychological aspects of human learning, pattern, culture and tradition. By dealing with the optical properties of materials it is seen that there are, perhaps, four headings under which possible measures might be made: colour, gloss, translucency and surface texture. It is recognised that these measures are not necessarily independent; colour may influence gloss, colour will certainly influence translucency, and surface texture is probably a function of all three of the other measures.

REFERENCES

Albers, J. *Interaction of Color*, Yale University Press, New Haven, CT, USA (1975).

Arai, H., and Baba, G., Micro-brilliance of anisotropic paint surfaces, *Proceedings of the 10th Congress of the International Colour Association Color '05*, pp. 813–816, Grenada, Spain (2005).

ASTM, *Annual Book of ASTM Standards: Textiles*, Vol. 7-01, ASTM International, West Conshohocken, PA, USA (2010a).

ASTM D523, *Standard test method for specular gloss*. ASTM International, West Conshohocken, PA, USA (2008).

ASTM D1003, *Standard test method for haze and luminous transmittance of transparent plastics*. ASTM International, West Conshohocken, PA, USA (2007).

ASTM E284, *Standard terminology of appearance*, ASTM International, West Conshohocken, PA, USA (2010b).

Barkas, W. W., Analysis of light scattered from a surface of low gloss into its specular and diffuse components, *Proc. Phys. Soc.*, **51**, 274–295 (1939).

Bäuml, K.-H., and Wandell, B. A., Color appearance of mixture gratings, *Vision Res.*, **36**, 2849–2864 (1996).

Beck, J., Prazdny, K., and Rosenfeld, A., A theory of textural segmentation, In Beck, J., Hope, B., and Rosenfeld, A., eds., *Human and Machine Vision*, pp. 1–38, Academic Press, New York, NY, USA (1983).

Béland, M.-C., and Bennett, J. M., Effect of local microroughness on the gloss uniformity of printed paper surfaces, *Appl. Optics*, **39**, 2719–2726 (2000).

Billmeyer, F. W., and Chen, Y., On the measurement of haze, *Color Res. Appl.*, **10**, 219–224 (1985).

Blakemore, C., and Campbell, F. W., On the existence of neurons in the human visual system selectively sensitive to the orientation and size of retinal images, *J. Physiol.*, **203**, 237–226 (1969).

BS1134, *Assessment of surface texture, Part 1, Methods and instrumentation*, British Standards Institution, London, England (1998).

BS1134, *Assessment of surface texture, Part 2, Guidance and general information*, British Standards Institution, London, England (1990).

BS EN ISO 2813, *Paints and varnishes. Determination of specular gloss of non-metallic paint films at 20°, 60° and 85°*, British Standards Institution, London, England (2000).

Byk Gardner Technical Note, see www.bykgardner.com. (2000).

Campbell, F. W., and Robson, J. G., Applications of Fourier analysis to the visibility of gratings, *J. Physiol. (Lond.)*, **181**, 576–593 (1968).

Chantler, M. J., Why illuminant direction is fundamental to texture analysis, *IEE Proc.-Vis. Image Signal Process.*, **142**, 199–205 (1995).

Chantler, M. J., and McGunnigle, G., The response of texture features to illuminant rotation, ICPR 2000, pp. 943–945, *15th International Conference on Pattern Recognition*, Barcelona, Spain (2000).

Chevreul, M. E., *De la loi du contraste simultané des couleurs et de l'assortiment des object colorés*, Paris, France (1839).

Chinga, G., Gregersen, Ø., and Dougherty, B., Paper surface characterisation by laser profilometry and image analysis, *Microscopy and Analysis*, **17**, 21–23, July (2003).

CIE Research Note, *Evaluation of the attribute of appearance called gloss*, In *CIE Journal*, **5**, 41–56, Commission Internationale de l'Éclairage, Vienna, Austria (1986).

CIE Publication 175:2006, *A framework for the measurement of visual appearance*, Commission Internationale de l'Éclairage, Vienna, Austria (2006).

CIE Publication 15:2004, *Colorimetry*, Commission Internationale de l'Éclairage, Vienna, Austria (2004).

Dana, K. J., van Ginneken, B., Nayar, S. K., and Koenderink, J. J., Reflectance and texture of real-world surfaces, *ACM Transactions on Graphics*, **18**, 1–34 (1999).

DeValois, R. L., Morgan, H., and Snodderly, D. M., Psychophysical studies of monkey vision – III. Spatial luminance contrast sensitivity tests of macaque and human observers, *Vision Res.*, **14**, 75–81 (1974).

Adelston, E. H., Perceptual organization and the judgment of brightness, Science, **262**, 2042–2044 (1993).

Edwards, M., MacPherson, L. M. D., Simmons, D. R., Harper Gilmour, W., and Stephan, K. W., The simulation of dental fluorosis, *Community Dent. Oral Epidemiol.*, **33**, 298–306 (2005).

Evans, R. M., *An Introduction to Color*, page 192, Wiley, New York, NY, USA (1948).

Gonzalez, A., *Model-based texture classification under varying illumination*, Heriot Watt University, Department of Computing and Electrical Engineering, Research Memorandum RM/02/8, Edinburgh, Scotland, September (2002). See: www.cee.hw.ac.uk.

Gonzalez, R. C., and Woods, R. E., *Digital Image Processing*, Addison-Wesley, Reading, MA, USA (1993).

Han, B., Cui, G. H., and Luo, M. R., Texture effect on evaluation of colour difference, *AIC Mid Term Meeting*, pp. 176–180, Bangkok, Thailand (2003).

Haralick, R. M., Statistical and structural approaches to texture, *Proc. IEEE*, **67**, 786–804 (1979).

Hunt, R. W. G., The strange journey from retina to brain, *J. Roy. Television Soc.*, **11**, 220–229 (1967).

Hunter, R. S., and Harold, R. W., *The Measurement of Appearance*, 2nd *Ed.*, Wiley, New York, NY, USA (1987).

Hunter, R. S., and Judd, D. B., Development of a method of classifying paints according to gloss, *ASTM Bulletin*, *No.39*, pp 11–18, ASTM International, West Conshohocken, PA, USA, March (1939).

Hutchings, J. B., *Food Colour and Appearance*, 2nd *Ed.*, Aspen Publishers, New York, NY, USA (1999).

Hutchings, J. B., *Expectation and the Food Industry*, Kluwer Academic/Plenum Publishers, New York, NY, USA (2003).

Hutchings, J. B., and Luo, M. R., Translucency, its perception and measurement, *AIC Congress - Color '05*, 10th Congress of the International Colour Association, pp. 835–838. Grenada, Spain (2005).

ISO 4287, *Geometrical Product Specification (GPS) – Surface texture – Profile method – Terms, definitions and surface texture parameters*, International Organisation for Standardisation, Geneva (1997).

Ji, W., Luo, M. R., and Kirchner, E., Assessing colour appearance and colour difference for automobile coatings – methods for assessing coarseness, *AIC Congress - Color '05*, pp. 631–634, 10th Congress of the International Colour Association, Grenada, Spain (2005).

Ji, W., Luo, M. R., Li, C., Cui. G., and Pointer, M. R., Methods of assessing gloss, *Proceedings of the 3rd Conference on Colour, Graphics, Imaging and Vision, CGIV 2004*, Aachen, Germany (2004).

Ji, W., Pointer, M. R., Luo, M. R., and Dakin, J., Gloss as an aspect of the measurement of appearance, *J. Opt. Soc. Amer. A*, **23**, 22–33 (2006).

Johnson, G. M., and Fairchild, M. D., A top down description of S-CIELAB and CIED2000, *Color Res. Appl*., **28**, 425–435 (2003).

Julesz, B., Texture and visual perception, *Scientific American*, **212**, 38–48 (1965).

Julesz, B., Experiments in the visual perception of texture, *Scientific American*, **232**, 34–43 (1975).

Julesz, B., Textons, the elements of texture perception, and their interactions, *Nature*, **290**, 91–97 (1981).

Kitaguchi, S., Westland, S., and Luo, M. R., Suitability of texture analysis for perceptual texture, *Proceeding of the 10th Congress of the International Colour Association Color '05*, pp. 923–926, Grenada, Spain (2005).

Kitaguchi, S., Westland, S., Owens, H., Luo, M. R., and Pointer, M. R., *Surface Texture: A Review*, National Physical Laboratory Report DQL-OR 006. National Physical Laboratory, Teddington, England (2004).

Konda, A., and Harold, R., Measuring the effect of substrate properties on color inkjet images, *GATF World*, Sep/Oct, 31–36 (1999).

Landy, M. S., and Graham, N., Visual perception of texture, In Chalupa, L. M. and Werner, J. S. (Eds.), *The Visual Neurosciences*, pp. 1106–1118, MIT Press, Cambridge, MA, USA (2004).

Lee, W., and Sato, M., Visual perception of texture of textiles, *Color Res. Appl*., **26**, 469–477 (2001).

Lindstrand, M., Instrumental gloss characterisation – in the light of visual evaluation – a review, *J. Imag. Sci. Tech*., **49**, 61–70 (2004).

Lu, S. Y., and Fu, K. S., A syntactic approach to texture analysis, *Computer Graphics and Image Processing*, **7**, 303–330 (1978).

McGunnigle, G., and Chantler, M. J., Rough surface classification using point statistics from photometric stereo, *Pattern Recognition Letters*, **21**, 593–604 (2000).

Nicodemus, F. E., Directional reflectance and emissivity of an object surface, *Appl. Optics*, **4**, 767–773 (1965).

Noorlander, C., and Koenerdick, J. J., Spatial and temporal discrimination ellipsoids in color space, *J. Opt. Soc. Amer*., **73**, 1533–1543 (1983).

Obein, G., Knoblauch, K., Chrisment, A., and Viénot, F., Perceptual scaling of the gloss of a one-dimensional series of painted black samples, *Perception*, **31**, 65 (2002).

Obein, G., Knoblauch, K., Viénot, F., Difference scaling of gloss: nonlinearity, binocularity and constancy, *J. of Vision*, **4**, 711–720 (2004).

Obein, G., Leroux, T., Knoblauch, K., and Viénot, F., Visually relevant gloss parameters, *Proceedings of the 11th International Metrology Conference*, pp. 20–24, Toulon, France (2003).

Obein, G., Leroux, T., and Viénot, F., Bi-directional reflectance distribution factor and gloss scales, *Proceedings of SPIE Human Vision and Electronic Imaging VI*, **4229**, 279–290 (2001).

Pfund, A. H., The measurement of gloss, *J. Opt. Soc. Amer.*, **20**, 23–26 (1930).

Poirson, A. B., and Wandell, B. A., Appearance of colored patterns: pattern-color separability, *J. Opt. Soc. Amer.*, **10**, 2458–247 (1993).

Sève, R., Problems associated with the concept of gloss, *Color Res. Appl.*, **18**, 241–252 (1993).

Shao, S., Xin, J. H., and Chung Fu-Lai, K., The effect of perceptual texture features on color variation of texture image, *Proceedings of the AIC Mid Term Meeting*, Bangkok, Thailand, pp. 209–214 (2003).

Sharma, M., Markou, M., and Singh, S., Evaluation of texture methods for image analysis, *Proceedings of the 7th Australia and New Zealand intelligent information systems conference*, pp. 117–121, Perth, Australia, September (2001).

Singh, M., Markou, M., and Singh, S., Colour image texture analysis: dependence on colour spaces, *Proceedings of the 5th International Conference on Pattern Recognition*, pp. 672–675, Quebec, Canada, August (2002).

Singh, M., and Singh, S., Spatial texture analysis: a comparative study, *Proceedings of the 5th International Conference on Pattern Recognition*, pp. 676–679, Quebec, Canada, August (2002).

Smith, K. B., A sharper look at gloss, *Surface Coatings International*, **80**, 573–576 (1997).

Suziki, K., and Baba, G., Prediction of color and appearance of textiles, *Proceedings of the 10th Congress of the International Colour Association Color '05*, pp. 817–820, Grenada, Spain (2005).

Tiddeman, B., Burt, M., and Perrett, D., Prototyping and transforming facial features for perception research, *IEEE Computer Graphics and Applications*, **21**, 42–50 (2001).

Tighe, B. J., Subjective and objective assessment of surfaces, In *Polymer Surfaces*, D. T. Clark and W. J. Feast (Eds.), Wiley, Chichester, England (1978).

Tingle, W. H., and Potter, F. R., New instrument grades polished metal surfaces, *Product Engineering*, 52–53, March 27 (1961).

Tuceryan, M., and Jain, A. K., Texture analysis, In *The Handbook of Pattern Recognition and Computer Vision, 2nd Ed.*, by Chen, C. H., Pau, L. F., and Wang, P. S. P. (Eds.), pp. 235–276, World Scientific Publishing, Hackensack, NJ, USA (1998).

Webber, A. C., Method for the measurement of transparency of sheet materials, *J. Op. Soc. Amer.*, **47**, 785–789 (1957).

Weidner, V. R., and Hsia, J. J., NBS reference hazemeter: its development and testing, *Appl. Optics*, **18**, 1619–1626 (1979).

Willmouth, F. M., Quantitative assessment of transparency, *Plastics and Rubber: Materials and Applications*, **1**, 101–108 (1976).

Willmouth, F. M., Transparency, translucency and gloss, In *Optical Properties of Polymers*, Meeten, G. H., ed., pp. 265–333, Elsevier, London (1986).

Wright, W. D., *The Measurement of Colour, 4th Ed.*, page 86, Hilger, London, England (1969).

Xin, J. H., Shen, H.-L., and Lam, C. C., Investigation of texture effect on visual colour difference evaluation, *Color Res. Appl.*, **30**, 341–347 (2005).

Xu, B., An overview of applications of image analysis to objectively evaluate fabric appearance, *Textile Chemist and Colorist*, **28**, 18–23 (1996).

Zhang, X., Farrell, J. E., and Wandell, B. A., Applications of a spatial extension to CIELAB, *Proc SPIEC conference: Very High Resolution and Quality Imaging II*, **3025**, 154–157 (1997).

Zhang, X., and Wandell, B. A., A spatial extension of CIELAB for digital image reproduction, *SID Digest '96*, **27**, 731–734 (1996).

Zhang, X., and Wandell, B. A., Color image fidelity metrics evaluated using image distortion maps, *Signal Processing*, **70**, 210–214 (1998).

15

The CIE Colour Appearance Model CIECAM02

15.1 INTRODUCTION

Most of this book has been concerned with internationally accepted procedures for colorimetry, as defined by the CIE. However, the first chapter was different in that it provided a description of the nature of the colour vision provided by the human eye and brain. The last chapters are devoted to describing models of colour vision which can be used for predicting the appearance of colours under a very wide range of viewing conditions. Models have been proposed by various workers (Seim and Valberg, 1986; Nayatani, Takahama, and Sobagaki, 1986; Nayatani, Hashimoto, Takahama, and Sobagaki, 1987; Nayatani, Takahama, Sobagaki, and Hashimoto, 1990; Nayatani, Sobagaki, Hashimoto, and Yano, 1997; Fairchild and Berns, 1993, Fairchild, 1996; Hunt and Pointer, 1985; Hunt, 1982, 1985, 1987, 1989, 1991, and 1994; Hunt and Luo, 1994; Luo, Lo, and Kuo, 1996). In these models, measures for related colours are derived that are intended to correlate not only with hue, saturation, lightness, and chroma, as in the CIELUV and CIELAB systems, but also with brightness and colourfulness. Such measures have already proved useful in some practical applications (Pointer, 1986; MacDonald, Luo, and Scrivener, 1990; Luo, Clarke, Rhodes, Schappo, Scrivener, and Tait, 1991; Attridge, Pointer, and Jacobson, 1997). The CIE has endorsed a model designated CIECAM02 (Moroney, Fairchild, Hunt, Li, Luo, and Newman, 2002; Li, Luo, Hunt, Moroney, Fairchild, and Newman, 2002; CIE, 2004), which is based on features drawn from all these models, and a description of this model is the main theme of this chapter. CIECAM02 is a development of the earlier model CIECAM97s, and versions of this were developed for unrelated colours, CAM97u, and for comprehensive use, CAM97c (Hunt, 2004).

Measuring Colour, Fourth Edition. R.W.G. Hunt and M.R. Pointer.

15.2 VISUAL AREAS IN THE OBSERVING FIELD

For related colours, five different visual fields are recognised in the model.

1. The colour element considered: typically a uniform patch of about 2° angular subtense.
2. The proximal field: the immediate environment of the colour element considered, extending typically for about 2° from the edge of the colour element considered in all or most directions.
3. The background: the environment of the colour element considered, extending typically for about 10° from the edge of the proximal field in all, or most directions. When the proximal field is the same colour as the background, the latter is regarded as extending from the edge of the colour element considered.
4. The surround: the field outside the background.
5. The adapting field: the total environment of the colour element considered, including the proximal field, the background, and the surround, and extending to the limit of vision in all directions.

The visual patterns of scenes viewed in practice are almost infinitely variable; but the phenomenon of colour constancy (see Sections 1.8 and 6.12) indicates that the effects of this variability on colour appearance are, to some extent, limited. The regime of fields described above is an attempt to simplify the situation sufficiently to make it feasible for modelling, while making it possible to include the most important factors that affect colour appearance. (The proximal field is not used in the CIECAM02 model described in this chapter, but is included for possible use in the future.)

If the colour element considered has an angular subtense of more than 4°, the proximal field can be thought of as extending for about the same angular subtense from the edge of the colour element, and tristimulus values for the CIE 1964 Standard Colorimetric Observer are used, the subscript 10 being attached to all the symbols; otherwise the tristimulus values used are for the CIE 1931 Standard Colorimetric Observer.

15.3 CHROMATIC ADAPTATION IN CIECAM02

The CAT02 chromatic adaptation transform described in Section 6.15 is used first to convert the tristimulus values, X, Y, Z, of test colours in a set of test conditions to tristimulus values, X_c, Y_c, Z_c, of corresponding colours, which define stimuli that have the same appearance in a set of reference conditions. Because it is necessary to model reduced levels of chromatic adaptation (Nayatani, 1997), the reference conditions must include a reference chromaticity relative to which test conditions can be considered more or less chromatic; this reference chromaticity is chosen to be the same as that of the equi-energy stimulus, S_E, because, it has been found that such stimuli appear achromatic to the dark-adapted eye (Hurvich and Jameson, 1951). In the model, the chromatic adaptation transform is therefore used to obtain corresponding colours in the equi-energy illuminant with the perfect diffuser as its reference white.

The CAT02 chromatic adaptation transform includes a parameter D, which is used to set the degree of adaptation taking place. D is calculated as:

$$D = F\left[1 - \left(\frac{1}{3.6}\right)e^{\left(\frac{-L_A-42}{92}\right)}\right]$$

where $F = 1$ for samples seen with a surround of luminance similar to the average luminance of the sample array; $F = 0.9$ for samples seen with a dim surround, and $F = 0.8$ for samples seen with a dark surround. A dim surround is regarded as being typical of when television displays are viewed in domestic environments; a dark surround would occur in typical cinema situations.

When observers attempt to identify the colours of surface objects, they can often make perceptual allowance for the colour of the prevailing illumination (McCann, McKee, and Taylor, 1976; McCann and Houston, 1983; Arend and Reeves, 1986). For instance, if an observer passes from an environment in which the illuminant is daylight to one in which the illuminant is tungsten light, then, although a piece of white paper generally appears to be yellowish in the tungsten light, it may still be correctly identified as a white, not as a yellow, object. This effect is sometimes referred to as discounting the colour of the illuminant. This mode of perception can be modelled by setting the D parameter in the CAT02 transform equal to unity. But if, on the other hand, no chromatic adaptation takes place, D can be set equal to zero. For general situations, D is formulated so that it increases towards unity as the adapting luminance increases. If partial discounting of the illuminant colour takes place, D can be set half way between its variable value and unity.

The R,G,B variables used in the CAT02 chromatic adaptation transform are not a set of cone responses, because it has been found that the use of a set of sharpened color-matching functions having some negative lobes (as shown in Figure 6.6) provides better predictions of experimental chromatic adaptation results.

The X, Y, Z tristimulus values obtained for the corresponding colours are then transformed into a set of cone responses.

15.4 SPECTRAL SENSITIVITIES OF THE CONES IN CIECAM02

The set of cone spectral sensitivities chosen is shown by the full lines in Figure 15.1; these curves are a linear combination of the colour-matching functions for the CIE 1931 Standard Colorimetric Observer, $\bar{x}(\lambda)$, $\bar{y}(\lambda)$, $\bar{z}(\lambda)$. For both the 1931 and the 1964 Standard Observers, the spectral sensitivities for the cones are obtained from the appropriate colour-matching functions by means of the following set of transformation equations:

$$\begin{aligned}
\rho &= 0.38971X_c + 0.68898Y_c - 0.07868Z_c \\
\gamma &= -0.22981X_c + 1.18340Y_c + 0.04641Z_c \\
\beta &= 1.00000Z_c
\end{aligned}$$

$\rho = \gamma = \beta = 100$ for the perfect diffuser white in illuminant S_E (the equi-energy illuminant) and this is referred to later as the PDE white.

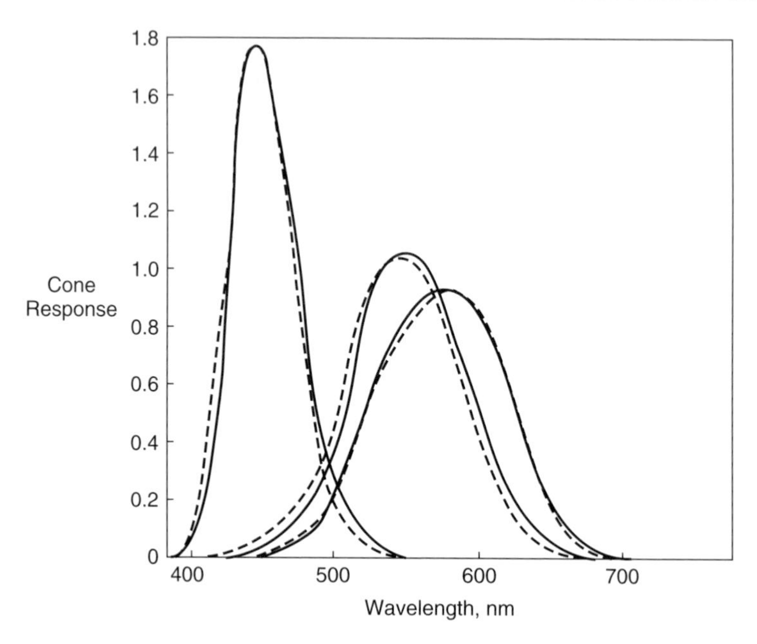

Figure 15.1 Spectral sensitivity functions (Hunt-Pointer-Estévez) used in the model for cone vision (full lines), compared with those obtained by Estévez (broken lines). These functions are for radiation incident on the cornea of the eye

The corresponding reverse set of transformation equations is:

$$X_c = 1.910197\rho - 1.112124\gamma + 0.201908\beta$$

$$Y_c = 0.370950\rho + 0.629054\gamma - 0.000008\beta$$

$$Z_c = 1.000000\beta$$

(The similar sets of equations given in Section 6.14 have different coefficients because they are normalised for D65 instead of for S_E.)

The broken lines in Figure 15.1 show the spectral sensitivities derived in a study by Estévez for 2° observations (Estévez, 1979). To reproduce these exactly would have required the use of colour-matching functions different from those of one of the CIE Standard Observers; this would have been very inconvenient for practical applications, and hence the approximation to these curves provided by the full lines has been used instead. The coefficients in the equations could have been chosen to generate a set of curves with the right-hand curve peaking at about 565 nm (instead of at about 575 nm), as suggested by some studies (Smith and Pokorny, 1972); this would make little difference to the predictions given by the model, but the unique hue criteria in Section 15.9 would become less simple.

The above set of transformation equations is used to derive the values of ρ, γ, and β, not only for colour-matching functions, but for any colour. The values of ρ, γ, and β, may be considered to be the amounts of radiation usefully absorbed per unit area of the retina by the three different types of cone in a given state of adaptation, for light incident on the cornea, after allowing for chromatic adaptation.

15.5 CONE DYNAMIC RESPONSE FUNCTIONS IN CIECAM02

Under a given set of viewing conditions, there will be a predictable relationship between the responses of the cones and the intensity of the stimulus (intensity denoting here simply the magnitude of the stimulus, not necessarily the flux per unit solid angle). There is much evidence to suggest that this relationship is non-linear. If the cone responses are taken as being proportional to the cube-root of the stimulus intensity, the curvatures of lines of constant hue in chromaticity diagrams can be predicted well using a simple criterion for constant hue (a square-root relationship has also been suggested, Hunt, 1982). A cube-root relationship would also result in a reduction in the dynamic range of the signals that have to be transmitted from the retina to the brain, and this seems likely on general grounds; thus, for example, a change in stimulus intensity of 1000 to 1 would only produce a change in cone response of 10 to 1. A simple cube-root relationship, however, cannot be correct for all stimulus intensities. When the intensity of the stimulus is very low, noise in the system must prevent extremely small cone responses from being significant; and, when the intensity of the stimulus is very high, the response must eventually reach a maximum level beyond which no further increase is possible (Baylor, 1987). These limits are illustrated by our inability to see modulations of colour in extremely dark objects, and by the tendency for very bright colours, such as lamp filaments seen through coloured filters, or coloured flares seen at close quarters, to appear pale or white. A hyperbolic function is therefore chosen to represent the response for the cones (as suggested by Seim and Valberg, 1986, and for which there is physiological evidence as described by Boynton and Whitten, 1970, and by Valeton and Van Norren, 1983). The responses given by the three different types of cone in a given state of adaptation are then formulated as:

$$\mathrm{f_n}(\rho) + 0.1 = [400(\rho/100)^{0.42}]/[27.13 + (\rho/100)^{0.42}] + 0.1$$

$$\mathrm{f_n}(\gamma) + 0.1 = [400(\gamma/100)^{0.42}]/[27.13 + (\gamma/100)^{0.42}] + 0.1$$

$$\mathrm{f_n}(\beta) + 0.1 = [400(\beta/100)^{0.42}]/[27.13 + (\beta/100)^{0.42}] + 0.1$$

The factor 100 positions the responses appropriately on the $\mathrm{f_n}$ function.

The +0.1 terms represent the noise. These responses give maximum values of 400.1 and minimum values of 0.1. In Figure 15.2, $\mathrm{f_n}(\rho) + 0.1$ is plotted against log $5L_A\rho/100$, where L_A is the adapting luminance, and $5L_A$ is used as representing the probable luminance of a white. In the lower figure, where $\log[\mathrm{f_n}(\rho) + 0.1]$ is plotted, it is clear that, over the central part of the curve, the response has a slope of 0.36 in log-log space, and therefore approximates a cube root relationship, as shown by the broken line. The responses $\log[\mathrm{f_n}(\gamma) + 0.1]$ and $\log[\mathrm{f_n}(\beta) + 0.1]$ would be represented by similar graphs. These functions are S-shaped, all having a curved toe and a curved shoulder.

15.6 LUMINANCE ADAPTATION IN CIECAM02

The actual response produced by the cones is dependent, not only on the intensity of the stimulus, but also on the state of adaptation of the eye. Adaptation usually provides an approximate compensation for the effects of changes in the level and colour of the

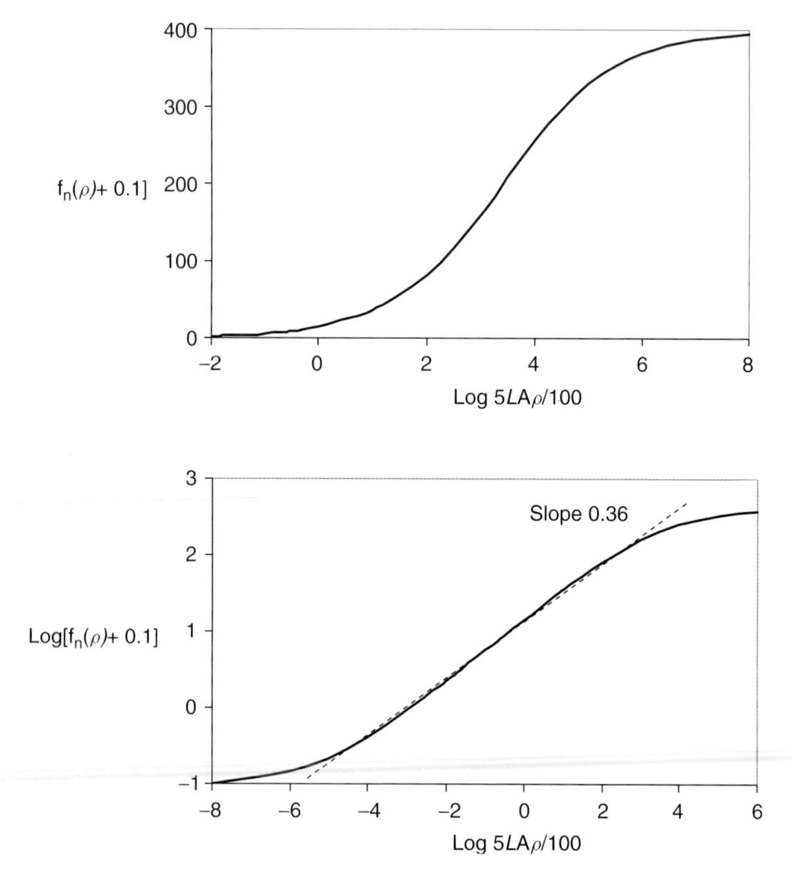

Figure 15.2 Cone response function. The function, $f_n(\rho) + 0.1$, is plotted against the log of $5L_A\rho/100$, where ρ is the radiation usefully absorbed by the ρ cones; $f_n(\rho) = [400(\rho/100)^{0.42}]/[27.13 + (\rho/100)^{0.42}]$. L_A is the adapting luminance, and $5L_A$ is used as representing the probable luminance of a white. Upper figure: $f_n(\rho) + 0.1$; lower figure: $\log_{10}[f_n(\rho) + 0.1]$

illumination, and this results in the phenomenon of colour constancy (see Sections 1.8 and 6.12). Compensation for the effects of changes in the colour of the illumination has already been made by means of the CAT02 chromatic adaptation transform (see Section 15.3), and a factor F_L is now introduced to model adaptation to changes in the level of the illumination. The cone responses after luminance adaptation are formulated as:

$$\rho_a = f_n(F_L\rho) + 0.1 = [400(F_L\rho/100)^{0.42}]/[27.13 + (F_L\rho/100)^{0.42}] + 0.1$$

$$\gamma_a = f_n(F_L\gamma) + 0.1 = [400(F_L\gamma/100)^{0.42}]/[27.13 + (F_L\gamma/100)^{0.42}] + 0.1$$

$$\beta_a = f_n(F_L\beta) + 0.1 = [400(F_L\beta/100)^{0.42}]/[27.13 + (F_L\beta/100)^{0.42}] + 0.1$$

where the function $f_n(I)$ is again of the form:

$$f_n(I) = [400(I/100)^{0.42}]/[27.13 + (I/100)^{0.42}]$$

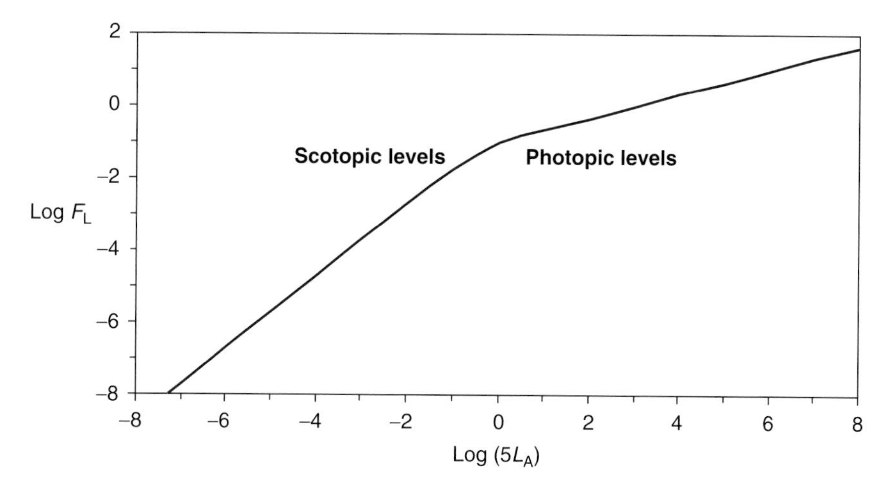

Figure 15.3 The log of the luminance-level adaptation-factor, log F_L, is plotted against log ($5L_A$), the log of five times the luminance, L_A, of the adapting field ($5L_A$ is taken to be the luminance of a typical white)

Objects look less bright in dim lighting. The factor, F_L, in the above equations, models this reduced luminance adaptation, and is defined as:

$$F_L = 0.2k^4(5L_A) + 0.1(1 - k^4)^2(5L_A)^{1/3}$$

where L_A is the luminance of the adapting field ($5L_A$ being regarded as the luminance of a typical white) and

$$k = 1/(5L_A + 1).$$

In Figure 15.3, log F_L is plotted against log ($5L_A$). The figure shows that, at photopic levels ($5L_A$ greater than 1, log ($5L_A$) greater than 0), F_L is approximately proportional to the cube root of $5L_A$ (slope of curve equal to 1/3), thus giving partial compensation for changes in adapting luminance; full compensation would occur if F_L were constant. At scotopic levels ($5L_A$ less than about 0.1, log ($5L_A$) less than −1) F_L is proportional to $5L_A$ (slope of curve equal to 45°) so that no compensation occurs.

15.7 CRITERIA FOR ACHROMACY AND FOR CONSTANT HUE IN CIECAM02

As mentioned in Section 1.6, there is a great deal of evidence that the responses from the three different types of cone are compared by neurons in the retina that result in colour difference signals being formed for subsequent transmission along the optic nerve fibres to the brain. We may represent these colour difference signals as:

$$C_1 = \rho_a - \gamma_a$$
$$C_2 = \gamma_a - \beta_a$$
$$C_3 = \beta_a - \rho_a$$

Almost certainly, their complements:

$$C'_1 = \gamma_a - \rho_a$$
$$C'_2 = \beta_a - \gamma_a$$
$$C'_3 = \rho_a - \beta_a$$

also exist, but, for simplicity, we will usually consider only C_1, C_2, and C_3.

Achromatic colours are those that do not exhibit a hue (such as whites, greys, and blacks). As suggested in Section 1.7, the criterion adopted for achromacy is:

$$\rho_a = \gamma_a = \beta_a$$

and hence

$$C_1 = C_2 = C_3 = 0$$

and colourfulness increases as C_1, C_2, and C_3, become increasingly different from zero.

The criterion for constant hue, as also suggested in Section 1.7, is:

$$C_1 \text{ to } C_2 \text{ to } C_3 \text{ in constant ratios.}$$

15.8 EFFECTS OF LUMINANCE ADAPTATION IN CIECAM02

In Figure 15.4, ρ_a is plotted against $\log(5L_A\,\rho/100) = \log(I)$, 100 being the value of ρ for PDE (the perfect diffuser white in the equi-energy illuminant S_E). If $5L_A$ is the luminance of the PDE, and the sample has the same chromaticity as the PDE (so that $\rho = \gamma = \beta$), then $5L_A\rho/100$ is equal to the luminance of the sample. The curves shown in Figure 15.4 are for values of log $(5L_A)$ equal to 5, 4, 3, 2, 1, and 0 log cd m^{-2} (full lines), and for dark adaptation (broken line).

Consider the curve labelled 3; this is for $\log(5L_A)$ equal to 3, and the open circle on this curve is for a colour having the same value of $\rho/100$ as for the PDE white, and the filled circle for a colour having $\rho/100$ equal to 0.03162 times that of the PDE white (that is, 1.5 less on the log scale). Relationships similar to those shown in the graph in Figure 15.4 also apply for γ_a and β_a. Hence, when $\log(5L_A) = 3$, the PDE white would be represented by points at the open-circle positions on curve 3 in all three graphs, and a colour having the same chromaticity as the PDE white, but a luminance 0.03162 times (1.5 log units) less, by points at the filled circles on curve 3 in all three graphs. The part of curve 3 between the open and filled points therefore represents the range of colours between a white (the PDE white) and a black (of luminance 3.162% of that of the PDE white), when log $(5L_A)$ is equal to 3 log cd m^{-2}. The position of the adapting field, L_A (taken as 1/5, that is 20%, of the luminance of the white, or 0.7 less on the log scale) is shown by the plus sign (+) on curve 3.

The other curves of the figure similarly represent the same range of colours for values of log $(5L_A)$ that become progressively smaller as the curve is displaced towards the left, and higher towards the right. Psychophysical and physiological studies show similar families of curves (Stevens, 1961; Valeton and Van Norren, 1983); progressive reduction of the

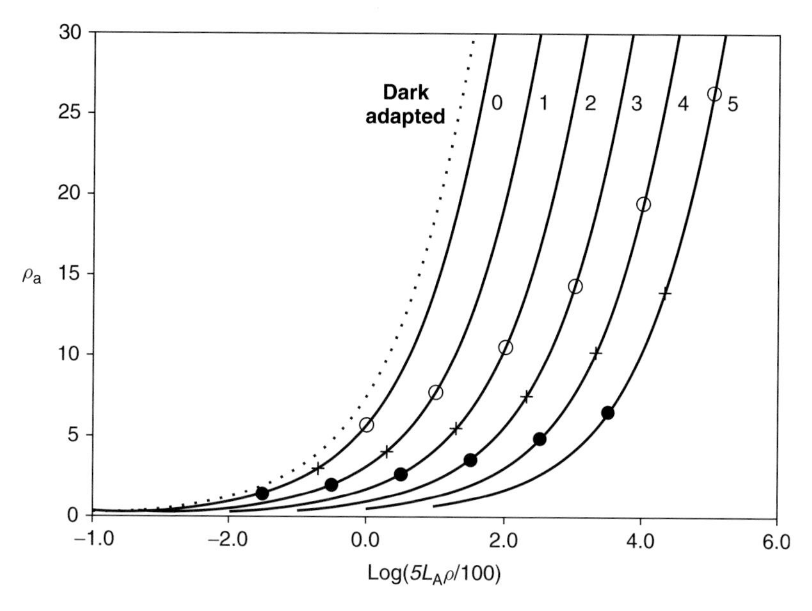

Figure 15.4 Response functions for the ρ cones. ρ_a is plotted against $\log[5L_A\rho/100] = \log(I)$, and L_A is the luminance of the adapting field in cd m^{-2}, for levels of $\log(5L_A)$ equal to 5, 4, 3, 2, 1, and 0 log cd m^{-2}, full lines, and for dark adaptation, broken line. Similar functions occur for the γ_a and β_a responses. Open circles: PDE white; filled circles: 3.162% black; plus signs: adapting field (luminance one fifth of that of the PDE white)

brightness of light colours and increase in the brightness of dark colours, as the level of illumination falls, is known as the *Stevens effect*.

The S-shaped nature of these curves predicts that, for colours of a given chromaticity, as the luminance factor is decreased, the colourfulness will usually decrease. This can be seen as follows. In Figure 15.5, curve 3 of Figure 15.4 is shown again, in this figure representing all three cone types, with vertical lines indicating three red colours of the same chromaticity but different luminance factors. The right-hand red and blue-green pair of vertical lines indicates a red colour having log I for ρ at the white level, and for γ and β at 0.4 less on the log scale. The centre pair of red and blue-green vertical lines has the same separation along the log I axis but is shifted 0.55 to the left, and therefore represents a colour of the same chromaticity but lower luminance factor. The left-hand pair of red and blue-green vertical lines has the same separation along the log I axis but is shifted a further 0.55 to the left, and therefore represents a colour of the same chromaticity but even lower luminance factor. As the luminance factor is decreased, the set of positions on the log I axis for each colour moves to the left; as a result, the responses come from parts of the curve having lower slopes. The difference between the ρ_a, and the γ_a responses therefore decreases, and hence C_1 decreases, as shown on the left, indicating reduced colourfulness (C_3 would also decrease). Thus, for a given chromaticity, as the luminance factor decreases, the colourfulness decreases. This is illustrated in Figure 8.26; in this figure, each vertical column of colours has the same chromaticity, but the luminance factor decreases from top to bottom of each colour; this is clearly accompanied by reduced colourfulness.

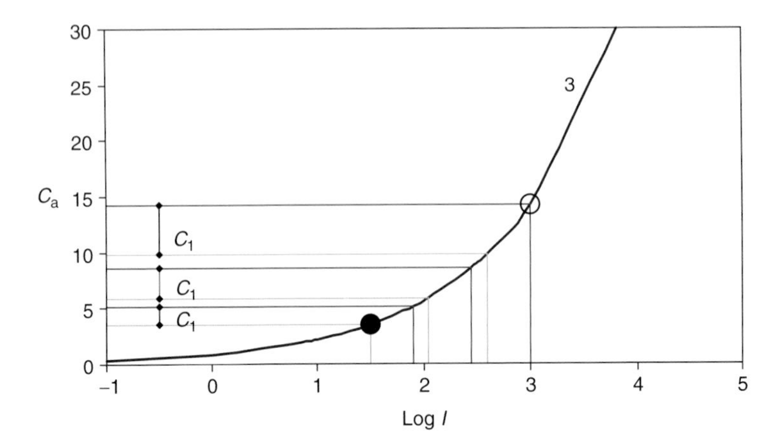

Figure 15.5 Curve 3 of Figure 15.4 representing the responses of all three cone types, indicated by C_a, together with representations of three stimuli having the same chromaticity but three different luminance factors. As the luminance factor decreases, the resulting colour difference signals also decrease

Returning to Figure 15.4 again, the following general features can be seen. As the luminance of the adapting field, L_A, decreases, the curves move to the left, indicating increasing sensitivity. But this movement is insufficient to provide full compensation, and hence the positions of the points representing white (O), the adapting field (+), and black (•) gradually move down each curve to regions of lower slope. This results in reductions in the differences in response between whites, adapting fields, and blacks. For colours, this results in reduced colourfulness. This is illustrated in Figure 15.6, where curves 1, 3, and

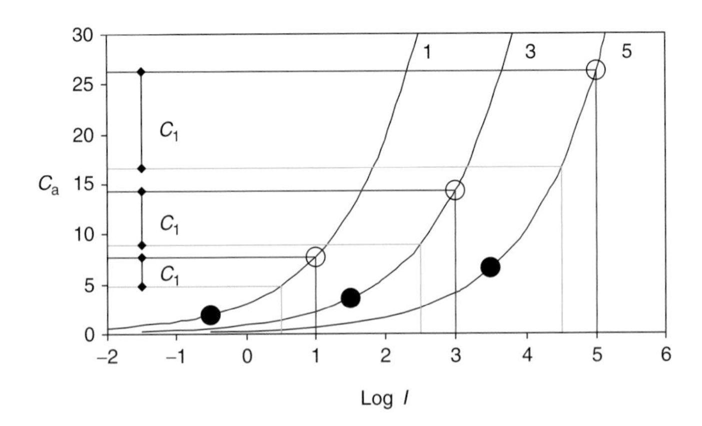

Figure 15.6 Curves 1, 3, and 5 of Figure 15.4, representing the responses of all three cone types, indicated by C_a, together with representations of a colour of the same chromaticity and luminance factor in the three different levels of adapting luminance (assuming the white has a luminance of five times that of the adapting luminance in each case). As the adapting luminance falls, the resulting colour difference signals also fall

5 of Figure 15.4 are reproduced. The red and blue-green vertical lines meeting curve 5 indicate a red colour having log I for ρ at the white level, and for γ and β at 0.4 less on the log scale; the corresponding value of C_1 is shown at the left. Similar vertical lines are shown meeting curves 3 and 1, and the corresponding values of C_1 are clearly smaller (as would also be the case for C_3). Hence the model predicts that, as the level of illumination is decreased, colours become of lower colourfulness, as is found in practice; this is sometimes referred to as the *Hunt effect* (Hunt, 1950, 1952, 1953.)

It is also clear from Figure 15.6 that the slopes of the curves near the black points (•) become very low for the curves towards the left; and this predicts that dark colours are difficult to distinguish in dim lighting, as is also found in practice.

For stimuli of luminances very much higher than that of the white, the responses may approach the maximum level, in which case they will also tend to be reduced in colourfulness. As already mentioned, there is a tendency for very bright colours, such as lamp filaments seen through coloured filters, or coloured flares seen at close quarters, to appear pale or white.

15.9 CRITERIA FOR UNIQUE HUES IN CIECAM02

As discussed in Section 8.6, there are four unique hues, red, green, yellow, and blue. The model predicts these as occurring at the following ratios of C_1 to C_2 or C_3 (because $C_1 + C_2 + C_3 = 0$, if one of these ratios is constant, the other will also be constant, and need not be specified in addition):

Unique red	$C_1 = C_2$
Unique green	$C_1 = C_3$
Unique yellow	$C_1 = C_2/11$
Unique blue	$C_1 = C_2/4$

The predictions given by these criteria are shown in Figure 15.7 by the full lines (for colours seen at high levels of adapting illumination, such as when L_A is around 200 cd m^{-2}, and illuminant S_E is used); the broken lines show the results obtained experimentally in the Natural Colour System (NCS, see Section 8.7), which are very similar. (The model does not include a representation of the *Bezold-Brücke phenomenon*, the change of hue produced by changing the luminance of a colour stimulus while keeping its chromaticity constant, but the magnitudes of these changes in hue are usually quite small.)

15.10 REDNESS-GREENNESS, *a*, AND YELLOWNESS-BLUENESS, *b*, IN CIECAM02

In the case of reddish colours, since the criterion for unique red is $C_1 = C_2$, it is to be expected that increasing departures from the unique hue condition, that is, increasing yellowness or blueness, would be indicated by increasing inequality of C_1 and C_2, that is, by $C_2 - C_1$ being increasingly different from zero. ($C_2 - C_1$ is used instead of $C_1 - C_2$ so that positive values indicate yellowness.) Similarly, because the criterion for unique green is $C_1 = C_3$, increasing yellowness or blueness of greenish colours would be indicated by $C_1 - C_3$ being increasingly different from zero (positive values indicating yellowness).

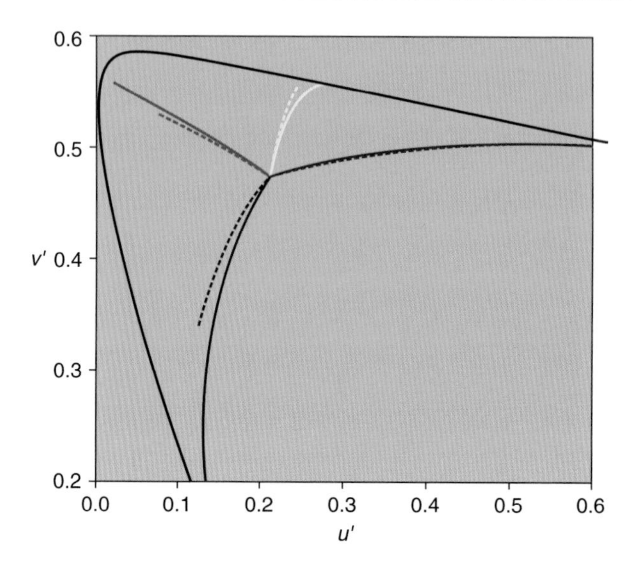

Figure 15.7 Unique hue loci predicted by the model (full lines) compared with those of the NCS (broken lines), for the equi-energy illuminant S_E. For the model, the hue loci shown are for an adapting luminance, L_A, of 200 cd m^{-2}. Because, as shown in Figure 15.2, the hyperbolic function approximates a simple power function over its central range, the chromaticities corresponding to loci of constant hue are not significantly affected by the level of the stimuli considered provided that they are in the central range. For the NCS, the luminance factors were the highest available in the system, and the tristimulus values were converted from those published for Illuminant C to the corresponding values for illuminant S_E using the CAM02 chromatic adaptation transform, with D set equal to 1.0

A measure of the yellowness or blueness of both reddish and greenish colours is therefore taken as the average of these two differences:

$$(C_2 - C_1 + C_1 - C_3)/2$$

and this is equal to

$$(C_2 - C_3)/2$$

(In Section 1.7, $C_2 - C_3$ was taken to indicate yellowness or blueness, and the use here of $(C_2 - C_3)/2$ is the same apart from the scaling factor of $1/2$.)

By similar arguments, redness or greenness of yellowish colours would be indicated by $C_1 - (C_2/11)$, and of bluish colours by $C_1 - (C_2/4)$; but, in this case, because the unique yellow hue is more sharply apparent than the unique blue hue, an average is not taken, and redness or greenness is taken to be indicated by:

$$C_1 - (C_2/11) = a.$$

(In Section 1.7, C_1 was used to indicate redness or greenness, and the use here of $C_1 - (C_2/11)$ is similar because $C_2/11$ is usually fairly small compared with C_1.)

It is now necessary to combine these correlates of yellowness-blueness and redness-greenness to obtain a measure of hue. But, because the number of β cones is only about 1/20th that of the ρ or γ cones (Walraven and Bouman, 1966), it is to be expected, on signal-to-noise ratio grounds, that the yellowness-blueness signal should have less weight than the redness-greenness signal; a factor of 1/4.5 (which is approximately equal to $1/20^{0.5}$) is used for this purpose, so that yellowness-blueness is taken to be indicated by:

$$[(C_2 - C_3)/4.5]/2 = b$$

Because, as can be seen from Figure 15.7, the red and green unique hue loci are not collinear, and the yellow and blue loci are also not collinear, the measures a and b are only approximate correlates of redness-greenness and yellowness-blueness. Accurate correlates of these attributes would require different formulations of a for yellowish and for bluish colours, and different formulations of b for red and for greenish colours (Hunt, 2007), but this elaboration is not included in CIECAM02.

15.11 HUE ANGLE, *h*, IN CIECAM02

A measure of hue is then obtained as the hue-angle, h, defined as:

$$h = \arctan(b/a)$$

where 'arctan' means 'the angle whose tangent is'. h lies between 0° and 90° if a and b are both positive; between 90° and 180° if a is negative and b is positive; between 180° and 270° if a and b are both negative; and between 270° and 360° if a is positive and b is negative.

15.12 ECCENTRICITY FACTOR, *e*, IN CIECAM02

The position of the achromatic point in contours of small constant saturation is eccentric (Hunt, 1985). The achromatic point becomes progressively nearer the contour as the hue considered is changed from yellow to red to green to blue; this is regarded as indicating increasing weight of perceptual colorization in the order yellow, red, green, and blue. To reflect this, an eccentricity factor, e_t is evaluated as:

$$e_t = [1/4][\cos(h\pi/180 + 2) + 3.8]$$

In Figure 15.8, e_t is shown plotted against h. The values of e_t for the unique hues are given here.

	Red	Yellow	Green	Blue	Red
i	1	2	3	4	5
h	20.14	90.00	164.25	237.53	380.14
e_t	0.774	0.723	0.988	1.198	0.774
e_i	0.8	0.7	1.0	1.2	0.8
H_i	0.0	100.0	200.0	300.0	400.0

The hue-angles, h, for the unique hues are determined from their ratios of C_1 to C_2 to C_3.

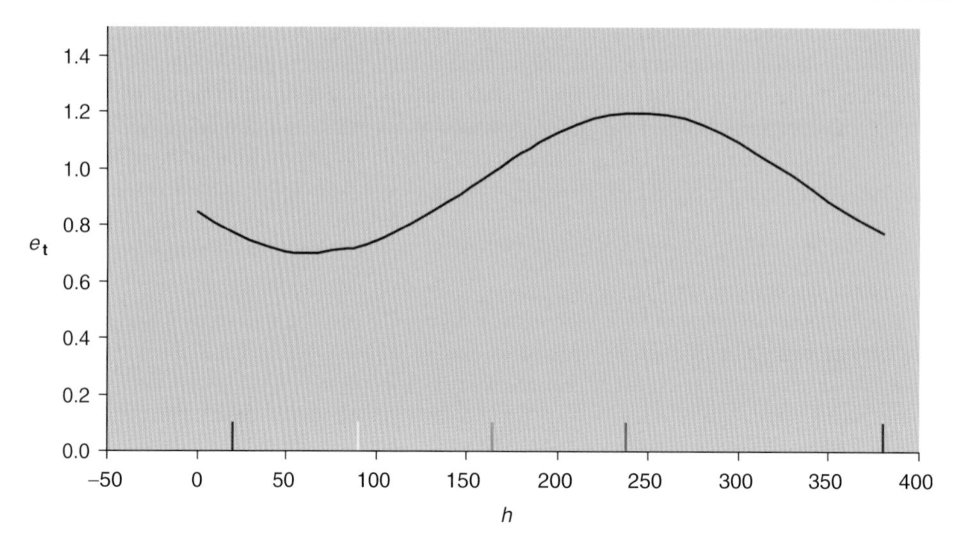

Figure 15.8 Eccentricity factor, e_t, plotted against hue angle (in degrees), h

15.13 HUE QUADRATURE, *H*, AND HUE COMPOSITION, H_c, IN CIECAM02

Hue can also be expressed in terms of the proportions of the unique hues perceived to be present, and the model provides a correlate of hue expressed in this way, Hue Quadrature, H. Hue Quadrature is computed by linear interpolation of the values of h and e_i shown in the table above; note that the values of e_i are slightly different from those for e_t. If h is less than h_1, then $h' = h + 360$, otherwise $h' = h$. A value of i is then chosen so that h_i is equal to or greater than h' and less than h_{i+1}. Hue Quadrature is then evaluated as:

$$H = H_i + 100[(h' - h_i)/e_i]/[(h' - h_i)/e_i + (h_{i+1} - h')/e_{i+1}]$$

The difference between hue angle, h, and hue quadrature, H, is illustrated in Figure 15.9, where the former is shown on the left, and the latter on the right. In the left half, is shown a plot of $(C_2 - C_3)/9 = b$ against $C_1 - (C_2/11) = a$. In this figure, the value of h is the angle between a horizontal line drawn from the origin towards the right and the line joining the origin to the point representing the colour considered. The positions of the unique hue lines are shown in this diagram by the full lines, R, Y, G, and B. The angular positions of the lines representing colours that are perceptually midway between adjacent pairs of unique hues (that is, appearing to contain 50% of each of the two hues) are shown by the broken lines; these broken lines are not at equal angular spacing between the lines representing the unique hues in the figure on the left, because of the effect the different colorising weights of the red, yellow, green, and blue unique hues. (In the case of the red-blue quadrant, because of its larger size in the diagram, broken lines are shown that divide it into four perceptually equal parts.) In the case of hue quadrature, H (on the right), the effects of these weights have been included in the derivation of H, and hence the broken lines are spaced at regular intervals. However, because unique red and green are placed

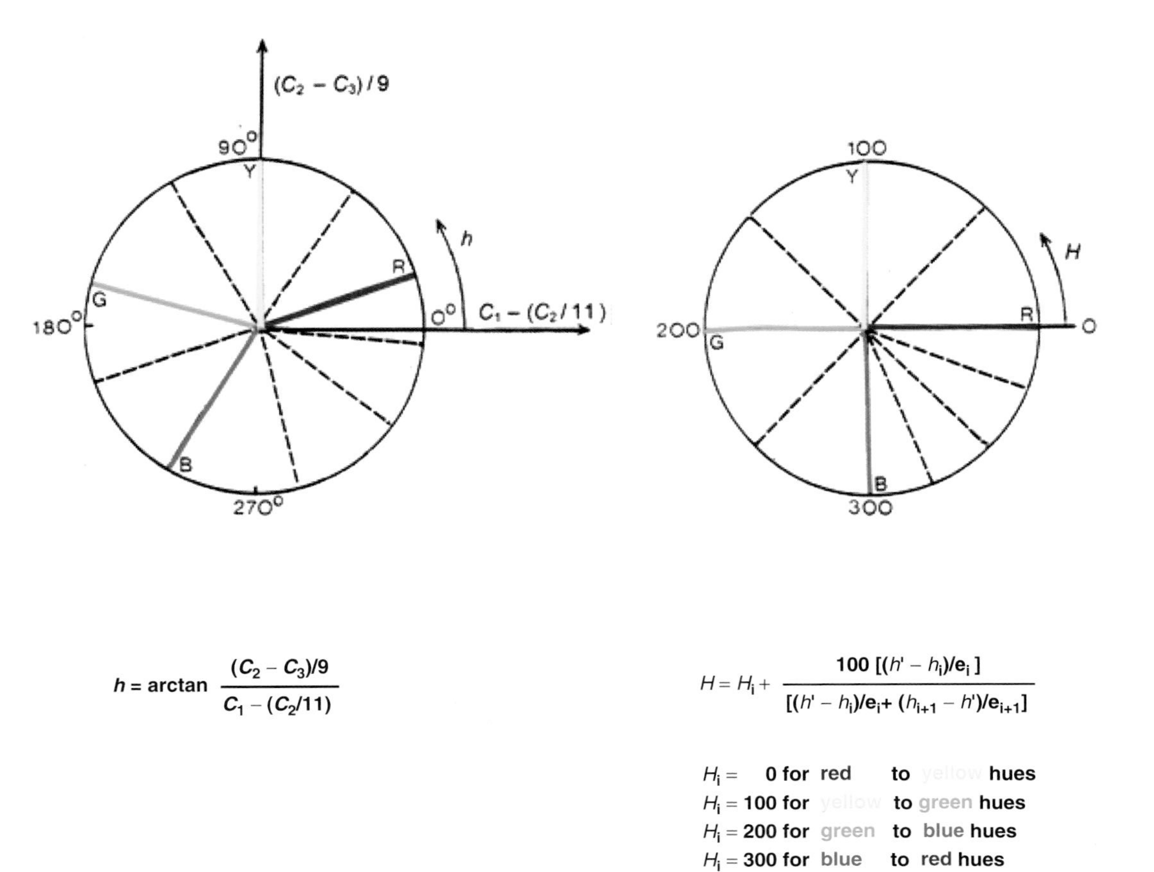

Figure 15.9 Left: hue angle, h, shown in a plot of $[(C_2 - C_3)/4.5]/2 = b$, yellowness-blueness, against $C_1 - (C_2/11) = a$, redness-greenness. Right: hue quadrature, H, shown in a plot where unique red and green are opposite one another, and unique yellow and blue are also opposite one another and at right-angles to the red-green directions

opposite one another, with unique yellow and blue also opposite one another and at right angles to the red-green axis, the four quadrants do not represent equal differences in hue; while the perceptual difference between the pairs of unique hues red and yellow, yellow and green, and green and blue, are not too different, the perceptual difference between blue and red is about twice as large, and this is represented by the crowding of the three broken lines in this quadrant. However, hue angle (on the left) represents differences in hue more uniformly. The spacing of hue angle (shown on the left) is similar to that of the hues in the Munsell system, while the spacing of Hue Quadrature (shown on the right) is similar to that of the hues in the NCS system.

H can be expressed both as a number, and as hue composition, H_c, in terms of the percentages of the component hues. When the two right-hand digits are more than 50, they indicate the main hue percentage, and the remaining percentage is that of the minor hue; for example if $H = 262$, the component hues in percentages are 62 Blue and 38 Green, which is abbreviated to 62B 38G. If the two right hand digits are less than 50, they represent the minor hue percentage, and the remaining percentage is that of the major hue; for example, if $H = 231$, the hue composition is 69 Green 31 Blue, or 69G 31B.

15.14 THE ACHROMATIC RESPONSE, *A*, IN CIECAM02

As mentioned in Section 1.6, in addition to the colour difference signals, the retina sends to the brain an achromatic signal. The achromatic signal is given by

$$A = [2\rho_a + \gamma_a + (1/20)\beta_a - 0.305]N_{bb}$$

assuming the relative ρ, γ, and β cone abundances to be in the ratios 2:1:1/20, respectively (Walraven and Bouman, 1966). The number, 0.305, is the sum $(0.2 + 0.1 + 0.005)$ of the separate noises of ρ_a, γ_a, and β_a and is subtracted so as to result in A being zero when X, Y, and Z are zero. N_{bb} is a factor to allow for the brightness induction of the background. No provision is made in CIECAM02 for any contribution from the rods.

$$N_{bb} = 0.725(1/n)^{0.2} \quad \text{where} \quad n = Y_b/Y_w$$

where Y_b is the Y tristimulus value of the background, and Y_w is that of the adopted white. The adopted white is the white that is appropriate for the application considered; it is usually different from PDE (the perfect diffuser in illuminant S_E). $N_{bb} = 1$ if $Y_b/Y_w = 0.2$ (background luminance 20% of that of the adopted white). If the chromaticity of the background is different from that of the adopted white, the corresponding-colour values of Y_b and Y_w are used, denoted by Y_{bc} and Y_{wc} so that then $n = Y_{bc}/Y_{wc}$.

15.15 CORRELATE OF LIGHTNESS, *J*, IN CIECAM02

Lightness is brightness judged relative to the brightness of the adopted white, and it is therefore formulated as a function of the ratio of the achromatic signals for the colour considered and for the adopted white. It is also made to be dependant on the luminance factor of the background, and on the type of surround. Three alternative surrounds are available depending on the ratio of the surround luminance to that of the colours being viewed:

Average when this ratio is about 1 (as is common when reflection prints are being viewed);

Dim when the ratio is about 0.15 (as is common when television displays are viewed in domestic environments);

Dark when the ratio is nearly zero (as is common in cinemas).

The correlate of lightness is then:

$$J = 100(A/A_{\mathrm{w}})^{cz}$$

where A and A_{w} are the values of the achromatic signal for the colour considered and for the adopted white respectively; c has values of 0.69 for an average surround, 0.59 for a dim surround, and 0.525 for a dark surround; and z is given by:

$$z = 1.48 + n^{0.5}$$

n being (as before) equal to $Y_{\mathrm{b}}/Y_{\mathrm{w}}$ (or $Y_{\mathrm{bc}}/Y_{\mathrm{wc}}$). When A is equal to A_{w} the value of J is 100.

15.16 CORRELATE OF BRIGHTNESS, *Q*, IN CIECAM02

The correlate of brightness is given by:

$$Q = (4/c)(J/100)^{0.5}(A_{\mathrm{w}} + 4)F_{\mathrm{L}}^{0.25}$$

In this expression, to give the necessary dependence on the luminance level, the correlate of lightness, J, is multiplied by $(A_{\mathrm{w}} + 4)F_{\mathrm{L}}^{0.25}$, where A_{w} is the achromatic signal for the adopted white. As the adapting luminance increases, A_{w}, increases, and this increases Q, as required; but the increase is insufficient, so the formula includes the power of F_{L}, the luminance-level adaptation factor. (However, if F_{L} were altered, in future, so as to increase the separation of the dynamic-response function curves along the log I axis, it might be possible to avoid having to use F_{L} in the formula for Q, so that its increase with adapting luminance then depended only on the dynamic-response function; this would be a more physiologically plausible result.)

In Figure 15.10 log Q is plotted against log luminance, log L, for an average surround. Results are for values of $5L_{\mathrm{A}}$ equal to 100 000, 1000, and 10, cd m^{-2}, that is, 5, 3, and 1, on the log scale, and for values of L/L_{W} (where L_{W} is the value of L for the adopted white) of 1.0, 0.3162, 0.1, 0.03162, and 0.01, that is, densities of 0, 0.5, 1.0, 1.5, and 2.0. The following features are evident from this figure. First, brightnesses are all reduced as the level of illumination is reduced. Second, the slopes of the lines are approximately equal, indicating that lightness is approximately constant with illumination level. Results similar to those shown in Figure 15.10, were obtained by Jameson and Hurvich and by Bartleson from both experimental scaling and modelling (Jameson and Hurvich, 1964; Bartleson, 1965, 1980). In this type of figure, equal differences in log Q represent approximately equal differences in brightness.

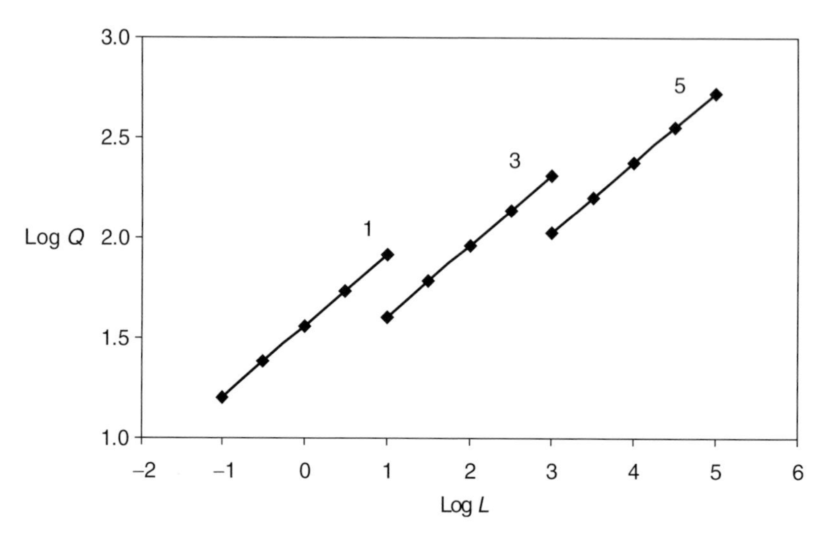

Figure 15.10 Brightness-luminance relationships for an average surround. Log Q is plotted against log L, where Q is the brightness response, and L is the luminance in cd m^{-2}. The relationships are shown for different levels of log adapting luminance in cd m^{-2}, L_A, such that log($5L_A$) is equal to 5, 3, and 1. On each curve the top point is for a white, and the lower points for luminance factors relative to the white of 31.62%, 10%, 3.162%, and 1% (densities of 0.5, 1.0, 1.5, and 2.0). The value of the relative luminance factor of the background, $n = Y_b/Y_w$, was set at 1.0, thus the background was white

15.17 CORRELATE OF CHROMA, *C*, IN CIECAM02

To obtain a correlate of chroma, C, it is necessary to combine the correlates of redness-greenness, a, and yellowness-blueness, b. But to do this effectively several factors have to be introduced.

First, a factor of 10/13 to allow for cross-channel noise in the system is incorporated (Hunt, 1982). Second, a chromatic surround induction factor, N_c, is used which makes allowance for the fact that dark or dim surrounds to colours can reduce their colourfulness; N_c is equal to 1.0 for average surrounds, 0.9 for dim surrounds, and 0.8 for dark surrounds. Third, to allow for the fact that, compared with their appearance when seen against a grey background, the colourfulness of colours tends to be reduced for light backgrounds, and increased for dark backgrounds (MacDonald, Luo, and Scrivener, 1990; Hunt, 1994), a chromatic background induction factor, N_{cb}, is introduced where:

$$N_{cb} = 0.725/n^{0.2} \quad \text{and} \quad n = Y_b/Y_w$$

Y_b and Y_w being the luminance factors of the background and of the adopted white, respectively. If $Y_b/Y_w = 1/5$ (background luminance 20% of that of the adopted white) then $N_{cb} = 1$. (If the chromaticity of the background is different from that of the adopted white, then the values for the corresponding colours Y_{cb} and Y_{cw} have to be used instead of Y_b and Y_w.) Fourth, the eccentricity factor, e_t, is incorporated. Finally, a factor 5000 is included to give convenient final numbers. Hence:

$$C = t^{0.9}(J/100)^{0.5}(1.64 - 0.29^n)^{0.73}$$

where

$$t = [(5000)(10/13)N_c N_{cb} e_t (a^2 + b^2)^{0.5}]/[\rho_a + \gamma_a + (21/20)\beta_a]$$

The inclusion of J causes C to diminish as the luminance factor of the sample decreases; N_{cb} causes C to increase as the luminance factor of the background decreases, and the $1.64 - 0.29^n$ term causes this effect to be reversed for light colours (Hunt, 1994).

Because a and b increase with the level of illumination, and chroma is independent of that variable, it is necessary to include the divisor $\rho_a + \gamma_a + (21/20)\beta_a$ in the formula for t; (if ρ_a, γ_a, and β_a are all multiplied by the same constant, as tends to happen when the adapting luminance changes, the value of C is not changed). An alternative plausible divisor would be $\rho_a + \gamma_a + \beta_a$, but $\rho_a + \gamma_a + (21/20)\beta_a$ is very similar, and facilitates reversing the model.

15.18 CORRELATE OF COLOURFULNESS, *M*, IN CIECAM02

The correlate of colourfulness, M, is given by:

$$M = C F_L^{0.25}$$

When the level of illumination increases, the value of C is not changed, so the F_L term is used to make M increase with level of illumination as is required. But if, in future, M were made to depend on $(a^2 + b^2)^{0.5}$ without the $\rho_a + \gamma_a + (21/20)\beta_a$ divisor that is incorporated in C, then M would increase as required, depending only on the dynamic response function, without using F_L; this would be a more physiologically plausible result.

15.19 CORRELATE OF SATURATION, *s*, IN CIECAM02

The correlate of saturation, s, is given by:

$$s = 100(M/Q)^{0.5}$$

Saturation is colourfulness judged in proportion to brightness; hence the correlate of saturation, s, depends, correctly, on the ratio of the correlate of colourfulness, M, to the correlate of brightness, Q.

15.20 COMPARISON OF CIECAM02 WITH THE NATURAL COLOUR SYSTEM

In Figure 15.11, the full lines show the loci of constant hue and saturation predicted by the model for a high level of adapting luminance, such as L_A equal to 200 cd m^{-2}; the broken lines show the results obtained experimentally in the Natural Colour System (NCS), and the two sets of lines are seen to be broadly similar.

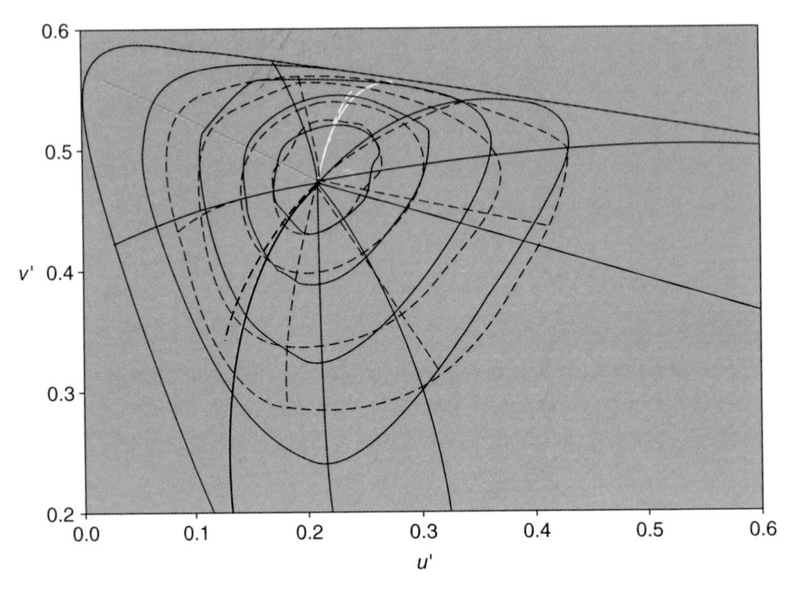

Figure 15.11 Constant hue loci, and constant saturation contours, predicted by the model (full lines) compared with those of the NCS (broken lines), the tristimulus values for which were converted from those published for Illuminant C to the corresponding values for illuminant S_E using the CAT02 chromatic adaptation transform, with D set equal to 1.0. The hues intermediate between the unique hues correspond to values of Hue Quadrature in CIECAM02 of 50, 150, 250, 325, 350, and 375, and of Hue notations in the NCS of Y50R, G50Y, B50G, R50B, an average of R25B and R30B, and an average of R70B and R80B. The loci intermediate between the unique hue loci divide each quadrant into segments of hue that are approximately equally different. The contours shown are for values of the saturation, s, of the model equal to 33. 43, 53, and 63, and for values of NCS chromaticness, c, equal to 40, 60, 80, and 90. The conditions for the model were as follows. The adapting luminance was 200 cd m^{-2}; N_c and N_{cb} were put equal to 1, and the lightness of the samples corresponded with a value of 50 for the Y tristimulus value. Because, as shown in Figure 15.2, the hyperbolic function approximates a simple power function over its central range, the chromaticities corresponding to hue and saturation are not significantly affected by the lightness of the samples or by typical photopic levels of illumination. For the NCS, the luminance factors were the highest available in the system

15.21 TESTING MODEL CIECAM02

Prior to its adoption by the CIE, the CIECAM02 model was tested against data that represented colour appearance in terms of magnitude estimation.

15.21.1 Colour appearance data sets

Colour appearance can be scaled in terms of hue, lightness, brightness, colourfulness, and saturation, by allocating appropriate numbers to each of these percepts. Hue is expressed

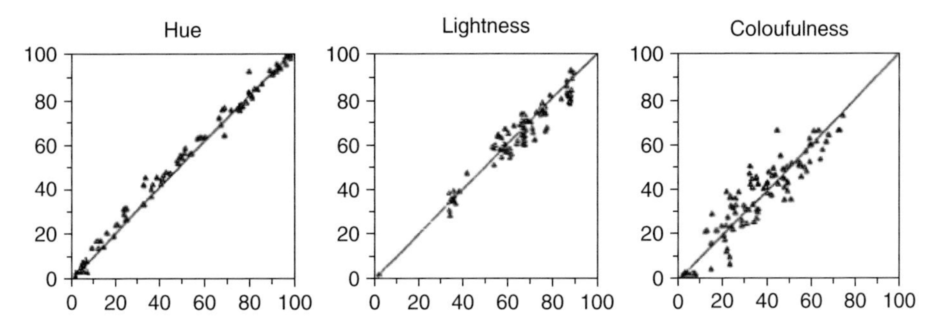

Figure 15.12 Typical examples of experimentally scaled magnitudes for hue (left), lightness (centre), and colourfulness (right), plotted as ordinates, against the predictions plotted as abscissas given by a colour appearance model for the same conditions of viewing

as the percentage contents of adjacent pairs of the unique hues (red, yellow, green, and blue). Lightness is expressed as a number on an imaginary scale having a reference white at 100 and the blackest black imaginable at zero. Brightness is expressed as a number on an imaginary scale having the blackest black imaginable at zero, and any suitable positive number. Colourfulness is expressed as the extent to which the hue is apparent, numbers ranging from zero for achromatic colours (white, grey, and black) to any suitable positive number. Saturation is expressed as the colourfulness as a proportion of the brightness with numbers ranging from zero for white, grey, and black, to any suitable positive number. Some typical magnitude-estimation results for hue, lightness, and colourfulness, plotted against predictions by a model, are shown in Figure 15.12.

The sets of data used for colour-appearance by magnitude estimation were those accumulated at LUTCHI (Luo, Clarke, Rhodes, Schappo, Scrivener, and Tait, 1991; Luo, Gao, Rhodes, Xin, Clarke, and Scrivener, 1993; Hunt, Li, Juan, and Luo, 2002) and by Kuo and Luo (Kuo, Luo, and Bez, 1995). These consisted of magnitude scaled values of hue, lightness, brightness, colourfulness, and saturation, averaged for groups of about six observers, together with the colorimetric data of the stimuli used. Over 100,000 estimations were made in over 50 phases of experimental conditions. The viewing conditions included D65, D50, and A simulators, White Fluorescent, TL84, and xenon light sources; small (2°) and large (10°) samples; illuminances ranging from 1.3 to 6600 lux; and white, grey, and black backgrounds. The data were divided into various sets: reflection samples seen at high (R-HL), low (R-LL), and varied luminances (R-VLH and R-VLL), large reflection samples (LT), self-luminous samples on monitors (CRT), textile samples (R-Tex), coloured cubes (Juan and Luo, 2002), and projected 35 mm transparencies (35 mm).

15.21.2 Testing methods

For the colour-appearance percept data (consisting of average estimations of hue, lightness, brightness, colourfulness, and saturation, together with the tristimulus values of the stimuli used), the testing procedure was as follows. The tristimulus values of the stimuli were used to obtain the predictions of hue, lightness, brightness, colourfulness, and saturation, for the model. The difference between the experimental results (V) and the predictions

(P) were expressed as coefficients of variation (CV) by calculating $CV = 100[\Sigma(V_i - P_i)^2/n]^{0.5}/[\Sigma(V_i)/n]$, where n is the number of samples, and i is the sample considered in a particular data set. The hue and lightness predictions of the models were tested directly against the experimental hue and lightness results, respectively. The brightness predictions were adjusted by multiplying them by a single scaling factor that resulted in the predictions fitting the experimental data best; this was also done for the colourfulness and saturation predictions. Finally a weighted average CV value was obtained by multiplying the average CV in each data set by the number of phases in that set, summing the results, and then dividing by the total number of phases. For a perfect performance by the model, the CV value should be zero.

15.21.3 Results for CIECAM02

The results obtained were as follows (Li, Luo, Hunt, Moroney, Fairchild, and Newman, 2002).

Data group	Hue	Lightness	Brightness	Colourfulness	Saturation
LT	6.0	11.1		14.7	
35 mm	7.0	18.9		16.6	
CRT	6.7	11.6		19.6	
R-Tex	7.4	14.8		24.1	
Juan	6.8	14.2		20.6	22
R-HL	7.1	10.6		18.1	
R-LL	8.1	17.5		19.7	
R-VHL	6.0	11.7	16	16.0	
R-VLL	7.8	16.4	24	30.1	
Weighted mean	6.9	13.6	20.4	19.2	22
Observer variation	8	13	10	18	15

The model may be considered to be giving satisfactory results if the CVs are similar to the extents to which the results for a single observer differ from the average for a group of about six observers; CVs for these observer variations are also shown. On this criterion, the model gives satisfactory predictions for hue, lightness, and colourfulness, but for brightness and saturation some improvement in the predictions given by the model are evidently needed.

15.22 FILTRATION OF PROJECTED SLIDES AND CIECAM02

If a uniformly coloured filter is placed over the whole of a colour slide, then its appearance when projected can be similar to that of the slide when projected without the filter; this is because the observer adapts very considerably to the different overall colour of the projected image. But, if the same filter is placed over the depiction on the slide of an object comprising only a small part of the area of the slide, then the apparent colour of that object in the projected image can change very dramatically (Hunt, 1991).

An example of these phenomena is a slide in which the picture contains a yellow cushion and a person wearing a white blouse. If a cyan filter is placed over the whole slide, the projected image, although slightly 'colder', looks generally similar to its unfiltered projected appearance. But, if the cyan filter is placed over only the depiction on the slide of the cushion, then the apparent colour of the cushion in the projected image changes from yellow to green. The CIECAM02 model predicts this type of effect, as shown by the following results.

	Hue	Colourfulness
Cushion: in tungsten light unfiltered	96Y 4R	35
in tungsten light filtered	75G 25Y	32
with filter over whole slide	75Y 25G	28
White: in tungsten light	75Y 25R	8
with filter over whole slide	91G 9B	11

15.23 COMPARISON OF CIECAM02 WITH CIECAM97S

The CIECAM02 model differs from its predecessor, the CIECAM97s model (Hunt, 2004), in the following ways. It incorporates a simpler Chromatic Adaptation Transform, CAT02; it changes the value of N_c for dim surrounds from 1.1 to 0.9. It uses a cone response function with a longer log-log straight-line section; the noise constant in the equation for the achromatic response, A, is chosen so that A is zero for stimuli having zero Y tristimulus values; revised correlates for chroma and colourfulness avoid excessive values for near neutral colours; a revised correlate for saturation avoids values, for colours of constant chromaticity, varying with changes in their luminance factors, and gives much improved agreement with experimental results.

15.24 UNIFORM COLOUR SPACE BASED ON CIECAM02

Spaces based on CIECAM02 have been proposed for evaluating colour differences (Luo, Cui, and Li, 2006). A space for universal use is obtained by plotting the following three variables at right-angles to one another:

$$a'_M = M' \cos(h)$$

$$b'_M = M' \sin(h)$$

$$J' = (1.7)J/(1 + 0.007J)$$

where

$$M' = (1/0.0228)\log_{10}(1 + 0.0228M)$$

The colour difference is given by:

$$\Delta E' = (\Delta J'^2 + \Delta a'^2_M + \Delta b'^2_M)^{0.5}$$

15.25 SOME PROBLEMS WITH CIECAM02

Problems in four areas have been reported in connection with CIECAM02.

First, the HPE matrix for deriving the cone signals has a top row that does not sum to unity. This can be corrected by changing the top right-hand entry from -0.7868 to -0.7869.

Second, there is mathematical failure for some colours. This arises because occasionally the value of A can become negative. A solution to this problem has been proposed (Li, Chorro-Calderon, Luo, and Pointer, 2009).

Third, the CIECAM02 colour domain is smaller than that of the ICC (International Color Consortium). A solution to this problem has been proposed (Gill, 2008).

Fourth, it has been reported that the brightness function does not correlate well with some experimental results. Further research in this area is required.

15.26 STEPS FOR USING THE CIECAM02 MODEL

It is first necessary to select from Table 15.1 the values of c, N_c, and F for the appropriate surround. (It is possible to use linear interpolation of the values of c, N_c, and F for intermediate surrounds.)

Starting data:

Sample in test conditions	x	y	Y
Adopted white in test conditions:	x_w	y_w	Y_w
Background in test conditions:	x_b	y_b	Y_b
Reference white in reference conditions:	$x_{wr} = 1/3$	$y_{wr} = 1/3$	$Y_{wr} = 100$
Luminance of test adapting field (cd m^{-2})	L_A		

L_A is normally taken as 1/5 of the luminance of the adopted test white.
Background parameters:

$$N_{bb} = N_{cb} = 0.725(1/n)^{0.2}$$

$$z = 1.48 + n^{0.5} \text{ where } n = Y_b/Y_w$$

If the test background chromaticity is different from that of the adopted test white, then, in the above expressions, the Y tristimulus values of the corresponding colours in the reference conditions, Y_{bc}, and Y_{wc}, have to be used instead of Y_b and Y_w.

Table 15.1 Values of c, N_c, and F for different surrounds

Surround	c	N_c	F
Average	0.69	1.0	1.0
Dim	0.59	0.9	0.9
Dark	0.525	0.8	0.8

Step 1. For the sample, calculate:

$$X = xY/y,\ Y,\ Z = (1 - x - y)Y/y$$

Convert the sample tristimulus values, X, Y, Z to sharpened R, G, B responses by using the CAT02 forward matrix, M_{CAT02}:

$$\begin{bmatrix} R \\ G \\ B \end{bmatrix} = M_{\mathrm{CAT02}} \begin{bmatrix} X \\ Y \\ Z \end{bmatrix}$$

$$M_{\mathrm{CAT02}} = \begin{bmatrix} 0.7328 & 0.4296 & -0.1624 \\ -0.7036 & 1.6975 & 0.0061 \\ 0.0030 & 0.0136 & 0.9834 \end{bmatrix}$$

Step 2. Compute the D factor, the degree of adaptation to the white point. The equation for D gives its default value. (L_A is the luminance of the adapting field in cd m^{-2}.) If full adaptation occurs, D should be set equal to 1; if no adaptation occurs D should be set equal to zero. A realistic minimum value is about 0.8.

$$D = F\left[1 - \left(\frac{1}{3.6}\right) e^{\left(\frac{-L_A - 42}{92}\right)}\right]$$

Note: if D is greater than 1 or less than 0, set it at 1 or 0 respectively.

Step 3. Obtain R_c, G_c, B_c, the R, G, B values for the corresponding colour. The subscript w indicates that the value is for the adopted white in the test situation, and the subscript wr indicates that the value is for the reference white (the perfect diffuser, $Y_{wr} = 100$) in the reference illuminant (the equi-energy stimulus, S_E, $x_{wr} = 1/3$, $y_{wr} = 1/3$).

$$R_c = D_R R \quad D_R = (Y_w/Y_{wr})(R_{wr}/R_w)D + (1 - D)$$

$$G_c = D_G G \quad D_G = (Y_w/Y_{wr})(G_{wr}/G_w)D + (1 - D)$$

$$B_c = D_B B \quad D_B = (Y_w/Y_{wr})(B_{wr}/B_w)D + (1 - D)$$

Step 4. Compute (where the subscript b indicates that the value is for the background):

$$k = 1/(5L_A + 1) \quad F_L = 0.2k^4(5L_A) + 0.1(1 - k^4)^2(5L_A)^{1/3}$$

$$n = Y_b/Y_w \quad N_{bb} = N_{cb} = 0.725(1/n)^{0.2} \quad z = 1.48 + n^{0.5}$$

Step 5. Convert to X,Y,Z corresponding colours:

$$\begin{bmatrix} X_c \\ Y_c \\ Z_c \end{bmatrix} = M_{\mathrm{CAT02}}^{-1} \begin{bmatrix} R_c \\ G_c \\ B_c \end{bmatrix} \quad M_{\mathrm{CAT02}}^{-1} = \begin{bmatrix} 1.096124 & -0.278869 & 0.182745 \\ 0.454369 & 0.473533 & 0.072098 \\ -0.009628 & -0.005698 & 1.015326 \end{bmatrix}$$

Step 6. Convert to Hunt-Pointer-Estevez cone responses:

$$\begin{bmatrix} \rho \\ \gamma \\ \beta \end{bmatrix} = M_{\mathrm{HPE}} \begin{bmatrix} X_c \\ Y_c \\ Z_c \end{bmatrix} \quad M_{\mathrm{HPE}} = \begin{bmatrix} 0.38971 & 0.68898 & -0.07868 \\ -0.22981 & 1.18340 & 0.04641 \\ 0.00000 & 0.00000 & 1.00000 \end{bmatrix}$$

Step 7. Apply luminance-level adaptation (F_L), and non-linear compression:

$$\rho_a = \{[400(F_L\rho/100)^{0.42}]/[27.13 + (F_L\rho/100)^{0.42}]\} + 0.1$$

$$\gamma_a = \{[400(F_L\gamma/100)^{0.42}]/[27.13 + (F_L\gamma/100)^{0.42}]\} + 0.1$$

$$\beta_a = \{[400(F_L\beta/100)^{0.42}]/[27.13 + (F_L\beta/100)^{0.42}]\} + 0.1$$

If any of the values of ρ, γ, or β are negative, then their positive equivalents must be used, and then the expressions in the {} brackets must be made negative. Calculate ρ_{aw}, γ_{aw}, and β_{aw} similarly.

Step 8. Calculate colour-difference signals, a and b, and hue-angle, h_r:

$$a = \rho_a - 12\gamma_a/11 + \beta_a/11 \quad b = (1/9)(\rho_a + \gamma_a - 2\beta_a) \quad h = \tan^{-1}(b/a)$$

If h is obtained in radians, multiplying by $180/\pi$ converts it into relative degrees, h_r. If a and b are both positive, h_r is positive, and the angle in absolute degrees, h, is the same as h_r. If a is negative and b is positive, h_r is negative, and $h = h_r + 180$. If a and b are both negative, h_r is positive, and $h = h_r + 180$. If a is positive and b is negative, h_r is negative, and $h = h_r + 360$.

Step 9. Hue quadrature, H, is calculated from the following unique hue data:

	Red	Yellow	Green	Blue	Red
i	1	2	3	4	5
h_i	20.14°	90.00°	164.25°	237.53°	380.14°
e_i	0.8	0.7	1.0	1.2	0.8
H_i	0.0	100.0	200.0	300.0	400.0

Set $h' = h + 360$ if $h < h_1$; otherwise $h' = h$.
Choose an appropriate value of i (1, 2, 3, or 4) so that $h_i \leq h' < h_{i+1}$

$$H = H_i + [100(h' - h_i)/e_i]/[(h' - h_i)/e_i + (h_{i+1} - h')/e_{i+1}]$$

Step 10. Calculate eccentricity factor, e_t:

$$e_t = [1/4][\cos(h\pi/180 + 2) + 3.8]$$

The cosine function uses radians; it is negative for angles from $\pi/2$ to $3\pi/2$. (For the unique red, yellow, green, and blue hues, the values of e_t are nearly, but not exactly, the same as those given in Step 9 for e_i.)

Step 11. Calculate the achromatic responses, A and A_w:

$$A = [2\rho_a + \gamma_a + (1/20)\beta_a - 0.305]N_{bb}$$

$$A_w = [2\rho_{aw} + \gamma_{aw} + (1/20)\beta_{aw} - 0.305]N_{bb}$$

Step 12. Calculate the correlate of lightness, J, from the achromatic responses of the stimulus, A, and of the white, A_w:

$$J = 100(A/A_w)^{cz}$$

Step 13. Calculate the correlate of brightness, Q:

$$Q = (4/c)(J/100)^{0.5}(A_w + 4)F_L^{0.25}$$

Step 14. Calculate the value of a temporary quantity, t. This quantity is used for calculating the correlate of chroma, C, and, by extension, the correlate of colourfulness, M.

$$t = [(50000/13)N_c N_{cb}][e_t(a^2 + b^2)^{0.5}]/[\rho_a + \gamma_a + (21/20)\beta_a]$$

Step 15. Calculate the correlate of chroma, C:

$$C = t^{0.9}(J/100)^{0.5}(1.64 - 0.29^n)^{0.73}$$

Step 16. Calculate the correlate of colourfulness, $M = CF_L^{0.25}$

Step 17. Calculate the correlate of saturation, $s = 100(M/Q)^{0.5}$

Step 18. Calculate corresponding Cartesian coordinates as necessary:

$$a_C = C\cos(h) \quad b_C = C\sin(h)$$

$$a_M = M\cos(h) \quad b_M = M\sin(h)$$

$$a_s = s\cos(h) \quad b_s = s\sin(h)$$

15.27 STEPS FOR USING THE CIECAM02 MODEL IN REVERSE MODE

The reverse model uses the same viewing-condition constants of Table 15.1 and the same values for the background parameters, n, N_{bb}, N_{cb}, and z.

Step R1. The value for A_w is computed for the adopted white point using the forward model. Then, if starting from Q, J can be computed:

$$J = 6.25\{[cQ]/[(A_w + 4)F_L^{0.25}]\}^2$$

If starting from M, then C can be computed:

$$C = M/F_L^{0.25}$$

If starting from J and s, then Q and C can be computed:

$$Q = (4/c)(J/100)^{0.5}(A_w + 4)F_L^{0.25}$$

$$C = (s/100)^2(Q/F_L^{0.25})$$

Step R2. Starting from H or h and the table of values in Step 9 of the forward model, and choosing an appropriate value of i (1, 2, 3, or 4) so that $H_i \leq H < H_{i+1}$, compute:

$$h' = [(H - H_i)(e_{i+1}h_i - e_i h_{i+1}) - 100h_i e_{i+1}]/[(H - H_i)(e_{i+1} - e_i) - 100e_{i+1}]$$

Set $h = (h' - 360)$ if $h' > 360$, otherwise $h = h'$.

Step R3. Compute t, e_t, A, p_1, p_2, and p_3:

$$t = \{C/[(J/100)^{0.5}(1.64 - 0.29^n)^{0.73}]\}^{1/0.9}$$

Note: if t is equal to zero, set a and b equal to zero, calculate A and p_2, and go directly to Step R5.

$$e_t = [1/4][\cos(2 + h\pi/180) + 3.8]$$

$$A = A_w(J/100)^{1/cz}$$

$$p_1 = (50000/13)(N_c N_{cb})e_t/t \quad p_2 = (A/N_{bb}) + 0.305 \quad p_3 = 21/20$$

Step R4. Calculate a and b:

$$h_r = h\pi/180$$

If $|\sin(h_r)| \geq |\cos(h_r)|$,

$$p_4 = p_1/\sin(h_r)$$

$$b = \frac{p_2(2+p_3)(460/1403)}{p_4+(2+p_3)(220/1403)[\cos(h_r)/\sin(h_r)] - (27/1403)+p_3(6300/1403)}$$

$$a = b[\cos(h_r)/\sin(h_r)]$$

If $|\sin(h_r)| < |\cos(h_r)|$,

$$p_5 = p_1/\cos(h_r)$$

$$a = \frac{p_2(2+p_3)(460/1403)}{p_5+(2+p_3)(220/1403) - [(27/1403) - p_3(6300/1403][(\sin(h_r)/\cos(h_r)]}$$

$$b = a[\sin(h_r)/\cos(h_r)]$$

Step R5. Compute ρ_a, γ_a and β_a:

$$\rho_a = (460/1403)p_2 + (451/1403)a + (288/1403)b$$

$$\gamma_a = (460/1403)p_2 - (891/1403)a - (261/1403)b$$

$$\beta_a = (460/1403)p_2 - (220/1403)a - (6300/1403)b$$

Step R6. Compute ρ, γ and β:

$$\rho = \{100/F_L\}\{[(27.13(|\rho_a - 0.1|)]/[(400 - (|\rho_a - 0.1|)]\}^{(1/0.42)}$$

$$\gamma = \{100/F_L\}\{[(27.13(|\gamma_a - 0.1|)]/[(400 - (|\gamma_a - 0.1|)]\}^{(1/0.42)}$$

$$\beta = \{100/F_L\}\{[(27.13(|\beta_a - 0.1|)]/[(400 - (|\beta_a - 0.1|)]\}^{(1/0.42)}$$

If any of the values of $\rho_a - 0.1$, $\gamma_a - 0.1$, and $\beta_a - 0.1$ are negative, then ρ, γ, or β must be made negative.

Step R7. Compute R_c, G_c and B_c:

$$\begin{bmatrix} R_c \\ G_c \\ B_c \end{bmatrix} = M_{CAT02} M_{HPE}^{-1} \begin{bmatrix} \rho \\ \gamma \\ \beta \end{bmatrix}$$

$$M_{HPE}^{-1} = \begin{bmatrix} 1.910197 & -1.112124 & 0.201908 \\ 0.370950 & 0.629054 & -0.000008 \\ 0.000000 & 0.000000 & 1.000000 \end{bmatrix}$$

Step R8. Compute R, G, and B and finally X, Y and Z:

$$R = R_c/[(Y_w/Y_{wr})(R_{wr}/R_w)D + (1 - D)]$$

$$G = G_c/[(Y_w/Y_{wr})(G_{wr}/G_w)D + (1 - D)]$$

$$B = B_c/[(Y_w/Y_{wr})(B_{wr}/B_w)D + (1 - D)]$$

$$\begin{bmatrix} X \\ Y \\ Z \end{bmatrix} = M_{CAT02}^{-1} \begin{bmatrix} R \\ G \\ B \end{bmatrix}$$

15.28 WORKED EXAMPLE FOR THE MODEL CIECAM02

The CIECAM02 model gives the following results for a sample in Standard Illuminant A (SA) at a level of adapting luminance, L_A, equal to 200.

Starting data:

Sample in test conditions	$X = 19.31$	$Y = 23.93$	$Z = 10.14$	
Adopted white in test conditions:	$X_w = 98.88$	$Y_w = 90.00$	$Z_w = 32.03$	
Parameters:	$Y_b = 18.00$	$F = 1.0$	$c = 0.69$	$N_c = 1.0$

Table 15.2 Details of forward model calculations for the $L_A = 200$ example

X	Y	Z	19.31	23.93	10.14
X_w	Y_w	Z_w	98.88	90.0	32.03
Y_b	c	N_c	18.0	0.69	1.0
F	n	z	1.0	0.2000	1.9272
N_{bb}	N_{cb}	F_L	1.0003	1.0003	1.0000
R_w	G_w	B_w	105.9216	83.3984	33.0189
R	G	B	22.7840	27.0965	10.3551
D_R	D_G	D_B	0.85269	1.07757	2.69117
R_c	G_c	B_c	19.4277	29.1985	27.8672
R_{wc}	G_{wc}	B_{wc}	90.3186	89.8679	88.8596
X_c	Y_c	Z_c	18.2452	24.66297	27.94090
X_{cw}	Y_{cw}	Z_{cw}	90.1777	90.0000	88.8398
ρ	γ	β	21.9043	26.2900	27.9409
ρ_w	γ_w	β_w	90.1614	89.9053	88.8398
ρ_a	γ_a	β_a	7.7428	8.3392	8.5481
ρ_{aw}	γ_{aw}	β_{aw}	13.7351	13.7193	13.6536
a	b	h^0	−0.5773	−0.1127	191.0452^0
e_t	A_w	A	1.09566	41.5797	23.9545
J	C	H	48.0314	38.7789	240.8885
Q	M	s	183.1240	38.7789	46.0177
H_c	t	D	41B 59G	98.9576	0.9800

Table 15.3 Details of reverse model calculations for the $L_A = 200$ example

J	C	H	48.0314098	38.7789012	240.8884534
X_w	Y_w	Z_w	98.88	90.0	32.03
Y_b	c	N_c	18.0	0.69	1.0
F	n	z	1.0	0.2000	1.9272
N_{bb}	N_{cb}	F_L	1.0003	1.0003	1.0000
R_w	G_w	B_w	105.9216	83.3984	33.0189
R_{wc}	G_{wc}	B_{wc}	90.3186	89.8679	88.8596
A_w	h^0	e_t	41.5797	191.0452^0	1.09566
D	t	A	0.9800	98.9576	23.9545
p_1	p_2	p_3	42.5977	24.2522	1.0500
a	b		−0.5773	−0.1127	
ρ_a	γ_a	β_a	7.7428	8.3392	8.5481
ρ	γ	β	21.9043	26.2900	27.9409
R_c	G_c	B_c	19.4279	29.1992	27.8672
R	G	B	22.7842	27.0972	10.3551
X	Y	Z	19.3101	23.9304	10.1400

REFERENCES

Arend, L., and Reeves, A., Simultaneous color constancy, *J Opt. Soc. Amer. A*, **3**, 1743–1751 (1986).

Attridge, G. G., Pointer, M. R., and Jacobson, R. E., A summary of applications of colour reproduction indices to imaging systems, *J. Phot. Sci.*, **44**, 187–192 (1997).

Bartleson, C. J., Interrelations among screen luminance, camera exposure and quality of projected color transparencies, *Phot. Sci. Eng.*, **9**, 174–183 (1965).

Bartleson, C. J., Measures of brightness and lightness, *Die Farbe*, **28**, 132–148 (1980).

Baylor, D. A., Photoreceptor cells and vision, *Investig. Ophthalmol.*, **28**, 34–47 (1987).

Boynton, R. M., and Whitten, D. N., Visual adaptation in monkey cones: recordings of late receptor potentials, *Science*, **170**, 1423–1426 (1970).

CIE Publication 159: 2004, *A colour appearance model for colour management systems: CIECAM02*, Commission Internationale de l'Éclairage, Vienna, Austria (2004).

Estévez, O., *On the Fundamental Data-Base of Normal and Dichromatic Colour Vision*, Ph.D. Thesis, University of Amsterdam, Holland (1979).

Fairchild, M. D., Refinement of the RLAB color space, *Color Res. Appl.*, **21**, 338–346 (1996).

Fairchild, M. D., and Berns, R. S., Image color-appearance specification through extension of CIELAB, *Color Res. Appl.*, **18**, 178–190 (1993).

Gill, G. W., A solution to CIECAM02 numerical and range issues, *IS&T and SID's 16th Color Imaging Conference: Color Science and Engineering Systems, Technologies, and Applications*, pp. 327–331. IS&T, Springfield, VA, USA (2008).

Hunt, R. W. G., The effects of daylight and tungsten light adaptation on colour perception, *J. Opt. Soc. Amer.*, **40**, 362–371 (1950).

Hunt, R. W. G., Light and dark adaptation and the perception of colour, *J Opt. Soc. Amer.*, **42**, 190–199 (1952).

Hunt, R. W. G., The perception of colour in 1^0 fields for different states of adaptation, *J. Opt. Soc. Amer.*, **43**, 479–484 (1953).

Hunt, R. W. G., A model of colour vision for predicting colour appearance, *Color Res. Appl.*, **7**, 95–112 (1982).

Hunt, R. W. G., Perceptual factors affecting colour order systems, *Color Res. Appl.*, **10**, 12–19 (1985).

Hunt, R. W. G., A visual model for predicting colour appearance under various viewing conditions, *Color Res. Appl.*, **12**, 297–314 (1987).

Hunt, R. W. G., Hue shifts in unrelated and related colours, *Color Res. Appl.*, **14**, 235–239 (1989).

Hunt, R. W. G., Revised colour appearance model for related and unrelated colours, *Color Res. Appl.*, **16**, 146–165 (1991).

Hunt, R. W. G., An improved predictor of colourfulness in a model of colour vision, *Color Res. Appl.*, **19**, 23–33 (1994).

Hunt, R. W. G., *The Reproduction of Colour*, 6*th* *Ed.*, Chapters 35 and 36, Wiley, Chichester, England (2004).

Hunt, R. W. G., In *Colorimetry, Understanding the CIE System*, Chapter 14, Schanda, J., ed., Wiley, New York, NY, USA (2007).

Hunt, R. W. G., Li, C. J., Juan, L. Y., and Luo, M. R., Further improvements to CIECAM97s, *Color Res. Appl.*, **27**, 164–170 (2002).

Hunt, R. W. G., and Luo, M. R., Evaluation of a model of colour vision by magnitude scalings: discussion of collected results, *Color Res. Appl.*, **19**, 27–33 (1994).

Hunt, R. W. G., and Pointer, M. R., A colour-appearance transform for the CIE 1931 standard colorimetric observer, *Color Res. Appl.*, **10**, 165–179 (1985).

Hurvich, L. M., and Jameson, D., A psychophysical study of white. I. Neutral adaptation. *J. Opt. Soc. Amer.*, **41**, 521–527 (1951).

Jameson, D., and Hurvich, L. M., Theory of brightness and color contrast in human vision, *Vision Res.*, **4**, 135–154 (1964).

Kuo, W.-G., Luo, M. R., and Bez, H. E., Various chromatic-adaptation transformations tested using new colour appearance data in textiles, *Color Res. Appl.*, **20**, 313–327 (1995)

Li, C., Chorro-Calderton, E., Luo, M. R., and Pointer, M. R., Recent progress with extensions to CIECAM02, *IS&T and SID's 17th Color Imaging Conference: Color Science and Engineering Systems, Technologies, and Applications*, pp. 69–74, IS&T, Springfield, VA, USA (2009).

Li, C., Luo, M. R., Hunt, R. W. G., Moroney, N., Fairchild, M. D., and Newman, T., The performance of CIECAM02, *IS&T and SID's 10th Color Imaging Conference: Color Science and Engineering Systems, Technologies, and Applications*, pp. 28–32, IS&T Springfield, VA, USA (2002).

Luo, M. R., Clarke, A. A., Rhodes, P. A., Schappo, A., Scrivener, S. A. R., and Tait, C. J., Quantifying colour appearance, Part I, LUTCHI colour appearance data, and Part II, testing colour models performance using LUTCHI colour appearance data, *Color Res. Appl.*, **16**, 166–180 and 181–197 (1991).

Luo, M. R., Gao, X. W., Rhodes, P. A., Xin, H. J., Clarke, A. A., and Scrivener, S. A. R. Quantifying colour appearance, Part III, supplementary colour appearance data, and Part IV transmissive media, *Color Res. Appl.*, **18**, 98–113 and 191–209 (1993).

Luo, M. R., Cui, G., and Li, C., Uniform colour spaces based on CIECAM02 Colour Appearance Model, *Color Res. Appl.*, **31**, 320–330 (2006).

Luo, M. R., Lo, M.-C., and Kuo, W.-G., The LLAB(*l*:c) colour model, *Color Res. Appl.*, **21**, 412–429 (1996).

MacDonald, L. W., Luo, M. R., and Scrivener, S. A. R., Factors affecting the appearance of coloured images on a video display monitor, *J. Phot. Sci.*, **38**, 177–186 (1990).

McCann, J. J., and Houston, K. L., In *Colour Vision*, Mollon, J. D., and Sharpe, L. T., eds., pp. 535–544, Academic Press, London (1983).

McCann, J. J., McKee, S. P., and Taylor, T. H., Quantitative studies in Retinex theory: a comparison between theoretical predictions and observer responses to the "color mondrian" experiments, *Vision Res.*, **16**, 445–448 (1976).

Moroney, N., Fairchild, M. D., Hunt, R. W. G., Li, C., Luo, M. R., and Newman, T., The CIECAM02 color appearance model, *IS&T and SID's 10th Color Imaging Conference: Color Science and Engineering Systems, Technologies, and Applications*, pp. 23–27, IS&T, Springfield, VA, USA (2002).

Nayatani, Y., Examination of adaptation coefficients for incomplete chromatic adaptation, *Color Res. Appl.*, **22**, 156–164 and A simple estimation method for effective adaptation coefficient, *Color Res. Appl.*, **22**, 259–268 (1997).

Nayatani, Y., Hashimoto, K., Takahama, K., and Sobagaki, H., Whiteness-blackness and brightness response in a non-linear color-appearance model, *Color Res. Appl.* **12**, 121–127, and A non-linear color-appearance model using Estévez-Hunt-Pointer primaries, *Color Res. Appl.*, **12**, 231–242 (1987).

Nayatani, Y., Sobagaki, H., Hashimoto, K., and Yano, T., Field trials of a non-linear color- appearance model, *Color Res. Appl.*, **22**, 240–258 (1997).

Nayatani, Y., Takahama, K., and Sobagaki, H., Prediction of color appearance under various adaptation conditions, *Color Res. Appl.*, **11**, 62–71 (1986).

Nayatani, Y., Takahama, K., Sobagaki, H., and Hashimoto, K., Color appearance model and chromatic adaptation transform, *Color Res. Appl.*, **15**, 210–221 (1990).

Pointer, M. R., Measuring colour reproduction, *J. Phot. Sci.*, **34**, 81–90 (1986).

Seim, T., and Valberg, A., Towards a uniform color space: a better formula to describe the Munsell and OSA color scales, *Color Res. Appl.*, **11**, 11–24 (1986).

Smith, V. C., and Pokorny, J., Spectral sensitivity of color blind observers and the cone pigments, *Vision Res.*, **12**, 2059–2071 (1972).

Stevens, S. S., To honor Fechner and repeal his laws, *Science*, **133**, 80–86 (1961).

Valeton, J. M., and Van Norren, D., Light adaptation of primate cones: an analysis based on extracellular data, *Vision Res.*, **23**, 1539–1547 (1983).

Walraven, P. L., and Bouman, M. A., Fluctuation theory of colour discrimination of normal trichromats, *Vision Res.*, **6**, 567–586 (1966).

16

Models of Colour Appearance for Stimuli of Different Sizes

16.1 INTRODUCTION

The colour appearance model CIECAM02 is applicable to related colours of angular subtense up to about 10°. Modifications of this model have been derived for stimuli of different sizes and for room colours. These will be described in this chapter.

16.2 STIMULI OF DIFFERENT SIZES

Revised correlates of lightness, J'', and chroma, C'', in CIECAM02, have been suggested for stimuli of different angular sizes, θ, in degrees (Xiao, Luo, Li, Cui, and Park, 2011; Xiao, Luo, and Li, 2011).

$$J'' = 100 + K_{\mathrm{J}}(J - 100)$$

$$C'' = K_{\mathrm{C}}C$$

$$H'' = H$$

where

$$K_{\mathrm{J}} = -0.007\theta + 1.1014$$

$$K_{\mathrm{C}} = 0.008\theta + 0.94$$

16.3 ROOM COLOURS

Paint manufacturers usually provide samples of their products in the form of painted swatches of size similar to that of postage stamps. It is a common experience that, when

Measuring Colour, Fourth Edition. R.W.G. Hunt and M.R. Pointer.

such paints are used to cover the walls of a room, the appearance of the colour is rather different from that on the swatch. There are at least two reasons for this.

First, the appearance of colour stimuli is dependent on their angular subtense. This is caused partly by the non-uniformity of the sensitivity of different parts of the retina of the eye, and partly by the effect of the surround on the colour. Second, the effective colour of the walls is changed by light reflected from wall to wall.

These phenomena are of considerable practical interest not only to paint manufacturers, but also to companies providing other components of furnishing, such as carpets and curtains, and to interior designers. A study was therefore carried out (Xiao, Luo, Li, and Hong, 2010) using a room the four walls of which were sequentially painted in twelve different colours, but with a ceiling that was white, and a floor that was grey. Two different light sources were used: a CIE D65 simulator and a Cool White fluorescent lamp. Measurements were made of the colours of the walls, of the light source, and also of a white card placed in the centre of the floor of the room to determine the extent to which the illuminant colour was modified by the coloured walls; all the measurements were evaluated using the CIE 10° standard observer. Observers then sat in the room and assessed the appearance of the colour of the walls using two different methods: in the first method an area on a cathode-ray tube subtending 10° to the observer's eye, with a black surround, was adjusted in colour to have the same appearance as that of the wall colour. In the second method, the observers selected a chip in an NCS atlas (see Section 8.8) that had the same appearance as the colour of the wall; the atlas was placed in a viewing cabinet fitted with similar fluorescent lamps to those in the room. The results obtained by the two methods were in general agreement.

It was found that the apparent colours of the walls were usually lighter and of higher chroma than those measured outside the room, but without any change in hue. The increase in chroma is not surprising because, as the light from the walls is inter-reflected in the room, the spectral absorption bands of the colorant would operate several times thus increasing the absorptions; although the adaptation of the observers to the prevailing colour would tend to counter this effect, it evidently did not eliminate it. The increase in lightness might be the result of the absence of any effective surround to the colour.

16.4 A MODEL FOR PREDICTING ROOM COLOURS

To predict the appearance of room colours, it is necessary to allow not only for the effects of the angular subtense, but also for the effects of the modification of the colour of the illuminant. The model to be described is based on the experimental results obtained by the observers.

The effects of the angular subtense are allowed for as follows. The correlates of lightness, J_R, chroma, C_R, and hue, H_R, for the room colours are obtained as:

$$J_R = \alpha_J J_S^2 + \beta_J J_S + 100(1 - 100\alpha_J - \beta_J)$$

$$C_R = k_C C_S$$

$$H_R = H_S$$

where the subscript R indicates the room-size colour, and the subscript S indicates the small-size colour, and

$$\alpha_J = -0.0026 \qquad \beta_J = 1.19 \qquad k_C = 1.41$$

The effects of the modification of the illuminant are allowed for as follows. The tristimulus values, $X_{10n}, Y_{10n}, Z_{10n}$, of the modified illuminant in different wall colours are obtained as:

$$X_{10n} = fX_{10W} + (1 - f)X_{10}$$

$$Y_{10n} = fY_{10W} + (1 - f)Y_{10}$$

$$Z_{10n} = fZ_{10W} + (1 - f)Z_{10}$$

where $X_{10W}, Y_{10W}, Z_{10W}$ are the tristimulus values of the unmodified light source, the value of f being 0.7.

16.5 STEPS IN USING THE MODEL FOR PREDICTING ROOM COLOURS

The input data required are:

The tristimulus values of the unmodified illuminant:	X_{10W},	Y_{10W},	Z_{10W}
The tristimulus values of the small-size colour of the walls:	X_{10S},	Y_{10S},	Z_{10S}
The following data for operation of the CIECAM02 model:	$c = 0.69$	$N_c = 1.0$	$F = 1.0$
Sample in test conditions:	X_{10S},	Y_{10S},	Z_{10S}
Adopted white in test conditions:	X_{10n},	Y_{10n},	Z_{10n}
Background in test conditions:	X_{10S},	Y_{10S},	Z_{10S}
Reference white in reference conditions:	$x_{10wr} = 1/3$	$y_{10wr} = 1/3$	$Y_{10wr} = 100$
Luminance of test adapting field (cd m^{-2})	L_A		

L_A is normally taken as 1/5 of the luminance of the adopted test white.
Background parameters:

$$N_{bb} = N_{cb} = 0.725(1/n)^{0.2}$$

$$z = 1.48 + n^{0.5} \text{ where } n = Y_{10S}/Y_{10W}$$

Step 1. Illuminant prediction. Prediction of the tristimulus values, $X_{10n}, Y_{10n}, Z_{10n}$, of the modified illuminant, from the tristimulus values, $X_{10W}, Y_{10W}, Z_{10W}$, of the unmodified illuminant, and $X_{10S}, Y_{10S}, Z_{10S}$, of the small-size colour of the walls of the room.

$$X_{10n} = fX_{10W} + (1 - f)X_{10S}$$

$$Y_{10n} = fY_{10W} + (1 - f)Y_{10S}$$

$$Z_{10n} = fZ_{10W} + (1 - f)Z_{10S}$$

where $f = 0.7$.

Step 2. Small-size colour-appearance prediction. With $X_{10S}, Y_{10S}, Z_{S10}$, as the input for the colour considered, and $X_{10n}, Y_{10n}, Z_{10n}$, as the input for the adopted white, CIECAM02 is used to derive J_S, C_S, and H_S, the correlates of lightness, chroma, and hue, respectively, of the small-size colour.

Step 3. Size effect correction. The correlates for lightness, J_R, chroma, C_R, and hue, H_R, for the room colour are calculated from those for the small colour, J_S, C_S, H_S.

$$J_R = -0.0026J_S^2 + 1.19J_S + 100[1 - 100(-0.0026) - 1.19]$$

$$C_R = 1.41C_S$$

$$H_R = H_S$$

Step 4. Tristimulus values for the room colour. With J_R, C_R, and H_R, as initial input, and $X_{10n}, Y_{10n}, Z_{10n}$, as the adopted white, the reverse CIECAM02 model is used to derive the tristimulus values of the room colour, $X_{10R}, Y_{10R}, Z_{10R}$.

REFERENCES

Xiao, K., Luo, M. R., Li, C., and Hong, G., Colour appearance of room colours, *Color Res. Appl.* **35**, 284–293 (2010).

Xiao, K., Luo, M. R., Li, C., Cui, G., and Park, D., Investigation of colour size effect for colour appearance assessment, *Color Res. Appl.*, **36**, 201–209 (2011).

Xiao, K., Luo, M. R., and Li, C. Colour size effect modelling, *Color Res. Appl.*, submitted (2011).

17

Model of Colour Appearance for Unrelated Colours in Photopic and Mesopic Illuminances

17.1 INTRODUCTION

The CIECAM02 model is applicable only to related colours. The terms related colour and unrelated colour are defined as follows:

Related colour. Colour perceived to belong to an area seen in relation to other colours.

Unrelated colour. Colour perceived to belong to an area seen in isolation from other colours.

Most colours seen in everyday life are, in this sense, related colours, and have the attributes hue, lightness, brightness, colourfulness, saturation, and chroma. Unrelated colours, in this sense, include light sources, self-luminous signage, and signal lights seen at night; they have only the attributes hue, brightness, colourfulness, and saturation, because lightness and chroma require comparison with a similarly illuminated white or highly transmitting area. The colour appearance of signal lights is of vital importance, and the specification of their colorimetry was one of the first uses of the CIE colorimetric system (CIE, 1935). However, colorimetry does not define appearance, and it is this that is of paramount importance in signal lights. There is therefore a need for a model of colour appearance for unrelated colours, particularly those of small angular subtense seen in the dark (*aperture colours*) at a range of luminances that includes mesopic values. A model for such purposes will now be described. This model is based on experiments involving the magnitude estimation of brightness, colourfulness, and hue, for unrelated colours seen

Measuring Colour, Fourth Edition. R.W.G. Hunt and M.R. Pointer.

in a completely darkened room; the stimuli had angular subtenses of 10°, 2°, 1°, and 0.5°, and luminances ranging from 60 cd m^{-2} down to 0.1 cd m^{-2} (Fu, Li, Luo, Hunt, and Pointer, 2007; Fu, Li, Luo, Hunt, and Pointer, 2010).

17.2 A MODEL FOR PREDICTING UNRELATED COLOURS

The CIECAM02 model does not include any contribution from rod vision. When luminances in the mesopic range are involved, it is necessary to include appropriate inputs from the rods in the formulation of the achromatic signal. To do this accurately it is necessary to know the scotopic luminances, L_S, of the stimuli considered, but these are not usually available. For illuminants they can be approximated (Hunt, 1998a) by:

$$L_S/2.26 = L(T/4000 - 0.4)^{1/3}$$

where L is the photopic luminance, and T is the correlated colour temperature of the illuminant; $L_S/2.26$ is used instead of L_S because for the equi-energy stimulus, S_E, $L_S/2.26 = L$, its photopic luminance. The value of T for S_E is about 5600, so in this case

$$L_S/2.26 = L.$$

The precise value used for the scotopic luminances of the stimuli is not critical in this application, so the above formula is used for all the stimuli, not just for those having the same spectral power distribution as S_E. All the scotopic luminances were therefore taken in the model as:

$$L_S = 2.26L$$

The CIECAM02 model is used to calculate the usual (cone) achromatic signal, A_a, to which is added a proportion of the rod contribution, A_S, to provide a total achromatic signal, A. As with the cone signal, the scotopic response (representing the rod signal) is compressed by a power function. The exponent, 0.42, in the power function is the same as that used in the dynamic adaptation function in CIECAM02. The whole achromatic signal A is calculated as shown below.

$$A_S = L_S^{0.42}$$

$$A = A_a + kA_S$$

where the constant k is used to determine the ratio between cone and rod contributions, this being different for different viewing conditions. The constant k was determined by minimising the difference between model predictions and the magnitude-estimation experimental data.

As the luminances of the stimuli decrease, the contributions of the rods increase, so that the value of k increases. However, there is also an effect of field size; as the field size becomes smaller the rod contribution decreases, because of the smaller proportion of rods in small field sizes. In the model these effects are allowed for as follows.

When $L \geq 1\,\text{cd}\,\text{m}^{-2}$

$$k = -5.3 \times \log_{10}(L) + 44.5 \text{ for } 0.5^\circ \text{ stimuli}$$

$$k = -5.9 \times \log_{10}(L) + 50.3 \text{ for } 10^\circ \text{ stimuli}$$

For intermediate field sizes the values of the constants should be interpolated.

When $L < 1\,\text{cd}\,\text{m}^{-2}$

$$k = 12.7 \times \log_{10}(\theta) + 22.7$$

where θ is the stimulus size in degrees.

Colourfulness decreases as the angular subtense of the stimulus decreases, and this is allowed for by the constant K_M in calculating the unrelated colourfulness correlate, M_{unr}, from the CIECAM02 colourfulness correlate, M.

$$M_{unr} = K_M \times M$$

When $L \geq 1\,\text{cd}\,\text{m}^{-2}$

$$K_M = 0.9 \quad \text{for } 0.5^\circ \text{ stimuli}$$

$$K_M = 1 \quad \text{for } 10^\circ \text{ stimuli}$$

When $L < 1\,\text{cd}\,\text{m}^{-2}$

$$K_M = 0.1 \times \log_{10}(\theta) + 0.27$$

where θ is the stimulus size in degrees.

Brightness also decreases as the angular subtense of the stimulus decreases, and this is allowed for by using M_{unr} in calculating the unrelated brightness correlate, Q_{unr}.

$$Q_{unr} = A + M_{unr}/100$$

In a previous model, CAM97u (Hunt, 1998b), a conditioning field could be introduced to represent adaptation to a field of significant luminance viewed immediately before observing the unrelated colour; this model does not include such a conditioning field.

17.3 INPUT DATA REQUIRED FOR THE MODEL

The input data required for the model is as follows:

The chromaticity co-ordinates, x, y of the colour considered.

The luminance, L, of the colour considered.

The adapting luminance L_A. This is taken as $L_W/5$ where L_W is the luminance of a white stimulus in the group of stimuli being considered.

The luminance factor of the background, Y_b; although, for unrelated colours, this should logically be taken as zero, it was found that this resulted in very poor predictions and that a value of 20 gave much better results.

The stimulus size θ, in degrees.

17.4 STEPS IN USING THE MODEL FOR UNRELATED COLOURS

Step 1. Measure or calculate absolute tristimulus values, X_L, Y_L, Z_L, for the test colour stimulus corresponding to CIE colour-matching functions (2° or 10°).

$$X_L = xL/y \qquad Y_L = L \qquad Z_L = (1 - x - y)L/y$$

Step 2. Calculate the cone achromatic signal (A_a), colourfulness (M), and hue (H_c) based on the forward CIECAM02. Since there is no adopted white involved with unrelated colours, a white point of ($Y = 100$, $x = 1/3$, $y = 1/3$) is used as the adopted white in the calculation when CIECAM02 is used. The luminance of the adapting field is taken as 1/5 of the luminance of a white stimulus in the group of stimuli being considered; the surround parameters are set as those under dark viewing conditions: $c = 0.525$, $N_c = 0.8$, $F = 0.8$. The value of n is set at 0.2.

Step 3. The CIECAM02 correlates of hue (hue angle, h, and hue quadrature, H_c) are used unchanged.

Step 4. Calculate the scaling factor K_M.

When $L \geq 1\,\text{cd}\,\text{m}^{-2}$

$$K_M = 0.9 \quad \text{for } 0.5° \text{ stimuli}$$

$$K_M = 1 \quad \text{for } 10° \text{ stimuli}$$

When $L < 1\,\text{cd}\,\text{m}^{-2}$

$$K_M = 0.1 \times \log_{10}(\theta) + 0.27$$

where θ is the stimulus size in degrees.

Step 5. Calculate the unrelated colourfulness correlate, M_{unr}.

$$M_{unr} = K_M \times M$$

Step 6. Calculate the rod contribution A_S in the model.

$$\text{Scotopic luminances } L_S \qquad L_S = 2.26L$$

$$\text{Rod contribution } A_S \qquad A_S = L_S^{0.42}$$

Step 7. Calculate the whole achromatic signal A in the model (A_a is obtained from Step 2).

$$A = A_a + kA_S$$

when $L \geq 1\,\text{cd}\,\text{m}^{-2}$

$$k = -5.3 \times \log_{10}(L) + 44.5 \quad \text{for } 0.5° \text{ stimuli}$$

$$k = -5.9 \times \log_{10}(L) + 50.3 \quad \text{for } 10° \text{ stimuli}$$

when $L < 1\,\text{cd}\,\text{m}^{-2}$

$$k = 12.7 \times \log_{10}(\theta) + 22.7$$

Step 8. Calculate the unrelated brightness correlate (Q_{unr}).

$$Q_{unr} = A + M_{unr}/100$$

17.5 WORKED EXAMPLE IN THE MODEL FOR PREDICTING UNRELATED COLOURS

Note: in these examples the symbols $R'G'B'$ are used instead of $\rho\gamma\beta$

	x	y	L (cd m^{-2})	$\theta°$
Sample 1	0.4318	0.5018	5.00	10
Sample 2	0.4318	0.5018	0.10	1
L_w	1/3	1/3	100	

Sample 1					
X	Y	Z	4.302	5.000	0.662
X_w	Y_w	Z_w	100.0	100.0	100.0
Y_b	c	N_c	20.0	0.53	0.8
F	n	z	0.8	0.20	1.93
$\theta°$	L_A	D	10	1	0.661
N_{bb}	N_{cb}	F_L	1.00	1.00	0.17
R	G	B	103.86	109.29	14.63
R_w	G_w	B_w	100.00	100.00	100.00
R_c	G_c	B_c	103.86	109.29	14.63
R_{cw}	G_{cw}	B_{cw}	100.0	100.0	100.0
X_c	Y_c	Z_c	86.0	100.0	13.2
X_{cw}	Y_{cw}	Z_{cw}	100.0	100.0	100.0
R'	G'	B'	101.4	99.2	13.2
R'_w	G'_w	B'_w	100.0	100.0	100.0
R'_a	G'_a	B'_a	7.0	7.0	3.1
R'_{aw}	G'_{aw}	B'_{aw}	7.0	7.0	7.0
a	b	h	−0.29181	0.874057	108.4618
e	H_c	A_w	2361.704	Y68G32	21.0801
A_a	s	L_s	20.9392	60.3257	11.30
A_s	k	A	2.77	46.18	148.79
M_{unr}	Q_{unr}	H	44.60	149.24	132.0999
Sample 2					
X	Y	Z	0.086	0.100	0.013
X_w	Y_w	Z_w	100.0	100.0	100.0
Y_b	c	N_c	20.0	0.53	0.8
F	n	z	0.8	0.20	1.93
$\theta°$	L_A	D	1	0.02	0.659

(Continued)

(*Continued*)

	x		y	L (cd m^{-2})	$\theta°$
N_{bb}	N_{cb}	F_L	1.00	1.00	0.02
R	G	B	103.86	109.29	14.63
R_w	G_w	B_w	100.00	100.00	100.00
R_c	G_c	B_c	103.86	109.29	14.63
R_{cw}	G_{cw}	B_{cw}	100.0	100.0	100.0
X_c	Y_c	Z_c	86.0	100.0	13.2
X_{cw}	Y_{cw}	Z_{cw}	100.0	100.0	100.0
R'	G'	B'	101.4	99.2	13.2
R'_w	G'_w	B'_w	100.0	100.0	100.0
R'_a	G'_a	B'_a	2.8	2.8	1.3
R'_{aw}	G'_{aw}	B'_{aw}	2.8	2.8	2.8
a	b	h	−0.11565	0.346889	108.4383
e	H_c	A_w	2361.489	Y68G32	8.3280
A_a	s	L_s	8.2722	85.2238	0.23
A_s	k	A	0.54	22.70	20.43
M_{unr}	Q_{unr}	H	6.75	20.49	132.063

REFERENCES

CIE Compte Rendu des Séances. Bureau Central de la CIE, Paris, France (1935).

Fu, C., Li, C., Luo, M.R., Hunt, R.W.G., and Pointer, M.R., Quantifying colour appearance for unrelated colour under photopic and mesopic vision, *IS&T and SID's 15th Color Imaging Conference: Color Science and Engineering Systems, Technologies, and Applications*, pp. 319–324, IS&T, Springfield, VA, USA (2007).

Fu, C., Li, C., Luo, M.R., Hunt, R.W.G., and Pointer, M.R., An investigation of colour appearance for unrelated colours under photopic and mesopic vision, *Color Res. Appl.* submitted (2010).

Hunt, R.W.G., *Measuring Colour, 3rd* Ed., page 238, Fountain Press, Kingston- upon-Thames, England (1998a).

Hunt, R.W.G., *Measuring* Colour, *3rd* Ed., pages 239–241, Fountain Press, Kingston- upon-Thames, England (1998b).

Appendices

The Appendices contain numerical and other information that enable colorimetric computations to be carried out.

Some of the data included in these Appendices are from CIE Publications, including CIE Publication 15:2004, and are reproduced with the permission of the CIE Central Bureau, Vienna.

Appendix 1

Radiometric and Photometric Terms and Units

A1.1 INTRODUCTION

Photometry is normally based on the photopic spectral luminous efficiency function, $V(\lambda)$, tabulated in Appendix 2, which was standardised by the CIE in 1924. The $V(\lambda)$ function is used as a weighting function for evaluating the total amount of light in a mixture of radiations of different wavelengths. If the radiant flux at wavelength λ_1 is P_1, that at wavelength λ_2 is P_2, etc., and the values of the $V(\lambda)$ function at these wavelengths are V_1, V_2, etc., then the *luminous flux* is given by:

$$L = K_m(P_1V_1 + P_2V_2 + \ldots P_nV_n)$$

The values, P, represent the radiant fluxes (powers) per small constant-width wavelength intervals centred on the wavelengths, λ, the intervals together covering all of the visible spectrum. The value of the constant, K_m, will be discussed later. (Although the values of the $V(\lambda)$ function at wavelengths below about 450 nm are now known to be too low, the resulting errors are negligible in most practical cases. More accurate values, $V_M(\lambda)$, are given in Appendix 2.)

When the level of illumination is so low that vision is mediated by the rods instead of by the cones, photometry is based on the scotopic spectral luminance efficiency function, $V'(\lambda)$, also tabulated in Appendix 2, which was standardised by the CIE in 1951. Unless otherwise stated, it is always assumed that the photopic function, $V(\lambda)$, is being used.

A1.2 PHYSICAL DETECTORS

Practical *photometry* is now almost always carried out using some kind of photo-electric physical detector, because this provides more precise and quicker measurements than

Measuring Colour, Fourth Edition. R.W.G. Hunt and M.R. Pointer.

are possible with visual photometry. But these physical detectors do not have spectral sensitivities that correspond to the $V(\lambda)$ or $V'(\lambda)$ functions of the eye. They therefore have to be used with filters that modify their spectral sensitivities to match these functions as closely as possible. Such filtered detectors then measure *luminous flux* (or *scotopic luminous flux* in the case of the $V'(\lambda)$ function).

Detectors used for *radiometry* should be equally sensitive to radiant flux of all wavelengths; this is usually only the case for thermopiles and bolometers, and other detectors require correcting filters to obtain equal sensitivity at all wavelengths as closely as possible. Such filtered detectors then measure *radiant flux*.

Photochemical and photobiological processes often depend on the number of quanta, irrespective of their energies, above a threshold. For these applications, measurements of the rate of flow of quanta are required, and physical detectors then need yet another type of correcting filter. Such filtered detectors then measure *quantum flux*.

For physical detectors to produce measurements that are true functions of these fluxes, it is necessary for them to have responses that are linearly proportional to the radiation incident upon them, and for the effects of radiation of different wavelengths to add together linearly.

An alternative to using filtered detectors, is to measure the radiant power at each small constant-width wavelength interval throughout the spectrum, and then to use the appropriate weighting function to convert the spectro-radiometric results obtained to luminous or quantum measures as required.

A1.3 PHOTOMETRIC UNITS AND TERMS

Lumen, lm

Unit of luminous flux (radiant flux weighted by the $V(\lambda)$ function). The luminous flux of a beam of monochromatic radiation whose frequency is 540×10^{12} hertz, and whose radiant flux is 1/683 watt, is 1 lumen. The frequency of 540×10^{12} hertz is closely equal to a wavelength of 555 nm, the wavelength for which the $V(\lambda)$ function has its maximum value of 1.0.

> The constant, K_{m}, in the equation given in Section A1.1 for evaluating the amount of light, L, is put equal to 683 to obtain the luminous flux in lumens when the radiant flux is expressed in watts. (If the scotopic function $V'(\lambda)$ is being used, K'_{m} is used instead of K_{m}, where K'_{m} is equal to 1700.)

Candela, cd

Unit of luminous intensity (luminous flux per unit solid angle). The luminous intensity, in a given direction, of a point source emitting 1 lumen per steradian is 1 candela.

Lux, lx

Unit of *illuminance* (luminous flux per unit area incident on a surface). The illuminance produced by a luminous flux of 1 lumen uniformly distributed over a surface of area 1 square metre is 1 lux.

Foot-candle or Lumen per square-foot, fc

Unit of illuminance. The illuminance produced by a luminous flux of 1 lumen uniformly distributed over a surface of area of 1 square foot is 1 foot-candle. 1 lumen per square foot = 10.76 lux.

Candela per square metre, cd m^{-2}

Unit of luminance (luminous flux per unit solid angle and per unit projected area, in a given direction, at a point on a surface). The luminance produced by a luminous intensity of 1 candela uniformly distributed over a surface area of 1 square metre is 1 candela per square metre. (Obsolete name: *nit*).

Other units of luminance are the lambert, the millilambert, the foot-lambert (sometimes called the equivalent foot-candle), and the apostilb.

$$\begin{aligned}
&1\ \text{lambert} &&= 10^4/\pi\ \text{cd}\,\text{m}^{-2} \\
&1\ \text{millilambert} &&= 10/\pi\ \text{cd}\,\text{m}^{-2} \\
&1\ \text{foot-lambert} &&= 10.76/\pi\ \text{cd m}^{-2} = 3.426\,\text{cd m}^{-2} \\
&1\ \text{apostilb} &&= 1/\pi\ \text{cd m}^{-2} = 0.3183\,\text{cd m}^{-2}
\end{aligned}$$

Luminance factor, β

Ratio of the luminance of an area to that of the perfect diffuser (an ideal isotropic diffuser with a reflectance equal to unity) identically illuminated.

Calculation of luminance from illuminance

An isotropic diffuser of luminance factor, β, under an illuminance, E lux, has a luminance, L, given by:

$$L = E\beta/\pi\ \text{cd}\,\text{m}^{-2}$$

Lumen per square metre, lm m^{-2}

Unit of *luminous exitance* (luminous flux per unit area leaving a surface).

Radiant efficiency

Ratio of the radiant flux emitted to the power consumed by a source.

Luminous efficacy

a. Of a source. Ratio of luminous flux emitted to power consumed. Unit: lm W^{-1}.
b. Of a radiation. Ratio of luminous flux to radiant flux. Unit lm W^{-1}.

The maximum spectral luminous efficacy for photopic vision occurs at a wavelength of 555 nm and is equal to 683 lm W^{-1}; and for scotopic vision occurs at a wavelength of 510 nm and is equal to 1700 scotopic lm W^{-1}.

Luminous efficiency

Ratio of radiant flux weighted according to the $V(\lambda)$ function, to the radiant flux.

Point brilliance

The illuminance produced by a source on a plane at the observer's eye, when the apparent diameter of the source is inappreciable. Unit: lx.

Troland

Unit used to express a quantity proportional to retinal illuminance produced by a light stimulus. When the eye views a surface of uniform luminance, the number of trolands is equal to the product of the area in square millimetres of the limiting pupil, natural or artificial, times the luminance of the surface in candelas per square metre.

A1.4 RADIANT AND QUANTUM UNITS AND TERMS

Most of the above terms and units have radiant and quantum equivalents, in which radiant flux and quantum flux, respectively, are used instead of luminous flux. All three types of term and unit are listed in Table A1.1.

Table A1.1 Luminous, radiant, and quantum terms and units

Luminous	Radiant	Quantum
luminous flux, F lm	radiant flux, F_e W	quantum flux, F_q s^{-1}
luminous intensity, I cd (lm sr^{-1})	radiant intensity, I_e W sr^{-1}	quantum intensity, I_q s^{-1} sr^{-1}
luminance, L cd m^{-2}	radiance, L_e W sr^{-1} m^{-2}	quantum radiance, L_q s^{-1} sr^{-1} m^{-2}
illuminance, E lx (lm m^{-2})	irradiance, E_e W m^{-2}	quantum irradiance, E_q s^{-1} m^{-2}
light exposure, H lx s	radiant exposure, H_e J m^{-2}	quantum exposure, H_q m^{-2}
luminous exitance, M lm m^{-2}	radiant exitance, M_e W m^{-2}	quantum exitance, M_q s^{-1} m^{-2}

Abbreviations:
lm: lumen W: watt s: second
cd: candela J: joule (watt second) m: metre
lx: lux sr: steradian (unit solid angle)

A1.5 RADIATION SOURCES

Incandescent sources

Incandescent sources emit radiation in accordance with their temperature and emissive properties.

Planckian (black-body) sources

Planckian sources emit radiation in accordance with Planck's law: this gives the spectral concentration of radiant exitance, M_e, in watts per square metre per wavelength interval, as a function of wavelength, λ, in metres, and temperature, T, in kelvins, by the formula:

$$M_e = c_1\lambda^{-5}(e^{c_2/\lambda T} - 1)^{-1}$$

in $W\,m^{-3}$, where $c_1 = 3.74183 \times 10^{-16}\,W\,m^2$, $c_2 = 1.4388 \times 10^{-2}\,m\,K$, and $e = 2.718282$.

A1.6 TERMS FOR MEASURES OF REFLECTION AND TRANSMISSION

The terms *reflection* and *transmission* refer to the processes of radiation being returned by a medium, or passed on through a medium, respectively. The amount of radiation reflected or transmitted by the medium relative to that by the perfect diffusing or non-diffusing reflector (perfect diffuser or perfect mirror), or perfect diffusing or non-diffusing transmitter, is expressed by means of different terms according to the geometrical arrangement for collecting the radiation. *Reflectance factor* and *transmittance factor* are used when the radiations reflected or transmitted by the sample and by the perfect diffuser or transmitter lie within a defined cone. If this cone is a hemisphere, the terms *reflectance* and *transmittance* are used. If the cone is very small, the term *luminance factor* or *radiance factor* is used. *Reflectometer value* is used for measurement by means of a particular instrument for measuring reflected radiation. The following adjectives are used to denote the type of radiation involved:

Spectral, when the radiation is monochromatic. In this case the symbol is followed by (λ), thus $\rho(\lambda)$, or $\tau(\lambda)$, for example. If the quantity is an absolute amount, this is denoted by an additional subscript λ; for example, $S_\lambda(\lambda)$.

Radiant, when the radiation is evaluated in terms of its total power or energy. In this case the symbol has a subscript e, thus ρ_e, or τ_e, for example.

Luminous, when the radiation is evaluated by using the $V(\lambda)$ function as a weighting function.

The terms themselves are defined as follows:

Reflectance, ρ [transmittance, τ]

Ratio of the reflected [transmitted] radiant or luminous flux to the incident flux under specified conditions of irradiation.

Note 1. Regular reflectance, ρ_r [*regular transmittance,* τ_r] is the ratio of the part of the regularly reflected [transmitted] flux to the incident flux.

Note 2. Diffuse reflectance, ρ_d [*diffuse transmittance,* τ_d] is the ratio of the part of the flux reflected [transmitted] by diffusion to the incident flux.

Note 3. $\rho = \rho_r + \rho_d$; $\tau = \tau_r + \tau_d$

Note 4. In the case of transmitting samples, if the irradiation and the collection are both diffuse, the ratio is referred to as doubly diffuse transmittance, τ_{dd}.

The perfect reflecting [transmitting] diffuser

An ideal isotropic diffuser with a reflectance [transmittance] equal to unity.

In this definition by *isotropic*, is meant that the radiation is reflected [transmitted] equally strongly in all directions, so that the perfect diffuser has a luminance that is independent of the direction of viewing; by a reflectance [transmittance] *equal to unity* is meant that the perfect diffuser reflects [transmits] all the light that is incident on it at every wavelength throughout the visible spectrum.

Reflectance factor, R [transmittance factor, T]

(at a representative surface element, for the part of the radiation contained in a given cone with apex at the representative surface element, and for incident radiation of given spectral composition and geometrical distribution):

Ratio of the radiant or luminous flux reflected [transmitted] in the directions limited by the cone to that reflected [transmitted] in the same directions by the perfect diffuser identically irradiated.

Note 1. For regularly reflecting surfaces that are irradiated by a source of small solid angle, the reflectance factor may be much larger than 1 if the cone contains the mirror image of the source.

Note 2. If the solid angle of the cone approaches a hemisphere, the reflectance [transmittance] factor approaches the reflectance [transmittance]. In this case, a regular component (if present) must be included; instruments employing integrating spheres may approximate this condition. A regular component can be excluded if it is sufficiently defined and a suitable trap is used; then the reflectance [transmittance] factor approaches the diffuse reflectance [transmittance]; instruments employing integrating spheres with gloss traps may approximate this condition.

Note 3. If the solid angle of the cone approaches zero, the reflectance [transmittance] factor approaches the radiance factor or luminance factor.

Radiance factor, β_e [luminance factor, β]

(at a representative surface element of a non-self-radiating medium, in a given direction, under specified conditions of irradiation)

Ratio of the radiance [luminance] of the medium to that of the perfect diffuser identically irradiated.

Note. In the case of photoluminescent media, the radiance [luminance] factor is the sum β_T, of two portions, the *reflected radiance [luminance] factor*, β_S, and the *luminescent radiance [luminance] factor*, β_L; $\beta_T = \beta_S + \beta_L$.

Density, D

Logarithm to the base 10 of the reciprocal of the reflectance factor or the transmittance factor, or other similar measure.

Absorptance, α

Ratio of the absorbed radiant or luminous flux to the incident flux under specified conditions.

A1.7 OTHER SPECTRAL LUMINOUS EFFICIENCY FUNCTIONS

The spectral luminous efficiency functions based on brightness matching for monochromatic 2^0 and 10^0 fields, $V_{b,2}(\lambda)$ and $V_{b,10}(\lambda)$, respectively, are also given in Appendix 2. These functions cannot be used as weighting functions for stimuli consisting of a variety of wavelengths.

A1.8 MESOPIC PHOTOMETRY

The recommended system for visual performance-based mesopic photometry describes spectral luminous efficiency, $V_{mes}(\lambda)$, in the mesopic region, as a linear combination of the photopic spectral luminous efficiency function, $V(\lambda)$, and the scotopic spectral luminous efficiency function, $V'(\lambda)$, and establishes a gradual transition between these two functions throughout the mesopic region.

The mesopic spectral luminous efficiency, $V_{mes}(\lambda)$, is of the form:

$$V_{mes}(\lambda) = [mV(\lambda) + (1-m)V'(\lambda)] \qquad \text{for } 0 \leq m \leq 1$$

where $V_{mes}(\lambda)$ should be normalised such that the function attains a maximum value of 1.

The coefficient, m, and the mesopic luminance, L_{mes}, can be calculated using an iterative approach as follows:

$$m_0 = 0.5,$$

$$L_{mes,n} = [m_{(n-1)}L_p + (1-m_{(n-1)})L_s V'(\lambda_0)]/[m_{(n-1)} + (1-m_{(n-1)})V'(\lambda_0)]$$

$$m_n = 0.7670 + 0.3334\log_{10}(L_{mes,n}) \quad \text{for } 0 \leq m_n \leq 1$$

where L_p is the photopic luminance, L_s the scotopic luminance, $V'(\lambda_0) = 683/1699$ is the value of scotopic spectral luminous efficiency function at $\lambda_0 = 555$ nm, and n is the iteration step.

The scotopic luminance is then defined as:

$$L_{mes} = \frac{683}{V_{mes}(\lambda_0)} \int V_{mes}(\lambda) L_e(\lambda) d\lambda$$

where:

L_{mes} is the mesopic luminance
$V_{mes}(\lambda_0)$ is the value of $V_{mes}(\lambda)$ at 555 nm

$L_e(\lambda)$ is the spectral radiance in W m^{-2} sr^{-1} nm^{-1}
if $L_{mes} \geq 5.0$ cd m^{-2}, then $m = 1$
if $L_{mes} \leq 0.005$ cd m^{-2}, then $m = 0$

REFERENCE

CIE Publication No.191:2010: *Recommended System for Visual Performance Based Mesopic Photometry*, CIE, Vienna (2010).

Appendix 2

Spectral Luminous Efficiency Functions

Wavelength, nm	$V(\lambda)$	$V_M(\lambda)$	$V'(\lambda)$	$V_{b,2}(\lambda)$	$V_{b,10}(\lambda)$
380	0.000039	0.0002	0.000589		
390	0.00012	0.0008	0.002209		
400	0.000396	0.0028	0.00929	0.0087	0.0087
410	0.00121	0.0074	0.03484	0.023	0.027
420	0.004	0.0175	0.0966	0.045	0.072
430	0.0116	0.0273	0.1998	0.065	0.115
440	0.023	0.0379	0.3281	0.085	0.17
450	0.038	0.0468	0.455	0.10	0.21
460	0.06	0.06	0.567	0.13	0.28
470	0.091	0.09098	0.676	0.18	0.36
480	0.139	0.13902	0.793	0.23	0.47
490	0.208	0.20802	0.904	0.30	0.56
500	0.323	0.323	0.982	0.44	0.72
510	0.503	0.503	0.997	0.71	0.93
520	0.71	0.71	0.935	0.98	1.15
530	0.862	0.862	0.811	1.20	1.20
540	0.954	0.954	0.65	1.26	1.26
550	0.995	0.99495	0.481	1.29	1.29
560	0.995	0.995	0.3288	1.15	1.15
570	0.952	0.952	0.2076	1.00	1.00

(*Continued*)

Measuring Colour, Fourth Edition. R.W.G. Hunt and M.R. Pointer.

(*Continued*)

Wavelength, nm	$V(\lambda)$	$V_M(\lambda)$	$V'(\lambda)$	$V_{b,2}(\lambda)$	$V_{b,10}(\lambda)$
580	0.87	0.87	0.1212	0.93	0.93
590	0.757	0.757	0.0655	0.89	0.89
600	0.631	0.631	0.03315	0.81	0.81
610	0.503	0.503	0.01593	0.72	0.72
620	0.381	0.381	0.00737	0.62	0.62
630	0.265	0.265	0.003335	0.48	0.48
640	0.175	0.175	0.001497	0.33	0.33
650	0.107	0.107	0.000677	0.21	0.21
660	0.061	0.061	0.0003129	0.129	0.129
670	0.032	0.032	0.000148	0.071	0.071
680	0.017	0.017	0.0000715	0.036	0.036
690	0.0082	0.00821	0.00003533	0.018	0.018
700	0.0041	0.004102	0.0000178	0.0093	0.0093
710	0.0021	0.002091	0.00000914	0.0045	0.0045
720	0.00105	0.001047	0.00000478	0.0022	0.0022
730	0.00052	0.00052	0.000002546	0.0011	0.0011
740	0.00025	0.0002492	0.000001379		
750	0.00012	0.00012	0.00000076		
760	0.00006	0.00006	0.000000425		
770	0.00003	0.00003	0.000000241		
780	0.000015	0.00001499	0.000000139		

Appendix 3

CIE Colour-Matching Functions

Wavelength, nm	$\bar{x}(\lambda)$	$\bar{y}(\lambda)$	$\bar{z}(\lambda)$	$\bar{x}_{10}(\lambda)$	$\bar{y}_{10}(\lambda)$	$\bar{z}_{10}(\lambda)$
380	0.0014	0.0000	0.0065	0.0002	0.0000	0.0007
385	0.0022	0.0001	0.0105	0.0007	0.0001	0.0029
390	0.0042	0.0001	0.0201	0.0024	0.0003	0.0105
395	0.0076	0.0002	0.0362	0.0072	0.0008	0.0323
400	0.0143	0.0004	0.0679	0.0191	0.0020	0.0860
405	0.0232	0.0006	0.1102	0.0434	0.0045	0.1971
410	0.0435	0.0012	0.2074	0.0847	0.0088	0.3894
415	0.0776	0.0022	0.3713	0.1406	0.0145	0.6568
420	0.1344	0.0040	0.6456	0.2045	0.0214	0.9725
425	0.2148	0.0073	1.0391	0.2647	0.0295	1.2825
430	0.2839	0.0116	1.3856	0.3147	0.0387	1.5535
435	0.3285	0.0168	1.6230	0.3577	0.0496	1.7985
440	0.3483	0.0230	1.7471	0.3837	0.0621	1.9673
445	0.3481	0.0298	1.7826	0.3867	0.0747	2.0273
450	0.3362	0.0380	1.7721	0.3707	0.0895	1.9948
455	0.3187	0.0480	1.7441	0.3430	0.1063	1.9007
460	0.2908	0.0600	1.6692	0.3023	0.1282	1.7454
465	0.2511	0.0739	1.5281	0.2541	0.1528	1.5549
470	0.1954	0.0910	1.2876	0.1956	0.1852	1.3176
475	0.1421	0.1126	1.0419	0.1323	0.2199	1.0302
480	0.0956	0.1390	0.8130	0.0805	0.2536	0.7721
485	0.0580	0.1693	0.6162	0.0411	0.2977	0.5701
490	0.0320	0.2080	0.4652	0.0162	0.3391	0.4153
495	0.0147	0.2586	0.3533	0.0051	0.3954	0.3024

(*Continued*)

Measuring Colour, Fourth Edition. R.W.G. Hunt and M.R. Pointer.

(Continued)

Wavelength, nm	$\bar{x}(\lambda)$	$\bar{y}(\lambda)$	$\bar{z}(\lambda)$	$\bar{x}_{10}(\lambda)$	$\bar{y}_{10}(\lambda)$	$\bar{z}_{10}(\lambda)$
500	0.0049	0.3230	0.2720	0.0038	0.4608	0.2185
505	0.0024	0.4073	0.2123	0.0154	0.5314	0.1592
510	0.0093	0.5030	0.1582	0.0375	0.6067	0.1120
515	0.0291	0.6082	0.1117	0.0714	0.6857	0.0822
520	0.0633	0.7100	0.0782	0.1177	0.7618	0.0607
525	0.1096	0.7932	0.0573	0.1730	0.8233	0.0431
530	0.1655	0.8620	0.0422	0.2365	0.8752	0.0305
535	0.2257	0.9149	0.0298	0.3042	0.9238	0.0206
540	0.2904	0.9540	0.0203	0.3768	0.9620	0.0137
545	0.3597	0.9803	0.0134	0.4516	0.9822	0.0079
550	0.4334	0.9950	0.0087	0.5298	0.9918	0.0040
555	0.5121	1.0000	0.0057	0.6161	0.9991	0.0011
560	0.5945	0.9950	0.0039	0.7052	0.9973	0.0000
565	0.6784	0.9786	0.0027	0.7938	0.9824	0.0000
570	0.7621	0.9520	0.0021	0.8787	0.9556	0.0000
575	0.8425	0.9154	0.0018	0.9512	0.9152	0.0000
580	0.9163	0.8700	0.0017	1.0142	0.8689	0.0000
585	0.9786	0.8163	0.0014	1.0743	0.8256	0.0000
590	1.0263	0.7570	0.0011	1.1185	0.7774	0.0000
595	1.0567	0.6949	0.0010	1.1343	0.7204	0.0000
600	1.0622	0.6310	0.0008	1.1240	0.6583	0.0000
605	1.0456	0.5668	0.0006	1.0891	0.5939	0.0000
610	1.0026	0.5030	0.0003	1.0305	0.5280	0.0000
615	0.9384	0.4412	0.0002	0.9507	0.4618	0.0000
620	0.8544	0.3810	0.0002	0.8563	0.3981	0.0000
625	0.7514	0.3210	0.0001	0.7549	0.3396	0.0000
630	0.6424	0.2650	0.0000	0.6475	0.2835	0.0000
635	0.5419	0.2170	0.0000	0.5351	0.2283	0.0000
640	0.4479	0.1750	0.0000	0.4316	0.1798	0.0000
645	0.3608	0.1382	0.0000	0.3437	0.1402	0.0000
650	0.2835	0.1070	0.0000	0.2683	0.1076	0.0000
655	0.2187	0.0816	0.0000	0.2043	0.0812	0.0000
660	0.1649	0.0610	0.0000	0.1526	0.0603	0.0000
665	0.1212	0.0446	0.0000	0.1122	0.0441	0.0000
670	0.0874	0.0320	0.0000	0.0813	0.0318	0.0000
675	0.0636	0.0232	0.0000	0.0579	0.0226	0.0000
680	0.0468	0.0170	0.0000	0.0409	0.0159	0.0000
685	0.0329	0.0119	0.0000	0.0286	0.0111	0.0000
690	0.0227	0.0082	0.0000	0.0199	0.0077	0.0000
695	0.0158	0.0057	0.0000	0.0138	0.0054	0.0000

(Continued)

(Continued)

Wavelength, nm	$\bar{x}(\lambda)$	$\bar{y}(\lambda)$	$\bar{z}(\lambda)$	$\bar{x}_{10}(\lambda)$	$\bar{y}_{10}(\lambda)$	$\bar{z}_{10}(\lambda)$
700	0.0114	0.0041	0.0000	0.0096	0.0037	0.0000
705	0.0081	0.0029	0.0000	0.0066	0.0026	0.0000
710	0.0058	0.0021	0.0000	0.0046	0.0018	0.0000
715	0.0041	0.0015	0.0000	0.0031	0.0012	0.0000
720	0.0029	0.0010	0.0000	0.0022	0.0008	0.0000
725	0.0020	0.0007	0.0000	0.0015	0.0006	0.0000
730	0.0014	0.0005	0.0000	0.0010	0.0004	0.0000
735	0.0010	0.0004	0.0000	0.0007	0.0003	0.0000
740	0.0007	0.0002	0.0000	0.0005	0.0002	0.0000
745	0.0005	0.0002	0.0000	0.0004	0.0001	0.0000
750	0.0003	0.0001	0.0000	0.0003	0.0001	0.0000
755	0.0002	0.0001	0.0000	0.0002	0.0001	0.0000
760	0.0002	0.0001	0.0000	0.0001	0.0000	0.0000
765	0.0001	0.0000	0.0000	0.0001	0.0000	0.0000
770	0.0001	0.0000	0.0000	0.0001	0.0000	0.0000
775	0.0001	0.0000	0.0000	0.0000	0.0000	0.0000
780	0.0000	0.0000	0.0000	0.0000	0.0000	0.0000

Appendix 4

CIE Spectral Chromaticity Co-Ordinates

Wavelength, nm	$x(\lambda)$	$y(\lambda)$	$u'(\lambda)$	$v'(\lambda)$	$x_{10}(\lambda)$	$y_{10}(\lambda)$	$u'_{10}(\lambda)$	$v'_{10}(\lambda)$
380	0.1741	0.0050	0.2569	0.0165	0.1813	0.0197	0.2524	0.0617
385	0.1740	0.0050	0.2567	0.0165	0.1809	0.0195	0.2519	0.0612
390	0.1738	0.0049	0.2564	0.0163	0.1803	0.0194	0.2512	0.0606
395	0.1736	0.0049	0.2560	0.0163	0.1795	0.0190	0.2502	0.0597
400	0.1733	0.0048	0.2558	0.0159	0.1784	0.0187	0.2498	0.0587
405	0.1730	0.0048	0.2553	0.0159	0.1771	0.0184	0.2472	0.0578
410	0.1726	0.0048	0.2545	0.0159	0.1755	0.0181	0.2449	0.0569
415	0.1721	0.0048	0.2537	0.0160	0.1732	0.0178	0.2417	0.0599
420	0.1714	0.0051	0.2522	0.0169	0.1706	0.0179	0.2376	0.0559
425	0.1703	0.0058	0.2496	0.0191	0.1679	0.0187	0.2325	0.0583
430	0.1689	0.0069	0.2461	0.0226	0.1650	0.0203	0.2266	0.0627
435	0.1669	0.0086	0.2411	0.0278	0.1622	0.0225	0.2202	0.0687
440	0.1644	0.0109	0.2347	0.0349	0.1590	0.0257	0.2127	0.0774
445	0.1611	0.0138	0.2267	0.0437	0.1554	0.0300	0.2038	0.0886
450	0.1566	0.0177	0.2161	0.0550	0.1510	0.0364	0.1926	0.1046
455	0.1510	0.0227	0.2033	0.0689	0.1459	0.0452	0.1796	0.1252
460	0.1440	0.0297	0.1877	0.0871	0.1389	0.0589	0.1620	0.1546
465	0.1355	0.0399	0.1690	0.1119	0.1295	0.0779	0.1410	0.1907
470	0.1241	0.0578	0.1441	0.1510	0.1152	0.1090	0.1130	0.2406
475	0.1096	0.0868	0.1147	0.2044	0.0957	0.1591	0.0812	0.3035
480	0.0913	0.1327	0.0828	0.2708	0.0728	0.2292	0.0519	0.3618

(*Continued*)

(Continued)

Wavelength, nm	$x(\lambda)$	$y(\lambda)$	$u'(\lambda)$	$v'(\lambda)$	$x_{10}(\lambda)$	$y_{10}(\lambda)$	$u'_{10}(\lambda)$	$v'_{10}(\lambda)$
485	0.0687	0.2007	0.0521	0.3427	0.0452	0.3275	0.0264	0.4310
490	0.0454	0.2950	0.0282	0.4117	0.0210	0.4401	0.0102	0.4807
495	0.0235	0.4127	0.0119	0.4698	0.0073	0.5625	0.0030	0.5200
500	0.0082	0.5384	0.0035	0.5131	0.0056	0.6745	0.0020	0.5477
505	0.0039	0.6548	0.0014	0.5432	0.0219	0.7526	0.0073	0.5650
510	0.0139	0.7502	0.0046	0.5638	0.0495	0.8023	0.0158	0.5763
515	0.0389	0.8120	0.0123	0.5770	0.0850	0.8170	0.0269	0.5820
520	0.0743	0.8338	0.0231	0.5837	0.1252	0.8102	0.0402	0.5847
525	0.1142	0.8262	0.0360	0.5861	0.1664	0.7922	0.0547	0.5857
530	0.1547	0.8059	0.0501	0.5868	0.2071	0.7663	0.0703	0.5854
535	0.1929	0.7816	0.0643	0.5865	0.2436	0.7399	0.0856	0.5846
540	0.2296	0.7543	0.0792	0.5856	0.2786	0.7113	0.1015	0.5831
545	0.2658	0.7243	0.0953	0.5941	0.3132	0.6813	0.1188	0.5812
550	0.3016	0.6923	0.1127	0.5821	0.3473	0.6501	0.1375	0.5789
555	0.3373	0.6589	0.1319	0.5796	0.3812	0.6182	0.1579	0.5762
560	0.3731	0.6245	0.1531	0.5766	0.4142	0.5858	0.1801	0.5730
565	0.4087	0.5896	0.1766	0.5732	0.4469	0.5531	0.2045	0.5693
570	0.4441	0.5547	0.2026	0.5694	0.4791	0.5209	0.2310	0.5653
575	0.4788	0.5202	0.2312	0.5651	0.5096	0.4904	0.2592	0.5611
580	0.5125	0.4866	0.2623	0.5604	0.5386	0.4614	0.2888	0.5567
585	0.5448	0.4544	0.2959	0.5554	0.5654	0.4346	0.3193	0.5521
590	0.5752	0.4242	0.3315	0.5501	0.5900	0.4100	0.3501	0.5457
595	0.6029	0.3965	0.3681	0.5446	0.6116	0.3884	0.3800	0.5430
600	0.6270	0.3725	0.4035	0.5393	0.6306	0.3694	0.4088	0.5387
605	0.6482	0.3514	0.4380	0.5342	0.6471	0.3529	0.4358	0.5346
610	0.6658	0.3340	0.4691	0.5296	0.6612	0.3388	0.4605	0.5309
615	0.6801	0.3197	0.4967	0.5254	0.6731	0.3269	0.4827	0.5276
620	0.6915	0.3083	0.5202	0.5219	0.6827	0.3173	0.5017	0.5247
625	0.7006	0.2993	0.5399	0.5190	0.6898	0.3120	0.5163	0.5225
630	0.7079	0.2920	0.5565	0.5165	0.6955	0.3045	0.5286	0.5207
635	0.7140	0.2859	0.5709	0.5144	0.7010	0.2990	0.5407	0.5189
640	0.7190	0.2809	0.5830	0.5125	0.7059	0.2941	0.5517	0.5172
645	0.7230	0.2770	0.5930	0.5110	0.7103	0.2898	0.5619	0.5157
650	0.7260	0.2740	0.6005	0.5099	0.7137	0.2863	0.5700	0.5145
655	0.7283	0.2717	0.6064	0.5090	0.7156	0.2844	0.5746	0.5138
660	0.7300	0.2700	0.6108	0.5084	0.7168	0.2832	0.5775	0.5134
665	0.7311	0.2689	0.6138	0.5079	0.7179	0.2821	0.5802	0.5130
670	0.7320	0.2680	0.6161	0.5076	0.7187	0.2813	0.5822	0.5127
675	0.7327	0.2673	0.6181	0.5073	0.7193	0.2807	0.5837	0.5124
680	0.7334	0.2666	0.6200	0.5070	0.7198	0.2802	0.5848	0.5123

(Continued)

(Continued)

Wavelength, nm	$x(\lambda)$	$y(\lambda)$	$u'(\lambda)$	$v'(\lambda)$	$x_{10}(\lambda)$	$y_{10}(\lambda)$	$u'_{10}(\lambda)$	$v'_{10}(\lambda)$
685	0.7340	0.2660	0.6216	0.5068	0.7200	0.2800	0.5854	0.5122
690	0.7344	0.2656	0.6226	0.5066	0.7202	0.2798	0.5858	0.5121
695	0.7346	0.2654	0.6231	0.5065	0.7203	0.2797	0.5861	0.5121
700	0.7347	0.2653	0.6234	0.5065	0.7204	0.2796	0.5863	0.5121
705	0.7347	0.2653	0.6234	0.5065	0.7203	0.2797	0.5862	0.5121
710	0.7347	0.2653	0.6234	0.5065	0.7202	0.2798	0.5859	0.5121
715	0.7347	0.2653	0.6234	0.5065	0.7201	0.2799	0.5856	0.5122
720	0.7347	0.2653	0.6234	0.5065	0.7199	0.2801	0.5851	0.5122
725	0.7347	0.2653	0.6234	0.5065	0.7197	0.2803	0.5846	0.5123
730	0.7347	0.2653	0.6234	0.5065	0.7195	0.2806	0.5840	0.5124
735	0.7347	0.2653	0.6234	0.5065	0.7192	0.2808	0.5834	0.5125
740	0.7347	0.2653	0.6234	0.5065	0.7189	0.2811	0.5827	0.5126
745	0.7347	0.2653	0.6234	0.5065	0.7186	0.2814	0.5819	0.5127
750	0.7347	0.2653	0.6234	0.5065	0.7183	0.2817	0.5811	0.5128
755	0.7347	0.2653	0.6234	0.5065	0.7180	0.2820	0.5803	0.5129
760	0.7347	0.2653	0.6234	0.5065	0.7176	0.2824	0.5795	0.5131
765	0.7347	0.2653	0.6234	0.5065	0.7172	0.2828	0.5786	0.5132
770	0.7347	0.2653	0.6234	0.5065	0.7169	0.2831	0.5777	0.5134
775	0.7347	0.2653	0.6234	0.5065	0.7165	0.2835	0.5767	0.5135
780	0.7347	0.2653	0.6234	0.5065	0.7161	0.2839	0.5757	0.5136

Appendix 5

Relative Spectral Power Distributions of Illuminants

A5.1 INTRODUCTION

Relative spectral power distributions are given as follows: in Section A5.2 for CIE Standard Illuminants A, and D65, and for CIE Illuminants B, C, D50, D55, D75, ID50, and ID65; in Section A5.3 for representative fluorescent lamps; in Section A5.4 for Planckian radiators; and in Section A5.5 for gas-discharge lamps. In Section A5.6, the method is given for deriving the relative spectral power distributions for CIE D Illuminants of different correlated colour temperatures.

A5.2 CIE ILLUMINANTS

Relative spectral power distributions of illuminants

Wavelength	A	B	C	D50	D55	D65	D75	ID50	ID65
300	0.93			0.02	0.02	0.03	0.04	0.00	0.00
305	1.13			1.03	1.05	1.66	2.59	0.00	0.00
310	1.36			2.05	2.07	3.29	5.13	0.00	0.00
315	1.62			4.91	6.65	11.77	17.47	0.00	0.00
320	1.93	0.02	0.0	7.78	11.22	20.2	29.8	0.03	0.01
325	2.27	0.26	0.20	11.26	15.94	28.64	42.37	0.33	0.13
330	2.66	0.50	0.40	14.75	20.65	37.05	54.93	1.60	0.64
335	3.10	1.45	1.55	16.35	22.27	38.50	56.09	4.17	1.77
340	3.59	2.40	2.70	17.95	23.88	39.95	57.26	8.76	3.94
345	4.14	4.00	4.85	19.48	25.85	42.43	60.00	15.61	7.16

(*Continued*)

Measuring Colour, Fourth Edition. R.W.G. Hunt and M.R. Pointer.

(*Continued*)

Wavelength	A	B	C	D50	D55	D65	D75	ID50	ID65
350	4.74	5.60	7.00	21.01	27.82	44.91	62.74	24.24	11.34
355	5.41	7.60	9.95	22.48	29.22	45.78	62.86	31.93	15.68
360	6.14	9.60	12.90	23.94	30.62	46.64	62.98	37.94	19.48
365	6.95	12.40	17.20	25.45	32.46	49.36	66.65	43.65	22.50
370	7.82	15.20	21.40	26.96	34.31	52.09	70.31	47.99	24.84
375	8.77	18.8	27.5	25.72	33.45	51.03	68.51	47.23	23.81
380	9.80	22.40	33.00	24.50	32.58	49.98	66.65	22.55	46.01
385	10.90	26.85	39.92	27.20	35.34	52.31	68.28	25.56	49.19
390	12.09	31.30	47.40	29.80	38.09	54.65	69.91	28.77	52.63
395	13.35	36.18	55.17	39.60	49.52	68.70	85.89	38.67	67.11
400	14.71	41.30	63.30	49.30	60.95	82.75	101.87	48.53	81.45
405	16.15	46.62	71.81	52.90	64.75	87.12	106.85	52.21	85.97
410	17.68	52.10	80.60	56.50	68.55	91.49	111.83	55.72	90.20
415	19.29	57.70	89.53	58.30	70.07	92.46	112.28	57.33	90.96
420	20.99	63.20	98.10	60.00	71.58	93.43	112.74	58.95	91.75
425	22.79	68.37	105.80	58.90	69.75	90.06	107.89	57.84	88.40
430	24.67	73.10	112.40	57.80	67.91	86.68	103.04	56.75	85.08
435	26.64	77.31	117.75	66.30	76.76	95.77	112.09	65.08	93.98
440	28.70	80.80	121.50	74.80	85.61	104.86	121.15	73.45	102.94
445	30.85	83.44	123.45	81.00	91.80	110.94	127.05	79.73	109.14
450	33.09	85.40	124.00	87.20	97.99	117.01	132.96	86.12	115.49
455	35.41	86.88	123.60	88.90	99.23	117.41	132.63	88.09	116.30
460	37.81	88.30	123.10	90.60	100.46	117.81	132.31	90.05	117.08
465	40.30	90.08	123.30	91.00	100.19	116.34	129.80	90.64	115.89
470	42.87	92.00	123.80	91.40	99.91	114.86	127.28	91.18	114.62
475	45.52	93.75	124.09	93.30	101.33	115.39	127.02	93.17	115.30
480	48.24	95.20	123.90	95.20	102.74	115.92	126.76	95.14	115.97
485	51.04	96.23	122.92	93.60	100.41	112.37	122.26	93.69	112.55
490	53.91	96.50	120.70	92.00	98.08	108.81	117.75	92.23	109.12
495	56.85	95.71	116.90	93.90	99.38	109.08	117.16	94.19	109.48
500	59.86	94.20	112.10	95.70	100.68	109.35	116.57	96.13	109.82
505	62.93	92.37	106.98	96.20	100.69	108.58	115.12	96.65	109.12
510	66.06	90.70	102.30	96.60	100.70	107.80	113.68	97.16	108.41
515	69.25	89.95	98.81	96.90	100.34	106.30	111.16	97.44	106.93
520	72.50	89.50	96.90	97.10	99.99	104.79	108.65	97.71	105.42
525	75.79	90.43	96.78	99.60	102.10	106.24	109.54	100.18	106.84
530	79.13	92.20	98.00	102.10	104.21	107.69	110.44	102.63	108.25
535	82.52	94.46	99.94	101.50	103.16	106.05	108.36	101.90	106.54
540	85.95	96.90	102.10	100.80	102.10	104.41	106.28	101.16	104.83
545	89.41	99.16	103.95	101.60	102.53	104.23	105.59	101.87	104.57

(*Continued*)

(*Continued*)

Wavelength	A	B	C	D50	D55	D65	D75	ID50	ID65
550	92.91	101.00	105.20	102.30	102.97	104.05	104.90	102.56	104.29
555	96.44	102.20	105.67	101.20	101.48	102.02	102.45	101.29	102.16
560	100.00	102.80	105.30	100.00	100.00	100.00	100.00	100.00	100.00
565	103.58	102.92	104.11	98.90	98.61	98.17	97.81	98.73	98.03
570	107.18	102.60	102.30	97.70	97.22	96.33	95.62	97.41	96.01
575	110.80	101.90	100.15	98.30	97.48	96.06	94.92	97.75	95.50
580	114.44	101.00	97.80	98.90	97.75	95.79	94.22	98.07	94.97
585	118.08	100.07	95.43	96.20	94.59	92.24	90.61	95.12	91.19
590	121.73	99.20	93.20	93.50	91.43	88.69	87.00	92.18	87.44
595	125.39	98.44	91.22	95.60	92.93	89.35	87.12	93.98	87.84
600	129.04	98.00	89.70	97.70	94.42	90.01	87.24	95.80	88.27
605	132.70	98.08	88.83	98.50	94.78	89.80	86.69	96.35	87.86
610	136.35	98.50	88.40	99.30	95.14	89.60	86.15	96.85	87.41
615	139.99	99.06	88.19	99.20	94.68	88.65	84.87	96.37	86.16
620	143.62	99.70	88.10	99.00	94.22	87.70	83.59	95.88	84.90
625	147.24	100.36	88.06	97.40	92.33	85.49	81.18	93.97	82.50
630	150.84	101.00	88.00	95.70	90.45	83.29	78.76	92.09	80.13
635	154.42	101.56	87.86	97.30	91.39	83.49	78.60	93.28	80.05
640	157.98	102.20	87.80	98.80	92.33	83.70	78.44	94.45	79.97
645	161.52	103.05	87.99	97.30	90.59	81.86	76.63	92.64	77.97
650	165.03	103.90	88.20	95.70	88.85	80.03	74.82	90.82	75.97
655	168.51	104.59	88.20	97.00	89.59	80.12	74.58	91.74	75.83
660	171.96	105.00	87.90	98.20	90.32	80.21	74.34	92.65	75.69
665	175.38	105.08	87.22	100.60	92.13	81.25	74.89	94.65	76.44
670	178.77	104.90	86.30	103.00	93.95	82.28	75.44	96.58	77.15
675	182.12	104.55	85.30	101.10	91.95	80.28	73.52	94.35	74.94
680	185.43	103.90	84.00	99.10	89.96	78.28	71.60	92.14	72.77
685	188.70	102.84	82.21	93.30	84.82	74.00	67.73	86.42	68.58
690	191.93	101.60	80.20	87.40	79.68	69.72	63.87	80.74	64.42
695	195.12	100.38	78.24	89.50	81.26	70.67	64.48	82.37	65.04
700	198.26	99.10	76.30	91.60	82.84	71.61	65.10	83.94	65.62
705	201.36	97.70	74.36	92.30	83.84	72.98	66.59	84.18	66.60
710	204.41	96.20	72.40	92.90	84.84	74.35	68.09	84.42	67.57
715	207.41	94.60	70.40	84.90	77.54	67.98	62.27	76.80	61.51
720	210.36	92.90	68.30	76.80	70.24	61.60	56.46	69.24	55.50
725	213.27	91.10	66.30	81.70	74.77	65.74	60.36	73.31	59.00
730	216.12	89.40	64.40	86.60	79.30	69.89	64.26	77.32	62.46
735	218.92	88.00	62.80	89.60	82.15	72.49	66.71	79.68	64.50
740	221.67	86.90	61.50	92.60	84.99	75.09	69.17	81.97	66.48
745	224.36	85.90	60.20	85.40	78.44	69.34	63.91	75.30	61.13

(*Continued*)

(Continued)

Wavelength	A	B	C	D50	D55	D65	D75	ID50	ID65
750	227.00	85.20	59.20	78.20	71.88	63.59	58.64	68.69	55.83
755	229.59	84.80	58.50	68.00	62.34	55.01	50.64	59.42	48.09
760	232.12	84.70	58.10	57.70	52.79	46.42	42.63	50.23	40.41
765	234.59	84.90	58.00	70.30	64.36	56.61	52.00	60.97	49.09
770	237.01	85.40	58.20	82.90	75.93	66.81	61.37	71.62	57.70
775	239.37	86.10	58.50	80.60	73.87	65.09	59.85	69.34	56.00
780	241.68	87.00	59.10	78.30	71.82	63.38	58.34	67.06	54.30
785	243.92		78.91	72.38	63.84	58.73			
790	246.12		79.55	72.94	64.30	59.14			
795	248.25		76.48	70.14	61.88	56.94			
800	250.33		73.40	67.35	59.45	54.73			
805	252.35		68.66	63.04	55.71	51.32			
810	254.31		63.92	58.73	51.96	47.92			
815	256.22		67.35	61.86	54.70	50.42			
820	258.07		70.78	64.99	57.44	52.92			
825	259.86		72.61	66.65	58.88	54.23			
830	261.60		74.44	68.31	60.31	55.54			

Tristimulus values, chromaticity co-ordinates and correlated colour temperatures (based on 5 nm intervals from 380 nm to 780 nm)

	X	Y	Z	x	y	u'	v'
A	109.85	100.00	35.58	0.4476	0.4074	0.2560	0.5243
B	99.09	100.00	85.31	0.3484	0.3516	0.2137	0.4852
C	98.07	100.00	118.23	0.3101	0.3162	0.2009	0.4609
D50	96.41	100.00	82.50	0.3457	0.3585	0.2091	0.4881
D55	95.68	100.00	92.14	0.3324	0.3474	0.2044	0.4807
D65	95.04	100.00	108.88	0.3127	0.3290	0.1978	0.4683
D75	94.97	100.00	122.57	0.2991	0.3149	0.1935	0.4586
ID50	95.28	100.00	82.33	0.3432	0.3602	0.2069	0.4885
ID65	93.95	100.00	108.46	0.3107	0.3307	0.1958	0.4689
	X_{10}	Y_{10}	Z_{10}	x_{10}	y_{10}	u'_{10}	v'_{10}
A	111.14	100.00	35.20	0.4512	0.4059	0.2590	0.5242
B	99.19	100.00	84.35	0.3498	0.3527	0.2142	0.4859
C	97.29	100.00	116.15	0.3104	0.3190	0.2000	0.4626
D50	96.71	100.00	81.41	0.3477	0.3596	0.2101	0.4889
D55	95.80	100.00	90.93	0.3341	0.3488	0.2051	0.4816
D65	94.81	100.00	107.33	0.3138	0.3310	0.1979	0.4695
D75	94.42	100.00	120.60	0.2997	0.3174	0.1931	0.4601
ID50	95.68	100.00	81.16	0.3456	0.3612	0.2081	0.4894
ID65	93.81	100.00	106.80	0.3121	0.3327	0.1960	0.4702

The correlated colour temperatures are as follows:

A	$T_c = 2848(1.4388/1.4380) = 2856$ K approximately
B	$T_c = 4874$ K
C	$T_c = 6774$ K
D50	$T_c = 5000(1.4388/1.4380) = 5003$ K approximately
D55	$T_c = 5500(1.4388/1.4380) = 5503$ K approximately.
D65	$T_c = 6500(1.4388/1.4380) = 6504$ K approximately
D75	$T_c = 7500(1.4388/1.4380) = 7504$ K approximately
ID50	$T_c = 5096$ K
ID65	$T_c = 6600$ K

A5.3 REPRESENTATIVE FLUORESCENT LAMPS

Spectral power distributions are given in this section for 27 types of fluorescent lamps. These distributions do not constitute CIE Standard Illuminants, but they have been compiled by the CIE for use as representative distributions for practical purposes. The first 12, FL1 to FL12, are in three different groups, normal, broad-band, and three-band. The distributions F2, F7, and F11, which are asterisked, are intended for use in preference to the others when the choice within each group is not critical. The subsequent 15, FL3.1 to FL3.15, are in five different groups: standard halophosphate, deluxe, three-band, multi-band, and D65 simulator.

Each of the distributions in the normal group consists of two semi-broad band emissions of antimony and manganese activations in calcium halo-phosphate phosphor. Those in the broad-band group are more or less enhanced in colour rendering properties as compared with those in the normal group, usually using multiple phosphors; this results in the distributions being flatter and having a wider range in the visible spectrum. The distributions in the three-band group consist mostly of three narrow-band emissions in red, green, and blue wavelength regions; in most cases the narrow band emissions are caused by ternary compositions of rare-earth phosphors.

Below 380 nm and above 780 nm, these spectral power distributions should be taken as zero.

Spectral power distributions of representative fluorescent lamps: FL1 to FL12.

Wavelength	FL1	*FL2	FL3	FL4	FL5	FL6	*FL7	FL8	FL9	FL10	*FL11	FL12
380	1.87	1.18	0.82	0.57	1.87	1.05	2.56	1.21	0.90	1.11	0.91	0.96
385	2.36	1.48	1.02	0.70	2.35	1.31	3.18	1.50	1.12	0.80	0.63	0.64
390	2.94	1.84	1.26	0.87	2.92	1.63	3.84	1.81	1.36	0.62	0.46	0.45
395	3.47	2.15	1.44	0.98	3.45	1.90	4.53	2.13	1.60	0.57	0.37	0.33
400	5.17	3.44	2.57	2.01	5.10	3.11	6.15	3.17	2.59	1.48	1.29	1.19
405	19.49	15.69	14.36	13.75	18.91	14.80	19.37	13.08	12.80	12.16	12.68	12.48
410	6.13	3.85	2.70	1.95	6.00	3.43	7.37	3.83	3.05	2.12	1.59	1.12
415	6.24	3.74	2.45	1.59	6.11	3.30	7.05	3.45	2.56	2.70	1.79	0.94
420	7.01	4.19	2.73	1.76	6.85	3.68	7.71	3.86	2.86	3.74	2.46	1.08

(*Continued*)

(Continued)

Wavelength	FL1	*FL2	FL3	FL4	FL5	FL6	*FL7	FL8	FL9	FL10	*FL11	FL12
425	7.79	4.62	3.00	1.93	7.58	4.07	8.41	4.42	3.30	5.14	3.33	1.37
430	8.56	5.06	3.28	2.10	8.31	4.45	9.15	5.09	3.82	6.75	4.49	1.78
435	43.67	34.98	31.85	30.28	40.76	32.61	44.14	34.10	32.62	34.39	33.94	29.05
440	16.94	11.81	9.47	8.03	16.06	10.74	17.52	12.42	10.77	14.86	12.13	7.90
445	10.72	6.27	4.02	2.55	10.32	5.48	11.35	7.68	5.84	10.40	6.95	2.65
450	11.35	6.63	4.25	2.70	10.91	5.78	12.00	8.60	6.57	10.76	7.19	2.71
455	11.89	6.93	4.44	2.82	11.40	6.03	12.58	9.46	7.25	10.67	7.12	2.65
460	12.37	7.19	4.59	2.91	11.83	6.25	13.08	10.24	7.86	10.11	6.72	2.49
465	12.75	7.40	4.72	2.99	12.17	6.41	13.45	10.84	8.35	9.27	6.13	2.33
470	13.00	7.54	4.80	3.04	12.40	6.52	13.71	11.33	8.75	8.29	5.46	2.10
475	13.15	7.62	4.86	3.08	12.54	6.58	13.88	11.71	9.06	7.29	4.79	1.91
480	13.23	7.65	4.87	3.09	12.58	6.59	13.95	11.98	9.31	7.91	5.66	3.01
485	13.17	7.62	4.85	3.09	12.52	6.56	13.93	12.17	9.48	16.64	14.29	10.83
490	13.13	7.62	4.88	3.14	12.47	6.56	13.82	12.28	9.61	16.73	14.96	11.88
495	12.85	7.45	4.77	3.06	12.20	6.42	13.64	12.32	9.68	10.44	8.97	6.88
500	12.52	7.28	4.67	3.00	11.89	6.28	13.43	12.35	9.74	5.94	4.72	3.43
505	12.20	7.15	4.62	2.98	11.61	6.20	13.25	12.44	9.88	3.34	2.33	1.49
510	11.83	7.05	4.62	3.01	11.33	6.19	13.08	12.55	10.04	2.35	1.47	0.92
515	11.50	7.04	4.73	3.14	11.10	6.30	12.93	12.68	10.26	1.88	1.10	0.71
520	11.22	7.16	4.99	3.41	10.96	6.60	12.78	12.77	10.48	1.59	0.89	0.60
525	11.05	7.47	5.48	3.90	10.97	7.12	12.60	12.72	10.63	1.47	0.83	0.63
530	11.03	8.04	6.25	4.69	11.16	7.94	12.44	12.60	10.78	1.80	1.18	1.10
535	11.18	8.88	7.34	5.81	11.54	9.07	12.33	12.43	10.96	5.71	4.90	4.56
540	11.53	10.01	8.78	7.32	12.12	10.49	12.26	12.22	11.18	40.98	39.59	34.40
545	27.74	24.88	23.82	22.59	27.78	25.22	29.52	28.96	27.71	73.69	72.84	65.40
550	17.05	16.64	16.14	15.11	17.73	17.46	17.05	16.51	16.29	33.61	32.61	29.48
555	13.55	14.59	14.59	13.88	14.47	15.63	12.44	11.79	12.28	8.24	7.52	7.16
560	14.33	16.16	16.63	16.33	15.20	17.22	12.58	11.76	12.74	3.38	2.83	3.08
565	15.01	17.56	18.49	18.68	15.77	18.53	12.72	11.77	13.21	2.47	1.96	2.47
570	15.52	18.62	19.95	20.64	16.10	19.43	12.83	11.84	13.65	2.14	1.67	2.27
575	18.29	21.47	23.11	24.28	18.54	21.97	15.46	14.61	16.57	4.86	4.43	5.09
580	19.55	22.79	24.69	26.26	19.50	23.01	16.75	16.11	18.14	11.45	11.28	11.96
585	15.48	19.29	21.41	23.28	15.39	19.41	12.83	12.34	14.55	14.79	14.76	15.32
590	14.91	18.66	20.85	22.94	14.64	18.56	12.67	12.53	14.65	12.16	12.73	14.27
595	14.15	17.73	19.93	22.14	13.72	17.42	12.45	12.72	14.66	8.97	9.74	11.86

(Continued)

(*Continued*)

Wavelength	FL1	*FL2	FL3	FL4	FL5	FL6	*FL7	FL8	FL9	FL10	*FL11	FL12
600	13.22	16.54	18.67	20.91	12.69	16.09	12.19	12.92	14.61	6.52	7.33	9.28
605	12.19	15.21	17.22	19.43	11.57	14.64	11.89	13.12	14.50	8.31	9.72	12.31
610	11.12	13.80	15.65	17.74	10.45	13.15	11.60	13.34	14.39	44.12	55.27	68.53
615	10.03	12.36	14.04	16.00	9.35	11.68	11.35	13.61	14.40	34.55	42.58	53.02
620	8.95	10.95	12.45	14.42	8.29	10.25	11.12	13.87	14.47	12.09	13.18	14.67
625	7.96	9.65	10.95	12.56	7.32	8.95	10.95	14.07	14.62	12.15	13.16	14.38
630	7.02	8.40	9.51	10.93	6.41	7.74	10.76	14.20	14.72	10.52	12.26	14.71
635	6.20	7.32	8.27	9.52	5.63	6.69	10.42	14.16	14.55	4.43	5.11	6.46
640	5.42	6.31	7.11	8.18	4.90	5.71	10.11	14.13	14.40	1.95	2.07	2.57
645	4.73	5.43	6.09	7.01	4.26	4.87	10.04	14.34	14.58	2.19	2.34	2.75
650	4.15	4.68	5.22	6.00	3.72	4.16	10.02	14.50	14.88	3.19	3.58	4.18
655	3.64	4.02	4.45	5.11	3.25	3.55	10.11	14.46	15.51	2.77	3.01	3.44
660	3.20	3.45	3.80	4.36	2.83	3.02	9.87	14.00	15.47	2.29	2.48	2.81
665	2.81	2.96	3.23	3.69	2.49	2.57	8.65	12.58	13.20	2.00	2.14	2.42
670	2.47	2.55	2.75	3.13	2.19	2.20	7.27	10.99	10.57	1.52	1.54	1.64
675	2.18	2.19	2.33	2.64	1.93	1.87	6.44	9.98	9.18	1.35	1.33	1.36
680	1.93	1.89	1.99	2.24	1.71	1.60	5.83	9.22	8.25	1.47	1.46	1.49
685	1.72	1.64	1.70	1.91	1.52	1.37	5.41	8.62	7.57	1.79	1.94	2.14
690	1.67	1.53	1.55	1.70	1.48	1.29	5.04	8.07	7.03	1.74	2.00	2.34
695	1.43	1.27	1.27	1.39	1.26	1.05	4.57	7.39	6.35	1.02	1.20	1.42
700	1.29	1.10	1.09	1.18	1.13	0.91	4.12	6.71	5.72	1.14	1.35	1.61
705	1.19	0.99	0.96	1.03	1.05	0.81	3.77	6.16	5.25	3.32	4.10	5.04
710	1.08	0.88	0.83	0.88	0.96	0.71	3.46	5.63	4.80	4.49	5.58	6.98
715	0.96	0.76	0.71	0.74	0.85	0.61	3.08	5.03	4.29	2.05	2.51	3.19
720	0.88	0.68	0.62	0.64	0.78	0.54	2.73	4.46	3.80	0.49	0.57	0.71
725	0.81	0.61	0.54	0.54	0.72	0.48	2.47	4.02	3.43	0.24	0.27	0.30
730	0.77	0.56	0.49	0.49	0.68	0.44	2.25	3.66	3.12	0.21	0.23	0.26
735	0.75	0.54	0.46	0.46	0.67	0.43	2.06	3.36	2.86	0.21	0.21	0.23
740	0.73	0.51	0.43	0.42	0.65	0.40	1.90	3.09	2.64	0.24	0.24	0.28
745	0.68	0.47	0.39	0.37	0.61	0.37	1.75	2.85	2.43	0.24	0.24	0.28
750	0.69	0.47	0.39	0.37	0.62	0.38	1.62	2.65	2.26	0.21	0.20	0.21
755	0.64	0.43	0.35	0.33	0.59	0.35	1.54	2.51	2.14	0.17	0.24	0.17
760	0.68	0.46	0.38	0.35	0.62	0.39	1.45	2.37	2.02	0.21	0.32	0.21
765	0.69	0.47	0.39	0.36	0.64	0.41	1.32	2.15	1.83	0.22	0.26	0.19
770	0.61	0.40	0.33	0.31	0.55	0.33	1.17	1.89	1.61	0.17	0.16	0.15
775	0.52	0.33	0.28	0.26	0.47	0.26	0.99	1.61	1.38	0.12	0.12	0.10
780	0.43	0.27	0.21	0.19	0.40	0.21	0.81	1.32	1.12	0.09	0.09	0.05

Tristimulus values, chromaticity co-ordinates and correlated colour temperatures (based on 5 nm intervals from 380 nm to 780 nm)

Group	Lamp	X	Y	Z	x	y	u'	v'	K	R_a
Normal	FL1	92.87	100.00	103.78	0.3131	0.3371	0.1951	0.4726	6430	76
	FL2	99.19	100.00	67.39	0.3721	0.3751	0.2202	0.4996	4230	64
	FL3	103.80	100.00	49.93	0.4091	0.3941	0.2368	0.5132	3450	57
	FL4	109.20	100.00	38.88	0.4402	0.4031	0.2531	0.5215	2940	51
	FL5	90.90	100.00	98.82	0.3138	0.3452	0.1927	0.4769	6350	72
	FL6	97.34	100.00	60.26	0.3779	0.3882	0.2190	0.5062	4150	59
Broadband	FL7	95.04	100.00	108.75	0.3129	0.3292	0.1979	0.4684	6500	90
	FL8	96.43	100.00	82.42	0.3458	0.3586	0.2092	0.4882	5000	95
	FL9	100.38	100.00	67.94	0.3741	0.3727	0.2225	0.4988	4150	90
Three-band	FL10	96.38	100.00	82.35	0.3458	0.3588	0.2091	0.4882	5000	81
	FL11	100.96	100.00	64.35	0.3805	0.3769	0.2251	0.5017	4000	83
	FL12	108.12	100.00	39.27	0.4370	0.4042	0.2506	0.5215	3000	83

Group	Lamp	X_{10}	Y_{10}	Z_{10}	x_{10}	y_{10}	u'_{10}	v'_{10}
Normal	FL1	94.82	100.00	103.26	0.3181	0.3355	0.1991	0.4725
	FL2	103.28	100.00	69.03	0.3793	0.3672	0.2282	0.4971
	FL3	109.01	100.00	52.00	0.4176	0.3831	0.2470	0.5099
	FL4	115.01	100.00	41.00	0.4492	0.3906	0.2647	0.5178
	FL5	93.39	100.00	98.70	0.3197	0.3424	0.1977	0.4763
	FL6	102.18	100.00	62.11	0.3866	0.3784	0.2285	0.5032
Broadband	FL7	95.79	100.00	107.69	0.3156	0.3295	0.1997	0.4690
	FL8	97.12	100.00	81.19	0.3490	0.3593	0.2110	0.4889
	FL9	102.13	100.00	67.87	0.3783	0.3704	0.2262	0.4984
Three-band	FL10	98.96	100.00	83.29	0.3506	0.3543	0.2141	0.4868
	FL11	103.86	100.00	65.61	0.3854	0.3711	0.2307	0.4998
	FL12	111.48	100.00	40.37	0.4427	0.3971	0.2574	0.5195

Spectral power distributions of representative fluorescent lamps: FL3.1 to FL3.6.

Wavelength	FL3.1	FL3.2	FL3.3	FL3.4	FL3.5	FL3.6
380	2.39	5.80	8.94	3.46	4.72	5.53
385	2.93	6.99	11.21	3.86	5.82	6.63
390	3.82	8.70	14.08	4.41	7.18	8.07
395	4.23	9.89	16.48	4.51	8.39	9.45

(Continued)

(Continued)

Wavelength	FL3.1	FL3.2	FL3.3	FL3.4	FL3.5	FL3.6
400	4.97	11.59	19.63	4.86	9.96	11.28
405	86.30	94.53	116.33	71.22	58.86	61.47
410	11.65	20.80	32.07	8.72	15.78	17.80
415	7.09	16.52	29.72	5.36	15.10	17.47
420	7.84	18.30	33.39	5.61	17.30	20.12
425	8.59	20.33	36.94	5.91	19.66	23.05
430	9.44	22.00	40.33	6.42	22.43	26.37
435	196.54	231.90	262.66	192.77	176.00	186.01
440	10.94	25.81	46.87	7.77	28.67	33.94
445	11.38	27.63	49.79	8.37	31.92	37.98
450	11.89	29.10	52.46	9.22	35.38	42.12
455	12.37	30.61	54.81	10.18	38.73	46.38
460	12.81	31.92	56.81	11.18	41.98	50.30
465	13.15	33.11	58.44	12.28	44.92	53.95
470	13.39	33.83	59.52	13.38	47.49	56.94
475	13.56	34.70	60.12	14.54	49.58	59.48
480	13.59	35.02	60.24	15.74	51.21	61.36
485	13.56	35.22	59.88	17.09	52.36	62.68
490	14.07	35.81	59.88	19.60	53.99	64.34
495	13.39	35.14	58.60	21.05	53.78	63.90
500	13.29	35.14	57.85	23.96	54.04	63.85
505	13.25	34.90	56.29	27.77	53.88	63.24
510	13.53	34.70	54.81	32.68	53.62	62.46
515	14.24	35.02	53.42	38.29	53.25	61.41
520	15.74	36.13	52.70	43.76	53.09	60.47
525	18.26	37.92	52.50	47.72	52.88	59.48
530	22.28	40.62	53.30	50.27	52.99	58.65
535	27.97	44.70	54.89	51.78	53.15	57.93
540	35.70	49.63	57.61	52.68	53.67	57.49
545	148.98	154.16	182.75	167.36	167.93	175.17
550	56.55	62.21	65.27	55.29	55.61	57.27
555	68.68	68.92	69.41	56.94	56.82	57.49
560	79.99	75.83	73.28	59.30	58.39	57.99
565	91.47	81.95	76.56	62.15	60.22	58.76
570	101.32	86.95	78.67	65.26	62.21	59.64
575	123.16	103.54	95.74	84.26	81.45	78.77
580	129.53	109.94	97.22	89.22	84.96	81.26
585	115.05	91.95	76.79	75.79	68.71	63.18
590	113.48	89.85	73.36	79.19	70.70	64.29
595	110.08	87.15	69.33	82.80	73.01	65.78

(Continued)

(Continued)

Wavelength	FL3.1	FL3.2	FL3.3	FL3.4	FL3.5	FL3.6
600	104.28	83.26	64.23	85.76	74.69	66.77
605	97.98	78.93	58.92	88.62	76.26	67.77
610	89.60	73.93	53.38	91.12	77.68	68.60
615	80.74	68.84	47.91	93.43	78.67	69.10
620	71.92	63.44	42.61	96.89	80.14	70.15
625	63.50	58.84	37.74	101.45	81.71	71.69
630	55.46	53.84	33.11	103.65	82.08	71.97
635	47.97	49.43	29.04	100.30	79.98	69.81
640	41.39	45.54	25.29	97.89	78.15	68.05
645	35.50	41.53	22.10	96.59	76.52	66.66
650	30.32	38.31	19.31	106.21	79.20	69.70
655	25.79	34.62	16.84	109.97	79.51	70.37
660	21.84	31.80	14.68	117.49	81.08	72.47
665	18.53	29.02	12.89	96.04	70.76	62.30
670	15.67	26.72	11.37	80.15	62.58	54.45
675	13.22	24.22	9.97	70.42	56.87	49.20
680	11.14	22.19	8.82	65.01	52.83	45.60
685	9.40	20.41	7.86	60.15	49.11	42.40
690	8.65	19.10	7.78	56.04	46.28	40.02
695	6.75	16.79	6.30	50.92	42.24	36.48
700	5.69	15.13	5.67	46.26	38.58	33.28
705	4.87	13.82	5.15	42.60	35.59	30.84
710	4.29	12.63	4.91	38.85	32.76	28.30
715	3.54	11.39	4.31	35.09	29.61	25.65
720	3.03	10.32	3.99	31.73	26.89	23.33
725	2.62	9.21	3.67	28.77	24.53	21.23
730	2.28	8.89	3.43	25.76	22.17	19.29
735	1.94	7.50	3.19	23.16	20.02	17.41
740	1.70	6.71	2.95	21.30	18.45	16.31
745	1.50	6.11	2.75	18.55	16.09	14.21
750	1.36	5.40	2.63	17.74	15.62	14.04
755	1.16	4.80	2.43	14.74	13.10	11.55
760	4.91	8.70	7.14	12.93	11.69	10.39
765	0.95	4.01	2.19	13.63	12.42	11.28
770	1.50	4.09	2.71	10.43	9.43	8.51
775	0.89	3.30	2.00	9.67	8.96	8.24
780	0.68	2.82	1.80	8.07	7.39	7.02

Tristimulus values, chromaticity co-ordinates and correlated colour temperatures (based on 5 nm intervals from 380 nm to 780 nm)

		X	Y	Z	x	y	u'	v'	K	R_a
Halophosphate	FL3.1	109.27	100.00	38.69	0.4407	0.4033	0.2533	0.5216	2932	51
	FL3.2	101.99	100.00	65.85	0.3808	0.3734	0.2267	0.5001	3965	70
	FL3.3	91.69	100.00	99.13	0.3153	0.3439	0.1941	0.4764	6280	72
DeLuxe	FL3.4	109.54	100.00	37.78	0.4429	0.4043	0.2543	0.5224	2904	87
	FL3.5	102.11	100.00	70.25	0.3749	0.3672	0.2253	0.4964	4086	95
	FL3.6	96.89	100.00	80.89	0.3488	0.3600	0.2107	0.4892	4894	96
		X_{10}	Y_{10}	Z_{10}	x_{10}	y_{10}	u'_{10}	v'_{10}		
Halophosphate	FL3.1	115.27	100.00	40.99	0.4498	0.3902	0.2653	0.5178		
	FL3.2	105.79	100.00	67.62	0.3869	0.3658	0.2340	0.4976		
	FL3.3	94.32	100.00	99.36	0.3212	0.3405	0.1994	0.4756		
DeLuxe	FL3.4	112.85	100.00	38.99	0.4481	0.3971	0.2610	0.5203		
	FL3.5	103.05	100.00	69.71	0.3778	0.3666	0.2275	0.4966		
	FL3.6	97.47	100.00	79.46	0.3520	0.3611	0.2124	0.4902		

Spectral power distributions of representative fluorescent lamps: FL3.7 to FL3.15.

Wavelength	FL3.7	FL3.8	FL3.9	FL3.10	FL3.11	FL3.12	FL3.13	FL3.14	FL3.15
380	3.79	4.18	3.77	0.25	3.85	1.62	2.23	2.87	300.00
385	2.56	2.93	2.64	0.00	2.91	2.06	2.92	3.69	286.00
390	1.91	2.29	2.06	0.00	2.56	2.71	3.91	4.87	268.00
395	1.42	1.98	1.87	0.00	2.59	3.11	4.55	5.82	244.00
400	1.51	2.44	2.55	0.69	3.63	3.67	5.46	7.17	304.00
405	73.64	70.70	71.68	21.24	74.54	74.60	77.40	72.21	581.00
410	7.37	10.19	12.05	2.18	14.69	8.88	11.25	13.69	225.00
415	4.69	9.79	13.57	1.86	17.22	4.77	7.69	11.12	155.00
420	5.33	13.21	19.60	3.10	24.99	4.72	8.29	12.43	152.00
425	6.75	17.79	27.33	5.00	34.40	4.72	8.98	13.90	170.00
430	8.51	22.98	35.39	7.03	44.57	4.94	10.01	15.82	295.00
435	181.81	191.43	211.82	45.08	228.08	150.29	204.45	200.99	1417.00
440	11.71	31.76	49.02	16.78	61.53	6.08	13.75	21.72	607.00
445	11.96	33.35	51.83	12.28	65.31	7.13	16.88	26.33	343.00
450	12.18	33.87	52.50	13.31	66.35	9.10	21.73	32.85	386.00
455	11.90	32.89	50.73	13.66	64.37	11.76	27.96	40.80	430.00
460	11.16	30.60	46.93	13.69	59.81	14.96	34.92	49.23	469.00
465	11.22	28.28	42.42	13.13	54.24	18.54	41.96	57.39	502.00
470	9.83	24.81	37.16	12.28	47.42	22.48	48.62	65.26	531.00

(Continued)

(Continued)

Wavelength	FL3.7	FL3.8	FL3.9	FL3.10	FL3.11	FL3.12	FL3.13	FL3.14	FL3.15
475	8.94	21.60	31.84	11.42	41.10	26.76	54.33	71.99	552.00
480	12.08	23.40	31.94	11.66	40.04	31.66	59.49	78.25	567.00
485	52.56	68.99	77.74	22.04	85.54	40.93	67.91	88.85	572.00
490	55.42	70.85	79.45	26.17	86.55	45.83	70.01	91.67	575.00
495	31.69	42.29	47.93	18.57	53.47	46.00	66.40	86.81	561.00
500	16.03	22.67	26.24	11.36	30.91	45.26	62.07	80.42	548.00
505	6.72	11.08	13.15	6.83	17.41	43.16	56.95	73.82	527.00
510	4.59	7.66	8.80	5.58	12.56	41.63	52.70	69.12	507.00
515	3.67	6.07	6.70	4.88	10.10	39.75	48.54	63.69	482.00
520	3.02	5.07	5.38	4.31	8.48	37.83	44.80	58.44	461.00
525	3.21	4.88	4.93	3.76	7.74	36.16	41.75	53.57	438.00
530	4.90	6.26	6.06	3.61	8.58	35.25	39.77	49.66	418.00
535	19.05	20.29	19.76	5.62	21.39	37.04	40.50	48.44	404.00
540	177.64	204.67	215.94	38.59	220.12	59.86	59.27	72.56	429.00
545	347.34	390.25	412.13	100.00	417.35	183.53	184.09	200.42	1016.00
550	116.80	135.69	142.39	36.54	146.13	59.03	59.06	65.00	581.00
555	31.87	34.57	34.74	10.57	36.67	47.93	49.95	47.49	370.00
560	16.37	15.71	14.76	2.98	16.51	48.67	50.90	44.14	368.00
565	14.92	12.60	10.99	2.05	12.56	52.69	54.51	44.71	371.00
570	14.12	11.05	9.25	1.84	10.81	57.24	58.33	46.01	377.00
575	29.50	25.05	23.50	6.09	25.31	77.75	77.49	63.52	490.00
580	61.40	54.98	53.05	17.27	53.31	87.81	85.78	71.73	525.00
585	85.05	82.84	81.90	21.77	80.75	80.55	76.20	63.52	402.00
590	64.86	58.22	54.92	18.72	53.56	84.83	78.73	64.13	404.00
595	65.01	53.06	47.80	10.15	44.02	86.84	78.95	63.74	412.00
600	53.17	41.44	36.65	7.26	33.05	91.44	81.48	66.82	418.00
605	34.22	25.26	21.82	5.17	20.26	96.51	84.57	70.65	425.00
610	427.27	329.89	285.69	56.66	233.61	105.25	87.75	79.29	428.00
615	201.10	161.29	139.94	49.39	118.20	106.74	89.56	80.77	432.00
620	58.63	54.19	53.37	18.57	51.66	108.53	91.36	83.59	433.00
625	72.01	66.30	64.30	14.21	61.27	106.92	89.00	82.59	431.00
630	88.19	71.43	64.04	14.01	55.15	101.54	83.67	77.60	427.00
635	20.07	15.74	13.79	5.99	12.95	95.20	78.26	72.47	420.00
640	13.10	10.22	9.06	2.68	8.93	89.34	73.19	68.34	410.00
645	12.92	10.68	9.83	3.14	9.77	82.95	67.61	63.82	399.00
650	24.54	20.32	18.60	6.25	17.12	75.78	61.42	58.57	385.00
655	15.94	14.13	13.38	5.78	13.01	68.65	55.49	53.18	370.00
660	13.56	11.72	10.99	6.75	10.45	61.70	49.78	47.97	352.00
665	13.38	11.75	10.77	5.16	10.33	55.23	44.46	43.14	336.00
670	8.42	7.87	7.57	3.03	7.70	48.58	39.13	38.19	317.00

(Continued)

(*Continued*)

Wavelength	FL3.7	FL3.8	FL3.9	FL3.10	FL3.11	FL3.12	FL3.13	FL3.14	FL3.15
675	6.57	6.38	6.19	1.57	6.34	42.90	34.45	33.85	298.00
680	7.18	7.23	7.09	1.72	7.35	37.74	30.28	29.94	277.00
685	9.90	8.94	8.54	1.54	8.22	32.93	26.37	26.24	260.00
690	11.47	9.79	8.77	1.71	7.93	29.65	23.88	23.90	242.00
695	8.88	7.26	6.41	1.10	5.70	25.19	20.10	20.33	223.00
700	3.05	2.59	2.26	0.28	2.23	21.69	17.40	17.42	202.00
705	22.04	17.03	15.02	3.65	12.43	19.28	15.29	15.64	187.00
710	42.79	33.69	29.39	7.54	24.24	17.36	13.62	14.34	167.00
715	14.40	12.02	10.22	2.34	8.74	14.74	11.68	12.21	152.00
720	1.88	1.68	1.42	0.05	1.39	12.86	10.31	10.65	136.00
725	1.60	1.50	1.23	0.04	1.23	11.28	9.11	9.43	125.00
730	1.42	1.31	1.10	0.04	1.10	9.97	8.03	8.34	113.00
735	1.05	1.01	0.84	0.03	0.84	8.88	7.13	7.52	103.00
740	1.23	1.16	0.97	0.03	0.94	7.78	6.31	6.73	93.00
745	1.76	1.59	1.35	0.02	1.23	7.04	5.67	6.08	84.00
750	0.74	0.79	0.65	0.02	0.68	6.30	5.11	5.52	75.00
755	0.52	0.67	0.13	0.01	0.52	5.55	4.55	5.00	66.00
760	4.10	4.82	4.22	0.01	4.60	10.15	9.06	9.47	58.00
765	0.46	0.61	0.10	0.00	0.45	4.50	3.74	4.08	51.00
770	0.99	1.25	0.68	0.00	1.04	4.81	4.04	4.43	46.00
775	0.43	0.79	0.16	0.00	0.45	3.72	3.14	3.39	41.00
780	0.00	0.58	0.00	0.00	0.00	3.28	2.75	3.17	37.00

Tristimulus values, chromaticity co-ordinates and correlated colour temperatures (based on 5 nm intervals from 380 nm to 780 nm)

		X	Y	Z	x	y	u'	v'	K	R_a
3-Band	FL3.7	108.38	100.00	38.82	0.4384	0.4045	0.2513	0.5218	2979	82
	FL3.8	99.69	100.00	61.29	0.3820	0.3832	0.2236	0.5046	4006	79
	FL3.9	97.43	100.00	81.05	0.3499	0.3591	0.2117	0.4890	4853	79
	FL3.10	97.07	100.00	83.87	0.3455	0.3560	0.2100	0.4868	5000	88
	FL3.11	94.51	100.00	96.72	0.3245	0.3434	0.2006	0.4775	5854	78
Multi-band	FL3.12	108.43	100.00	39.30	0.4377	0.4037	0.2512	0.5213	2984	93
	FL3.13	102.85	100.00	65.65	0.3831	0.3724	0.2286	0.5001	3896	96
	FL3.14	95.51	100.00	81.55	0.3447	0.3609	0.2076	0.4891	5045	95
D65 Simulator	FL3.15	95.11	100.00	109.07	0.3127	0.3288	0.1979	0.4682	6509	98

(*Continued*)

		X_{10}	Y_{10}	Z_{10}	x_{10}	y_{10}	u'_{10}	v'_{10}
3-Band	FL3.7	111.98	100.00	40.05	0.4443	0.3968	0.2586	0.5196
	FL3.8	103.00	100.00	62.75	0.3876	0.3763	0.2300	0.5024
	FL3.9	100.35	100.00	82.58	0.3547	0.3534	0.2172	0.4870
	FL3.10	98.41	100.00	83.29	0.3493	0.3550	0.2130	0.4869
	FL3.11	97.17	100.00	97.94	0.3293	0.3389	0.2055	0.4759
Multi-band	FL3.12	110.23	100.00	39.01	0.4422	0.4012	0.2553	0.5211
	FL3.13	103.20	100.00	63.98	0.3863	0.3743	0.2300	0.5014
	FL3.14	94.67	100.00	77.96	0.3473	0.3668	0.2071	0.4922
D65 Simulator	FL3.15	94.37	100.00	105.59	0.3146	0.3334	0.1975	0.4709

A5.4 PLANCKIAN RADIATORS

Relative spectral power distributions ($c_2 = 1.4388 \times 10^{-2}$ m K).

Wavelength	1000 K	2000 K	2500 K	3000 K	4000 K	5000 K	6000 K	7000 K	8000 K	10000 K
380	0.00	1.58	5.34	12.03	33.11	60.62	90.33	119.59	147.00	194.46
385	0.01	1.90	6.09	13.27	35.07	62.65	91.86	120.22	146.51	191.56
390	0.01	2.26	6.92	14.60	37.07	64.65	93.30	120.73	145.92	188.63
395	0.01	2.68	7.83	16.00	39.09	66.60	94.64	121.14	145.23	185.68
400	0.02	3.15	8.82	17.49	41.13	68.51	95.90	121.44	144.46	182.71
405	0.03	3.70	9.90	19.06	43.19	70.37	97.06	121.65	143.60	179.74
410	0.04	4.32	11.07	20.71	45.27	72.18	98.14	121.77	142.66	176.76
415	0.06	5.03	12.34	22.44	47.36	73.93	99.13	121.79	141.66	173.78
420	0.08	5.82	13.71	24.26	49.46	75.63	100.04	121.73	140.59	170.81
425	0.11	6.71	15.18	26.15	51.56	77.28	100.87	121.59	139.46	167.85
430	0.16	7.71	16.76	28.12	53.66	78.87	101.62	121.38	138.27	164.90
435	0.22	8.82	18.45	30.17	55.76	80.40	102.28	121.09	137.04	161.97
440	0.30	10.05	20.25	32.30	57.85	81.87	102.88	120.73	135.76	159.07
445	0.41	11.41	22.17	34.50	59.93	83.28	103.39	120.31	134.44	156.19
450	0.56	12.91	24.20	36.78	62.00	84.63	103.84	119.83	133.08	153.33
455	0.75	14.57	26.36	39.13	64.06	85.92	104.21	119.29	131.70	150.51
460	1.00	16.38	28.63	41.54	66.10	87.15	104.52	118.69	130.28	147.72
465	1.33	18.36	31.03	44.03	68.12	88.33	104.76	118.05	128.84	144.96
470	1.75	20.52	33.56	46.58	70.11	89.44	104.94	117.36	127.37	142.23
475	2.29	22.86	36.20	49.18	72.08	90.49	105.06	116.63	125.89	139.55
480	2.99	25.40	38.98	51.85	74.02	91.48	105.12	115.85	124.39	136.90
485	3.86	28.15	41.88	54.58	75.93	92.41	105.13	115.04	122.88	134.29
490	4.97	31.11	44.91	57.35	77.81	93.29	105.08	114.19	121.35	131.71
495	6.35	34.30	48.06	60.18	79.65	94.11	104.98	113.31	119.82	129.18

(Continued)

(Continued)

Wavelength	1000 K	2000 K	2500 K	3000 K	4000 K	5000 K	6000 K	7000 K	8000 K	10000 K
500	8.07	37.72	51.34	63.05	81.46	94.87	104.83	112.40	118.28	126.69
505	10.21	41.39	54.75	65.97	83.24	95.57	104.64	111.47	116.74	124.24
510	12.86	45.30	58.28	68.93	84.97	96.22	104.40	110.51	115.20	121.83
515	16.10	49.48	61.93	71.92	86.67	96.82	104.11	109.53	113.66	119.46
520	20.07	53.92	65.70	74.95	88.32	97.37	103.79	108.52	112.11	117.13
525	24.90	58.64	69.60	78.01	89.93	97.86	103.43	107.50	110.58	114.85
530	30.76	63.64	73.61	81.10	91.51	98.31	103.04	106.47	109.04	112.60
535	37.82	68.94	77.73	84.21	93.03	98.71	102.60	105.41	107.51	110.40
540	46.31	74.53	81.97	87.34	94.52	99.05	102.14	104.35	105.99	108.24
545	56.47	80.43	86.32	90.48	95.95	99.36	101.65	103.28	104.48	106.12
550	68.59	86.63	90.78	93.65	97.35	99.61	101.12	102.19	102.98	104.04
555	82.98	93.16	95.34	96.82	98.70	99.83	100.58	101.10	101.48	102.00
560	100.00	100.00	100.00	100.00	100.00	100.00	100.00	100.00	100.00	100.00
565	120.07	107.17	104.76	103.19	101.26	100.13	99.40	98.90	98.53	98.04
570	143.66	114.67	109.62	106.37	102.47	100.22	98.78	97.79	97.07	96.12
575	171.27	122.50	114.56	109.56	103.63	100.27	98.14	96.68	95.63	94.24
580	203.50	130.67	119.60	112.74	104.75	100.29	97.48	95.57	94.20	92.39
585	241.00	139.18	124.71	115.92	105.83	100.27	96.80	94.45	92.78	90.58
590	284.48	148.04	129.91	119.08	106.86	100.21	96.10	93.34	91.38	88.81
595	334.75	157.23	135.18	122.24	107.84	100.12	95.39	92.23	89.99	87.08
600	392.71	166.77	140.52	125.37	108.78	100.00	94.66	91.12	88.62	85.38
605	459.32	176.66	145.94	128.50	109.67	99.85	93.92	90.01	87.27	83.72
610	535.68	186.90	151.41	131.60	110.52	99.67	93.17	88.91	85.93	82.09
615	622.96	197.48	156.94	134.68	111.32	99.46	92.41	87.81	84.61	80.49
620	722.46	208.40	162.53	137.73	112.09	99.22	91.64	86.72	83.30	78.93
625	835.60	219.67	168.17	140.76	112.80	98.96	90.86	85.63	82.01	77.40
630	963.93	231.28	173.86	143.76	113.48	98.67	90.07	84.55	80.74	75.91
635	1109.1	243.24	179.58	146.73	114.12	98.36	89.28	83.47	79.49	74.44
640	1273.0	255.53	185.35	149.67	114.71	98.02	88.48	82.40	78.25	73.01
645	1457.5	268.15	191.15	152.57	115.27	97.66	87.67	81.34	77.03	71.61
650	1664.8	281.11	196.98	155.44	115.78	97.28	86.86	80.29	75.83	70.23
655	1897.2	294.39	202.83	158.27	116.26	96.88	86.05	79.25	74.64	68.89
660	2157.1	308.00	208.71	161.06	116.69	96.47	85.23	78.21	73.47	67.57
665	2447.2	321.93	214.60	163.81	117.09	96.03	84.41	77.18	72.32	66.29
670	2770.3	336.16	220.50	166.52	117.46	95.58	83.59	76.17	71.19	65.03
675	3129.4	350.71	226.42	169.19	117.78	95.11	82.76	75.16	70.07	63.79
680	3527.8	365.56	232.33	171.81	118.08	94.63	81.94	74.16	68.97	62.59
685	3968.9	380.70	238.25	174.39	118.33	94.13	81.12	73.17	67.89	61.41
690	4456.3	396.14	244.17	176.92	118.56	93.62	80.29	72.19	66.82	60.25
695	4994.0	411.86	250.08	179.41	118.75	93.09	79.47	71.22	65.77	59.12

(Continued)

(*Continued*)

Wavelength	1000 K	2000 K	2500 K	3000 K	4000 K	5000 K	6000 K	7000 K	8000 K	10000 K
700	5586.0	427.85	255.98	181.85	118.91	92.56	78.65	70.26	64.74	58.01
705	6236.7	444.11	261.87	184.24	119.04	92.01	77.83	69.32	63.72	56.93
710	6950.7	460.63	267.74	186.58	119.13	91.45	77.02	68.38	62.72	55.87
715	7732.8	477.41	273.59	188.87	119.20	90.88	76.20	67.45	61.73	54.83
720	8588.0	494.43	279.42	191.11	119.24	90.30	75.39	66.54	60.76	53.81
725	9521.8	511.69	285.22	193.30	119.25	89.72	74.58	65.63	59.81	52.82
730	10539.7	529.17	290.99	195.44	119.23	89.12	73.78	64.74	58.87	51.85
735	11647.5	546.88	296.72	197.53	119.19	88.52	72.98	63.85	57.95	50.90
740	12851.5	564.80	302.43	199.57	119.12	87.91	72.18	62.98	57.04	49.96
745	14158.1	582.92	308.09	201.55	119.03	87.30	71.39	62.12	56.15	49.05
750	15573.8	601.23	313.71	203.48	118.91	86.68	70.61	61.27	55.27	48.16
755	17105.7	619.73	319.28	205.36	118.77	86.05	69.82	60.43	54.41	47.28
760	18761.1	638.41	324.81	207.19	118.60	85.42	69.05	59.60	53.56	46.43
765	20547.4	657.25	330.30	208.97	118.42	84.78	68.28	58.78	52.73	45.59
770	22472.4	676.25	335.72	210.69	118.21	84.15	67.51	57.97	51.90	44.77
775	24544.3	695.39	341.10	212.36	117.98	83.50	66.75	57.17	51.10	43.97
780	26771.4	714.68	346.42	213.98	117.73	82.86	66.00	56.39	50.30	43.18

Tristimulus values and chromaticity co-ordinates (based on 5 nm intervals from 380 to 780 nm)

	X	Y	Z	x	y	u'	v'
1000 K	189.56	100.00	0.81	0.6528	0.3444	0.4481	0.5319
2000 K	127.44	100.00	14.52	0.5267	0.4133	0.3051	0.5386
2500 K	115.31	100.00	26.43	0.4770	0.4137	0.2722	0.5311
3000 K	108.13	100.00	39.34	0.4369	0.4041	0.2506	0.5214
4000 K	100.98	100.00	64.44	0.3805	0.3768	0.2251	0.5016
5000 K	98.15	100.00	86.24	0.3451	0.3516	0.2114	0.4847
6000 K	97.08	100.00	104.31	0.3221	0.3318	0.2033	0.4712
7000 K	96.79	100.00	119.11	0.3064	0.3166	0.1981	0.4606
8000 K	96.85	100.00	131.22	0.2952	0.3048	0.1946	0.4521
10000 K	97.33	100.00	149.47	0.2807	0.2884	0.1903	0.4399
	X_{10}	Y_{10}	Z_{10}	x_{10}	y_{10}	u'_{10}	v'_{10}
1000 K	184.59	100.00	0.63	0.6472	0.3506	0.4378	0.5337
2000 K	128.56	100.00	14.00	0.5300	0.4123	0.3078	0.5387
2500 K	116.66	100.00	25.97	0.4808	0.4122	0.2754	0.5311
3000 K	109.38	100.00	38.99	0.4404	0.4026	0.2534	0.5213
4000 K	101.80	100.00	64.20	0.3827	0.3759	0.2269	0.5016
5000 K	98.54	100.00	85.91	0.3464	0.3516	0.2123	0.4848
6000 K	97.11	100.00	103.76	0.3228	0.3324	0.2035	0.4716
7000 K	96.52	100.00	118.27	0.3066	0.3177	0.1978	0.4612
8000 K	96.33	100.00	130.08	0.2951	0.3064	0.1940	0.4530
10000 K	96.43	100.00	147.74	0.2802	0.2905	0.1891	0.4412

A5.5 GAS DISCHARGE LAMPS

Relative spectral power distributions

Below 380 nm and above 780 nm, these spectral power distributions should be taken as zero.

	Low pressure sodium lamp SOX	Standard high pressure sodium lamp HP1	Colour Enhanced High Pressure Sodium Lamp HP2	High pressure mercury lamp type MB	High pressure mercury lamp type MBF	High pressure mercury lamp type MBTF	High pressure mercury lamp type HMI	Xenon
380	0.1	1.90	2.64	5.14	4.53	4.73	116.39	93.03
385	0.0	2.20	2.77	4.28	3.91	4.01	114.92	94.59
390	0.0	2.50	3.42	7.49	6.68	6.40	115.62	96.33
395	0.1	2.70	3.68	5.26	4.81	4.85	107.07	100.56
400	0.0	3.10	4.33	7.71	7.73	6.77	108.19	102.81
405	0.2	4.30	5.50	121.05	107.15	102.94	123.74	100.40
410	0.0	3.80	5.94	61.57	54.92	54.53	125.42	100.84
415	0.0	4.20	7.20	5.73	6.08	5.97	123.88	101.45
420	0.0	4.80	9.02	3.67	4.33	4.19	125.56	102.57
425	0.1	5.19	10.27	4.29	4.97	4.82	116.39	101.94
430	0.1	5.89	12.48	6.26	7.15	6.75	89.57	101.29
435	0.0	7.39	16.82	168.75	146.23	144.91	122.69	101.54
440	0.0	7.89	16.04	140.57	123.51	131.86	120.03	103.74
445	0.0	5.69	15.26	5.84	6.82	7.00	71.29	103.67
450	0.0	12.89	22.58	3.86	4.66	4.96	65.93	110.30
455	0.0	6.69	20.07	3.22	3.79	4.28	73.25	112.78
460	0.0	4.30	15.13	2.84	3.46	4.08	80.53	116.52
465	0.1	20.78	25.27	2.72	3.27	4.05	72.48	129.56
470	0.1	12.99	28.04	2.74	3.49	4.34	66.65	141.07
475	0.0	6.69	15.99	3.00	3.70	4.68	67.82	126.45
480	0.0	1.40	10.40	2.77	3.58	4.68	66.60	115.94
485	0.1	1.50	11.10	3.10	3.94	5.20	59.40	118.42
490	0.0	3.20	13.44	5.24	6.04	7.27	60.43	111.75
495	0.3	18.18	22.62	6.11	6.39	8.03	62.92	113.67
500	0.7	56.24	49.71	2.88	3.73	5.59	64.03	105.17
505	0.0	2.90	17.21	2.58	3.49	5.55	68.89	103.88
510	0.0	2.10	17.12	2.39	3.57	5.73	57.96	102.90
515	0.2	13.39	27.26	2.58	3.75	6.16	57.28	102.71

(*Continued*)

(*Continued*)

	Low pressure sodium lamp SOX	Standard high pressure sodium lamp HP1	Colour Enhanced High Pressure Sodium Lamp HP2	High pressure mercury lamp type MB	High pressure mercury lamp type MBF	High pressure mercury lamp type MBTF	High pressure mercury lamp type HMI	Xenon
520	0.1	2.10	20.02	2.28	3.67	6.24	54.24	102.29
525	0.0	2.00	21.54	2.43	3.88	6.70	50.74	101.90
530	0.1	2.20	23.36	2.53	4.38	7.29	52.25	101.54
535	0.1	2.30	25.66	3.67	6.74	9.43	53.26	101.47
540	0.1	2.60	29.69	4.29	9.93	12.21	56.32	101.12
545	0.1	5.10	43.12	195.18	198.10	195.59	114.85	101.43
550	0.1	11.39	98.3	181.63	171.97	178.84	119.19	100.98
555	0.2	15.48	125.6	6.93	10.57	14.46	59.24	100.75
560	0.2	20.78	134.57	3.74	7.55	11.60	56.91	100.44
565	2.1	55.64	149.7	3.50	6.61	11.26	62.96	100.28
570	8.1	254.03	166.12	3.79	6.76	11.92	65.18	100.19
575	1.3	56.14	98.77	100.00	100.00	100.00	100.00	100.00
580	1.4	111.78	30.47	292.87	273.65	265.84	158.96	99.86
585	131.8	297.98	1.17	39.62	42.59	47.39	81.58	100.47
590	1000.0	142.55	0.39	4.23	15.05	19.07	82.42	100.33
595	150.6	334.84	1.65	3.38	27.64	28.62	72.41	99.04
600	2.0	189.40	21.41	2.74	13.65	18.67	74.44	97.17
605	1.5	117.78	76.11	2.97	10.75	16.68	62.19	96.65
610	1.3	79.92	126.16	2.78	24.34	26.49	61.75	96.84
615	3.5	108.09	161.96	3.10	83.81	69.19	61.00	98.50
620	1.9	46.85	160.06	2.65	149.80	119.79	60.76	100.21
625	0.6	38.16	158.19	3.51	58.07	53.70	62.91	99.91
630	0.6	32.47	153.69	2.67	17.99	23.74	55.78	97.80
635	0.8	28.37	147.40	2.91	10.51	18.90	50.61	96.84
640	1.8	25.37	140.60	2.66	8.09	18.23	52.97	98.92
645	0.6	22.98	134.92	2.76	7.86	18.56	49.83	99.18
650	0.7	20.38	127.59	2.66	10.98	21.14	46.82	101.99
655	0.4	19.78	124.65	2.78	10.76	21.58	47.96	98.78
660	0.3	17.78	118.02	2.68	7.25	19.31	56.13	97.14
665	0.2	16.78	113.94	2.81	6.29	18.98	53.38	98.43
670	0.5	19.18	118.1	2.96	6.06	19.18	67.89	100.30
675	0.2	17.98	115.16	3.18	5.91	19.68	73.25	101.29
680	0.0	13.69	102.85	2.75	5.18	19.43	47.32	101.97
685	0.0	9.99	90.54	2.95	5.48	20.20	50.08	109.55

(*Continued*)

(*Continued*)

	Low pressure sodium lamp SOX	Standard high pressure sodium lamp HP1	Colour Enhanced High Pressure Sodium Lamp HP2	High pressure mercury lamp type MB	High pressure mercury lamp type MBF	High pressure mercury lamp type MBTF	High pressure mercury lamp type HMI	Xenon
690	0.1	8.19	83.34	6.51	9.83	24.27	44.96	110.88
695	0.5	7.59	79.44	5.19	18.35	30.79	56.34	102.08
700	0.2	6.99	76.97	2.82	46.48	51.26	69.68	94.49
705	0.6	6.79	74.85	3.17	35.63	43.56	36.16	93.05
710	0.1	6.49	73.12	4.29	15.82	29.82	28.41	98.10
715	0.0	6.39	71.51	3.22	5.92	23.00	27.15	106.20
720	0.1	6.09	70.13	2.76	4.61	22.22	28.21	98.24
725	0.2	5.99	69.04	2.97	4.48	22.86	34.95	95.69
730	0.0	5.79	67.48	2.74	4.11	22.84	34.94	101.73
735	0.0	5.79	66.70	2.97	4.23	23.63	27.68	108.38
740	0.2	5.79	66.31	2.71	3.92	23.41	24.19	102.62
745	0.0	5.79	65.14	2.97	4.08	24.35	22.61	100.28
750	0.4	6.39	65.70	2.63	3.84	24.30	21.79	97.33
755	0.2	5.99	64.79	3.00	4.13	26.32	30.47	97.45
760	0.1	5.59	64.10	2.67	3.67	25.18	33.75	101.71
765	23.6	31.97	83.04	3.10	3.99	26.40	36.16	137.21
770	55.8	27.87	86.25	2.99	3.79	26.22	32.39	105.22
775	11.1	5.89	63.93	3.86	4.61	27.96	26.55	84.41
780	0.0	6.69	64.92	2.64	3.41	27.08	24.38	80.55

Tristimulus values, chromaticity co-ordinates and correlated colour temperatures (based on 5 nm intervals from 380 nm to 780 nm)

	X	Y	Z	x	y	u'	v'	K	R_a
LP Sodium	135.24	100.00	0.24	0.5743	0.4246	0.3307	0.5501	1725	
HP1	128.45	100.00	12.54	0.5330	0.4150	0.3084	0.5402	1959	8
HP2	114.90	100.00	25.58	0.4778	0.4158	0.2717	0.5320	2506	83
MB	86.97	100.00	75.95	0.3308	0.3803	0.1917	0.4959	5563	15
MBF	104.94	100.00	57.00	0.4006	0.3818	0.2364	0.5068	3531	48
MBTF	103.16	100.00	56.98	0.3965	0.3844	0.2326	0.5073	3649	49
HMI	103.75	100.00	116.80	0.3237	0.3120	0.2124	0.4606	5983	88
Xenon	99.78	100.00	110.21	0.3219	0.3226	0.2068	0.4662	6043	94

	X_{10}	Y_{10}	Z_{10}	x_{10}	y_{10}	u'_{10}	v'_{10}
LP Sodium	143.44	100.00	0.09	0.5890	0.4106	0.3491	0.5475
HP1	134.06	100.00	12.68	0.5433	0.4053	0.3207	0.5382
HP2	117.49	100.00	25.93	0.4826	0.4108	0.2772	0.5309
MB	96.95	100.00	84.63	0.3443	0.3551	0.2095	0.4863
MBF	112.44	100.00	63.12	0.4080	0.3629	0.2496	0.4995
MBTF	110.44	100.00	62.78	0.4042	0.3660	0.2456	0.5003
HMI	105.45	100.00	120.71	0.3233	0.3066	0.2144	0.4574
Xenon	98.97	100.00	108.74	0.3216	0.3250	0.2056	0.4675

A5.6 METHOD OF CALCULATING D ILLUMINANT DISTRIBUTIONS

For daylight illuminants, CIE Standard Illuminant D65 should be used whenever possible; if D65 is not appropriate, then either ID65, D50, ID50, D55, or D75, should be used if possible. If none of these daylight illuminants is appropriate, then one of the other CIE D illuminants, defined below, should be used.

Chromaticity

The chromaticity co-ordinates must be such that:

$$y_D = -3.000x_D^2 + 2.870x_D - 0.275$$

with x_D being within the range of 0.250 to 0.380. The correlated colour temperature T_c (calculated with c_2 of Planck's Law being equal to 1.4388×10^{-2} mK) of daylight D is related to x_D by the following formulae based on normals to the Planckian locus on a chromaticity diagram in which v is plotted against u (not v' against u'):

a. for correlated colour temperatures from 4000 K to 7000 K:

$$x_D = -4.6070(10^9/T_c^3) + 2.9678(10^6/T_c^2) + 0.09911(10^3/T_c) + 0.244063$$

b. for correlated colour temperatures from 7000 K to 25000 K

$$x_D = -2.0064(10^9/T_c^3) + 1.9018(10^6/T_c^2) + 0.24748(10^3/T_c) + 0.237040$$

The relative spectral power distributions, $S(\lambda)$, of the D illuminants are given by:

$$S(\lambda) = S_0(\lambda) + M_1 S_1(\lambda) + M_2 S_2(\lambda)$$

where $S_0(\lambda)$, $S_1(\lambda)$, $S_2(\lambda)$ are functions of wavelength, λ, as given in the table below, and M_1, M_2 are factors whose values are related to the chromaticity co-ordinates x_D, y_D as follows:

$$M_1 = (-1.3515 - 1.7703x_D + 5.9114y_D)/(0.0241 + 0.2562x_D - 0.7341y_D)$$

$$M_2 = (0.0300 - 31.4424x_D + 30.0717y_D)/(0.0241 + 0.2562x_D - 0.7341y_D)$$

Values of x_D, y_D, M_1, and M_2 for correlated colour temperatures in the range 4000 K to 25000 K are given below.

Wave-length	S_0	S_1	S_2	Wave-length	S_0	S_1	S_2
300	0.04	0.02	0.00	575	95.55	−2.55	0.35
305	3.02	2.26	1.00	580	95.10	−3.50	0.50
310	6.00	4.50	2.00	585	92.10	−3.50	1.30
315	17.80	13.45	3.00	590	89.10	−3.50	2.10
320	29.60	22.40	4.00	595	89.80	−4.65	2.65
325	42.45	32.20	6.25	600	90.50	−5.80	3.20
330	55.30	42.00	8.50	605	90.40	−6.50	3.65
335	56.30	41.30	8.15	610	90.30	−7.20	4.10
340	57.30	40.60	7.80	615	89.35	−7.90	4.40
345	59.55	41.10	7.25	620	88.40	−8.60	4.70
350	61.80	41.60	6.70	625	86.20	−9.05	4.90
355	61.65	39.80	6.00	630	84.00	−9.50	5.10
360	61.50	38.00	5.30	635	84.55	−10.20	5.90
365	65.15	40.20	5.70	640	85.10	−10.90	6.70
370	68.80	42.40	6.10	645	83.50	−10.80	7.00
375	66.10	40.45	4.55	650	81.90	−10.70	7.30
380	63.40	38.50	3.00	655	82.25	−11.35	7.95
385	64.60	36.75	2.10	660	82.60	−12.00	8.60
390	65.80	35.00	1.20	665	83.75	−13.00	9.20
395	80.30	39.20	0.05	670	84.90	−14.00	9.80
400	94.80	43.40	−1.10	675	83.10	−13.80	10.00
405	99.80	44.85	−0.80	680	81.30	−13.60	10.20
410	104.80	46.30	−0.50	685	76.60	−12.80	9.25
415	105.35	45.10	−0.60	690	71.90	−12.00	8.30
420	105.90	43.90	−0.70	695	73.10	−12.65	8.95
425	101.35	40.50	−0.95	700	74.30	−13.30	9.60
430	96.80	37.10	−1.20	705	75.35	−13.10	9.05
435	105.35	36.90	−1.90	710	76.40	−12.90	8.50
440	113.90	36.70	−2.60	715	69.85	−11.75	7.75
445	119.75	36.30	−2.75	720	63.30	−10.60	7.00
450	125.60	35.90	−2.90	725	67.50	−11.10	7.30
455	125.55	34.25	−2.85	730	71.70	−11.60	7.60
460	125.50	32.60	−2.80	735	74.35	−11.90	7.80
465	123.40	30.25	−2.70	740	77.00	−12.20	8.00
470	121.30	27.90	−2.60	745	71.10	−11.20	7.35

(*Continued*)

(Continued)

Wave-length	S_0	S_1	S_2	Wave-length	S_0	S_1	S_2
475	121.30	26.10	−2.60	750	65.20	−10.20	6.70
480	121.30	24.30	−2.60	755	56.45	−9.00	5.95
485	117.40	22.20	−2.20	760	47.70	−7.80	5.20
490	113.50	20.10	−1.80	765	58.15	−9.50	6.30
495	113.30	18.15	−1.65	770	68.60	−11.20	7.40
500	113.10	16.20	−1.50	775	66.80	−10.80	7.10
505	111.95	14.70	−1.40	780	65.00	−10.40	6.80
510	110.80	13.20	−1.30	785	65.50	−10.50	6.90
515	108.65	10.90	−1.25	790	66.00	−10.60	7.00
520	106.50	8.60	−1.20	795	63.50	−10.15	6.70
525	107.65	7.35	−1.10	800	61.00	−9.70	6.40
530	108.80	6.10	−1.00	805	57.15	−9.00	5.95
535	107.05	5.15	−0.75	810	53.30	−8.30	5.50
540	105.30	4.20	−0.50	815	56.10	−8.80	5.80
545	104.85	3.05	−0.40	820	58.90	−9.30	6.10
550	104.40	1.90	−0.30	825	60.40	−9.55	6.30
555	102.20	0.95	−0.15	830	61.90	−9.80	6.50
560	100.00	0.00	0.00				
565	98.00	−0.80	0.10				
570	96.00	−1.60	0.20				

Chromaticity co-ordinates x_D, y_D and factors M_1, M_2 used in the calculation of the relative spectral power distributions of CIE D Illuminants. The corresponding chromaticity co-ordinates u'_D, v'_D are also given.

T_c	x_D	y_D	u'_D	v'_D	M_1	M_2
4000	0.3823	0.3838	0.2236	0.5049	−1.505	2.827
4100	0.3779	0.3812	0.2217	0.5031	−1.464	2.460
4200	0.3737	0.3786	0.2200	0.5014	−1.422	2.127
4300	0.3697	0.3760	0.2183	0.4997	−1.378	1.825
4400	0.3658	0.3734	0.2168	0.4979	−1.333	1.550
4500	0.3621	0.3709	0.2153	0.4962	−1.286	1.302
4600	0.3585	0.3684	0.2139	0.4946	−1.238	1.076
4700	0.3551	0.3659	0.2126	0.4929	−1.190	0.871
4800	0.3519	0.3634	0.2114	0.4913	−1.140	0.686
4900	0.3487	0.3610	0.2102	0.4897	−1.090	0.518

(Continued)

(*Continued*)

T_c	x_D	y_D	u'_D	v'_D	M_1	M_2
5000	0.3457	0.3587	0.2091	0.4882	−1.040	0.367
5100	0.3429	0.3564	0.2081	0.4866	−0.989	0.230
5200	0.3401	0.3541	0.2071	0.4851	−0.939	0.106
5300	0.3375	0.3619	0.2062	0.4837	−0.888	−0.005
5400	0.3349	0.3497	0.2053	0.4822	−0.837	−0.105
5500	0.3325	0.3476	0.2044	0.4808	−0.786	−0.195
5600	0.3302	0.3455	0.2036	0.4795	−0.736	−0.276
5700	0.3279	0.3435	0.2028	0.4781	−0.685	−0.348
5800	0.3258	0.3416	0.2021	0.4768	−0.635	−0.412
5900	0.3237	0.3397	0.2014	0.4755	−0.586	−0.469
6000	0.3217	0.3378	0.2007	0.4743	−0.536	−0.519
6100	0.3198	0.3360	0.2001	0.4730	−0.487	−0.563
6200	0.3179	0.3342	0.1995	0.4719	−0.439	−0.602
6300	0.3161	0.3325	0.1989	0.4707	−0.391	−0.635
6400	0.3144	0.3308	0.1983	0.4695	−0.343	−0.664
6500	0.3128	0.3292	0.1978	0.4684	−0.296	−0.688
6600	0.3112	0.3276	0.1973	0.4673	−0.250	−0.709
6700	0.3097	0.3260	0.1968	0.4663	−0.204	−0.726
6800	0.3082	0.3245	0.1963	0.4652	−0.159	−0.739
6900	0.3067	0.3231	0.1959	0.4642	−0.114	−0.749
7000	0.3054	0.3216	0.1955	0.4632	−0.070	−0.757
7100	0.3040	0.3202	0.1950	0.4623	−0.026	−0.762
7200	0.3027	0.3189	0.1946	0.4613	0.017	−0.765
7300	0.3015	0.3176	0.1943	0.4604	0.060	−0.765
7400	0.3003	0.3163	0.1939	0.4595	0.102	−0.763
7500	0.2991	0.3150	0.1935	0.4586	0.144	−0.760
7600	0.2980	0.3138	0.1932	0.4578	0.184	−0.755
7700	0.2969	0.3126	0.1928	0.4569	0.225	−0.748
7800	0.2958	0.3115	0.1925	0.4561	0.264	−0.740
7900	0.2948	0.3103	0.1922	0.4553	0.303	−0.730
8000	0.2938	0.3092	0.1919	0.4545	0.342	−0.720
8100	0.2928	0.3081	0.1916	0.4537	0.380	−0.708
8200	0.2919	0.3071	0.1913	0.4530	0.417	−0.695
8300	0.2910	0.3061	0.1911	0.4523	0.454	−0.682
8400	0.2901	0.3051	0.1908	0.4515	0.490	−0.667
8500	0.2892	0.3041	0.1906	0.4508	0.526	−0.652
9000	0.2853	0.2996	0.1894	0.4475	0.697	−0.566
9500	0.2818	0.2956	0.1884	0.4446	0.856	−0.471
10000	0.2788	0.2920	0.1876	0.4419	1.003	−0.369
10500	0.2761	0.2887	0.1868	0.4395	1.139	−0.265

(*Continued*)

(Continued)

T_c	x_D	y_D	u'_D	v'_D	M_1	M_2
11000	0.2737	0.2858	0.1861	0.4373	1.266	−0.160
12000	0.2697	0.2808	0.1850	0.4335	1.495	0.045
13000	0.2664	0.2767	0.1841	0.4303	1.693	0.239
14000	0.2637	0.2732	0.1834	0.4275	1.868	0.419
15000	0.2614	0.2702	0.1828	0.4252	2.021	0.586
17000	0.2578	0.2655	0.1818	0.4214	2.278	0.878
20000	0.2539	0.2603	0.1809	0.4172	2.571	1.231
25000	0.2499	0.2548	0.1798	0.4126	2.907	1.655
5003	0.3457	0.3585	0.2092	0.4881	−1.039	0.363
5503	0.3324	0.3474	0.2044	0.4807	−0.785	−0.198
6504	0.3127	0.3290	0.1978	0.4683	−0.295	−0.689
7504	0.2990	0.3149	0.1935	0.4585	0.145	−0.760

Appendix 6

Colorimetric Formulae

A6.1 CHROMATICITY RELATIONSHIPS

$$u' = 4X/(X + 15Y + 3Z)$$

$$v' = 9Y/(X + 15Y + 3Z)$$

$$w' = (-3X + 6Y + 3Z)/(X + 15Y + 3Z)$$

$$u' = 4x/(-2x + 12y + 3)$$

$$v' = 9y/(-2x + 12y + 3)$$

$$w' = (-6x + 3y + 3)/(-2x + 12y + 3)$$

$$x = 9u'/(6u' - 16v' + 12)$$

$$y = 4v'/(6u' - 16v' + 12)$$

$$z = (-3u' - 20v' + 12)/(6u' - 16v' + 12)$$

$$u = u'$$

$$v = (2/3)v'$$

A6.2 CIELUV, CIELAB, AND U*V*W* RELATIONSHIPS

$$L^* = 116\mathrm{f}(Y/Y_\mathrm{n}) - 16$$

where $\mathrm{f}(Y/Y_\mathrm{n}) = (Y/Y_\mathrm{n})^{1/3}$ for $Y/Y_\mathrm{n} > (6/29)^3$

$\mathrm{f}(Y/Y_\mathrm{n}) = (841/108)(Y/Y_\mathrm{n}) + 4/29$ for $Y/Y_\mathrm{n} \leq (6/29)^3$

Measuring Colour, Fourth Edition. R.W.G. Hunt and M.R. Pointer.

$$u^* = 13L^*(u' - u'_n)$$

$$v^* = 13L^*(v' - v'_n)$$

$$a^* = 500[f(X/X_n) - f(Y/Y_n)]$$

$$b^* = 200[f(Y/Y_n) - f(Z/Z_n)]$$

where $f(X/X_n) = (X/X_n)^{1/3}$ for $X/X_n > (6/29)^3$

$f(X/X_n) = (841/108)(X/X_n) + 4/29$ for $X/X_n \leq (6/29)^3$

and $f(Y/Y_n) = (Y/Y_n)^{1/3}$ for $Y/Y_n > (6/29)^3$

$f(Y/Y_n) = (841/108)(Y/Y_n) + 4/29$ for $Y/Y_n \leq (6/29)^3$

and $f(Z/Z_n) = (Z/Z_n)^{1/3}$ for $Z/Z_n > (6/29)^3$

$f(Z/Z_n) = (841/108)(Z/Z_n) + 4/29$ for $Z/Z_n \leq (6/29)^3$

$$U^* = 13W^*(u - u_n)$$

$$V^* = 13W^*(v - v_n)$$

$$W^* = 25Y^{1/3} - 17$$

The subscript n indicates that the value is for the reference white. In the formula for W^* it is necessary to express Y as a percentage.

$$h_{uv} = \arctan[(v' - v'_n)/(u' - u'_n)]$$

$$s_{uv} = 13[(u' - u'_n)^2 + (v' - v'_n)^2]^{1/2}$$

$$h_{uv} = \arctan(v^*/u^*)$$

$$h_{ab} = \arctan(b^*/a^*)$$

h_{uv} lies between 0° and 90° if v^* and u^* are both positive; between 90° and 180° if v^* is positive and u^* is negative; between 180° and 270° if v^* and u^* are both negative; and between 270° and 360° if v^* is negative and u^* is positive (and similarly for h_{ab} and a^* and b^*).

$$C^*_{uv} = (u^{*2} + v^{*2})^{1/2} = L^* s_{uv}$$

$$C^*_{ab} = (a^{*2} + b^{*2})^{1/2}$$

$$\Delta E^*_{uv} = [(\Delta L^*)^2 + (\Delta u^*)^2 + (\Delta v^*)^2]^{1/2}$$

$$\Delta E^*_{ab} = [(\Delta L^*)^2 + (\Delta a^*)^2 + (\Delta b^*)^2]^{1/2}$$

$$\Delta E^*_{uv} = [(\Delta L^*)^2 + (\Delta H^*_{uv})^2 + (\Delta C^*_{uv})^2]^{1/2}$$

$$\Delta E^*_{ab} = [(\Delta L^*)^2 + (\Delta H^*_{ab})^2 + (\Delta C^*_{ab})^2]^{1/2}$$

where

$$\Delta H^*_{uv} = [(\Delta E^*_{uv})^2 - (\Delta L^*)^2 - (\Delta C^*_{uv})^2]^{1/2}$$
$$\Delta H^*_{ab} = [(\Delta E^*_{ab})^2 - (\Delta L^*)^2 - (\Delta C^*_{ab})^2]^{1/2}$$

ΔH^* is positive if indicating an increase in h and negative if indicating a decrease in h. ΔH^*_{ab} can also be defined as follows:

$$\Delta H^*_{ab} = 2(C^*_{ab,1} - C^*_{ab,2})^{1/2} \sin(\Delta h_{ab}/2)$$

where $C^*_{ab,1}$ and $C^*_{ab,2}$ refer to the values of C* of the two samples being compared, and Δh_{ab} is the difference in the values of the hue angles (in radians) of those two samples. For small colour differences away from the achromatic axis the above equation can be written as:

$$\Delta H^*_{ab} = 2(C^*_{ab,1} - C^*_{ab,2})^{1/2} \Delta h_{ab}$$

Corresponding equations can be applied to CIE 1976 CIELUV colour difference calculations.

Appendix 7

Calculation of the CIE Colour Rendering Indices

A7.1 SPECTRAL RADIANCE FACTORS OF TEST COLOURS

Spectral Radiance Factors of CRI Test Colours 1–8

Wavelength	1 7.5R6/4	2 5Y6/4	3 5GY6/8	4 2.5G6/6	5 10BG6/4	6 5PB6/8	7 2.5P6/8	8 10P6/8
380	0.219	0.070	0.065	0.074	0.295	0.151	0.378	0.104
385	0.239	0.079	0.068	0.083	0.306	0.203	0.459	0.129
390	0.252	0.089	0.070	0.093	0.310	0.265	0.524	0.170
395	0.256	0.101	0.072	0.105	0.312	0.339	0.546	0.240
400	0.256	0.111	0.073	0.116	0.313	0.410	0.551	0.319
405	0.254	0.116	0.073	0.121	0.315	0.464	0.555	0.416
410	0.252	0.118	0.074	0.124	0.319	0.492	0.559	0.462
415	0.248	0.120	0.074	0.126	0.322	0.508	0.560	0.482
420	0.244	0.121	0.074	0.128	0.326	0.517	0.561	0.490
425	0.240	0.122	0.073	0.131	0.330	0.524	0.558	0.488
430	0.237	0.122	0.073	0.135	0.334	0.531	0.556	0.482
435	0.232	0.122	0.073	0.139	0.339	0.538	0.551	0.473
440	0.230	0.123	0.073	0.144	0.346	0.544	0.544	0.462
445	0.226	0.124	0.073	0.151	0.352	0.551	0.535	0.450
450	0.225	0.127	0.074	0.161	0.360	0.556	0.522	0.439
455	0.222	0.128	0.075	0.172	0.369	0.556	0.506	0.426

(*Continued*)

Measuring Colour, Fourth Edition. R.W.G. Hunt and M.R. Pointer.

(Continued)

Wavelength	1 7.5R6/4	2 5Y6/4	3 5GY6/8	4 2.5G6/6	5 10BG6/4	6 5PB6/8	7 2.5P6/8	8 10P6/8
460	0.220	0.131	0.077	0.186	0.381	0.554	0.488	0.413
465	0.218	0.134	0.080	0.205	0.394	0.549	0.469	0.397
470	0.216	0.138	0.085	0.229	0.403	0.541	0.448	0.382
475	0.214	0.143	0.094	0.254	0.410	0.531	0.429	0.366
480	0.214	0.150	0.109	0.281	0.415	0.519	0.408	0.352
485	0.214	0.159	0.126	0.308	0.418	0.504	0.385	0.337
490	0.216	0.174	0.148	0.332	0.419	0.488	0.363	0.325
495	0.218	0.190	0.172	0.352	0.417	0.469	0.341	0.310
500	0.223	0.207	0.198	0.370	0.413	0.450	0.324	0.299
505	0.225	0.225	0.221	0.383	0.409	0.431	0.311	0.289
510	0.226	0.242	0.241	0.390	0.403	0.414	0.301	0.283
515	0.226	0.253	0.260	0.394	0.396	0.395	0.291	0.276
520	0.225	0.260	0.278	0.395	0.389	0.377	0.283	0.270
525	0.225	0.264	0.302	0.392	0.381	0.358	0.273	0.262
530	0.227	0.267	0.339	0.385	0.372	0.341	0.265	0.256
535	0.230	0.269	0.370	0.377	0.363	0.325	0.260	0.251
540	0.236	0.272	0.392	0.367	0.353	0.309	0.257	0.250
545	0.245	0.276	0.399	0.354	0.342	0.293	0.257	0.251
550	0.253	0.282	0.400	0.341	0.331	0.279	0.259	0.254
555	0.262	0.289	0.393	0.327	0.320	0.265	0.260	0.258
560	0.272	0.299	0.380	0.312	0.308	0.253	0.260	0.264
565	0.283	0.309	0.365	0.296	0.296	0.241	0.258	0.269
570	0.298	0.322	0.349	0.280	0.284	0.234	0.256	0.272
575	0.318	0.329	0.332	0.263	0.271	0.227	0.254	0.274
580	0.341	0.335	0.315	0.247	0.260	0.225	0.254	0.278
585	0.367	0.339	0.299	0.229	0.247	0.222	0.259	0.284
590	0.390	0.341	0.285	0.214	0.232	0.221	0.270	0.295
595	0.409	0.341	0.272	0.198	0.220	0.220	0.284	0.316
600	0.424	0.342	0.264	0.185	0.210	0.220	0.302	0.348
605	0.435	0.342	0.257	0.175	0.200	0.220	0.324	0.384
610	0.442	0.342	0.252	0.169	0.194	0.220	0.344	0.434
615	0.448	0.341	0.247	0.164	0.189	0.220	0.362	0.482
620	0.450	0.341	0.241	0.160	0.185	0.223	0.377	0.528
625	0.451	0.339	0.235	0.156	0.183	0.227	0.389	0.568
630	0.451	0.339	0.229	0.154	0.180	0.233	0.400	0.604
635	0.451	0.338	0.224	0.152	0.177	0.239	0.410	0.629
640	0.451	0.338	0.220	0.151	0.176	0.244	0.420	0.648
645	0.451	0.337	0.217	0.149	0.175	0.251	0.429	0.663

(Continued)

(*Continued*)

Wavelength	1 7.5R6/4	2 5Y6/4	3 5GY6/8	4 2.5G6/6	5 10BG6/4	6 5PB6/8	7 2.5P6/8	8 10P6/8
650	0.450	0.336	0.216	0.148	0.175	0.258	0.438	0.676
655	0.450	0.335	0.216	0.148	0.175	0.263	0.445	0.685
660	0.451	0.334	0.219	0.148	0.175	0.268	0.452	0.693
665	0.451	0.332	0.224	0.149	0.177	0.273	0.457	0.700
670	0.453	0.332	0.230	0.151	0.180	0.278	0.462	0.705
675	0.454	0.331	0.238	0.154	0.183	0.281	0.466	0.709
680	0.455	0.331	0.251	0.158	0.186	0.283	0.468	0.712
685	0.457	0.330	0.269	0.162	0.189	0.286	0.470	0.715
690	0.458	0.329	0.288	0.165	0.192	0.291	0.473	0.717
695	0.460	0.328	0.312	0.168	0.195	0.296	0.477	0.719
700	0.462	0.328	0.340	0.170	0.199	0.302	0.483	0.721
705	0.463	0.327	0.366	0.171	0.200	0.313	0.489	0.720
710	0.464	0.326	0.390	0.170	0.199	0.325	0.496	0.719
715	0.465	0.325	0.412	0.168	0.198	0.338	0.503	0.722
720	0.466	0.324	0.431	0.166	0.196	0.351	0.511	0.725
725	0.466	0.324	0.447	0.164	0.195	0.364	0.518	0.727
730	0.466	0.324	0.460	0.164	0.195	0.376	0.525	0.729
735	0.466	0.323	0.472	0.165	0.196	0.389	0.532	0.730
740	0.467	0.322	0.481	0.168	0.197	0.401	0.539	0.730
745	0.467	0.321	0.488	0.172	0.200	0.413	0.546	0.730
750	0.467	0.320	0.493	0.177	0.203	0.425	0.553	0.730
755	0.467	0.318	0.497	0.181	0.205	0.436	0.559	0.730
760	0.467	0.316	0.500	0.185	0.208	0.447	0.565	0.730
765	0.467	0.315	0.502	0.189	0.212	0.458	0.570	0.730
770	0.467	0.315	0.505	0.192	0.215	0.469	0.575	0.730
775	0.467	0.314	0.510	0.194	0.217	0.477	0.578	0.730
780	0.467	0.314	0.516	0.197	0.219	0.485	0.581	0.730

Spectral Radiance Factor of CRI Test Colours 9–14

Wavelength	9 4.5R4/13	10 5Y8/10	11 4.5G5/8	12 3PB3/11	13 5YR8/4	14 5GY4/4
380	0.066	0.050	0.111	0.120	0.104	0.036
385	0.062	0.054	0.121	0.103	0.127	0.036
390	0.058	0.059	0.127	0.090	0.161	0.037
395	0.055	0.063	0.129	0.082	0.211	0.038
400	0.052	0.066	0.127	0.076	0.264	0.039
405	0.052	0.067	0.121	0.068	0.313	0.039

(*Continued*)

(*Continued*)

Wavelength	9 4.5R4/13	10 5Y8/10	11 4.5G5/8	12 3PB3/11	13 5YR8/4	14 5GY4/4
410	0.051	0.068	0.116	0.064	0.341	0.040
415	0.050	0.069	0.112	0.065	0.352	0.041
420	0.050	0.069	0.108	0.075	0.359	0.042
425	0.049	0.070	0.105	0.093	0.361	0.042
430	0.048	0.072	0.104	0.123	0.364	0.043
435	0.047	0.073	0.104	0.160	0.365	0.044
440	0.046	0.076	0.105	0.207	0.367	0.044
445	0.044	0.078	0.106	0.256	0.369	0.045
450	0.042	0.083	0.110	0.300	0.372	0.045
455	0.041	0.088	0.115	0.331	0.374	0.046
460	0.038	0.095	0.123	0.346	0.376	0.047
465	0.035	0.103	0.134	0.347	0.379	0.048
470	0.033	0.113	0.148	0.341	0.384	0.050
475	0.031	0.125	0.167	0.328	0.389	0.052
480	0.030	0.142	0.192	0.307	0.397	0.055
485	0.029	0.162	0.219	0.282	0.405	0.057
490	0.028	0.189	0.252	0.257	0.416	0.062
495	0.028	0.219	0.291	0.230	0.429	0.067
500	0.028	0.262	0.325	0.204	0.443	0.075
505	0.029	0.305	0.347	0.178	0.454	0.083
510	0.030	0.365	0.356	0.154	0.461	0.092
515	0.030	0.416	0.353	0.129	0.466	0.100
520	0.031	0.465	0.346	0.109	0.469	0.108
525	0.031	0.509	0.333	0.090	0.471	0.121
530	0.032	0.546	0.314	0.075	0.474	0.133
535	0.032	0.581	0.294	0.062	0.476	0.142
540	0.033	0.610	0.271	0.051	0.483	0.150
545	0.034	0.634	0.248	0.041	0.490	0.154
550	0.035	0.653	0.227	0.035	0.506	0.155
555	0.037	0.666	0.206	0.029	0.526	0.152
560	0.041	0.678	0.188	0.025	0.553	0.147
565	0.044	0.687	0.170	0.022	0.582	0.140
570	0.048	0.693	0.153	0.019	0.618	0.133
575	0.052	0.698	0.138	0.017	0.651	0.125
580	0.060	0.701	0.125	0.017	0.680	0.118
585	0.076	0.704	0.114	0.017	0.701	0.112
590	0.102	0.705	0.106	0.016	0.717	0.106
595	0.136	0.705	0.100	0.016	0.729	0.101

(*Continued*)

(Continued)

Wavelength	9 4.5R4/13	10 5Y8/10	11 4.5G5/8	12 3PB3/11	13 5YR8/4	14 5GY4/4
600	0.190	0.706	0.096	0.016	0.736	0.098
605	0.256	0.707	0.092	0.016	0.742	0.095
610	0.336	0.707	0.090	0.016	0.745	0.093
615	0.418	0.707	0.087	0.016	0.747	0.090
620	0.505	0.708	0.085	0.016	0.748	0.089
625	0.581	0.708	0.082	0.016	0.748	0.087
630	0.641	0.710	0.080	0.018	0.748	0.086
635	0.682	0.711	0.079	0.018	0.748	0.085
640	0.717	0.712	0.078	0.018	0.748	0.084
645	0.740	0.714	0.078	0.018	0.748	0.084
650	0.758	0.716	0.078	0.019	0.748	0.084
655	0.770	0.718	0.078	0.020	0.748	0.084
660	0.781	0.720	0.081	0.023	0.747	0.085
665	0.790	0.722	0.083	0.024	0.747	0.087
670	0.797	0.725	0.088	0.026	0.747	0.092
675	0.803	0.729	0.093	0.030	0.747	0.096
680	0.809	0.731	0.102	0.035	0.747	0.102
685	0.814	0.735	0.112	0.043	0.747	0.110
690	0.819	0.739	0.125	0.056	0.747	0.123
695	0.824	0.742	0.141	0.074	0.746	0.137
700	0.828	0.746	0.161	0.097	0.746	0.152
705	0.830	0.748	0.182	0.128	0.746	0.169
710	0.831	0.749	0.203	0.166	0.745	0.188
715	0.833	0.751	0.223	0.210	0.744	0.207
720	0.835	0.753	0.242	0.257	0.743	0.226
725	0.836	0.754	0.257	0.305	0.744	0.243
730	0.836	0.755	0.270	0.354	0.745	0.260
735	0.837	0.755	0.282	0.401	0.748	0.277
740	0.838	0.755	0.292	0.446	0.750	0.294
745	0.839	0.755	0.302	0.485	0.750	0.310
750	0.839	0.756	0.310	0.520	0.749	0.325
755	0.839	0.757	0.314	0.551	0.748	0.339
760	0.839	0.758	0.317	0.577	0.748	0.353
765	0.839	0.759	0.323	0.599	0.747	0.366
770	0.839	0.759	0.330	0.618	0.747	0.379
775	0.839	0.759	0.334	0.633	0.747	0.390
780	0.839	0.759	0.338	0.645	0.747	0.399

A7.2 WORKED EXAMPLE OF THE CIE COLOUR RENDERING INDICES

1. As an example consider that the test lamp has the same spectral power distribution, $S(\lambda)$, as CIE illuminant F2 (See Appendix 5).

Wave-length	$S(\lambda)$	Wave-length	$S(\lambda)$	Wave-length	$S(\lambda)$	Wave-length	$S(\lambda)$
380	1.18	480	7.65	580	22.79	680	1.89
385	1.48	485	7.62	585	19.29	685	1.64
390	1.84	490	7.62	590	18.66	690	1.53
395	2.15	495	7.45	595	17.73	695	1.27
400	3.44	500	7.28	600	16.54	700	1.10
405	15.69	505	7.15	605	15.21	705	0.99
410	3.85	510	7.05	610	13.80	710	0.88
415	3.74	515	7.04	615	12.36	715	0.76
420	4.19	520	7.16	620	10.95	720	0.68
425	4.62	525	7.47	625	9.65	725	0.61
430	5.06	530	8.04	630	8.40	730	0.56
435	34.98	535	8.88	635	7.32	735	0.54
440	11.81	540	10.01	640	6.31	740	0.51
445	6.27	545	24.88	645	5.43	745	0.47
450	6.63	550	16.64	650	4.68	750	0.47
455	6.93	555	14.59	655	4.02	755	0.43
460	7.19	560	16.16	660	3.45	760	0.46
465	7.40	565	17.56	665	2.96	765	0.47
470	7.54	570	18.62	670	2.55	770	0.40
475	7.62	575	21.47	675	2.19	775	0.33
						780	0.27

2. Calculate the colorimetry of the test lamp using the CIE 2° observer and hence the correlated colour temperature:

X	99.1863	x	0.3721	u	0.2202	T	4224.4
Y	100.0000	y	0.3751	v	0.3331	Δ_{uv}	0.0018
Z	67.3942						

The correlated colour temperature of the test lamp is found by inspection of lines normal to the daylight or Planckian locus in a u,v chromaticity diagram. Robertson (Robertson, 1968) describes a method that can be easily programmed in a computer. The spectral power distribution of the reference illuminant can be calculated from the value of the correlated colour temperature of the test lamp. If the value is less than 5000 K then a Planckian radiator is used, if equal to or greater than 5000 K then a D-illuminant is used. The former is calculated using Planck's Law (Section 4.8) and the latter using the CIE method (Appendix 5.6).

The correlated colour temperature of the test lamp is 4224.4 K and the data below represent a Planckian radiator at this colour temperature.

Wave-length	$S(\lambda)$	Wave-length	$S(\lambda)$	Wave-length	$S(\lambda)$	Wave-length	$S(\lambda)$
380	38.90	480	78.32	580	103.54	680	111.29
385	40.93	485	80.01	585	104.31	685	111.31
390	42.99	490	81.67	590	105.04	690	111.30
395	45.05	495	83.28	595	105.72	695	111.26
400	47.12	500	84.84	600	106.36	700	111.20
405	49.19	505	86.36	605	106.95	705	111.11
410	51.26	510	87.84	610	107.51	710	110.99
415	53.33	515	89.27	615	108.02	715	110.85
420	55.38	520	90.65	620	108.49	720	110.69
425	57.43	525	91.99	625	108.92	725	110.50
430	59.46	530	93.27	630	109.32	730	110.29
435	61.47	535	94.51	635	109.67	735	110.06
440	63.46	540	95.71	640	109.99	740	109.81
445	65.43	545	96.85	645	110.27	745	109.54
450	67.37	550	97.95	650	110.51	750	109.26
455	69.28	555	99.00	655	110.72	755	108.95
460	71.16	560	100.00	660	110.90	760	108.62
465	73.00	565	100.95	665	111.04	765	108.28
470	74.81	570	101.86	670	111.16	770	107.92
475	76.59	575	102.73	675	111.24	775	107.55
						780	107.16

3. The colorimetry of the 15 CIE test colours is calculated using the spectral power distribution of the test lamp: the spectral radiance factors of each of the 15 test colours can be found in Section A7.1.

Test colour	x	y	u	v	Y
1	0.4282	0.3837	0.2539	0.3412	31.1863
2	0.4340	0.4359	0.2358	0.3552	30.2858
3	0.4179	0.4908	0.2076	0.3656	31.9163
4	0.3536	0.4494	0.1841	0.3508	28.1160
5	0.3232	0.3678	0.1910	0.3261	29.1288
6	0.2994	0.3014	0.1990	0.3005	27.0425
7	0.3403	0.3001	0.2299	0.3041	28.2231
8	0.3765	0.3191	0.2478	0.3151	30.3309
9	0.5686	0.3393	0.3832	0.3431	10.3247
10	0.4660	0.4812	0.2377	0.3682	63.6958
11	0.3200	0.4541	0.1639	0.3489	17.9934

Test colour	x	y	u	v	Y
12	0.1825	0.1789	0.1526	0.2245	4.3447
13	0.4262	0.4029	0.2442	0.3462	59.8609
14	0.4028	0.4707	0.2054	0.3601	12.1915
15	0.4296	0.3931	0.2506	0.3439	32.5579
Source	0.3721	0.3751	0.2202	0.3331	100.0000

4. The colorimetry of the 15 CIE test colours is calculated using the spectral power distribution of the reference illuminant:

Test colour	x	y	u	v	Y
1	0.4395	0.3725	0.2667	0.3391	31.0654
2	0.4379	0.4237	0.2430	0.3527	29.6747
3	0.4146	0.4817	0.2086	0.3635	30.5715
4	0.3336	0.4468	0.1734	0.3484	28.4318
5	0.3085	0.3690	0.1812	0.3251	29.6440
6	0.2942	0.3084	0.1926	0.3027	28.5372
7	0.3511	0.3029	0.2368	0.3063	29.4348
8	0.3992	0.3177	0.2655	0.3170	32.2824
9	0.6136	0.3205	0.4368	0.3422	13.3770
10	0.4751	0.4637	0.2496	0.3654	61.2613
11	0.2954	0.4607	0.1489	0.3482	19.1591
12	0.1707	0.1972	0.1359	0.2355	5.4574
13	0.4326	0.3913	0.2533	0.3437	59.0055
14	0.3992	0.4620	0.2061	0.3579	11.7150
15	0.4421	0.3802	0.2648	0.3416	33.8406
Source	0.3711	0.3707	0.2213	0.3317	100.0000

5. To account for the adaptive shift due to the different state of chromatic adaptation under the test lamp, k, and the reference illuminant, r, the following formulae are applied:

$$u'_{\mathrm{k,i}} = \frac{10.872 + 0.404\dfrac{c_{\mathrm{r}}}{c_{\mathrm{k}}}c_{\mathrm{k,i}} - 4\dfrac{d_{\mathrm{r}}}{d_{\mathrm{k}}}d_{\mathrm{k,i}}}{16.518 + 1.481\dfrac{c_{\mathrm{r}}}{c_{\mathrm{k}}}c_{\mathrm{k,i}} - \dfrac{d_{\mathrm{r}}}{d_{\mathrm{k}}}d_{\mathrm{k,i}}}$$

$$v'_{\mathrm{k,i}} = \frac{5.520}{16.518 + 1.481\dfrac{c_{\mathrm{r}}}{c_{\mathrm{k}}}c_{\mathrm{k,i}} - \dfrac{d_{\mathrm{r}}}{d_{\mathrm{k}}}d_{\mathrm{k,i}}}$$

The values of $u'_{k,i}$ and $v'_{k,i}$ are the chromaticity coordinates of a test colour sample i after consideration of the adaptive colour shift. The functions c and d are calculated for the test lamp u_k, v_k giving c_k and d_k, and the test colours under the test lamp $u_{k,i}$, $v_{k,i}$ giving $c_{k,i}$ and $d_{k,i}$ using the following formulae:

$$c = \frac{1}{v}(4 - u - 10v)$$

$$d = \frac{1}{v}(1.708v - 0.404 - 1.481u)$$

Test	u_k	v_k	T	Δuv	c_k	d_k
lamp	0.2202	0.3331	4224.42	0.00179	1.347884	1.94162
Reference	u_r	v_r			c_r	d_r
illuminant	0.2213	0.3317	4224.09	8.4E-06	1.39307	1.93778
	Δu	Δv				
	−0.00108	0.00142				

where Δu, Δv represent the distance of the test lamp from the Planckian locus

6. Calculate the respective values of $c_{k,I}$ and $d_{k,I}$ from the u,v coordinates, $u_{k,i}$, $v_{k,i}$ of the test lamp, then the corresponding values $u'_{k,i}$, $v'_{k,I}$ after correction for chromatic adaptation.

Test colour	$u_{k,i}$	$v_{k,i}$	$c_{k,i}$	$d_{k,i}$	$u'_{k,i}$	$v'_{k,i}$
1	0.2539	0.3412	0.9805	1.7902	0.2547	0.3401
2	0.2358	0.3552	0.5971	1.8623	0.2367	0.3545
3	0.2076	0.3656	0.3719	1.9721	0.2086	0.3651
4	0.1841	0.3508	0.8766	2.0825	0.1853	0.3498
5	0.1910	0.3261	1.6800	2.0794	0.1923	0.3244
6	0.1990	0.3005	2.6498	2.0717	0.2004	0.2983
7	0.2299	0.3041	2.3970	1.9170	0.2309	0.3021
8	0.2478	0.3151	1.9072	1.8253	0.2487	0.3134
9	0.3832	0.3431	0.5427	1.2313	0.3836	0.3424
10	0.2377	0.3682	0.2195	1.8493	0.2386	0.3678
11	0.1639	0.3489	0.9946	2.1701	0.1653	0.3477
12	0.1526	0.2245	7.1381	2.5007	0.1551	0.2213
13	0.2442	0.3462	0.8481	1.8304	0.2451	0.3452
14	0.2054	0.3601	0.5378	1.9851	0.2065	0.3594
15	0.2506	0.3439	0.9018	1.8037	0.2515	0.3429
	u_k	v_k	c_k	d_k	u'_k	v'_k
Source	0.2202	0.3331	1.3479	1.9416	0.2213	0.3317

7. Convert to U*V*W* space and calculate colour differences

	Test lamp			Reference illuminant		
Test colour	$W^*_{r,i}$	$U^*_{r,i}$	$V^*_{r,i}$	$W^*_{k,i}$	$U^*_{k,i}$	$V^*_{k,i}$
1	61.5897	36.3399	5.9600	61.6915	26.7916	6.7389
2	60.3990	16.9941	16.4788	60.9267	12.2051	18.0497
3	61.1710	−10.1320	25.3069	62.3008	−10.2771	27.0833
4	59.3030	−36.9377	12.9118	59.0194	−27.6409	13.9000
5	60.3723	−31.4961	−5.1809	59.9215	−22.5941	−5.6305
6	59.3971	−22.2223	−22.3549	58.0393	−15.8147	−25.1917
7	60.1899	12.0794	−19.8217	59.1159	7.3845	−22.7422
8	62.6029	35.9323	−11.9559	60.9654	21.7213	−14.5084
9	42.3461	118.6376	5.8174	37.4376	78.9656	5.2426
10	81.5527	29.9951	35.7639	82.8413	18.6010	38.9103
11	49.8958	−47.0065	10.7567	48.5105	−35.3360	10.1295
12	27.0150	−30.0197	−33.7830	23.7937	−20.4882	−34.1504
13	80.3279	33.4275	12.5955	80.7960	24.9693	14.2477
14	39.7789	−7.8536	13.5587	40.5385	−7.8128	14.6029
15	63.8635	36.0899	8.2541	62.8286	24.6087	9.1732
Source	99.0397	0.0000	0.0000	99.0397	0.0000	0.0000

8. Calculate the special colour rendering indices, R_i

Test colour	ΔE_i	R_i	
1	9.58	R_1	55.93
2	5.07	R_2	76.69
3	2.11	R_3	90.29
4	9.35	R_4	56.97
5	8.92	R_5	58.95
6	7.14	R_6	67.17
7	5.63	R_7	74.09
8	14.53	R_8	33.16
9	39.98	R_9	−83.90
10	11.89	R_{10}	45.30
11	11.77	R_{11}	45.86
12	10.07	R_{12}	53.69
13	8.63	R_{13}	60.30
14	1.29	R_{14}	94.06
15	11.56	R_{15}	46.80

9. Calculate the general colour rendering index R_a

R_a 64.16

Appendix 8

Illuminant-Observer Weights for Calculating Tristimulus Values

This appendix gives illuminant-observer weights for calculating tristimulus values from data of spectral reflectance factor or spectral transmittance factor. These weights were published as Table 6 in ASTM Document E 308–2006, and are reproduced by permission. These values should ideally be used only for data uncorrected for bandpass errors, but they have been found to be beneficial for general use (see Section 8.14). The occasional negative values are caused by the interpolation method used.

ASTM Document E 2022 – 2008. Standard practice for calculation of weighting factors for tristimulus integration. ASTM International, West Conshohocken, PA (2008).

Table A8.1 Illuminant A, 1931 Observer

10 nm Interval			
nm	W_X	W_Y	W_Z
360	0.000	0.000	0.000
370	0.000	0.000	0.001
380	0.001	0.000	0.005
390	0.004	0.000	0.018
400	0.017	0.000	0.081
410	0.057	0.002	0.272
420	0.246	0.007	1.178
430	0.660	0.025	3.214
440	0.942	0.059	4.710

(*Continued*)

Measuring Colour, Fourth Edition. R.W.G. Hunt and M.R. Pointer.

Table A8.1 (*Continued*)

10 nm Interval			
nm	W_X	W_Y	W_Z
450	1.039	0.113	5.454
460	1.043	0.205	5.969
470	0.790	0.353	5.209
480	0.416	0.608	3.602
490	0.148	1.012	2.277
500	0.016	1.749	1.493
510	0.028	3.047	0.963
520	0.388	4.778	0.505
530	1.187	6.345	0.305
540	2.288	7.625	0.157
550	3.702	8.594	0.071
560	5.484	9.255	0.034
570	7.562	9.496	0.020
580	9.739	9.265	0.018
590	11.644	8.567	0.013
600	12.811	7.563	0.010
610	12.782	6.365	0.004
620	11.460	5.076	0.002
630	8.991	3.689	0.001
640	6.536	2.543	0.000
650	4.296	1.616	0.000
660	2.583	0.954	0.000
670	1.405	0.514	0.000
680	0.780	0.283	0.000
690	0.388	0.140	0.000
700	0.200	0.072	0.000
710	0.106	0.038	0.000
720	0.054	0.020	0.000
730	0.028	0.010	0.000
740	0.014	0.005	0.000
750	0.007	0.002	0.000
760	0.003	0.001	0.000
770	0.002	0.001	0.000
780	0.001	0.000	0.000
Check Sum	109.848	99.997	35.586
White Point	109.850	100.000	35.585

Table A8.2 Illuminant A, 1931 Observer

20 nm Interval			
nm	W_X	W_Y	W_Z
360	0.000	0.000	0.000
380	0.013	0.000	0.060
400	−0.026	0.000	−0.123
420	0.483	0.009	2.306
440	1.955	0.106	9.637
460	2.145	0.385	12.257
480	0.848	1.119	7.301
500	−0.112	3.247	2.727
520	0.611	9.517	1.035
540	4.407	15.434	0.274
560	10.804	18.703	0.055
580	19.601	18.746	0.034
600	26.256	15.233	0.018
620	23.295	10.105	0.003
640	12.853	4.939	0.000
660	4.863	1.784	0.000
680	1.363	0.495	0.000
700	0.359	0.129	0.000
720	0.100	0.036	0.000
740	0.023	0.008	0.000
760	0.006	0.002	0.000
780	0.002	0.001	0.000
Check Sum	109.849	99.998	35.584
White Point	109.850	100.000	35.585

Table A8.3 Illuminant A, 1964 Observer

10 nm Interval			
nm	$W_{10,X}$	$W_{10,Y}$	$W_{10,Z}$
360	0.000	0.000	0.000
370	0.000	0.000	0.000
380	0.000	0.000	0.000
390	0.002	0.000	0.007
400	0.018	0.002	0.078
410	0.118	0.012	0.540
420	0.372	0.038	1.760
430	0.686	0.082	3.374
440	0.982	0.154	5.024
450	1.094	0.255	5.876
460	1.024	0.414	5.882
470	0.747	0.688	5.023
480	0.326	1.073	3.236
490	0.061	1.589	1.926
500	0.003	2.397	1.129
510	0.189	3.503	0.638
520	0.717	4.857	0.377
530	1.617	6.096	0.205
540	2.823	7.290	0.100
550	4.296	8.116	0.028
560	6.177	8.799	−0.003
570	8.285	9.039	0.001
580	10.218	8.758	0.000
590	12.041	8.350	0.000
600	12.850	7.492	0.000
610	12.441	6.337	0.000
620	10.872	5.025	0.000
630	8.604	3.753	0.000
640	5.951	2.469	0.000
650	3.846	1.537	0.000
660	2.259	0.891	0.000
670	1.242	0.485	0.000
680	0.643	0.250	0.000
690	0.324	0.126	0.000
700	0.160	0.062	0.000
710	0.078	0.030	0.000
720	0.039	0.015	0.000
730	0.019	0.007	0.000
740	0.010	0.004	0.000

(*Continued*)

Table A8.3 (*Continued*)

10 nm Interval			
nm	$W_{10,X}$	$W_{10,Y}$	$W_{10,Z}$
750	0.005	0.002	0.000
760	0.002	0.001	0.000
770	0.001	0.001	0.000
780	0.001	0.000	0.000
Check Sum	111.143	99.999	35.201
White Point	111.144	100.000	35.200

Table A8.4 Illuminant A, 1964 Observer

20 nm Interval			
nm	$W_{10,X}$	$W_{10,Y}$	$W_{10,Z}$
360	0.000	0.000	0.000
380	0.007	0.000	0.037
400	−0.016	0.000	−0.088
420	0.691	0.066	3.226
440	2.025	0.285	10.278
460	2.158	0.796	12.345
480	0.642	2.043	6.555
500	−0.160	4.630	1.966
520	1.284	9.668	0.721
540	5.445	14.621	0.171
560	12.238	17.766	−0.013
580	20.755	17.800	0.004
600	26.325	15.129	−0.001
620	22.187	10.097	0.000
640	11.816	4.858	0.000
660	4.221	1.643	0.000
680	1.154	0.452	0.000
700	0.282	0.109	0.000
720	0.068	0.026	0.000
740	0.017	0.007	0.000
760	0.004	0.002	0.000
780	0.001	0.000	0.000
Check Sum	111.144	99.998	35.201
White Point	111.144	100.000	35.200

Table A8.5 Illuminant C, 1931 Observer

10 nm Interval			
nm	W_X	W_Y	W_Z
360	0.000	0.000	0.000
370	0.001	0.000	0.003
380	0.004	0.000	0.017
390	0.015	0.000	0.069
400	0.074	0.002	0.350
410	0.261	0.007	1.241
420	1.170	0.032	5.605
430	3.074	0.118	14.967
440	4.066	0.259	20.346
450	3.951	0.437	20.769
460	3.421	0.684	19.624
470	2.292	1.042	15.153
480	1.066	1.600	9.294
490	0.325	2.332	5.115
500	0.025	3.375	2.788
510	0.052	4.823	1.481
520	0.535	6.468	0.669
530	1.496	7.951	0.381
540	2.766	9.193	0.187
550	4.274	9.889	0.081
560	5.891	9.898	0.036
570	7.353	9.186	0.019
580	8.459	8.008	0.015
590	9.036	6.621	0.010
600	9.005	5.302	0.007
610	8.380	4.168	0.003
620	7.111	3.147	0.001
630	5.300	2.174	0.000
640	3.669	1.427	0.000
650	2.320	0.873	0.000
660	1.333	0.492	0.000
670	0.683	0.250	0.000
680	0.356	0.129	0.000
690	0.162	0.059	0.000
700	0.077	0.028	0.000
710	0.038	0.014	0.000
720	0.018	0.006	0.000
730	0.008	0.003	0.000
740	0.004	0.001	0.000

(*Continued*)

Table A8.5 (*Continued*)

10 nm Interval			
nm	W_X	W_Y	W_Z
750	0.002	0.001	0.000
760	0.001	0.000	0.000
770	0.000	0.000	0.000
780	0.000	0.000	0.000
Check Sum	98.074	99.999	118.231
White Point	98.074	100.000	118.232

Table A8.6 Illuminant C, 1931 Observer

20 nm Interval			
nm	W_X	W_Y	W_Z
360	0.000	0.000	0.000
380	0.066	0.000	0.311
400	−0.164	0.001	−0.777
420	2.373	0.044	11.296
440	8.595	0.491	42.561
460	6.939	1.308	39.899
480	2.045	3.062	18.451
500	−0.217	6.596	4.728
520	0.881	12.925	1.341
540	5.406	18.650	0.319
560	11.842	20.143	0.059
580	17.169	16.095	0.028
600	18.383	10.537	0.013
620	14.348	6.211	0.002
640	7.148	2.743	0.000
660	2.484	0.911	0.000
680	0.600	0.218	0.000
700	0.136	0.049	0.000
720	0.031	0.011	0.000
740	0.006	0.002	0.000
760	0.002	0.001	0.000
780	0.000	0.000	0.000
Check Sum	98.073	99.998	118.231
White Point	98.074	100.000	118.232

Table A8.7 Illuminant C, 1964 Observer

10 nm Interval			
nm	$W_{10.X}$	$W_{10.Y}$	$W_{10.Z}$
360	0.000	0.000	0.000
370	0.000	0.000	0.000
380	0.000	0.000	0.000
390	0.006	0.001	0.025
400	0.071	0.007	0.317
410	0.519	0.054	2.362
420	1.690	0.173	7.995
430	3.050	0.364	15.015
440	4.055	0.638	20.751
450	3.974	0.936	21.364
460	3.207	1.316	18.457
470	2.067	1.938	13.957
480	0.792	2.693	7.968
490	0.123	3.489	4.126
500	0.008	4.395	2.006
510	0.297	5.276	0.935
520	0.939	6.275	0.480
530	1.944	7.299	0.244
540	3.259	8.401	0.114
550	4.739	8.926	0.030
560	6.340	8.995	−0.003
570	7.694	8.357	0.001
580	8.479	7.236	0.000
590	8.929	6.171	0.000
600	8.630	5.020	0.000
610	7.794	3.966	0.000
620	6.446	2.978	0.000
630	4.848	2.114	0.000
640	3.191	1.323	0.000
650	1.986	0.793	0.000
660	1.114	0.439	0.000
670	0.577	0.226	0.000
680	0.280	0.109	0.000
690	0.130	0.050	0.000
700	0.059	0.023	0.000
710	0.027	0.010	0.000
720	0.012	0.005	0.000
730	0.005	0.002	0.000
740	0.003	0.001	0.000

(Continued)

Table A8.7 (*Continued*)

10 nm Interval			
nm	$W_{10.X}$	$W_{10.Y}$	$W_{10.Z}$
750	0.001	0.000	0.000
760	0.001	0.000	0.000
770	0.000	0.000	0.000
780	0.000	0.000	0.000
Check Sum	97.286	99.999	116.144
White Point	97.285	100.000	116.145

Table A8.8 Illuminant C, 1964 Observer

20 nm Interval			
nm	$W_{10.X}$	$W_{10.Y}$	$W_{10.Z}$
360	0.000	0.000	0.000
380	0.043	0.002	0.213
400	−0.122	−0.004	−0.622
420	3.216	0.301	15.025
440	8.476	1.239	43.144
460	6.668	2.577	38.431
480	1.430	5.320	15.661
500	−0.249	8.742	3.219
520	1.734	12.466	0.897
540	6.364	16.891	0.187
560	12.790	18.284	−0.014
580	17.338	14.617	0.004
600	17.597	10.019	−0.001
620	13.045	5.925	0.000
640	6.283	2.581	0.000
660	2.055	0.800	0.000
680	0.488	0.191	0.000
700	0.100	0.039	0.000
720	0.021	0.008	0.000
740	0.004	0.002	0.000
760	0.001	0.000	0.000
780	0.000	0.000	0.000
Check Sum	97.282	100.000	116.144
White Point	97.285	100.000	116.145

Table A8.9 Illuminant D50, 1931 Observer

10 nm Interval			
nm	W_X	W_Y	W_Z
360	0.000	0.000	0.000
370	0.001	0.000	0.005
380	0.003	0.000	0.014
390	0.008	0.000	0.039
400	0.058	0.002	0.277
410	0.191	0.005	0.906
420	0.751	0.021	3.603
430	1.592	0.060	7.747
440	2.519	0.158	12.593
450	2.824	0.310	14.834
460	2.556	0.511	14.659
470	1.717	0.776	11.344
480	0.832	1.246	7.240
490	0.250	1.783	3.934
500	0.025	2.892	2.447
510	0.047	4.610	1.432
520	0.538	6.586	0.688
530	1.590	8.435	0.403
540	2.770	9.185	0.186
550	4.210	9.733	0.080
560	5.662	9.503	0.035
570	7.092	8.882	0.019
580	8.681	8.225	0.016
590	9.175	6.728	0.010
600	9.966	5.884	0.008
610	9.556	4.752	0.003
620	8.099	3.584	0.002
630	5.835	2.392	0.000
640	4.199	1.633	0.000
650	2.539	0.954	0.000
660	1.517	0.560	0.000
670	0.831	0.304	0.000
680	0.423	0.153	0.000
690	0.178	0.064	0.000
700	0.096	0.035	0.000
710	0.049	0.018	0.000
720	0.020	0.007	0.000
730	0.012	0.004	0.000
740	0.006	0.002	0.000

(Continued)

Table A8.9 (*Continued*)

10 nm Interval			
nm	W_X	W_Y	W_Z
750	0.002	0.001	0.000
760	0.001	0.000	0.000
770	0.001	0.000	0.000
780	0.000	0.000	0.000
Check Sum	96.422	99.998	82.524
White Point	96.422	100.000	82.521

Table A8.10 Illuminant D50, 1931 Observer

20 nm Interval			
nm	W_X	W_Y	W_Z
360	0.000	0.000	0.000
380	0.021	0.000	0.100
400	−0.013	0.003	−0.060
420	1.297	0.023	6.170
440	5.218	0.290	25.788
460	5.326	0.984	30.489
480	1.554	2.291	13.965
500	−0.191	5.461	4.224
520	0.915	13.421	1.430
540	5.528	18.956	0.313
560	11.324	19.226	0.057
580	17.119	16.204	0.028
600	20.222	11.611	0.014
620	16.400	7.117	0.002
640	7.922	3.030	0.000
660	2.835	1.043	0.000
680	0.741	0.268	0.000
700	0.150	0.054	0.000
720	0.044	0.016	0.000
740	0.009	0.003	0.000
760	0.002	0.001	0.000
780	0.001	0.000	0.000
Check Sum	96.424	100.002	82.520
White Point	96.422	100.000	82.521

Table A8.11 Illuminant D50, 1964 Observer

10 nm Interval			
nm	$W_{10,X}$	$W_{10,Y}$	$W_{10,Z}$
360	0.000	0.000	0.000
370	0.000	0.000	0.000
380	0.001	0.000	0.002
390	0.002	0.000	0.009
400	0.059	0.006	0.263
410	0.385	0.040	1.751
420	1.087	0.112	5.154
430	1.598	0.190	7.864
440	2.556	0.398	13.066
450	2.888	0.675	15.511
460	2.437	1.000	14.023
470	1.574	1.469	10.623
480	0.630	2.130	6.312
490	0.096	2.715	3.227
500	0.006	3.842	1.796
510	0.284	5.138	0.919
520	0.965	6.500	0.501
530	2.101	7.872	0.263
540	3.317	8.532	0.114
550	4.745	8.931	0.031
560	6.194	8.780	−0.003
570	7.547	8.214	0.001
580	8.847	7.557	0.000
590	9.218	6.375	0.000
600	9.712	5.663	0.000
610	9.035	4.597	0.000
620	7.465	3.447	0.000
630	5.426	2.366	0.000
640	3.713	1.541	0.000
650	2.208	0.882	0.000
660	1.289	0.509	0.000
670	0.714	0.279	0.000
680	0.338	0.131	0.000
690	0.144	0.056	0.000
700	0.075	0.029	0.000
710	0.035	0.014	0.000
720	0.014	0.005	0.000
730	0.008	0.003	0.000
740	0.004	0.002	0.000

(*Continued*)

Table A8.11 (*Continued*)

10 nm Interval			
nm	$W_{10,X}$	$W_{10,Y}$	$W_{10,Z}$
750	0.002	0.001	0.000
760	0.001	0.000	0.000
770	0.000	0.000	0.000
780	0.000	0.000	0.000
Check Sum	96.720	100.001	81.427
White Point	96.720	100.000	81.427

Table A8.12 Illuminant D50, 1964 Observer

20 nm Interval			
nm	$W_{10,X}$	$W_{10,Y}$	$W_{10,Z}$
360	0.000	0.000	0.000
380	0.001	−0.001	0.010
400	0.035	0.009	0.131
420	1.856	0.174	8.631
440	5.234	0.748	26.634
460	5.206	1.975	29.874
480	1.104	4.046	12.054
500	−0.238	7.459	2.948
520	1.816	13.203	0.969
540	6.614	17.441	0.186
560	12.430	17.746	−0.014
580	17.595	14.952	0.004
600	19.678	11.219	−0.001
620	15.166	6.902	0.000
640	7.075	2.898	0.000
660	2.387	0.931	0.000
680	0.612	0.240	0.000
700	0.111	0.043	0.000
720	0.030	0.012	0.000
740	0.006	0.002	0.000
760	0.001	0.000	0.000
780	0.001	0.000	0.000
Check Sum	96.720	99.999	81.426
White Point	96.720	100.000	81.427

Table A8.13 Illuminant D55, 1931 Observer

10 nm Interval			
nm	W_X	W_Y	W_Z
360	0.000	0.000	0.000
370	0.001	0.000	0.006
380	0.004	0.000	0.019
390	0.011	0.000	0.051
400	0.072	0.002	0.343
410	0.232	0.006	1.105
420	0.897	0.026	4.303
430	1.872	0.071	9.113
440	2.881	0.181	14.405
450	3.169	0.348	16.648
460	2.831	0.567	16.238
470	1.874	0.849	12.388
480	0.896	1.346	7.807
490	0.266	1.902	4.187
500	0.026	3.042	2.570
510	0.050	4.806	1.490
520	0.554	6.779	0.707
530	1.624	8.605	0.411
540	2.807	9.303	0.188
550	4.236	9.789	0.080
560	5.660	9.497	0.035
570	7.052	8.829	0.018
580	8.575	8.123	0.015
590	8.968	6.574	0.010
600	9.626	5.681	0.008
610	9.151	4.550	0.003
620	7.698	3.406	0.002
630	5.508	2.258	0.000
640	3.916	1.523	0.000
650	2.356	0.885	0.000
660	1.393	0.514	0.000
670	0.757	0.277	0.000
680	0.383	0.139	0.000
690	0.162	0.059	0.000
700	0.087	0.031	0.000
710	0.045	0.016	0.000
720	0.018	0.007	0.000
730	0.011	0.004	0.000
740	0.005	0.002	0.000

(*Continued*)

Table A8.13 (*Continued*)

10 nm Interval			
nm	W_X	W_Y	W_Z
750	0.002	0.001	0.000
760	0.001	0.000	0.000
770	0.001	0.000	0.000
780	0.000	0.000	0.000
Check Sum	95.678	99.998	92.150
White Point	95.682	100.000	92.149

Table A8.14 Illuminant D55, 1931 Observer

20 nm Interval			
nm	W_X	W_Y	W_Z
360	0.000	0.000	0.000
380	0.027	0.000	0.127
400	−0.016	0.004	−0.072
420	1.578	0.029	7.506
440	5.983	0.334	29.586
460	5.881	1.094	33.691
480	1.663	2.481	15.012
500	−0.202	5.771	4.413
520	0.950	13.833	1.471
540	5.611	19.197	0.314
560	11.328	19.214	0.057
580	16.931	16.001	0.028
600	19.527	11.196	0.013
620	15.581	6.759	0.002
640	7.384	2.823	0.000
660	2.600	0.956	0.000
680	0.669	0.242	0.000
700	0.137	0.049	0.000
720	0.040	0.014	0.000
740	0.008	0.003	0.000
760	0.001	0.001	0.000
780	0.001	0.000	0.000
Check Sum	95.682	100.001	92.148
White Point	95.682	100.000	92.149

Table A8.15 Illuminant D55, 1964 Observer

10 nm Interval			
nm	$W_{10,X}$	$W_{10,Y}$	$W_{10,Z}$
360	0.000	0.000	0.000
370	0.000	0.000	0.000
380	0.001	0.000	0.003
390	0.003	0.000	0.012
400	0.073	0.008	0.326
410	0.466	0.048	2.122
420	1.291	0.133	6.120
430	1.870	0.222	9.203
440	2.910	0.454	14.875
450	3.224	0.755	17.323
460	2.686	1.104	15.458
470	1.710	1.599	11.543
480	0.675	2.289	6.773
490	0.101	2.882	3.418
500	0.007	4.021	1.876
510	0.296	5.329	0.952
520	0.989	6.657	0.513
530	2.134	7.993	0.267
540	3.345	8.600	0.115
550	4.751	8.939	0.031
560	6.162	8.732	−0.003
570	7.468	8.126	0.001
580	8.697	7.426	0.000
590	8.966	6.199	0.000
600	9.336	5.442	0.000
610	8.610	4.380	0.000
620	7.061	3.261	0.000
630	5.097	2.222	0.000
640	3.446	1.430	0.000
650	2.039	0.814	0.000
660	1.178	0.465	0.000
670	0.647	0.253	0.000
680	0.305	0.119	0.000
690	0.131	0.051	0.000
700	0.067	0.026	0.000
710	0.032	0.012	0.000
720	0.012	0.005	0.000
730	0.007	0.003	0.000
740	0.004	0.001	0.000

(*Continued*)

Table A8.15 (*Continued*)

10 nm Interval			
nm	$W_{10,X}$	$W_{10,Y}$	$W_{10,Z}$
750	0.001	0.001	0.000
760	0.001	0.000	0.000
770	0.000	0.000	0.000
780	0.000	0.000	0.000
Check Sum	95.799	100.001	90.928
White Point	95.799	100.000	90.926

Table A8.16 Illuminant D55, 1964 Observer

20 nm Interval			
nm	$W_{10,X}$	$W_{10,Y}$	$W_{10,Z}$
360	0.000	0.000	0.000
380	0.001	−0.001	0.013
400	0.044	0.010	0.165
420	2.237	0.210	10.414
440	5.965	0.856	30.366
460	5.721	2.183	32.860
480	1.170	4.359	12.878
500	−0.246	7.830	3.064
520	1.870	13.538	0.992
540	6.678	17.576	0.186
560	12.373	17.649	−0.013
580	17.314	14.694	0.004
600	18.909	10.768	−0.001
620	14.336	6.522	0.000
640	6.563	2.688	0.000
660	2.177	0.849	0.000
680	0.551	0.216	0.000
700	0.101	0.039	0.000
720	0.027	0.011	0.000
740	0.006	0.002	0.000
760	0.001	0.000	0.000
780	0.000	0.000	0.000
Check Sum	95.798	99.999	90.928
White Point	95.799	100.000	90.926

Table A8.17 Illuminant D65, 1931 Observer

10 nm Interval			
nm	W_X	W_Y	W_Z
360	0.000	0.000	0.000
370	0.002	0.000	0.009
380	0.006	0.000	0.029
390	0.016	0.000	0.077
400	0.097	0.003	0.460
410	0.311	0.009	1.477
420	1.164	0.033	5.581
430	2.400	0.092	11.684
440	3.506	0.221	17.532
450	3.755	0.413	19.729
460	3.298	0.662	18.921
470	2.141	0.973	14.161
480	1.001	1.509	8.730
490	0.293	2.107	4.623
500	0.028	3.288	2.769
510	0.054	5.122	1.584
520	0.581	7.082	0.736
530	1.668	8.833	0.421
540	2.860	9.472	0.191
550	4.257	9.830	0.081
560	5.632	9.446	0.034
570	6.960	8.709	0.018
580	8.344	7.901	0.015
590	8.676	6.357	0.009
600	9.120	5.379	0.007
610	8.568	4.259	0.003
620	7.119	3.149	0.001
630	5.049	2.070	0.000
640	3.522	1.370	0.000
650	2.112	0.794	0.000
660	1.229	0.454	0.000
670	0.658	0.240	0.000
680	0.331	0.120	0.000
690	0.142	0.051	0.000
700	0.074	0.027	0.000
710	0.039	0.014	0.000
720	0.016	0.006	0.000
730	0.009	0.003	0.000
740	0.005	0.002	0.000

(Continued)

Table A8.17 (*Continued*)

10 nm Interval			
nm	W_X	W_Y	W_Z
750	0.002	0.001	0.000
760	0.001	0.000	0.000
770	0.001	0.000	0.000
780	0.000	0.000	0.000
Check Sum	95.047	100.001	108.882
White Point	95.047	100.000	108.883

Table A8.18 Illuminant D65, 1931 Observer

20 nm Interval			
nm	W_X	W_Y	W_Z
360	0.000	0.000	0.000
380	0.040	0.000	0.187
400	−0.026	0.004	−0.120
420	2.114	0.041	10.065
440	7.323	0.411	36.235
460	6.815	1.281	39.090
480	1.843	2.797	16.753
500	−0.219	6.291	4.727
520	1.003	14.463	1.532
540	5.723	19.509	0.314
560	11.284	19.106	0.058
580	16.548	15.600	0.027
600	18.528	10.607	0.013
620	14.397	6.240	0.002
640	6.646	2.540	0.000
660	2.290	0.842	0.000
680	0.574	0.208	0.000
700	0.120	0.043	0.000
720	0.034	0.012	0.000
740	0.007	0.003	0,000
760	0.001	0.000	0.000
780	0.001	0.000	0.000
Check Sum	95.046	99.998	108.883
White Point	95.047	100.000	108.883

Table A8.19 Illuminant D65, 1964 Observer

	10 nm Interval		
nm	$W_{10.X}$	$W_{10.Y}$	$W_{10.Z}$
360	0.000	0.000	0.000
370	0.000	0.000	−0.001
380	0.001	0.000	0.004
390	0.005	0.000	0.020
400	0.097	0.010	0.436
410	0.616	0.064	2.808
420	1.660	0.171	7.868
430	2.377	0.283	11.703
440	3.512	0.549	17.958
450	3.789	0.888	20.358
460	3.103	1.277	17.861
470	1.937	1.817	13.085
480	0.747	2.545	7.510
490	0.110	3.164	3.743
500	0.007	4.309	2.003
510	0.314	5.631	1.004
520	1.027	6.896	0.529
530	2.174	8.136	0.271
540	3.380	8.684	0.116
550	4.735	8.903	0.030
560	6.081	8.614	−0.003
570	7.310	7.950	0.001
580	8.393	7.164	0.000
590	8.603	5.945	0.000
600	8.771	5.110	0.000
610	7.996	4.067	0.000
620	6.476	2.990	0.000
630	4.635	2.020	0.000
640	3.074	1.275	0.000
650	1.814	0.724	0.000
660	1.031	0.407	0.000
670	0.557	0.218	0.000
680	0.261	0.102	0.000
690	0.114	0.044	0.000
700	0.057	0.022	0.000
710	0.028	0.011	0.000
720	0.011	0.004	0.000
730	0.006	0.002	0.000
740	0.003	0.001	0.000

(*Continued*)

Table A8.19 (*Continued*)

10 nm Interval			
nm	$W_{10.X}$	$W_{10.Y}$	$W_{10.Z}$
750	0.001	0.000	0.000
760	0.000	0.000	0.000
770	0.000	0.000	0.000
780	0.000	0.000	0.000
Check Sum	94.813	99.997	107.304
White Point	94.811	100.000	107.304

Table A8.20 Illuminant D65, 1964 Observer

20 nm Interval			
nm	$W_{10.X}$	$W_{10.Y}$	$W_{10.Z}$
360	0.000	0.000	0.000
380	0.003	−0.001	0.025
400	0.056	0.013	0.199
420	2.951	0.280	13.768
440	7.227	1.042	36.808
460	6.578	2.534	37.827
480	1.278	4.872	14.226
500	−0.259	8.438	3.254
520	1.951	14.030	1.025
540	6.751	17.715	0.184
560	12.223	17.407	−0.013
580	16.779	14.210	0.004
600	17.793	10.121	−0.001
620	13.135	5.971	0.000
640	5.859	2.399	0.000
660	1.901	0.741	0.000
680	0.469	0.184	0.000
700	0.088	0.034	0.000
720	0.023	0.009	0.000
740	0.005	0.002	0.000
760	0.001	0.000	0.000
780	0.000	0.000	0.000
Check Sum	94.812	100.001	107.306
White Point	94.811	100.000	107.304

Table A8.21 Illuminant D75, 1931 Observer

10 nm Interval			
nm	W_X	W_Y	W_Z
360	0.000	0.000	0.000
370	0.003	0.000	0.012
380	0.008	0.000	0.038
390	0.021	0.001	0.098
400	0.120	0.003	0.567
410	0.378	0.010	1.798
420	1.403	0.040	6.728
430	2.820	0.108	13.727
440	4.028	0.254	20.146
450	4.244	0.467	22.301
460	3.677	0.739	21.106
470	2.350	1.071	15.552
480	1.087	1.642	9.485
490	0.313	2.262	4.951
500	0.029	3.484	2.929
510	0.058	5.371	1.657
520	0.599	7.281	0.754
530	1.702	9.005	0.430
540	2.890	9.564	0.192
550	4.265	9.845	0.081
560	5.592	9.375	0.034
570	6.853	8.571	0.018
580	8.161	7.725	0.015
590	8.429	6.174	0.009
600	8.777	5.176	0.007
610	8.176	4.064	0.003
620	6.737	2.980	0.001
630	4.728	1.938	0.000
0.34	3.279	1.275	0.000
650	1.956	0.735	0.000
660	1.128	0.417	0.000
670	0.599	0.219	0.000
680	0.301	0.109	0.000
690	0.128	0.046	0.000
700	0.067	0.024	0.000
710	0.036	0.013	0.000
720	0.014	0.005	0.000
730	0.009	0.003	0.000
740	0.004	0.002	0.000

Table A8.21 (*Continued*)

10 nm Interval			
nm	W_X	W_Y	W_Z
750	0.002	0.001	0.000
760	0.001	0.000	0.000
770	0.000	0.000	0.000
780	0.000	0.000	0.000
Check Sum	94.972	99.999	122.639
White Point	94.972	100.000	122.638

Table A8.22 Illuminant D75, 1931 Observer

20 nm Interval			
nm	W_X	W_Y	W_Z
360	0.000	0.000	0.000
380	0.050	0.000	0.235
400	−0.030	0.005	−0.142
420	2.571	0.051	12.243
440	8.429	0.475	41.731
460	7.578	1.434	43.498
480	1.982	3.045	18.114
500	−0.231	6.706	4.973
520	1.042	14.911	1.575
540	5.798	19.708	0.314
560	11.210	18.953	0.057
580	16.196	15.245	0.026
600	17.836	10.201	0.012
620	13.604	5.892	0.002
640	6.169	2.358	0.000
660	2.102	0.773	0.000
680	0.518	0.188	0.000
700	0.109	0.039	0.000
720	0.031	0.011	0.000
740	0.007	0.002	0.000
760	0.001	0.000	0.000
780	0.000	0.000	0.000
Check Sum	94.972	99.997	122.638
White Point	94.972	100.000	122.638

Table A8.23 Illuminant D75, 1964 Observer

10 nm Interval			
nm	$W_{10,X}$	$W_{10,Y}$	$W_{10,Z}$
360	0.000	0.000	0.000
370	0.000	0.000	−0.001
380	0.001	0.000	0.005
390	0.006	0.001	0.026
400	0.119	0.013	0.535
410	0.745	0.077	3.396
420	1.985	0.205	9.410
430	2.773	0.330	13.652
440	4.009	0.628	20.503
450	4.254	0.998	22.859
460	3.437	1.417	19.790
470	2.112	1.986	14.275
480	0.805	2.751	8.104
490	0.116	3.374	3.981
500	0.008	4.534	2.105
510	0.328	5.863	1.043
520	1.051	7.042	0.539
530	2.203	8.241	0.274
540	3.392	8.711	0.116
550	4.713	8.858	0.030
560	5.997	8.493	−0.003
570	7.149	7.773	0.001
580	8.154	6.959	0.000
590	8.303	5.736	0.000
600	8.386	4.885	0.000
610	7.580	3.855	0.000
620	6.088	2.811	0.000
630	4.312	1.879	0.000
640	2.843	1.179	0.000
650	1.669	0.666	0.000
660	0.940	0.371	0.000
670	0.504	0.197	0.000
680	0.236	0.092	0.000
690	0.102	0.040	0.000
700	0.051	0.020	0.000
710	0.025	0.010	0.000
720	0.010	0.004	0.000
730	0.006	0.002	0.000
740	0.003	0.001	0.000

(Continued)

Table A8.23 (*Continued*)

10 nm Interval			
nm	$W_{10,X}$	$W_{10,Y}$	$W_{10,Z}$
750	0.001	0.000	0.000
760	0.000	0.000	0.000
770	0.000	0.000	0.000
780	0.000	0.000	0.000
Check Sum	94.416	100.002	120.640
White Point	94.416	100.000	120.641

Table A8.24 Illuminant D75, 1964 Observer

20 nm Interval			
nm	$W_{10,X}$	$W_{10,Y}$	$W_{10,Z}$
360	0.000	0.000	0.000
380	0.003	−0.002	0.029
400	0.071	0.015	0.252
420	3.555	0.339	16.605
440	8.252	1.195	42.050
460	7.268	2.815	41.829
480	1.358	5.270	15.257
500	−0.266	8.912	3.401
520	2.006	14.363	1.045
540	6.791	17.776	0.182
560	12.060	17.154	−0.013
580	16.311	13.796	0.004
600	17.015	9.671	−0.001
620	12.327	5.601	0.000
640	5.403	2.212	0.000
660	1.733	0.676	0.000
680	0.421	0.165	0.000
700	0.080	0.031	0.000
720	0.021	0.008	0.000
740	0.005	0.002	0.000
760	0.001	0.000	0.000
780	0.000	0.000	0.000
Check Sum	94.415	99.999	120.640
White Point	94.416	100.000	120.641

Table A8.25 Illuminant F2, 1931 Observer

10 nm Interval			
nm	W_X	W_Y	W_Z
360	0.000	0.000	0.000
370	0.000	0.000	0.000
380	0.001	0.000	0.004
390	−0.007	0.000	−0.038
400	0.082	0.002	0.390
410	0.175	0.005	0.836
420	−0.048	−0.010	−0.293
430	2.994	0.139	14.707
440	4.235	0.248	21.081
450	1.115	0.145	5.992
460	1.462	0.290	8.373
470	1.020	0.463	6.727
480	0.487	0.714	4.211
490	0.150	1.063	2.353
500	0.008	1.592	1.318
510	0.025	2.406	0.738
520	0.292	3.473	0.370
530	0.656	4.112	0.214
540	2.917	9.247	0.176
550	5.409	12.968	0.124
560	6.217	10.369	0.034
570	10.109	12.644	0.027
580	13.826	13.167	0.024
590	13.136	9.598	0.014
600	12.110	7.113	0.009
610	9.497	4.706	0.003
620	6.361	2.802	0.001
630	3.637	1.484	0.000
640	1.867	0.723	0.000
650	0.864	0.324	0.000
660	0.363	0.134	0.000
670	0.140	0.051	0.000
680	0.054	0.020	0.000
690	0.021	0.008	0.000
700	0.008	0.003	0.000
710	0.003	0.001	0.000
720	0.001	0.000	0.000
730	0.001	0.000	0.000
740	0.000	0.000	0.000

(Continued)

Table A8.25 (*Continued*)

10 nm Interval			
nm	W_X	W_Y	W_Z
750	0.000	0.000	0.000
760	0.000	0.000	0.000
770	0.000	0.000	0.000
780	0.000	0.000	0.000
Check Sum	99.188	100.004	67.395
White Point	99.186	100.000	67.393

Table A8.26 Illuminant F2, 1931 Observer

20 nm Interval			
nm	W_X	W_Y	W_Z
360	0.000	0.000	0.000
380	−0.015	−0.001	−0.075
400	0.126	0.006	0.604
420	0.723	0.016	3.459
440	7.638	0.413	37.775
460	2.320	0.518	13.826
480	0.931	1.364	8.340
500	−0.106	3.077	2.271
520	0.034	5.636	0.725
540	5.711	18.719	0.319
560	13.144	23.526	0.088
580	27.390	25.997	0.044
600	24.880	13.965	0.017
620	12.425	5.247	0.001
640	3.276	1.258	0,000
660	0.613	0.222	0.000
680	0.082	0.030	0.000
700	0.014	0.005	0.000
720	0.002	0.001	0.000
740	0.000	0.000	0.000
760	0.000	0.000	0.000
780	0.000	0.000	0.000
Check Sum	99.188	99.999	67.394
White Point	99.186	100.000	67.393

Table A8.27 Illuminant F2, 1964 Observer

10 nm Interval			
nm	$W_{10,X}$	$W_{10,Y}$	$W_{10,Z}$
360	0.000	0.000	0.000
370	0.000	0.000	0.000
380	0.001	0.000	0.003
390	−0.020	−0.001	−0.097
400	0.130	0.014	0.588
410	0.326	0.034	1.494
420	0.088	−0.005	0.303
430	3.107	0.407	15.491
440	4.387	0.658	22.288
450	1.169	0.312	6.415
460	1.441	0.587	8.294
470	0.954	0.895	6.428
480	0.383	1.257	3.794
490	0.060	1.664	1.973
500	0.002	2.156	0.989
510	0.151	2.752	0.492
520	0.528	3.519	0.267
530	0.934	3.956	0.148
540	3.551	8.791	0.108
550	6.295	12.243	0.056
560	6.984	9.828	−0.004
570	11.012	11.985	0.000
580	14.508	12.451	0.000
590	13.512	9.315	0.000
600	12.111	7.032	0.000
610	9.208	4.671	0.000
620	6.030	2.776	0.000
630	3.450	1.496	0.000
640	1.702	0.704	0.000
650	0.767	0.305	0.000
660	0.317	0.125	0.000
670	0.122	0.048	0.000
680	0.045	0.017	0.000
690	0.017	0.007	0.000
700	0.006	0.002	0.000
710	0.002	0.001	0.000
720	0.001	0.000	0.000
730	0.000	0.000	0.000
740	0.000	0.000	0.000

(*Continued*)

Table A8.27 (*Continued*)

10 nm Interval			
nm	$W_{10,X}$	$W_{10,Y}$	$W_{10,Z}$
750	0.000	0.000	0.000
760	0.000	0.000	0.000
770	0.000	0.000	0.000
780	0.000	0.000	0.000
Check Sum	103.281	100.002	69.030
White Point	103.279	100.000	69.027

Table A8.28 Illuminant F2, 1964 Observer

20 nm Interval			
nm	$W_{10,X}$	$W_{10,Y}$	$W_{10,Z}$
360	0.000	0.000	0.000
380	−0.038	−0.005	−0.171
400	0.234	0.028	1.066
420	1.022	0.100	4.782
440	7.898	1.121	39.933
460	2.301	1.042	13.716
480	0.686	2.475	7.408
500	−0.133	4.279	1.613
520	0.444	5.769	0.511
540	6.953	17.713	0.191
560	14.911	22.281	−0.001
580	28.878	24.639	0.002
600	24.810	13.883	0.000
620	11.708	5.211	0.000
640	3.014	1.241	0.000
660	0.516	0.197	0.000
680	0.073	0.030	0.000
700	0.010	0.004	0.000
720	0.001	0.001	0.000
740	0.000	0.000	0.000
760	0.000	0.000	0.000
780	0.000	0.000	0.000
Check Sum	103.288	100.009	69.050
White Point	103.279	100.000	69.027

Table A8.29 Illuminant F7, 1931 Observer

10 nm Interval			
nm	W_X	W_Y	W_Z
360	0.000	0.000	0.000
370	0.000	0.000	−0.001
380	0.001	0.000	0.005
390	−0.004	0.000	−0.021
400	0.105	0.003	0.499
410	0.266	0.007	1.265
420	0.203	−0.005	0.904
430	4.113	0.186	20.179
440	5.834	0.346	29.065
450	2.301	0.278	12.246
460	2.650	0.527	15.185
470	1.855	0.842	12.235
480	0.889	1.303	7.684
490	0.273	1.933	4.285
500	0.016	2.937	2.432
510	0.052	4.495	1.372
520	0.537	6.254	0.661
530	1.118	6.620	0.329
540	3.686	11.541	0.214
550	5.727	13.711	0.131
560	4.699	7.698	0.021
570	7.124	8.867	0.019
580	9.875	9.422	0.017
590	8.833	6.445	0.010
600	8.895	5.236	0.007
610	7.999	3.978	0.003
620	6.510	2.879	0.001
630	4.709	1.929	0.000
640	3.068	1.192	0.000
650	1.924	0.723	0.000
660	1.076	0.397	0.000
670	0.417	0.152	0.000
680	0.168	0.061	0.000
690	0.073	0.026	0.000
700	0.029	0.011	0.000
710	0.013	0.005	0.000
720	0.005	0.002	0.000
730	0.002	0.001	0.000
740	0.001	0.000	0.000

(*Continued*)

Table A8.29 (*Continued*)

10 nm Interval			
nm	W_X	W_Y	W_Z
750	0.000	0.000	0.000
760	0.000	0.000	0.000
770	0.000	0.000	0.000
780	0.000	0.000	0.000
Check Sum	95.042	100.002	108.748
White Point	95.041	100.000	108.747

Table A8.30 Illuminant F7, 1931 Observer

20 nm Interval			
nm	W_X	W_Y	W_Z
360	0.000	0.000	0.000
380	−0.007	−0.001	−0.033
400	0.121	0.007	0.578
420	1.323	0.028	6.323
440	10.790	0.584	53.336
460	4.665	0.963	27.365
480	1.708	2.492	15.213
500	−0.218	5.611	4.189
520	0.379	11.237	1.309
540	7.709	23.952	0.351
560	10.453	18.318	0.071
580	18.791	17.848	0.030
600	17.996	10.198	0.013
620	13.114	5.650	0.001
640	5.970	2.291	0.000
660	1.965	0.720	0.000
680	0.204	0.074	0.000
700	0.073	0.026	0.000
720	0.003	0.001	0.000
740	0.003	0.001	0.000
760	0.000	0.000	0.000
780	0.000	0.000	0.000
Check Sum	95.042	100.000	108.746
White Point	95.041	100.000	108.747

Table A8.31 Illuminant F7, 1964 Observer

10 nm Interval			
nm	$W_{10,X}$	$W_{10,Y}$	$W_{10,Z}$
360	0.000	0.000	0.000
370	0.000	0.000	0.000
380	0.000	0.000	0.001
390	−0.021	−0.001	−0.101
400	0.156	0.016	0.700
410	0.493	0.051	2.259
420	0.461	0.030	2.055
430	4.148	0.535	20.642
440	5.863	0.885	29.819
450	2.334	0.589	12.685
460	2.532	1.034	14.581
470	1.682	1.578	11.335
480	0.677	2.224	6.711
490	0.106	2.935	3.483
500	0.004	3.858	1.770
510	0.278	4.979	0.888
520	0.934	6.139	0.462
530	1.518	6.168	0.221
540	4.342	10.634	0.126
550	6.462	12.551	0.058
560	5.108	7.073	−0.006
570	7.520	8.149	0.000
580	10.048	8.637	0.000
590	8.809	6.065	0.000
600	8.627	5.018	0.000
610	7.521	3.826	0.000
620	5.988	2.766	0.000
630	4.332	1.886	0.000
640	2.715	1.126	0.000
650	1.659	0.663	0.000
660	0.912	0.359	0.000
670	0.354	0.138	0.000
680	0.134	0.052	0.000
690	0.058	0.022	0.000
700	0.023	0.009	0.000
710	0.009	0.003	0.000
720	0.003	0.001	0.000
730	0.001	0.001	0.000
740	0.001	0.000	0.000

(Continued)

Table A8.31 (*Continued*)

10 nm Interval			
nm	$W_{10,X}$	$W_{10,Y}$	$W_{10,Z}$
750	0.000	0.000	0.000
760	0.000	0.000	0.000
770	0.000	0.000	0.000
780	0.000	0.000	0.000
Check Sum	95.791	99.999	107.689
White Point	95.792	100.000	107.686

Table A8.32 Illuminant F7, 1964 Observer

20 nm Interval			
nm	$W_{10,X}$	$W_{10,Y}$	$W_{10,Z}$
360	0.000	0.000	0.000
380	−0.036	−0.005	−0.161
400	0.246	0.031	1.106
420	1.824	0.177	8.525
440	10.807	1.533	54.683
460	4.506	1.899	26.455
480	1.222	4.373	13.104
500	−0.261	7.596	2.884
520	1.147	11.062	0.890
540	9.029	21.938	0.199
560	11.459	16.827	0.000
580	19.208	16.389	0.002
600	17.412	9.821	−0.001
620	12.049	5.451	0.000
640	5.311	2.182	0.000
660	1.641	0.638	0.000
680	0.169	0.067	0.000
700	0.055	0.021	0.000
720	0.001	0.000	0.000
740	0.002	0.001	0.000
760	0.000	0.000	0.000
780	0.000	0.000	0.000
Check Sum	95.791	100.001	107.686
White Point	95.792	100.000	107.686

Table A8.33 Illuminant F11, 1931 Observer

10 nm Interval			
nm	W_X	W_Y	W_Z
360	0.000	0.000	0.000
370	0.000	0.000	0.000
380	0.001	0.000	0.005
390	−0.009	0.000	−0.048
400	0.061	0.002	0.291
410	0.107	0.003	0.511
420	−0.205	−0.014	−1.044
430	2.800	0.130	13.758
440	4.264	0.251	21.231
450	1.277	0.164	6.849
460	1.367	0.280	7.848
470	0.695	0.255	4.495
480	0.435	0.754	3.956
490	0.341	2.063	4.778
500	−0.004	1.088	0.792
510	0.007	0.469	0.054
520	−0.001	0.229	0.032
530	−0.925	−2.067	0.000
540	9.613	29.254	0.535
550	11.438	28.030	0.300
560	0.196	−0.695	−0.031
570	0.602	0.870	0.002
580	7.021	6.565	0.012
590	9.070	6.866	0.011
600	4.247	2.617	0.004
610	29.903	14.812	0.010
620	13.567	6.132	0.003
630	3.446	1.329	0.000
640	0.630	0.240	0.000
650	0.534	0.199	0.000
660	0.297	0.110	0.000
670	0.084	0.031	0.000
680	0.043	0.016	0.000
690	0.028	0.010	0A00
700	0.013	0.005	0.000
710	0.020	0.007	0.000
720	0.001	0.000	0.000
730	0.000	0.000	0.000
740	0.000	0.000	0.000

(*Continued*)

Table A8.33 (*Continued*)

10 nm Interval			
nm	W_X	W_Y	W_Z
750	0.000	0.000	0.000
760	0.000	0.000	0.000
770	0.000	0.000	0.000
780	0.000	0.000	0.000
Check Sum	100.964	100.005	64.354
White Point	100.962	100.000	64.350

Table A8.34 Illuminant F11, 1931 Observer

20 nm Interval			
nm	W_X	W_Y	W_Z
360	0.000	0.000	0.000
380	−0.014	−0.001	−0.076
400	0.100	0.005	0.509
420	0.256	−0.001	1.093
440	8.207	0.419	40.877
460	1.559	0.623	9.228
480	0.600	0.507	8.258
500	1.524	7.107	4.371
520	−5.091	−14.004	−0.965
540	20.536	58.821	1.039
560	3.973	7.524	−0.034
580	9.894	9.370	0.032
600	24.253	13.848	0.011
620	37.637	17.208	0.009
640	−4.377	−2.270	−0.002
660	2.164	0.978	0.001
680	−0.411	−0.200	0.000
700	0.172	0.075	0.000
720	−0.025	−0.012	0.000
740	0.006	0.003	0.000
760	−0.001	−0.001	0.000
780	0.000	0.000	0.000
Check Sum	100.962	99.999	64.351
White Point	100.962	100.000	64.350

Table A8.35 Illuminant F11, 1964 Observer

10 nm Interval			
nm	$W_{10.X}$	$W_{10.Y}$	$W_{10.Z}$
360	0.000	0.000	0.000
370	0.000	0.000	0.000
380	0.001	0.000	0.004
390	−0.019	−0.001	−0.088
400	0.102	0.011	0.460
410	0.196	0.021	0.897
420	−0.134	−0.028	−0.756
430	2.908	0.381	14.502
440	4.426	0.666	22.492
450	1.339	0.355	7.327
460	1.348	0.566	7.783
470	0.657	0.513	4.313
480	0.329	1.316	3.539
490	0.176	3.206	4.053
500	−0.006	1.464	0.581
510	0.039	0.510	0.020
520	0.015	0.238	0.024
530	−1.070	−1.951	0.005
540	11.643	27.854	0.327
550	13.374	26.520	0.149
560	0.159	−0.660	−0.023
570	0.674	0.822	0.000
580	7.362	6.226	0.000
590	9.374	6.653	0.000
600	4.309	2.597	0.000
610	29.011	14.710	0.000
620	12.930	6.080	0.000
630	3.263	1.353	0.000
640	0.571	0.232	0.000
650	0.473	0.187	0.000
660	0.261	0.103	0.000
670	0.073	0.029	0.000
680	0.036	0.014	0.000
690	0.023	0.009	0.000
700	0.010	0.004	0.000
710	0.015	0.006	0.000
720	0.001	0.000	0.000
730	0.000	0.000	0.000
740	0.000	0.000	0.000

(*Continued*)

Table A8.35 (*Continued*)

10 nm Interval			
nm	$W_{10,X}$	$W_{10,Y}$	$W_{10,Z}$
750	0.000	0.000	0.000
760	0.000	0.000	0.000
770	0.000	0.000	0.000
780	0.000	0.000	0.000
Check Sum	103.869	100.006	65.609
White Point	103.863	100.000	65.607

Table A8.36 Illuminant F11, 1964 Observer

20 nm Interval			
nm	$W_{10,X}$	$W_{10,Y}$	$W_{10,Z}$
360	0.000	0.000	0.000
380	−0.029	−0.005	−0.142
400	0.181	0.026	0.869
420	0.414	0.019	1.729
440	8.515	1.220	43.348
460	1.544	0.977	9.002
480	0.319	1.693	7.470
500	1.673	8.341	3.484
520	−5.992	−13.547	−0.739
540	24.601	55.948	0.625
560	4.494	7.060	−0.051
580	10.526	8.885	0.014
600	24.099	13.702	−0.004
620	36.033	17.112	0.001
640	−4.279	−2.247	0.000
660	2.026	0.952	0.000
680	−0.397	−0.198	0.000
700	0.155	0.072	0.000
720	−0.025	−0.013	0.000
740	0.006	0.003	0.000
760	−0.001	−0.001	0.000
780	0.000	0.000	0.000
Check Sum	103.863	99.999	65.606
White Point	103.863	100.000	65.607

Appendix 9

Glossary of Terms

The following list of terms and their definitions follows broadly the recommendations made in the fourth edition of the CIE International Lighting Vocabulary (CIE Publication No. 17.4, 1987). The numbers in brackets after the terms indicate the sections in which these terms are discussed.

Abney phenomenon (3.4)

Change of hue produced by decreasing the purity of a colour stimulus while keeping its dominant wavelength and luminance constant.

Abney's Law (2.3)

An empirical law stating that if two colour stimuli, A and B, are perceived to be of equal brightness, and two other colour stimuli, C and D, are perceived to be of equal brightness, then the additive mixtures of A with C and B with D will also be perceived to be of equal brightness. (The validity of this law depends strongly on the observing conditions.)

absorptance, α (A1.6)

Ratio of the absorbed radiant or luminous flux to the incident flux under specified conditions.

absorption (13.4)

Process by which radiant energy is converted to a different form of energy by interaction with matter.

achromatic, colour (1.7)

Colour devoid of hue (The names white, grey, black, neutral, and colourless, are commonly used for these colours.)

achromatic, signal (1.6 and 15.14)

Visual signal from the retina composed of additions of the cone and rod responses.

Measuring Colour, Fourth Edition. R.W.G. Hunt and M.R. Pointer.

achromatic stimulus (3.4)

A stimulus that is chosen to provide a reference that is regarded as achromatic in colorimetry.

action spectra (1.5)

Spectral sensitivity of a visual mechanism in terms of the light incident on the cornea of the eye.

adaptation (1.8)

Visual process whereby approximate compensation is made for changes in the luminances and colours of stimuli, especially in the case of changes in illuminants.

adaptation, incomplete (6.12)

A phenomenon in which the reference (or adopted) white in a given viewing environment does not actually appear white to an observer.

Note. Familiar instances include colour images on newsprint or displays that appear "too yellow or too blue."

adaptive colorimetric shift (6.16)

Mathematical adjustment in chromaticity and luminance factor of an object colour stimulus to correct for a change in chromatic adaptation.

adaptive colour shift (6.12)

Change in the perceived colour of an object caused solely by change of chromatic adaptation.

additive mixing (2.3 and 2.4)

Addition of colour stimuli on the retina in such a way that they cannot be perceived individually.

alychne

Surface in tristimulus space that represents the locus of colour stimuli of zero luminance.

Note This surface passes through the origin of the space. It intersects any chromaticity diagram in a straight line which is also called the alychne; this line lies wholly outside the domain of chromaticities bounded by the spectrum locus and the purple boundary.

anomalous trichromatism (1.10)

Form of trichromatic vision in which colour discrimination is less than normal.

aperture colour (17.1)

Perceived colour for which there is no definite spatial localisation in depth, such as that perceived as filling a hole in a screen.

apostilb, asb (A1.3)

Unit of luminance equal to 0.3183 cd m^{-2}.

arctan (3.7)

Abbreviation meaning: angle whose tangent is; sometimes written as: $\tan^{-1}$.

assimilation (3.3)

The tendency of very small stimuli to change their appearance by becoming more like their surrounding colours (also termed *spreading effect*).

Bezold-Brücke phenomenon (15.9)

Change of hue produced by changing the luminance (within the range of photopic vision) of a colour stimulus while keeping its chromaticity constant.

Note With certain monochromatic stimuli, hue remains constant over a wide range of luminances (for a given condition of adaptation). The wavelengths of these stimuli are sometimes referred to as *invariant wavelengths*.

binary hue (8.6)

Hue that can be described as a combination of two unique hues.

Note For example: orange is a yellowish-red or reddish-yellow; violet is reddish-blue, etc.

blackness (8.7)

1. Attribute of a visual sensation according to which an area appears to contain more or less black content.

2. Measure of the black content of colours in the NCS colour order system.

bright (1.7)

Adjective denoting high brightness.

brightness (1.7 and Table 3.1)

Attribute of a visual perception according to which an area appears to exhibit more or less light.

candela, cd (A1.3)

Unit of luminous intensity. The candela is the luminous intensity, in a given direction, of a source emitting a monochromatic radiation of frequency 540×10^{12} hertz, the radiant intensity of which in that direction is 1/683 watt per steradian.

candela per square metre, cd m^{-2} (A1.3)

Unit of luminance.

chroma (1.9 and Table 3.1)

The colourfulness of an area judged in proportion to the brightness of a similarly illuminated area that appears to be white or highly transmitting.

chromatic adaptation (3.14, 6.13, and 15.3)

Visual process whereby approximate compensation is made for changes in the colours of stimuli, especially in the case of changes in illuminants.

chromatic colour (1.7)

Colour exhibiting hue (as distinct from those commonly called white, grey, black, neutral, and colourless).

chromatic induction (3.3, 14.3)

A modification of the visual response that occurs when two colour stimuli (of any spectral irradiance distribution) are viewed side-by-side in which each stimulus alters the appearance of the other.

Note 1 It is often referred to as *simultaneous contrast* or *spatial contrast* because the effect is one of enhancement of colour difference.

Note 2 The effect is virtually instantaneous, in contrast to chromatic adaptation which is considered to develop slowly in the visual system.

chromaticity (3.3 and Table 3.1)

Property of a colour stimulus defined by its chromaticity co-ordinates.

chromaticity co-ordinates (3.3, A4)

Ratio of each of a set of tristimulus values to their sum.

chromaticity diagram (3.3)

A two-dimensional diagram in which points specified by chromaticity co-ordinates represent the chromaticities of colour stimuli.

chromaticness (8.7)

1. Measure of the chromatic content of colours in the NCS colour order system.
2. An alternative term for colourfulness (obsolete).
3. Perceptual colour attribute consisting of the hue and saturation of a colour.

CIE (Commission Internationale de l'Éclairage) (2.2)

The International Commission on Illumination; the body responsible for international recommendations for photometry and colorimetry. French name: Commission Internationale de l'Éclairage.

CIE 1931 standard colorimetric observer (2.6)

Ideal observer whose colour matching properties correspond to the CIE colour-matching functions for the 2° field size.

CIE 1964 standard colorimetric observer (2.6)

Ideal observer whose colour matching properties correspond to the CIE colour-matching functions for the 10° field size.

CIE 1976 chroma, C^*_{uv}, C^*_{ab} (3.9 and Table 3.1)

Correlate of chroma in the CIELUV and CIELAB colour spaces.

CIE 1976 colour difference, ΔE^*_{uv}, ΔE^*_{ab} (3.10)

Correlate of colour difference in the CIELUV and CIELAB colour spaces.

CIE 1976 hue-angle, h_{uv}, h_{ab} (3.9 and Table 3.1)

Correlate of hue in the CIELUV and CIELAB colour spaces.

CIE 1976 hue-difference, ΔH^*_{uv}, ΔH^*_{ab} (3.10)

Correlate of hue difference in the CIELUV and CIELAB colour spaces.

CIE 1976 lightness, L^* (3.9 and Table 3.1)

Correlate of lightness in the CIELUV and CIELAB colour spaces.

CIE 1976 saturation, s_{uv} (3.9 and Table 3.1).

Correlate of saturation in the CIELUV colour space.

CIE 1976 uniform chromaticity scale (UCS) diagram (3.6 and Table 3.1)

Chromaticity diagram in which u' and v' are plotted; equal distances in this diagram represent more nearly, than in the x,y diagram, equal colour differences for stimuli having the same luminance.

CIE colour-matching functions (2.6)

Functions $\bar{x}(\lambda)$, $\bar{y}(\lambda)$, $\bar{z}(\lambda)$, in the CIE 1931 standard colorimetric system or $\bar{x}_{10}(\lambda)$, $\bar{y}_{10}(\lambda)$, $\bar{z}_{10}(\lambda)$, in the CIE 1964 standard colorimetric system

CIE colour rendering index, R_a (7.3 and A7)

A CIE method of assessing the degree to which a test illuminant renders colours similar in appearance to their appearance under a reference illuminant.

CIE illuminants (4.14, 4.15, 4.17)

Illuminants defined by the CIE that include CIE illuminants A, C, D, ID50, and ID65. (See also *CIE Standard Illuminants*).

CIE standard deviate observer (6.7)

Standard observer whose colour matching functions deviate from those of the CIE standard observer in a defined manner

CIE standard illuminants (4.14 and 4.17)

Illuminants A and D65 defined by the CIE in terms of relative spectral power distributions.

Note 1 These illuminants are intended to represent: A, Planckian radiation at a temperature of about 2856 K;
D65, a daylight illuminant, defined by its correlated colour temperature of 6500 K.
See CIE Publication 15:2004 Colorimetry

Note 2 Illuminants B, C and other D illuminants, previously denoted as standard illuminants, should now be termed CIE illuminants.

CIE standard photometric observer (2.2 and 2.3)

Ideal observer whose relative spectral sensitivity function conforms to the photopic, $V(\lambda)$, or scotopic, $V'(\lambda)$, spectral luminous efficiency function, and that complies with the summation law implied in the definition of luminous flux.

CIE standard sources (4.16)

Artificial sources specified by the CIE whose radiations approximate CIE standard illuminants.

CIELAB colour space (3.9)

Colour space in which L^*, a^*, b^* are plotted at right angles to one another. Equal distances in the space represent approximately equal colour differences.

CIELUV colour space (3.9)

Colour space in which L^*, u^*, v^* are plotted at right angles to one another. Equal distances in the space represent approximately equal colour differences.

colorimetric colour space (3.9)

Colour space defined by three colorimetric coordinates.

colorimetric purity, p_c (3.16)

Quantity defined by the expression $p_c = L_d/(L_d + L_n)$ where L_d and L_n are the respective luminances of the monochromatic stimulus and of a specified achromatic stimulus that match the colour stimulus considered in an additive mixture. (In the case of purple stimuli, the monochromatic stimulus is replaced by a stimulus whose chromaticity is represented by a point on the purple boundary.)

Coloroid system (8.10)

A colour order system.

colour constancy (1.8, 6.12, and 15.3)

Effect of visual adaptation whereby the appearance of colours remains approximately constant when the level and colour of the illuminant are changed.

colour difference signal (1.6 and 15.7)

Visual signal from the retina composed of differences between the cone responses.

colour-matching functions (2.4)

The tristimulus values of monochromatic stimuli of equal radiant power per small constant-width wavelength interval throughout the spectrum.

colour order system (Chapter 8)

Arrangement of samples according to a set of principles for the ordering and denotation of their colour, usually according to defined scales.

colour stimulus function, $\varphi_\lambda(\lambda)$ (2.6)

Description of a colour stimulus by an absolute measure of a radiant quantity per small constant-width wavelength interval throughout the spectrum.

colour rendering index (7.3 and A7)

A method of assessing the degree to which a test illuminant renders colours similar in appearance to their appearance under a reference illuminant.

colour temperature, T_c. Unit: kelvin, K (4.2)

The temperature of a Planckian radiator whose radiation has the same chromaticity as that of a given stimulus.

colourfulness (1.7, Table 3.1, 15.18)

Attribute of a visual perception according to which an area appears to exhibit more or less of its hue.

Note For a colour stimulus of a given chromaticity and, in the case of related colours, of a given luminance factor, this attribute usually increases as the luminance is raised except when the brightness is very high.

complementary colour stimulus (3.4)

Two colour stimuli are complementary when it is possible to reproduce the tristimulus values of a specified achromatic stimulus by an additive mixture of these two stimuli.

complementary wavelength, λ_c (3.4)

Wavelength of the monochromatic stimulus that, when additively mixed in suitable proportions with the colour stimulus considered, matches the specified achromatic stimulus.

cones (1.4)

Photoreceptors in the retina that contain light-sensitive pigments capable of initiating the process of photopic vision.

correlated colour temperature, T_{cp}. Unit: kelvin, K (4.8)

Temperature of the Planckian radiator whose perceived colour most closely resembles that of a given stimulus seen at the same brightness and under specified viewing conditions. The recommended method of calculating the correlated colour temperature of a stimulus is to determine on the u,v (not the u′,v′) chromaticity diagram the temperature corresponding to the point on the locus of Planckian radiators that is nearest to the point representing the stimulus.

corresponding colour stimuli (3.14)

Pairs of colour stimuli that look alike when one is seen in one set of adaptation conditions, and the other is seen in a different set.

D illuminants (4.17 and A5.6)

CIE Standard Illuminants having defined relative spectral power distributions that represent phases of daylight with different correlated colour temperatures.

dark (1.9)

Adjective denoting low lightness.

daylight illuminant (4.12. 4.15, and 4.17)

Illuminant having the same, or nearly the same, relative spectral power distribution as a phase of daylight.

daylight locus (4.17)

The locus of points in a chromaticity diagram that represent chromaticities of phases of daylight with different correlated colour temperatures.

defective colour vision (1.10)

Abnormal colour vision in which there is a reduced ability to discriminate between some or all colours.

density, (optical) D (A1.6)

Logarithm to the base 10 of the reciprocal of the reflectance factor or the transmittance factor or other similar measure.

deutan (1.10)

Adjective denoting deuteranopia or deuteranomaly.

deuteranomaly (1.10)

Defective colour vision in which discrimination of the reddish and greenish contents of colours is reduced, without any colours appearing abnormally dim.

deuteranopia (1.10)

Defective colour vision in which discrimination of the reddish and greenish contents of colours is absent, without any colours appearing abnormally dim.

dichromatism (1.10)

Defective colour vision in which all colours can be matched using additive mixtures of only two matching stimuli.

diffuse reflectance, ρ_d (5.9, Table 5.1, and A1.6)

The ratio of the part of the radiant or luminous flux reflected by diffusion to the incident flux, under specified conditions of irradiance.

diffuse transmittance, τ_d (5.9, Table 5.2, and A1.6)

The ratio of the part of the radiant or luminous flux transmitted by diffusion to the incident flux, under specified conditions of irradiance.

dim (1.7)

Adjective denoting low brightness.

DIN system (8.9)

A colour order system.

distribution temperature T_d. Unit: kelvin, K (4.8)

Temperature of a Planckian radiator whose relative spectral power distribution is the same as that of the radiation considered.

dominant wavelength, λ_d (3.4 and Table 3.1)

Wavelength of the monochromatic stimulus that, when additively mixed in suitable proportions with the specified achromatic stimulus, matches the colour stimulus considered.

doubly diffuse transmittance, τ_{dd} (5.9, Table 5.2, and A1.6)

The ratio of the transmitted to the incident radiant or luminous flux when the sample is irradiated and viewed diffusely.

efficacy (see luminous efficacy)

emission

The process of emitting radiation.

emissivity (4.8)

Ratio of the radiant exitance of a radiator to that of a Planckian radiator at the same temperature.

equi-energy stimulus, S_E (2.4)

Stimulus consisting of equal amounts of power per small constant-width wavelength interval throughout the spectrum.

erg

Unit of energy or work equal to 10^{-7} joules.

excitation purity, p_e (3.4 and Table 3.1)

Quantity defined by the ratio NC/ND of two collinear distances on the x,y, or on the x_{10},y_{10}, chromaticity diagram. NC is the distance between the point C representing the colour stimulus considered and the point N representing the specified achromatic stimulus; ND is the distance between the point N and the point D on the spectral locus at the dominant wavelength of the colour stimulus considered. In the case of purple stimuli, the point on the spectral locus is replaced by a point on the purple boundary.

exitance

At a point on a surface, the flux leaving the surface per unit area.

fluorescence (10.1)

Process whereby colours absorb radiant power at one wavelength and immediately re-emit it at another (usually longer) wavelength.

flux

Rate of flow per unit cross-section normal to the direction of flow.

foot-candle, fc (A1.3)

Unit of illuminance equal to 10.76 lux.

foot-lambert, fL (A1.3)

Unit of luminance equal to 3.426 cd m^{-2}.

fovea (1.3)

Central part of the retina that contains almost exclusively cones, and forming the site of most distinct vision. It subtends an angle of about 0.087 radians (5°) in the visual field.

foveola (1.3)

Central region of the fovea that contains only cones and is limited to a diameter of about 0.3 mm; it subtends an angle of about 0.017 radians (1°) in the visual field.

frequency (1.2)

The number of events occurring per unit of time. (The frequency of a radiation is independent of the medium though which it is passing, unlike wavelength which is dependent.)

gloss trap (5.9 and Tables 5.1 and 5.2)

Device used in spectrophotometry to eliminate specular components from measurements.

Grassmann's Laws (2.4)

Three empirical laws that describe the colour-matching properties of additive mixtures of colour stimuli:

1. To specify a colour match, three independent variables are necessary and sufficient.
2. For an additive mixture of colour stimuli, only their tristimulus values are relevant, not their spectral compositions.
3. In additive mixtures of colour stimuli, if one or more components of the mixture are gradually changed, the resulting tristimulus values also change gradually.

Helmholtz-Kohlrausch effect (2.3, 3.17)

Change in brightness of perceived colours produced by increasing the purity of a colour stimulus while keeping its luminance constant (within the range of photopic vision).

Helson-Judd effect (7.1)

Tendency, in coloured illumination, for light colours to be tinged with the hue of the illuminant, and for dark colours to be tinged with the complementary hue.

hertz, hz

Unit of frequency denoting the number of events per second.

hue (1.7 and Table 3.1)

Attribute of a visual perception according to which an area appears to be similar to one, or to proportions of two, of the perceived colours, red, yellow, green, and blue.

hue quadrature (15.13)

Correlate of hue expressed in terms of the proportions of the unique hues perceived to be present.

Hunt effect (15.8)

Progressive reduction of the colourfulness of colours as the level of illumination falls.

illuminance, E. Unit: lux, lx (A1.3 and Table A1.1)

Luminous flux per unit area incident on a surface.

illuminant colorimetric shift (6.16)

Change in colorimetric specification cause by a change of illuminant.

integrating sphere (5.9)

Hollow sphere, whitened inside, used in colour measuring instruments.

intensity, I (A1.3 and Table A1.1)

1. Luminous flux per unit solid angle. Unit: candela, cd.

2. General term used to indicate the magnitude of a variable.

irradiance, E. Unit: Watt per square metre, W m^{-2} (Table A1.1)

Radiant flux per unit area incident on a surface.

isotropic, (5.7)

Independent of direction.

joule, J

Unit of power expended or consumed, equal to 1 watt second.

Judd correction (2.3 and A2)

Modification of the $V(\lambda)$ function to correct the low values at wavelengths below 460 nm, called the CIE 1988 2° spectral luminous efficiency function for photopic vision, $V_M(\lambda)$.

kelvin, K (4.8)

Unit of temperature used for expressing colour temperatures. The temperature in kelvin is equal to that in Celsius plus 273.

lambert, L (A1.3)

unit of luminance equal to 3183 cd m^{-2}.

light (adjective) (1.9)

Adjective denoting high lightness.

light exposure, H. Unit: lux seconds, lx s. (Table A1.1)

Quantity of light received per unit area.

lightness (1.9 and Table 3.1)

The brightness of an area judged relative to the brightness of a similarly illuminated area that appears to be white or highly transmitting.

lumen, lm (A1.3 and Table A 1.1)

Unit of luminous flux. The lumen is the luminous flux emitted within unit solid angle (1 steradian) by a point source having an isotropic luminous intensity of 1 candela. It is the luminous flux of a beam of monochromatic radiation whose frequency is 540×10^{12} hertz and whose radiant flux is 1/683 watt.

lumen per square foot, lm ft^{-2} (A1.3)

Unit of illuminance equal to 10.76 lux.

lumen per square metre, $lm.m^{-2}$ (Table A1.1)

1. Unit of illuminance called lux, lx.

2. Unit of luminous exitance.

luminance, L. Unit: cd m^{-2} (Table 3.1, A1.3 and Table A1.1)

In a given direction, at a point in the path of a beam, the luminous intensity per unit projected area (the projected area being at right angles to the given direction).

luminance factor, β (2.6, 3.2, Table 3.1, and A1.6)

Ratio of the luminance to that of the perfect diffuser identically illuminated and viewed.

luminescence (10.2)

Emission of radiation in excess of that caused by thermal radiation.

luminous (2.6, A1.2, and A1.6)

1. Adjective denoting measures evaluated in terms of spectral power weighted by the $V(\lambda)$ function.

2. Adjective denoting stimuli that produce light or appear to do so.

luminous efficacy. Unit: lm W^{-1}(A1.3)

1. Of a source: ratio of luminous flux emitted to power consumed.

2. Of a radiation: ratio of luminous flux to radiant flux.

luminous efficiency (A1.3)

Ratio of radiant flux weighted according to the $V(\lambda)$ function, to the radiant flux.

luminous exitance, M. Unit: lm m^{-1} (Table A1.1)

At a point on a surface, the luminous flux leaving the surface per unit area.

luminous flux, F. Unit: lumen, lm (A1.3 and Table A1.1)

Radiant flux weighted by the $V(\lambda)$ function.

luminous intensity, I. Unit: candela, cd (A1.3 and Table A1.1)
Luminous flux per unit solid angle.

lux, lx (A1.3 and Table A1.1)
Unit of illumination equal to 1 lumen per square metre.

macula lutea (1.3)
Layer of photostable pigment covering parts of the retina in the foveal region.

mesopic, vision (1.4)
Vision intermediate between photopic and scotopic vision.

metameric colour stimuli (2.7 and 6.1)
Spectrally different colour stimuli that have the same tristimulus values.

metamerism index (6.6, 6.7, 6.8)
Measure of the extent to which two stimuli that match one another become different when the illuminant or the observer is changed.

millilambert, ml (A1.3)
Unit of luminance equal to 3.183 cd m^{-2}.

monochromat (1.10)
Observer who is completely unable to discriminate stimuli by their colours.

monochromatic stimulus (2.4)
A stimulus consisting of a very small range of wavelengths which can be adequately described by stating a single wavelength.

monochromatism (1.10)
Defective colour vision in which all colours can be matched using only a single matching stimulus.

Munsell system (8.4, 8.5)
A colour order system.

nanometre, nm (1.2)
Very small unit of length equal to 10^{-9} metre, commonly used for identifying wavelengths of the spectrum.

Natural Colour System (NCS) (8.7, 8.8)
A colour order system.

NCS (8.7, 8.8)
Abbreviation for Natural Colour System.

neuron (1.6)
Nerve cell.

nit, nt (A1.3)

Obsolete name for the unit of luminance, cd m^{-2}.

non-self-luminous colour stimulus (5.6)

Stimulus that consists of an illuminated object.

observer metamerism (6.7)

Variations of colour matches (of spectrally different stimuli) amongst different observers.

optical axis (of the eye) (1.3)

The direction defined by a line passing normally through the optical elements of the eye.

optical brightening agent (3.15, 10.1)

Compound added to coloured objects to increase their luminance factors by fluorescence, particularly in the case of whites.

Optical Society of America system (OSA system) (8.11)

A colour order system.

optimal colour stimuli (8.3)

Colour stimuli whose luminance factors have maximum possible values for each chromaticity when their spectral radiance factors do not exceed unity for any wavelength.

OSA system (8.11)

Abbreviation for Optical Society of America system.

paramers (6.11)

Spectrally different colour stimuli that have nearly the same tristimulus values. The corresponding property is called *paramerism*.

pearlescent (5.8)

Adjective to denote reflecting colours that contain metallic or other particles which impart reflective properties similar to those of pearls.

perfect diffuser (2.6, 5.7, A1.6)

An ideal diffuser with a reflectance (or transmittance) equal to unity at all wavelengths of the visible spectrum, and that looks equally bright for all directions of viewing.

Phosphorescence (10.2)

Photoluminescence that continues appreciably after the exciting radiation is removed.

Photoluminescence (10.2)

Luminescence caused by ultraviolet, visible, or infrared radiation.

photon

A quantum of light or of other electromagnetic radiation.

photopic vision (1.4)

Vision by the normal eye when it is adapted to levels of luminance of at least several candelas per square metre. (The cones are the principal photoreceptors that are active in photopic vision.)

Planckian locus (4.8, A5.4)

The locus of points in a chromaticity diagram that represent the chromaticities of the radiation of Planckian radiators at different temperatures.

Planckian radiator (4.8, A1.5, A5.4)

A body that emits radiation, because of its temperature, according to Planck's Law.

Planck's Law (4.8)

Formula for the spectral exitance of radiation from an enclosure with a small aperture.

point brilliance Unit: lux, lx (A1.3)

The illuminance produced by a source on a plane at the observer's eye, when the apparent diameter of the source is inappreciable.

point source (A1.3)

Source of radiation the dimensions of which are small enough, compared with the distance between source and detector, for them to be neglected in calculations.

power, P. Unit: watt, W

Energy per unit time.

protan (1.10)

Adjective denoting protanopia or protanomaly.

protanomaly (1.10)

Defective colour vision in which discrimination of the reddish and greenish contents of colours is reduced, with reddish colours appearing abnormally dim.

protanopia (1.10)

Defective colour vision in which discrimination of the reddish and greenish contents of colours is absent, with reddish colours appearing abnormally dim.

proximal field (15.2)

The immediate environment of the colour considered, extending typically about 10^0 from the edge of the colour element considered in all or most directions.

purity (3.4)

A measure of the proportions of the amounts of the monochromatic stimulus and of the specified achromatic stimulus that, when additively mixed, match the colour stimulus considered. (In the case of purple stimuli, the monochromatic stimulus is replaced by a stimulus whose chromaticity is represented by a point on the purple boundary.)

Purkinje phenomenon (1.4 and 2.3)

Reduction in the brightness of a predominantly long-wavelength colour stimulus relative to that of a predominantly short-wavelength colour stimulus, when the luminances are reduced in the same proportion from photopic to mesopic or scotopic levels, without changing the respective relative spectral power distributions of the stimuli involved.

purple boundary (3.3)

The line in a chromaticity diagram, or the surface in a colour space, that represents additive mixtures of monochromatic stimuli from the two ends of the spectrum, corresponding to wavelengths of approximately 380 and 780 nm.

purple stimulus (3.3)

Stimulus that is represented in any chromaticity diagram by a point lying within the triangle defined by the point representing the specified achromatic stimulus and the two ends of the spectral locus, which correspond approximately to the wavelengths 380 and 780 nm.

quantum (A1.2, A1.6, Table A1.1)

1. Extremely small indivisible unit of energy. (In the case of light or of other electromagnetic radiation, the term photon may be used.)

2. Adjective denoting measures evaluated in terms of number of quanta.

quantum exitance, M_q. Unit: s^{-1} m^{-2} (Table A1.1)

At a point on a surface, the quantum flux leaving the surface per unit area.

quantum exposure, H_q. Unit: m^{-2} (Table A1.1)

Number of quanta received per unit area.

quantum flux, F_q. Unit: s^{-1} (Table A1.1)

Number of quanta emitted, transferred, or received per unit of time.

quantum intensity, I_q. Unit: $s^{-1}.sr^{-1}$ (Table A1.1)

Quantum flux per unit solid angle.

quantum irradiance, E_q. Unit: s^{-1} m^{-2} (Table A1.1)

Quantum flux per unit area incident on a surface.

quantum radiance, L_q. Unit: s^{-1} sr^{-1} m^{-2} (Table A1.1)

In a given direction, at a point in the path of a beam, the quantum intensity per unit projected area (the projected area being at right angles to the given direction).

radiance, L_e. Unit: W sr^{-1} m^{-2} (Table A1.1)

In a given direction, at a point in the path of a beam, the radiant intensity per unit projected area (the projected area being at right angles to the given direction).

radiance factor, β_e (2.6, 5.9, Table 5.1, Table 5.2, A1.1)

Ratio of the radiance to that of the perfect diffuser identically irradiated.

radiant (2.6, A1.2, A1.6)

Adjective denoting measures evaluated in terms of power.

radiant efficiency (A1.3)

Ratio of the radiant flux emitted to the power consumed by a source.

radiant exitance, M_e. Unit: W m^{-2} (Table A1.1)

At a point on a surface, the radiant flux leaving the surface per unit area.

radiant exposure, H_e. Unit: J m^{-2} (Table A1.1)

Quantity of radiant energy received per unit area.

radiant flux, F_e. Unit: W (Table A1.1)

Power (energy per unit time) emitted, transferred, or received in the form of radiation.

radiant intensity, I_e. Unit: W sr^{-1} (Table A1.1)

Radiant flux per unit solid angle.

reflectance, ρ (2.6, 5.9, Table 5.1, A1.6)

Ratio of the reflected radiant or luminous flux to the incident flux under specified conditions of irradiation.

reflectance factor, R (2.6, 5.9, Table 5.1, A1.6)

Ratio of the radiant or luminous flux reflected in a given cone, whose apex is on the surface considered, to that reflected in the same directions by the perfect diffuser identically irradiated. In the colorant industries *reflectance* is often used as an abbreviation.

reflection (A1.6)

Return of radiation by a medium without change of frequency (that is, without fluorescence).

regular reflectance, ρ_r (5.9, Table 5.1, and A1.6)

Ratio of the regularly reflected radiant or luminous flux to the incident flux, under specified conditions of irradiance.

regular reflection (5.9, Table 5.1, A1.6)

Reflection as in a mirror without deviation by scattering, diffraction, or diffusion.

regular transmission (Table 5.2, A1.6)

Transmission without deviation by scattering, diffraction, or diffusion.

regular transmittance, τ_r (Table 5.2, A1.6)

Ratio of the regularly transmitted radiant or luminous flux to the incident flux, under specified conditions of irradiance.

related colours (1.9, 15.1, 15.2)

Colour perceived to belong to an area seen in relation to other colours.

relative colour stimulus function, $\varphi(\lambda)$ (2.6)

Relative measure of a radiant quantity per small constant-width wavelength interval throughout the spectrum.

relative spectral power distribution, $S(\lambda)$ (2.6)

Spectral power per small constant-width wavelength interval throughout the spectrum relative to a fixed reference value.

retina (1.3)

Light sensitive layer on the inside of the back of the eye; it contains the photoreceptors and nerve cells that transmit the visual signals to the optic nerve.

rhodopsin (1.5)

Visual pigment present in the rods of the retina.

rods (1.4)

Photoreceptors in the retina that contain a light-sensitive pigment capable of initiating the process of scotopic vision.

saturation (1.9, Table 3.1, 15.19)

Colourfulness of an area judged in proportion to its brightness.

Note For given viewing conditions and at luminance levels within the range of photopic vision, a colour stimulus of a given chromaticity exhibits approximately constant saturation for all luminance levels, except when the brightness is very high.

scotopic vision (1.4)

Vision by the normal eye when it is adapted to levels of luminance less than some hundredths of a candela per square metre.

self-luminous colour stimulus (5.5)

Stimulus that produces its own light.

shadow series (8.9)

A series of colours of constant chromaticity but varying luminance factor.

simultaneous contrast (14.3.2)

Change of appearance of a colour caused by adjacent colours.

solid angle (2.2)

The part of space that is bounded by lines radiating from a point and passing through a closed curve that does not contain the point.

spectral (2.6, A1.6)

Adjective denoting that monochromatic concepts are being considered.

spectral conventional reflectometer value, ρ_C (10.2)

The apparent spectral reflectance factor obtained when a fluorescent sample is measured relative to a non-fluorescent white sample, using monochromatic illumination and heterochromatic detection.

spectral luminescent radiance factor, β_L (10.2)

Ratio, at a given wavelength, of the radiance produced by luminescence by a sample to that produced by the perfect reflecting diffuser identically irradiated.

spectral luminous efficiency, $V(\lambda)$, $V'(\lambda)$ (2.2 and 2.3)

Weighting functions used to derive photometric measures from radiometric measures in photometry. Normally the $V(\lambda)$ function is used, but, if the conditions are such that the vision is scotopic, the $V'(\lambda)$ function is used and the measures are then distinguished by the adjective 'scotopic' and the symbols by the superscript $'$. (See also Judd correction).

spectral power distribution, $S_\lambda(\lambda)$ (2.6)

Spectral power per small constant-width wavelength interval throughout the spectrum.

spectral radiance factor, $\beta_e(\lambda)$ (10.2)

Ratio, at a given wavelength, of the spectral radiance to that of the perfect diffuser identically irradiated.

spectral reflectance, $\rho(\lambda)$ (5.9, Table 5.1, A1.6)

Ratio, at a given wavelength, of the spectral reflected radiant or luminous flux to the incident flux under specified conditions of irradiation.

spectral reflectance factor, $R(\lambda)$ (5.9, Table 5.1, A1.6)

Ratio, at a given wavelength, of the spectral radiant or luminous flux reflected in a given cone, whose apex is on the surface considered, to that reflected in the same directions by the perfect diffuser identically irradiated.

spectral reflected radiance factor, $\beta_S(\lambda)$ (10.2)

Ratio, at a given wavelength, of the spectral radiance produced by reflection by a sample to that produced by the perfect diffuser identically irradiated.

spectral total radiance factor, $\beta_T(\lambda)$ (10.2)

The sum, at a given wavelength, of the spectral reflected and spectral luminescent radiance factors.

spectral transmittance, $\tau(\lambda)$ (5.9, Table 5.1, A1.6)

Ratio, at a given wavelength, of the spectral transmitted radiant or luminous flux to the incident flux under specified conditions of irradiation.

spectral transmittance factor, $T(\lambda)$ (5.9, Table 5.1, A1.6)

Ratio, at a given wavelength, of the spectral radiant or luminous flux transmitted in a given cone, whose apex is on the surface considered, to that transmitted in the same directions by the perfect transmitter identically irradiated.

spreading effect (3.3)

The tendency of very small stimuli to change their appearance by becoming more like their surrounding colours (also termed *assimilation*).

spectrum locus (3.3)

Locus in a chromaticity diagram or colour space of the points that represent monochromatic stimuli throughout the spectrum.

steradian, sr (2.2)

Unit of solid angle defined as: the solid angle that, having its vertex in the middle of a sphere, cuts off an area on the surface of the sphere equal to that of a square with side of length equal to that of the radius of the sphere.

Stevens effect (15.8)

Progressive reduction of the brightness of light colours and increase in the brightness of dark colours, as the level of illumination falls.

Stiles-Crawford effect (1.3)

Decrease of the brightness of a light stimulus with increasing eccentricity of the entry of the light through the pupil of the eye. If the variation is in hue and saturation instead of in brightness, it is called the Stiles-Crawford effect of the second kind.

Stokes' Law (10.5)

In fluorescence, the wavelength of the exciting radiation is shorter than that of the emitted radiation.

strong (1.9)

Adjective denoting high chroma.

subtractive mixing of colorants (2.3, 13.1)

Production of colours by mixing colorants in such a way that they each subtract light from some parts of the spectral power distribution used to illuminate them.

successive contrast (14.3.1)

Change of appearance of a colour caused by the previous viewing of different colours.

tele-spectroradiometry (5.4)

Spectroradiometry carried out remotely from a sample with the aid of a telescopic optical system.

thermochromism (9.3)

The process by which an object changes colour because of a change in its temperature.

tint (of whites) (3.15)

Reddish or greenish hue of whites.

total spectral radiance factor, $\beta_T(\lambda)$ (10.2)

The sum of the spectral reflected and spectral luminescent radiance factors.

transformation equations (2.5 and 11.6)

Set of three simultaneous equations used to transform a colour specification from one set of matching stimuli to another.

transmission (A1.6)

Passage of radiation through a medium without change of frequency (that is, without fluorescence).

transmittance, τ (2.6, Table 5.2, A1.6)

Ratio of the transmitted radiant or luminous flux to the incident flux under specified conditions of irradiation.

transmittance factor, T (2.6, Table 5.2, A1.6)

Ratio of the radiant or luminous flux transmitted in a given cone, whose apex is on the surface considered, to that transmitted in the same directions by the perfect diffuser identically irradiated.

trichromatic matching (2.4)

Action of making a colour stimulus appear the same colour as a given stimulus by adjusting three components of an additive colour mixture.

tristimulus values (2.4)

Amounts of the three matching stimuli, in a given trichromatic system, required to match the stimulus considered.

tritan (1.10)

Adjective denoting tritanopia or tritanomaly.

tritanomaly (1.10)

Defective colour vision in which discrimination of the bluish and yellowish contents of colours is reduced.

tritanopia (1.10)

Defective colour vision in which discrimination of the bluish and yellowish contents of colours is absent.

troland, td (A1.3)

Unit used to express a quantity proportional to retinal illuminance produced by a light stimulus. When the eye views a surface of uniform luminance, the number of trolands is equal to the product of the area in square millimetres of the limiting pupil, natural or artificial, times the luminance of the surface in candelas per square metre.

u,v diagram (3.6)

Uniform chromaticity diagram introduced by the CIE in 1960, but now superseded by the u′,v′ diagram.

u′, v′ diagram (3.6)

Uniform chromaticity diagram introduced by the CIE in 1976.

uniform chromaticity diagram (3.6)

Chromaticity diagram in which equal distances approximately represent equal colour differences for stimuli having the same luminance.

uniform colour space (3.9)

Colour space in which equal distances approximately represent equal colour differences.

unique hue (8.6)

Perceived hue that cannot be further described by the use of hue names other than its own; there are four unique hues: red, green, yellow, and blue.

unrelated colour (1.9, 15.1, 15.2, 17.1, 17.2)

Colour perceived to belong to an area seen in isolation from other colours.

$V(\lambda)$ *function* (2.3)

Weighting function used to derive photometric measures from radiometric measures in photometry.

$V'(\lambda)$ *function* (2.2)

Weighting function used to derive scotopic photometric measures from radiometric measures in photometry. Normally the $V(\lambda)$ function is used, but if the conditions are such that the vision is scotopic the $V'(\lambda)$ is used and the measures are then distinguished by the adjective 'scotopic' and the symbols by the superscript $'$.

visual axis (1.3)

The direction defined by a line joining the centre of the fovea to the centre of the pupil of the eye; the visual axis is offset from the optical axis of the eye by about 4°.

Von Kries transformation (6.14)

Algebraic transformation whereby changes in adaptation are represented as adjustments of the sensitivities of the three cone systems such as to compensate fully for changes in the colour of illuminants.

watt, W

Unit of power equal to 1 joule per second or 10^7 ergs per second.

weak (1.9)

Adjective denoting low chroma.

white point (3.4)

An achromatic reference stimulus in a chromaticity diagram that corresponds to the stimulus that produces an image area that has the perception of white.

white point (colour appearance), adopted (15.14)

The computational reference white point (the white point used in the computation, e.g. the adopted white point or the display illuminant white point) used by a colour appearance model

Note 1 A white point used by a colour appearance model is an adopted white point.

Note 2 The adopted white point may or may not be the adapted white point, and one of the most common applications of an adopted white point is to achieve an optimally reproduced device white. For example, an adopted white point corresponding to a given medium may be used in place of an adapted white point in order that no colorants are used to create a white.

Note 3 The concept of adopted white point applies to both colour appearance models and reproduction models.

whiteness (3.15 and 8.7)

1. Attribute of a visual perception according to which an area appears to contain more or less white content.

2. Attribute that enables whites of different colours to be ranked in order of increasing similarity to some ideal white.

3. Measure of the white content of colours in the NCS colour order system.

x,y diagram (3.3)

Chromaticity diagram in which the x,y chromaticity co-ordinates of the CIE XYZ system are used.

yellow spot (1.3)

Layer of photostable pigment covering parts of the retina in the foveal region (also called the macula lutea).

REFERENCE

CIE Publication No. 17.4, *CIE International Lighting Vocabulary* (1987).

Index

Entries for authors are shown in italics.

A

Abney phenomenon, 45
Abney's Law, 26, 449
absorptance, 343, 449
absorption co-efficient, 262
acceptability of colour differences, 62
accuracy in colorimetry, 197
achromacy, criterion for, 300
achromatic colour, 10, 300
achromatic response, 308
achromatic signal, 8, 449
achromatic stimulus, 45, 432
action spectra, 7, 432
Adams-Nickerson formula, 59
adaptation, 10, 65, 432
adaptation, incomplete, 432
adaptation, luminance, effects of, 300
adaptation, luminance-level, 297
adapting field, 294
adaptive colorimetric shift, 134, 432
adaptive colour shift, 127, 432
additive mixing, 25, 432
additivity of colour matches, 29
additivity, brightness, 26
Albers, J., 273, 289
Alessi, P.J., 197, 198, 200, 216
Allebach, J.P., 246, 256
Allen, E., 226, 230
Alman, D.H., 202, 215, 225, 230
alychne, 432
Amano, K., 127, 141
Analoui, M., 246, 256
Anderson, P.G., 247, 256
anomalous trichromatism, 14, 432
aperture colour, 432
aperture vignetting, 200
apostilb, 339, 432
Arai, H., 287, 289
arctan, 51, 433
Arend, L., 295, 322
art restoration, 264
assimilation, 43, 271, 433
ASTM, 107, 115, 198, 199, 202, 210, 214, 215, 216, 274, 275, 280, 287, 289
Aston, S.M., 151, 152
attributes of colour, perceptual, 25, 26
Attridge, G.G., 251, 256, 293, 322

B

B.R.E.M.A., 231, 240
B.R.E.M.A., phosphors, 231
Baba, G., 287, 289, 292
background, 294
Bailey, J.R., 14, 16
bandwidth function, 203
bandwidth, 203
Barbur, J.L., 14, 16
Barkas, W.W., 277, 289
Bartleson, C.J., 37, 40, 115, 151, 152, 221, 222, 225, 230, 309, 322
Bäuml, K.-H., 286, 289
Baylor, D.A., 297, 322
beam-landing errors in cathode-ray tube displays, 239
Beck, J., 282, 289
Béland, M.-C., 273, 289

Bellchambers, H.E., 151, 152
Bennett, J.M., 273, 289
Berns, R.S., 40, 115, 118, 122, 141, 159, 187, 194, 197, 199, 204, 205, 206, 216, 240, 264, 265, 293, 322
Bez, H.E., 129, 142, 313, 323
Bezold-Brücke phenomenon, 303, 433
Billmeyer, F.W., 40, 93, 97, 115, 122, 141, 159, 186, 187, 194, 197, 198, 199, 200, 202, 204, 208, 215, 216, 222, 225, 230, 281, 289
binary hue, 169, 433
Birch, J., 13, 14, 16
black-body radiators, 83, 341
black-body sources, 83, 341
blackness, 170, 433
Blakemore C., 288, 290
blind spot, 4
Boddeke, F.R., 246, 256
Bouma, P.J., 148, 153
Bouman, M.A., 7, 17, 305, 308, 324
Bowmaker, J.K., 6, 7, 16
Boyce, P.R., 144, 153
Boynton, R.M., 17, 297, 322
Bradford transform, 130
Braun, K.M., 129, 141
BRDF (Bidirectional Reflectance Distribution Function), 279
BREMA 1969, 231, 240
Breneman, E.J., 129, 141
bright, 9, 10, 433
brightness additivity, 26
brightness, 9, 65, 433
brightness, correlate of, 309
brightness, stimuli of equal, 67
Brill, M.H., 31, 40
British Standard 6923;1988, 61, 72
British Standard 950, 150, 152
Brockes, A., 125, 141
BS 1134, 282, 289
BS EN ISO 2813, 275
Budde, W., 115, 198, 200, 216
Burgt, van der P.J.M., 145, 150, 153, 154
Burt, M., 288, 292
Butterworth, M., 242, 255
Byk Gardner, 275, 290

C

calculation of X,Y,Z from R,G,B tristimulus values, 32, 235
CAM (colour appearance model), 129, 293
CAM97s colour appearance model, 315
CAM97u colour appearance model for unrelated colours, 331
Cambridge (HRR) colour vision test, 14
camera characterisation, 242, 244
Campbell F.W., 288, 290
candela per square metre, 339, 433
candela, 338, 433
candle light, 73
Capilla, P., 246, 256
carbon arcs, 82
Carroll, J., 7, 13, 16
Carter, E.C., 197, 199, 216
CAT (chromatic adaptation transform), 131, 315
CAT02 chromatic adaptation transform, 131, 130, 131, 134, 315
CAT97, 129
categorical colour scaling, 151
cathodoluminescence, 74
Cayless, M.A., 97, 154
centre of gravity law of colour mixture, 48, 234
CGATS, 249, 255
Chantler, M.J., 284, 290, 291
chemiluminescence, 74
Chen, Y., 281, 289
Cheung, V., 248, 255
Chevreul, M.E., 269, 290
Chiba University, 249, 255
Chinga, G., 288, 290
Cho, M., 157, 194
Chong, T.-F., 222, 230
Chorro-Calderon, E., 316, 323
Chrisment, A., 275, 277, 291
chroma, 11, 12, 53, 57, 65, 433
chroma, correlate of, 310
chromatic adaptation transform (CAT), 131, 315
chromatic adaptation transform (CAT02), 131, 130, 131, 134, 315

chromatic adaptation transform (Von Kries), 129
chromatic adaptation transform, Bradford, 130
chromatic adaptation, 65, 294, 434
chromatic colour, 10, 434
chromatic induction, 43, 269, 434
chromaticity, 42, 65, 434
chromaticity co-ordinates for D Illuminants, 355
chromaticity co-ordinates for Planckian radiators, 368
chromaticity co-ordinates for spectral colours, 351
chromaticity co-ordinates for Standard Illuminants D65 and A, 355
chromaticity co-ordinates for Standard Illuminants, 355
chromaticity co-ordinates, 42, 233, 434
chromaticity diagram, 42, 434
chromaticity diagrams using r and g, 233
chromaticity relationships, 379
chromaticness, 170, 434
Chung Fu-Lai, K., 287, 292
CIE (Commission Internationale de l'Éclairage), 20, 434
CIE 1931 Standard Colorimetric Observer, 33, 434
CIE 1964 Standard Colorimetric Observer, 37, 434
CIE 1976 chroma, 53, 57, 435
CIE 1976 colour difference formulae, 59
CIE 1976 colour difference, 57, 435
CIE 1976 hue-angle, 51, 57, 435
CIE 1976 hue-difference, 59, 435
CIE 1976 lightness, 52, 435
CIE 1976 saturation, 51, 435
CIE 1976 uniform chromaticity scale (UCS) diagram, 49, 435
CIE 1994 colour difference formula, 62
CIE colour rendering indices, 145, 383, 435
CIE colour-matching functions, 33, 122, 347, 435
CIE DE2000 colour difference formula, 63, 64
CIE geometries of illumination and measurement, 105, 109
CIE Illuminant B, 92, 355
CIE Illuminant C, 92, 355
CIE Illuminants, 91, 92, 94, 355, 435
CIE indoor daylights, 94
CIE Journal, 130, 141
CIE modified two degree spectral luminous efficiency function for photopic vision, 39
CIE Proceedings 12th Session Stockholm, 39, 40
CIE Publication 119:1995, 62, 71
CIE Publication 142:2001, 62, 71
CIE Publication 159:2004, 131, 132, 141, 293, 322
CIE Publication 165:2005, 38, 40
CIE Publication 167:2005, 207
CIE Publication 170:1:2006, 38, 40
CIE Publication 175:2006, 267
CIE Publication 177:2007, 147,153
CIE Publication 184:2009, 96, 97
CIE Publication 191:2010, 344
CIE Publication 192:2010, 94, 97
CIE Publication No. 13, 145, 153
CIE Publication No. 15, 15.2, 15:2004, 37, 40, 72, 94, 97, 115, 122, 141, 200, 216, 286, 290
CIE Publication No. 17.4, 126, 141
CIE Publication No. 51, 94, 97
CIE Publication No. 75, 39, 40
CIE Publication No. 80, 123, 141
CIE Publication No. 86, 39, 40
CIE research note, 279
CIE S012-E:2004, 94, 97
CIE S014-1E:2006, 38, 40
CIE S014-2E:2006, 94, 97
CIE S014-3E:2011, 38, 40
CIE sources, 93
CIE spectral chromaticity co-ordinates, 351
CIE standard deviate observer, 122, 435
CIE Standard Illuminant A, 91
CIE Standard Illuminants D, 94, 355
CIE Standard Illuminants D, correlated colour temperatures of, 359

CIE standard illuminants, 435
CIE Standard Illuminants, D, chromaticity co-ordinates for, 358
CIE Standard Illuminants, relative spectral power distributions of, 355
CIE Standard on Colorimetric Illuminants, CIE S 014-2, 94, 97
CIE Standard on Colorimetric Observers, CIE S 014-1, 37, 40
CIE standard photometric observer, 23, 436
CIE standard scotopic photometric observer, 20
CIE standard sources, 93, 436
CIE, Compte Rendu, 12th Session, 39, 40
CIE, Compte Rendu, 1935, 329
CIECAM02 colour appearance model, 293
CIECAM02 uniform colour space, 315
CIECAM97s colour appearance model, 315
CIEDE2000, 61, 64, 69, 71
CIELAB colour space, 55, 436
CIELAB formulae, 379
CIELUV colour space, 53, 436
CIELUV formulae, 379
City University colour vision test, 14
Clarke, A.A., 129, 142, 293, 313, 323
Clarke, F.J.J., 61, 72, 200, 203, 216
Clarke, P.J., 198, 202, 203, 204, 216
CMC colour difference formula, 61
colorant mixtures, 257
colorimeters using filtered photocells, 114
colorimetric colour space, 436
colorimetric computations, 207
colorimetric formulae, 379
colorimetric purity, 67, 436
colorimetry, errors for different types of summation, 213
colorimetry, practical precautions in, 214
colorimetry, precision and accuracy, 197
colorimetry, sample preparation, 198
colorimetry, secondary standards, 203
Coloroid system, 182, 436
colour appearance model (CAM), 129, 293
colour appearance model CIECAM02, 293
colour appearance model CIECAM97s, 293, 315
colour appearance model for stimuli of different sizes, 325
colour appearance model for unrelated colours, 329
colour appearance model for unrelated colours, CAM97u, 331
colour appearance model, comprehensive, CAM97c, 293
colour appearance of objects, 267, 268
colour blindness, 13
colour blindness, tests for, 14
colour constancy, 10, 117, 127, 129, 436
colour deficiency, 13
colour deficiency, tests for, 14
colour difference formula, CIE (1976), 57
colour difference formula, CIE (1994), 61
colour difference formula, CIEDE2000, 62, 63, 69
colour difference formula, CMC, 61
colour difference formulae, allowing for chromatic adaptation, 65
colour difference signals, 9, 299, 436
colour inconstancy index (CON), 127, 134
colour matching equations, 236
colour matching functions, 26, 30, 234, 347, 436
colour mixture on chromaticity diagrams, 194
colour mixture, Centre of Gravity Law of, 48, 234
colour opponency, 168
colour order systems, 156, 436
colour order systems, advantages, 192
colour order systems, disadvantages, 192
colour order systems, variables in, 155
colour pseudo-stereopsis, 15
Colour Quality Scale (CQS), 151
colour rendering, 143
colour rendering index, 145, 383, 437
colour spaces, 53
colour stimuli, relations between, 41
colour stimulus function, 35, 37, 437
colour temperature, 85, 437
colour vector maps, 150

colour vision, 1
colour vision, models of, 293, 325, 329
colourfulness, 9, 65, 437
colourfulness, correlate of, 311
colour-matching functions in RGB systems, 234
colour-matching functions, 26, 436
colour-matching functions, CIE, 33, 347
complementary colour stimulus, 45, 437
complementary wavelength, 45, 437
Compton, J.A., 200, 216
computer displays, 239
CON (colour inconstancy index), 134
Condax, Louis M., 50
cone response functions, 297
cones, 4, 437
cones, spectral sensitivities of, 5, 295
Connah, D., 248, 255
contrast sensitivity function (CSF), 288
conventional reflectometer, 225
cornea, 3
correcting errors in spectral data, 204
correlated colour temperature, 86, 437
correlated colour temperatures of D Illuminants, 355
correlated colour temperatures of light sources, 96
corresponding colour stimuli, 6.13, 437
Costa, L.F., 222, 230
Cowan, W.B., 68, 72, 239, 240
Crawford, B.H., 5, 16, 150, 153
Cui, G.H., 287, 290
Cui. G., 57, 72, 242, 255, 277, 291, 315, 323, 325, 328

D

D illuminants, 94, 97, 437
D Illuminants, correlated colour temperatures of, 359
D Illuminants, method of calculating spectral powers of, 374
Dain, S.J., 14, 16
Dakin, J., 242, 255, 277, 291
Dana, K.J., 283, 290
dark, 11, 438
Dartnall, H.J.A., 6, 16
Davidson, H.R., 165, 194
Davidson, J.G., 151, 153, 249, 256
Davis, W., 151, 153
Day, D.C., 265
Day, E.A., 265
daylight, 89
daylight illuminant, 438
daylight illuminants, 94
daylight locus, 85, 94, 438
Decononck, G., 151, 153
deconvolution formula, 210
defective colour vision, 13, 438
DeMarsh, L.E., 121, 142
density, optical, 257, 343, 438
deutan, 51, 438
deuteranomaly, 14, 51, 438
deuteranopia, 13, 51, 438
DeValois R.L., 288, 290
device-independent co-ordinates, 242
dichromatism, 14, 438
diffuse reflectance, 108, 341, 438
diffuse transmittance, 109, 341, 438
diffusing colorants, 262
digital camera characterisation, 242
digital cameras, colorimetry with, 241
dim, 9, 10, 438
DIN system, 179, 438
discounting the colour of the illuminant, 295
distribution temperature, 85, 439
dominant wavelength, 44, 45, 65, 439
Donaldson, R., 221, 230
double monochromators, 220
doubly diffuse transmittance, 109, 341, 439
Dougherty, B., 288, 290
Drew, M.S., 246, 255

E

E.B.U. phosphors, 232
Early, E.A., 198, 216
eccentricity factor, 305
Eckerle, K.L., 203, 216
Adelston, E.H., 270, 290

Edwards, M., 289, 290
Eem, J.K., 246, 256
efficacy, 143, 439
Eitle, D., 226, 230
electroluminescence, 74
emission, 439
emissivity, 83, 439
Endoh, H., 151, 154
Eppeldauer, G.P., 198, 217
equi-energy stimulus, S_E, 27, 294, 439
Erb, W., 115, 198, 200, 213, 216
erg, 439
Estévez, O., 6, 16, 51, 72, 132, 141, 296, 322
Evans, R.M., 157, 194, 271, 290
excitation purity, 44, 46, 65, 67, 439
exitance, 439
eye, construction of, 3
eye-camera metamerism, 244

F

Fach, C. C., 6, 16
Fairchild, M.D., 124, 129, 141, 195, 199, 216, 287, 291, 293, 314, 322, 323, 324
Fairman, H.S., 31, 40, 125, 126, 141, 208, 210, 212, 216, 217
Farnsworth, D., 14, 16
Farnsworth-Munsell 100 Hue test, 14
Farrell, J.E., 286, 292
field size, 124, 294
Field, G.G., 191, 194
filtered-photocell colorimeters, 114
filter-reduction method for fluorescent colours, 226
filtration of projected slides, 314
Finlayson, G.D., 246, 248, 255
Fletcher, R., 14, 16
flicker photometry, 23
fluorent colours, 157
fluorescence, 74, 157, 219, 439
fluorescence, practical considerations, 227
fluorescent colours, colorimetry of, 219
fluorescent lamps, 78
fluorescent lamps, data for representative, 359
fluorescent lamps, spectral power distributions of, 78, 359
fluorescent whitening agents, 94
flux, 439
foot-candle, 339, 439
foot-lambert, 339, 440
Foster, D.H., 127, 141
fovea, 4, 440
foveola, 4, 440
frequency, 3, 440
Fu, C., 330, 334
Fu, K.S., 284, 291
full radiators, 83

G

gamut area, 150
Ganz, E., 226, 230
Gao, X.W., 129, 142, 313, 323
Gardner, J.L., 198, 217
gas discharge lamps, relative spectral power distributions of, 371
gas discharges, 74
gas mantles, 82
geometric metamerism, 124
geometries of illumination and measurement, 104, 124, 199
geometries of illumination and viewing, 103, 124, 199
Gerrity, A., 93, 97
Gill, G.W., 316, 322
Gloag, H.L., 178, 194
gloss meters, 275
gloss traps, 106, 440
gloss, 271
Godlove, M.N., 165, 194
Gold, M.J., 178, 194
gonio-spectrophotometers, 125
Gonzalez, A., 283, 284, 290
Gonzalez, R.C., 283, 290
Gorzinski, M.E., 240
Graham, N., 284, 291
Grassmann's Laws, 29, 440
Gregersen, Ø., 288, 290
Gregory, Richard L., 17
Grum, F., 37, 40, 115, 199, 200, 217, 221, 222, 225, 230

Guild, J., 27, 31, 40
Günay, M., 242, 255

H

half-tone images, 262, 286
Halstead, M.B., 147, 150, 153, 154
Han, B., 287, 290
Hanselaer, P., 151, 153
Hanson, A.H., 198, 216
Haralick, R.M., 285, 290
Hård, A., 170, 194
Hardeberg, J., 248, 255
Harlow, A.J., 14, 16
Harold, R.W., 271, 273, 280, 290, 291
Harper Gilmour, W., 289, 290
Hashimoto, K., 124, 142, 151, 153, 293, 324
haze, 280
Healey, G., 248, 256
Helmholtz, H.v., 168
Helmholtz-Kohlrausch effect, 23, 440
Helson, H., 129, 141
Helson-Judd effect, 143, 440
Hemmendinger, H., 31, 40, 165, 194, 200, 216
Henderson, S.T., 97, 150, 153
Hering, E., 168, 170
hertz, 440
Hesselgren, S., 155, 170, 194
Hofer, H., 7, 13, 16
Holst, G.C., 243, 256
Hong, G., 244, 245, 246, 256, 325, 326, 328
Houston, K.L., 295, 323
Hsia, J.J., 281, 292
hue, 9, 65, 440
hue composition, 306
hue quadrature, 306, 441
hue, correlates of, 305, 306
hue, criterion for constant, 299
hue-angle, 51, 53, 57, 65, 305
hue-difference, 57
hues, unique, 168, 302
hues, unique, criteria for, 302
Hunt effect, 303, 441
Hunt, R.W.G., 17, 103, 115, 119, 129, 130, 141, 142, 144, 151, 153, 155, 194, 232, 239, 240, 265, 271, 290, 293, 297, 303, 305, 310, 311, 313, 314, 315, 322, 323, 324, 330, 331, 334
Hunter Lab system, 187
Hunter, R.S., 187, 194, 271, 275, 280, 290
Hurlbert, A.C., 151, 153
Hurvich, L.M., 4, 16, 294, 309, 323
Hutchings, J.B., 279, 282, 290, 291

I

IEC 61966-2-1:1999, 232
Ikeda, K., 129, 142
illuminance, 340, 441
illuminant colorimetric shift, 134, 441
illuminant metamerism index, 122
illuminant metamerism, 122
illuminant spectral power distributions, 355
illumination level, effect of, 127, 300
illumination with white light for fluorescent colours, 221
illumination, geometries of, 103, 104, 124, 199
Imai, F.H., 265
Imura, K., 227, 230
incandescence, 74
incandescent light sources, 82, 368, 340
indigo, 2
indoor daylight CIE Illuminants, 94
infrared, 3
integrating spheres, 104, 441
intensity, 441
interpolation formulae, 208
iris, 4
irradiance, 340, 441
Ishihara test, 14
ISO 4467, 282, 291
isotropic, 102, 342, 441
Ito, M., 130, 142

J

Jacobson, R.E., 251, 256, 293, 322
Jagla, W., 39, 40

Jägle, H., 39, 40
Jain, A.K., 284, 292
Jameson, D., 294, 309, 323
Jerome, C.W., 145, 153
Ji, W., 277, 287, 291
Johansson, T., , 170
Johnson, G.M., 287, 291
joule, 441
Juan, L.Y., 313, 323
Judd correction, 39, 40, 441
Judd, D.B., 1, 16, 40, 115, 129, 141, 145, 153, 165, 183, 194, 203, 217, 271, 275, 290
Julesz, B., 284, 287, 291

K

Kaiser, P.K., 68, 72
Kang, H.R., 247, 256
Katoh, N., 130, 142
Keegan, H.J., 203, 217
kelvin, 83, 441
Kemenade, van J.T.C., 145, 150, 153, 154
Kim, H.S., 246, 256
Kim, J., 157, 194
King, P.A., 239, 240
Kirchner, E., 287, 291
Kitaguchi, S., 287, 288, 291
Knoblauch, K., 17, 275, 277, 291
Koenderick, J.J., 283, 286, 290, 291
Komatsubara, H., 129, 142
Konda, A., 273, 291
Kosztyán, Z.T., 198, 217
Krystek, M., 86, 97, 115, 198, 213, 216
Kubelka, P., 263, 265
Kubelka-Munk function, 263
Kuehni, R.G., 118, 125, 141, 142, 195
Kuo, W.-G., 129, 130, 142, 293, 313, 323

L

Lagrange interpolation formula, 208
Lam, C.C., 287, 292
Lam, K.M., 130, 132, 142, 226, 246
lambert, 339, 441
Land, E.H., 11, 16
Landy, M. S., 284, 291
Larkin, J.A., 203, 205, 217
LCDs (Liquid Crystal Displays), 202
LEDs (Light Emitting Diodes), 88
Lee, R.L., 247, 256
Lee, W., 247, 256, 287, 291
lens, of the eye, 3
Leroux, T., 275, 277, 291
Lester, A.A., 199, 216
Li, C., 57, 72, 151, 153, 157, 194, 248, 255, 277, 291, 293, 313, 314, 315, 316, 323, 324, 325, 326, 328, 330, 334
Li, X., 157, 194
light (adjective), 11, 441
light emitting diodes, 88
light exposure, 340, 441
light sources, 73
light, methods of producing, 74
light, speed of, 3
Lighting Handbook, 97
lightness, 11, 52, 65, 442
lightness, correlate of, 308
lime light, 82
Lindstrand, M., 279, 291
linearity of photodetectors, 202
Ling, Y., 151, 153
Lo, M.-C., 129, 130, 142, 293, 323
loss of colour vision, 128
Lovibond, J.W., 190
Lovibond-Schofield Tintometer, 191
Lu, S.Y., 284, 291
lumen, 338, 442
lumen per square foot, 339, 442
lumen per square metre, 339, 442
luminance, 23, 65, 340, 442
luminance adaptation, effects of, 297
luminance factor, 35, 41, 65, 339, 341, 342, 442
luminance, calculation from illuminance, 339
luminance-level adaptation, 300
luminescence, 219, 442
luminescence-weakening method for fluorescent colours, 226
luminescent luminance factor, 342
luminescent radiance factor, 220, 342

luminous, 35, 340, 442
luminous efficacy, 340, 339, 442
luminous efficiency functions, spectral, 19, 21, 343, 345
luminous efficiency, 339, 340, 442
luminous efficiency, photopic, 21
luminous efficiency, scotopic, 19
luminous exitance, 340, 442
luminous flux, 337, 338, 340, 442
luminous intensity, 340, 443
luminous reflectance factor, 35
luminous terms and units, 340
Luo, M.R., 57, 72, 129, 130, 142, 151, 153, 157, 194, 242, 244, 245, 246, 255, 256, 277, 279, 287, 288, 291, 293, 310, 313, 314, 315, 316, 323, 324, 325, 326, 328, 330, 334
LUTCHI colour-appearance data, 313
lux, 338, 443
Lynes, J.A., 144, 153

M

MacAdam, D.L., 115, 157, 187, 194
MacBeth ColorChecker chart, 151
MacDonald, L.W., 293, 310, 323
MacPherson, L.M.D., 289, 290
macula lutea, 4, 443
Malkin, E., 203, 205, 217
Maloney, L.T., 248, 256
Maplesden, N., 242, 255
Marcus, H., 151, 153, 249, 256
Marcus, R.T., 197, 217
Marin-Franch, I., 127, 141
Markou, M., 282, 285, 292
Marsden, A.M., 97, 154
Marshall, P.J., 239, 240
Martinez-Verdú, F., 246, 256
matching stimuli, 26
matching stimuli, choice of units for, 27, 233
matching stimuli, choice of, 231
matching stimuli, specification of, 27, 231
Maxwell, J.C., 168
McCamy, C.S., 151, 153, 155, 194, 249, 256
McCann, J.J., 11, 16, 295, 324
McDonald, R., 61, 72, 115, 262, 264, 265
McGunnigle, G., 284, 290, 291
McKee, S.P., 295, 324
McLaren, K., 2, 16, 61, 72, 115, 125, 142
measurement of optical properties, 267
memory colours, 151
Merbs, S.L., 6, 13, 16
mercury lamps, 76
mesopic photometry, 343
mesopic vision, 5, 443
metameric colour stimuli, 117, 443
metameric colours, 117, 126
metamerism, 38, 117, 244
metamerism in practice, 119
metamerism in practice, 119
metamerism index for change of field size, 124
metamerism index for change of geometry, 124
metamerism index for change of illuminant, 122
metamerism index for change of observer, 122
metamerism index of potential, 126
metamerism index, 443
metamerism, cause of, 117
metamerism, definition of, 118
metamerism, degree of, 121
metamerism, geometric, 124
metamerism, perceived, 119, 126
metamerism, psychophysical, 119
metamerism, terms used in connection with, 126
metamers, 38, 117, 126
millilambert, 339, 443
Minato, H., 221, 230
mireds, 96
mireks, 96
models for predicting colour appearance, 293, 325, 329
Mohammadi, M., 264, 265
Mollon, J.D., 6, 7, 8, 16, 17
monitors, characterization of, 239
monochromat, 13, 443
monochromatic stimuli, 27, 443

monochromatism, 13, 443
monochromators, double, 220
Morgan H., 288, 290
Mori, L., 129, 142
Moriyama, T., 151, 154
Morley, D.I., 147, 153
Moroney, N., 293, 314, 323, 324
Morovic, P., 248, 255
Motta, R.J., 240
Mullikan, J.C., 246, 256
multiple linear regression, 206
multi-spectral analysis, 264
Munk, F., 263, 265
Munsell Book of Colour, 165
Munsell Chroma, 162
Munsell colours for CRI, spectral data, 383
Munsell Hue, 160
Munsell Notation, 165
Munsell Renotation, 165
Munsell System, 159, 308, 443
Munsell Value, 159
Munsell, A.H., 159

N

Nadel, M.E., 198, 216
Naghshineh, S., 13, 16
Nakabayashi, K., 130, 142
Nanjo, M., 221, 230
nanometre, 3, 443
Nascimento, M.C., 127, 141
Nathans, L., 9, 13, 16
Natural Colour System (NCS), 170, 303, 308, 311, 443
Nayar, S.K., 283, 290
Nayatani, Y., 124, 130, 142, 151, 153, 221, 230, 293, 294, 324
NCS (Natural Colour System), 170, 303, 308, 311, 443
NCS atlas, 172
Neitz, J., 7, 13, 14, 16
Neitz, M., 7, 13, 14, 16
Nemcsics, A., 182, 194
neuron, 8, 443
Newhall, S.M., 165, 194
Newman, T., 293, 314, 323, 324
Newton, Isaac, 1
Nezamabadi, M., 264, 265
Nickerson, D., 159, 165, 183, 187, 194, 195
Nicodemus, F.E., 279, 291
Nimeroff, L., 122, 142
nit, 339, 444
Nobbs, J.H., 64, 72, 262, 265
non-diffusing colorants in a transmitting layer, 257
non-diffusing colorants in optical contact with a reflecting diffuser, 259
non-self-luminous colour stimulus, 101, 444
Noorlander, C., 286, 291
North, A.D., 124, 141
nuance, 178

O

O'Connor, M., 178, 195
Obein, G., 275, 277, 291
observer metamerism, 122, 444
obtaining tristimulus values, 99
Ohno Y., 151, 153
Ohno, S., 130, 142
Ohta, N., 124, 142
Okumura, Y., 264, 265
opponent hues, 10, 168, 303
optic nerve, 3
optical axis of the eye, 3, 4, 444
optical brightening agent, 66, 94, 219, 444
optical contact, 259
Optical Society of America (OSA) system, 183, 444
optimal colour stimuli, 157, 444
optimized weights for calculating tristimulus values, 210
OSA system, 183, 444
Oskoui, P., 130, 142
Owens, H., 288, 291
Ozturk, L.D., 192, 195

P

Palmer, D., 147, 153
Palmer, D.A., 26, 40
Pantone system, 191
parameric colours, 126, 444

Park, D., 325, 328
Park, J.M., 246, 256
Park, S.O., 246, 256
Park, Y., 148, 153
pearlescence, 125, 444
perceived metameric colours, 126
perceptibility of colour differences, 61
perceptual attributes of colour, 11
perfect diffuser, 34, 102, 200, 342, 444
Perrett, D., 288, 292
Petersen, K.H., 197, 204, 205, 206, 216
Pfund, A.H., 274, 292
Phillips, D.G., 225, 230
phosphorescence, 74, 219, 444
photoluminescence, 74, 219, 444
photometric terms and units, 337, 338
photometry, 99, 100, 337
photon, 444
photopic spectral luminous efficiency function, 22
photopic spectral luminous efficiency, 337
photopic vision, 4, 39, 445
physical detectors for radiometry and photometry, 227
Pinney, J.E., 121, 142
Pirotta, E., 130, 142
Planck's law, 83, 445
Planckian locus, 83, 85, 445
Planckian radiators, 83, 341, 445
Planckian radiators, chromaticity co-ordinates for, 368
Planckian radiators, relative spectral power distributions of, 368
Plant, G.T., 14, 16
point brilliance, 340, 445
point source, 39, 445
Pointer, M. R., 151, 153, 157, 159, 194, 195, 251, 256, 277, 288, 291, 293, 316, 322, 323, 324, 330, 334
Poirson, A.B., 286, 292
Pokorny, J., 6, 16, 17, 296, 324
polarization, 200
Potter, F.R., 272, 292
power, 445
Prazdny, K., 284, 289
precision in colorimetry, 197
protan, 51, 445
protanomaly, 13, 51, 445
protanopia, 13, 51, 445
proximal field, 294, 445
psychophysical metamerism, 119
Pujol, J., 246, 256
pupil, 4
purity, 44, 67, 445
purity, colorimetric, 67, 70
purity, excitation, 44
Purkinje phenomenon, 5, 22, 446
purple boundary, 43, 446
purple colours, 127
purple stimulus, 43, 446

Q

quantum, 446
quantum exitance, 340, 446
quantum exposure, 340, 446
quantum flux, 338, 340, 446
quantum intensity, 340, 446
quantum irradiance, 340, 446
quantum radiance, 340, 446
quantum terms and units, 340

R

r,g chromaticity diagram, 233
radiance factor, 35, 108, 220, 341, 342, 447
radiance, 340, 446
radiant, 35, 219, 340, 447
radiant efficiency, 339, 447
radiant exitance, 340, 447
radiant exposure, 340, 447
radiant flux, 252, 338, 340, 447
radiant intensity, 340, 447
radiant power, 99, 338
radiant reflectance factor, 219
radiant terms and units, 340
radiation sources, 254, 340
radiometer, 100
radiometric terms and units, 337, 338
radiometry, 99
RAL system, 191
receptors, retinal, 4

redness-greenness, 303
Reeves, A., 295, 322
reference illuminant, 144
reference whites, 102, 200, 294
reflectance factor, 35, 41, 108, 220, 341, 342, 447
reflectance, 35, 108, 341, 447
reflected luminance factor, 342
reflected radiance factor, 342
reflection, 35, 103, 341, 447
reflection, terms for measures of, 35, 341
reflection, total internal, 259
reflectometer value, 220, 341
regular reflectance, 341, 447
regular reflection, 103, 200, 447
regular transmission, 447
regular transmittance, 341, 447
related colour, 12, 65, 329, 448
relations between colour stimuli, 41
relative colour stimulus function, 5, 37, 448
relative spectral power distribution, 35, 448
relative spectral power distributions of illuminants, 355
Reniff, L., 204, 216
resting rate, of nerve signal, 8
retina, 4, 448
retinal receptors, 4
retinal receptors, spectral sensitivities of, 5, 295
Retinex theory, 11
reverse chromatic adaptation transform (CAT02), 133
reverse CIECAM02 model, 320
RGB colorimetry, 231
Rhodes, P.A., 129, 142, 242, 244, 245, 246, 255, 256, 293, 313, 323
rhodopsin, 5, 448
Rich, D.C., 200, 217
Richter, M., 179, 195
Rigg, B., 61, 72, 130, 151, 153, 242, 255
Ripamonti, C., 256
Robertson, A.R., 60, 72, 86, 97, 145, 153, 155, 195, 204, 217
Robson J.G., 288, 290
rod contribution, 308, 330
rods, 4, 448
room colours, 325
Rösch, S., 157, 195
Rosenfeld, A., 284, 289
Rowell, N., 240
Ruddock, K.H., 13, 16
Ryckaert, W.R., 151, 153

S

Saltzman, M., 115, 187, 194, 200, 217
sample preparation, 198
Sato, M., 287, 291
saturation, 12, 51, 65, 180, 448
saturation, correlate of, 311
Saunderson equation, 264
Saunderson, J.L., 264, 265
scattering co-efficient, 262
Schanda, J.D., 198, 217
Schappo, A., 129, 142, 293, 313, 323
Schleter, J.B., 203, 217
Schofield, S.K., 191, 195
SCOT-Munsell system, 193
scotopic luminance, 20
scotopic luminance, method of estimating, 330
scotopic luminous flux, 338
scotopic spectral luminous efficiency, 19, 337
scotopic vision, 4, 448
Scrivener, S.A.R., 129, 142, 293, 310, 313, 323
secondary standards, 203
Seim, T., 293, 297, 324
self-luminous colour, 12, 101, 448
Sève, R., 277, 279, 292
shadow series, 181, 448
Shao, S., 287, 292
Sharma, G., 256
Sharma, M., 285, 292
Sharpe, L. T., 6, 16, 39, 40
Shen, H.-L., 287, 292
Shi, M., 248, 256
Shimuzi, M., 151, 153
Shin, H.K., 199, 216
Shioiri, S., 151, 154

signal transmission, visual, 8
Simmons, D.R., 289, 290
Simon, F.T., 225, 230
simultaneous contrast, 43, 269, 448
Singh, M., 282, 285, 292
Singh, S., 282, 285, 292
Sivik, L., 170, 194
Smet, K., 151, 153
Smith, K.B., 272, 292
Smith, V.C., 6, 16, 296, 324
Snodderly D.M., 288, 290
Sobagaki, H., 124, 129, 130, 142, 293, 324
sodium lamps, 75, 128
solid angle, 20, 448
sources of radiation, 340
sources, measuring and desired, correcting for differences between, 222
spectral chromaticity co-ordinates, 351
spectral conventional reflectometer value, 220, 221, 449
spectral data, correcting for errors in, 204
spectral density, 258
spectral, 341, 448
spectral locus, 43
spectral luminescent radiance factor, 220, 342, 449
spectral luminous efficiency functions based on brightness matching, 39
spectral luminous efficiency functions, 345
spectral luminous efficiency, 20, 21, 39, 345, 449
spectral luminous efficiency, 449
spectral power distribution, 35, 449
spectral power distributions of illuminants, 355
spectral radiance factor, 220, 449
spectral reflectance factor, 35, 449
spectral reflectance, 35, 449
spectral reflected radiance factor, 219, 449
spectral sensitivities of the cones, 5, 296
spectral total radiance factor, 220, 342, 449
spectral transmittance factor, 449
spectral transmittance, 449
spectral weighting functions, 19
spectrophotometers, 101, 108
spectrophotometry, 101, 108
spectroradiometers, 100, 108
spectroradiometry, 100, 101
spectroradiometry of non-self-luminous colours, 101
spectroradiometry of self-luminous colours, 101
spectrum locus, 43, 450
spectrum, 1
specular reflection, 103, 200
SPEX, 106
Spillman, W., 155, 195
SPINC, 106
spreading effect, 271, 450
Sproson, W.N., 232, 240
Stainsby, A.G., 147, 153
standard deviate observer (SDO), 122
Standard Illuminant A, 91, 355
Standard Illuminants D, 93, 355
Standard Illuminants D, correlated colour temperatures of, 355
Standard illuminants, 91
Standard Illuminants, D, chromaticity co-ordinates for, 355
Standard Illuminants, relative spectral power distributions of, 355
Standard Observer, CIE, Colorimetric, 1931 (2°), 33, 347
Standard Observer, CIE, Colorimetric, 1964 (10°), 33, 347
standard observer, CIE, photopic photometric, 21
standard observer, CIE, scotopic photometric, 19
standard photometric observer, 21
standard scotopic photometric observer, 19
standard sources, 91, 93, 355
Stearns, E.I., 209, 210, 217
Stearns, R.E., 210, 217
Stephan, K.W., 289, 290
steradian, 20, 450
Stevens effect, 301, 450
Stevens, S.S., 300, 324
Stiles, W.S., 7, 31, 37, 40, 86, 97, 115, 165, 179, 187, 195
Stiles-Crawford effect, 4, 450

Stockman, A., 6, 16, 39, 40
Stokes' law, 223, 450
stray light, 202
strong, 12, 450
subtractive mixing of colorants, 25, 257, 450
successive contrast, 43, 268, 450
Sung, C.-H., 6, 13, 16
surface reflection, 261
surface texture, 281
surround, 294
Suziki, K., 287, 292
Swedish Standard, SS 01 91 02 (1982), 173, 195
Swenholt, B.K., 157, 194

T

Tait, C.J., 129, 142, 293, 313, 323
Tait, D.M., 14, 16
Takahama, K., 124, 130, 142, 293, 324
Taplin, L.A., 264, 265
Taylor, T.H., 295, 324
tele-spectroradiometers, 100
tele-spectroradiometry, 450
television displays, 101, 239
television gamuts, 60
Terstiege, H., 94, 97, 227, 230
texture, surface, 281
thermochromism, 199, 450
third degree polynomial interpolation formula, 208
Thomson, M.G.A., 245, 256
Thornton, W.A., 144, 145, 150, 153, 154
Tiddeman, B., 288, 292
Tighe, B.J., 277, 292
Tingle, W.H., 272, 292
tint, of whites, 66, 450
Tintometer, The, 190, 231
total internal reflection, 259
total spectral radiance factor, 220, 342, 450
transformation equations, 32, 235, 451
transformation from *R*, *G*, *B* to *X*, *Y*, *Z*, 32, 235
transformation from *X*, *Y*, *Z* to *R*, *G*, *B*, 235
translucency, 279
translucent blurring, 200
transmission, 331, 341, 451
transmission, terms for measures of, 341
transmittance factor, 35, 41, 341, 342, 451
transmittance, 35, 341, 451
transmitting diffuser, 34, 102, 342
transparency, 280
Trezona, P.W., 26, 40
triangular window of bandwidth function, 209
trichromatic matching, 26, 451
tristimulus values, 28, 66, 233, 451
tristimulus values, calculation of, from spectral data, 34, 111, 207, 393
tristimulus values, correcting for inequalities of, 125
tritan, 51, 451
tritanomaly, 14, 51, 451
tritanopia, 13, 51, 451
troland, 340, 451
Tuceryan, M., 284, 292
tungsten lamps, 86
tungsten-halogen lamps, 87
turbidity, 280
two-mode method for fluorescent colours, 225
two-monochromator method for fluorescent colours, 224

U

U*V*W* space, 54, 145
u,v diagram, 49, 451
u′,v′ diagram, 49, 451
ultraviolet, 3
uniform chromaticity diagram, 48, 452
uniform colour space, 53, 315, 452
unique hues, 10, 168, 303, 452
unique hues, criteria for, 303
unitary hues, 10, 452
units for photometry, 337, 338
units for radiometry, 340
units in colorimetry, 27, 233
unrelated colour, 12, 65, 329, 452
unrelated colours, model for, 329

V

V(λ) function, 21, 337, 452
V'(λ) function, 19, 337, 452
Valberg, A., 17, 293, 297, 324
Valeton, J.M., 297, 300, 324
Van der Burgt, P.J.M., 145, 150, 153, 154
Van Der Feltz, G., 246, 256
Van Ginneken, B., 283, 290
Van Kemenade, J.T.C., 145, 150, 153, 154
Van Norren, D., 297, 300, 324
Van-Viet, L.J., 246, 256
Venable, W.H., Jr., 203, 210, 212, 216, 217
Verrill, J.F., 198, 203, 205, 216, 217
Viénot, F., 275, 277, 291
viewing, geometries of, 103, 124, 199
visual axis, 3, 452
visual clarity, 144
visual fields, 294
$V_M(\lambda)$ function, 39, 337
Voke, J., 14, 16
Von Kries transformation, 130, 452
Von Kries, J.A., 129, 142

W

Wald, G., 5, 16
Walraven, P.L., 7, 17, 305, 308, 324
Wandell, B.A., 17, 248, 256, 286, 289, 292
Wang, Y., 6, 13, 16
Wardman, R.H., 203, 205, 217
Ware, C., 68, 72
Warren, M.H., 129, 141
watt, 452
wavelength calibration, 202
wavelength, 2
weak, 12, 452
Webber, A.C., 280, 292
Weidner, V.R., 281, 292
weight (of a colour), 178
weighting functions, spectral, 19
weights for computing tristimulus values, 393
Weitz, C.J., 6, 13, 16
Westland, S., 245, 247, 248, 255, 256, 287, 288, 291
Wharmby, D.O., 115, 198, 217
white light, illumination with, for fluorescent colours, 221
white point, 452
white point (colour appearance), adopted, reference, 294, 296, 300, 308, 309, 310, 313, 316, 320, 452
whiteness, 66, 170, 453
whites, reference, 100, 102, 200, 294
Whitfield, T.W.A., 178, 195
Whitten, D.N., 297, 322
Wightman, T.E., 199, 217
Williams, D.R., 7, 13, 17
Willmouth, F.M., 271, 279, 292
Wiltshire, T.L., 178, 195
Witt, K., 179, 195
Woods, R.E., 283, 290
Wright, W.D., 27, 31, 40, 48, 72, 115, 155, 195, 271, 292
Wu, W., 246, 256
Wyszecki, G., 1, 16, 37, 40, 86, 97, 115, 165, 179, 187, 195

X

x,y diagram, 42, 453
xenon lamps, 81
Xiao, K., 325, 326, 328
Xin, H.J., 129, 142, 287, 292, 313, 323
Xu, B., 287, 292

Y

Y tristimulus value, 41
Yaguchi, H., 151, 154
Yano, T., 151, 153, 293, 324
yellow spot, 4, 453
yellowness-blueness, 303
Young, I.T., 246, 256
Young, T., 168
Yurow, J.A., 122, 142

Z

zero level, 202
Zhang, X., 286, 292
Zhao, Y., 264, 265